Sports Economics, Management and Policy

Volume 24

Series Editor
Dennis Coates, Baltimore, MD, USA

The aim of this series is to provide academics, students, sports business executives, and policy makers with information and analysis on the cutting edge of sports economics, sport management, and public policy on sporting issues.

Volumes in this series can focus on individual sports, issues that cut across sports, issues unique to professional sports, or topics in amateur sports. Each volume will provide rigorous analysis with the purpose of advancing understanding of the sport and the sport business, improving decision making within the sport business and regarding policy toward sports, or both. Volumes may include any or all of the following: theoretical modelling and analysis, empirical investigations, or description and interpretation of institutions, policies, regulations, and law.

Derek Van Rheenen • Olivier Naria
Ricardo Melo • Claude Sobry
Editors

Sport Tourism, Island Territories and Sustainable Development

A Comparative Perspective

Springer

Editors
Derek Van Rheenen
Cultural Studies of Sport in Education
University of California Berkeley
Berkeley, CA, USA

Ricardo Melo
Coimbra Education School
Polytechnic of Coimbra
Coimbra, Portugal

Olivier Naria
Department STAPS
Université de La Réunion
Le Tampon, ïle de la Réunion
Réunion, France

Claude Sobry
Faculty of Sport and Physical Education
Université de Lille
Lille, Nord, France

ISSN 2191-298X ISSN 2191-2998 (electronic)
Sports Economics, Management and Policy
ISBN 978-3-031-51704-4 ISBN 978-3-031-51705-1 (eBook)
https://doi.org/10.1007/978-3-031-51705-1

This Springer imprint is published by the registered company Springer Nature Switzerland AG
The registered company address is: Gewerbestrasse 11, 6330 Cham, Switzerland

Foreword

Established in 2010, the International Research Network in Sport Tourism (IRNIST) has sought to study sustainable local development through sport tourism. Although the genesis of the establishment of the research network was to promote rigorous academic studies involving professionals from numerous universities across the world, it soon became apparent that the concept of sport tourism was poorly understood, with the conceptual boundaries not well circumscribed. This lack of clarity was due largely to a multitude of working definitions of sport tourism.

Acknowledging this lack of conceptual coherence in defining sport tourism (JS&T, 2014; Gammon et al., 2017), international scholars have sought for decades to better understand this dynamic construct as more than the sum of its parts (Downward, 2005; Pigeassou et al., 2003; Weed & Bull, 2012). Since its establishment, IRNIST scholars have attempted to develop a more empirically grounded and inclusive definition of sport tourism (Sobry et al., 2016; Van Rheenen et al., 2017; Melo et al., 2021). Van Rheenen et al. (2017), for example, conducted a systematic content analysis of published definitions between 1993 and 2014 ($n = 517$), narrowing these texts to 30 distinct definitions of sport tourism. Among these 30 definitions, these researchers found five paradigmatic dimensions underlying this dynamic construct: (i) time (the duration away from a home environment), (ii) space (encompassing travel away from a home environment), (iii) sport as motivation for travel (type, level, and extent of sport practice or activity), (iv) participant experience of the sport tourism activity, and (v) economic motivation (reference to the development of an economic supply or niche market).

Since then, IRNIST scholars have emphasized the need to develop a deeper understanding of the field's supply or production (Sobry et al., 2016; Bouchet & Bouhaouala, 2009; Bouchet & Sobry, 2019; Bouhaouala, 2018), the least cited paradigm in previous definitions (Van Rheenen et al., 2017). Sobry et al. (2016) refer to this supply as "opportunity sport tourism," akin to Gammon and Robinson's (2003) earlier reference to "tourism sport." As a modern addendum, Sobry et al. proposed a socioeconomic index system to more accurately measure this highly diversified and growing sport tourism economic market that ranges from mega-events to small-scale producers of activities that represent tertiary excursions for travelers.

Additionally, scholars have sought to understand sport tourism *relationally*, recognizing the systems or matrices of power that inform these socio-historical activities (Saavedra, 2018; Sykes, 2017; Van Rheenen & Roberson, 2023). One might argue that a focus on the social and historical relations inherent to the logic of sport tourism production and consumption is nothing more than another way of describing participant experience; Van Rheenen and Roberson (2023) contend, however, that structures of dominance or matrices of power (e.g., racism, sexism, classism, etc.) inform and differentially impact the sport tourism experience. They must therefore be made visible, revealed, and studied from a particular theoretical framework.

Finally, both of these conceptual developments have prompted recent work on sport tourism and sustainable development more broadly understood (Carlsen & Bruggemann, 2022; Melo et al., 2021; Naria et al., 2022). During the twenty-first century, then, environmental sustainability has become one of the most commonly discussed dimensions in terms of sport tourism impact (Carneiro et al., 2016; Van Rheenen & Melo, 2021) but not a particularly well-researched dimension of sustainability (Gibson et al., 2012; Hinch et al., 2016; Mascarenhas et al., 2021). Even in relatively small-scale sport tourism activities and events, potential negative environmental impacts may occur as a by-product of these activities, such as noise, water and air pollution, soil erosion, natural landscape destruction, fauna and flora destruction, and the deterioration of monuments and historic sites (Markwick, 2000; Melo & Gomes, 2016; Orams, 2005; Van Rheenen & Melo, 2021).

The concept of sustainability has also evolved significantly since the end of the twentieth century. While the three pillars of the triple-bottom-line articulated in *Our Common Future*, otherwise known as the Brundtland Report[1] (UNCED, 1987; Dwyer, 2015; Elkington, 1997), provided the initial conceptual framework for scholars interested in the intersection of sport tourism and sustainable development. But the economic, social, and environmental factors underlying sustainability were so broad as to limit the application of this conceptualization. Changes in the global socio-political context have led the United Nations (UN) to weigh in more recently on the issue of sustainable development, further expanding our understanding of the concept to include 17 distinct Sustainable Development Goals (SDGs) (Carlsen & Bruggemann, 2022; Chirico, 2018; United Nations, 2015, 2021). These SDGs map loosely onto the three overarching pillars referenced in the Brundtland Report almost a half century ago (Mangukiya & Sklarew, 2023; Rockström & Sukhdev, 2016; Sachs et al., 2022).

The 17 SDG's include decent work and economic growth, sustainable cities and communities, climate action, and gender equality, among others. The World Tourism Organisation, a specialized agency of the United Nations, has also committed itself to the 2030 Agenda for Sustainable Development, recognizing in particular Goals 8, 12, and 14. These SDGs target inclusive and sustainable economic growth,

[1] Norway's first female Prime Minister Gro Harlem Brundtland was Director-General of the World Health Organization (WHO) from 1998 to 2003. She also chaired the Brundtland Commission which presented the Report on Sustainable Development, entitled *Our Common Future*, in 1987. It has come to be referred to as the *Brundtland Report*.

sustainable consumption and production (SCP), and the sustainable use of oceans and marine resources (UNWTO, 2023).

In a document entitled "The Contribution of Sport to the Sustainable Development Goals and the post-2015 Development Agenda," the International Olympic Committee (IOC) "recognizes that this Agenda is an historical opportunity to ensure that sport and physical activity are integrated as a meaningful and cost-effective tool to achieve these sustainable development goals" (IOC, 2015, p. 2). In this document, the IOC promotes 6 of the 17 SDG's in particular as important to sport's potential for positive global change. These include promoting health and well-being (SDG # 3), achieving quality education (SDG # 4), promoting gender equality (SDG # 5), promoting sustainable cities and communities (SDG # 11), contributing to peaceful and just societies (SDG # 16), and finally, developing critical partnerships for the realization of these sustainable development goals globally (SDG # 17).

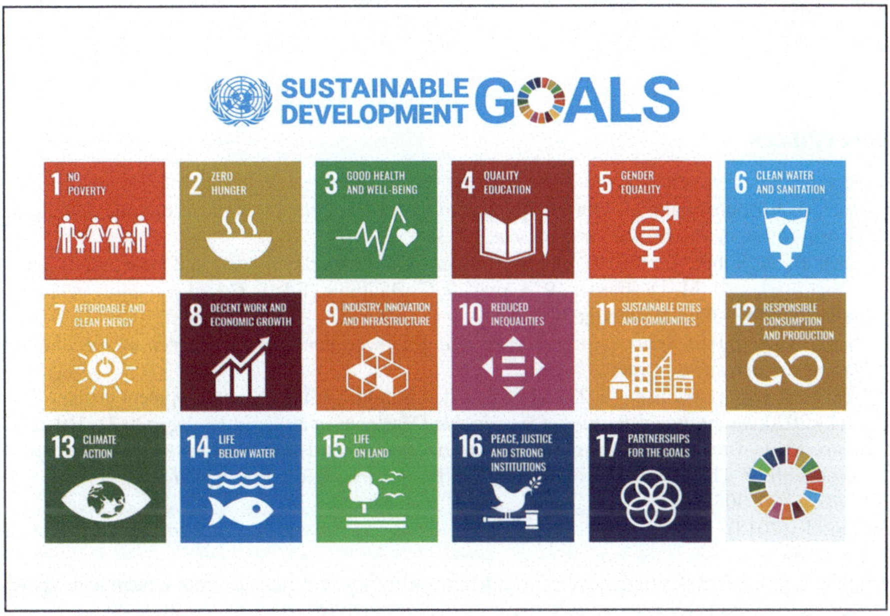

UNWTO (2023)

Many of these goals will be referenced in the following comparative research project examining the development of small island states or territories through sport tourism. Specifically, the conceptual model and comparative research methodology proposed in this book explicitly address these sustainable development goals. It would be naive to assume that sport tourism offers a panacea for the sustainable development of small island territories. Rather, this project begins with the premise that this economic and social activity could play a unique role in the development of certain small island economies as they make intentional efforts to foster sustainable growth, protect and preserve natural resources, and honor the dignity of indigenous peoples. In analyzing several island case studies, the following comparative research project hopes to articulate a potential model or heuristic that might be adopted by other like territories and adapted to their own local contexts. These contexts must assess the varying levels of vulnerability, adaptability or adaptive capacity and myriad types of public-private and core-peripheral partnerships that could successfully be forged to accelerate small island development through sport tourism. This significant contribution to the literature offers an opportunity for local to global (e.g., glocal) stakeholders to reimagine sustainable development through a more just and intentional ethic of care for the health and biodiversity of the environment.

President, International Research Network in Sport Tourism (IRNIST),
Professor Emeritus, Economics
Université of Lille,
Lille, France

Claude Sobry

References

Bouchet, P., & Bouhaouala, M. (2009). Sporting tourism. An attempt at socio-economic definition. *Téoros, Revue de Recherche en Tourisme, 28*, 3–8.

Bouchet, P., & Sobry, C. (2019). Sports tourism: Contemporary issues and new trends on a global market. In M. Desbordes, P. Aymar, & C. Hautbois (Eds.), *The global sport economy* (pp. 295–314, 330p). Routledge.

Bouhaouala, M. (2018). *Sport, économie et société. Un programme de recherche en socioéconomie du sport*. Univesité de Lille.

Carlsen, L., & Bruggemann, R. (2022). The 17 United Nations' sustainable development goals: A status by 2020. *International Journal of Sustainable Development & World Ecology, 29*(3), 219–229.

Carneiro, M. J., Breda, Z., & Cordeiro, C. (2016). Sports tourism development and destination sustainability: The case of the coastal area of the Aveiro region, Portugal. *Journal of Sport & Tourism, 20*, 305–334.

Chirico, J. (2018). *Global problems, global solutions: Prospects for a better world*. SAGE Publications.

Downward, P. (2005). Critical (realist) reflection on policy and management research in sport, tourism and sports tourism. *European Sport Management Quarterly, 5*(3), 303–320.

Dwyer, L. (2015). Triple bottom line reporting as a basis for sustainable tourism: Opportunities and challenges. *Acta Turistica, 27*(1), 33–62.

Elkington, J. (1997). The triple bottom line. In M. V. Russo (Ed.), *Environmental management: Readings and cases* (2nd ed., pp. 49–66). Sage.

Gammon, S., & Robinson, T. (2003). Sport and tourism: A conceptual framework. *Journal of Sport & Tourism, 8*, 21–26.

Gammon, S., Ramshaw, G., & Wright, R. (2017). Theory in sport tourism: Some critical reflections. *Journal of Sport & Tourism, 21*(2), 69–74.

Gibson, H. J., Kaplanidou, K., & Kang, S. J. (2012). Small-scale event sport tourism: A case study in sustainable tourism. *Sport Management Review, 15*, 160–170.

Hinch, T. D., Higham, J. E., & Moyle, B. D. (2016). Sport tourism and sustainable destinations: Foundations and pathways. *Journal of Sport & Tourism, 20*, 163–173.

International Olympic Committee (IOC). (2015). *The contribution of sport to the sustainable development goals and the post-2015 development agenda*. https://www.sportanddev.org/sites/default/files/downloads/sport_contribution_to_post_2015_agenda_eng_feb.pdf. Accessed December 2022.

Journal of Sport & Tourism (JS&T). (2014). Call for papers: After 20 years. *Journal of Sport & Tourism*. Retrieved from: http://explore.tandfonline.com/page/pgas/rjto-20

Mangukiya, R. D., & Sklarew, D. M. (2023). Analyzing three pillars of sustainable development goals at sub-national scales within the USA. *World Development Sustainability, 2*, 100058.

Markwick, M. C. (2000). Golf tourism development, stakeholders, differing discourses and alternative agendas: The case of Malta. *Tourism Management, 21*, 515–524.

Mascarenhas, M., Pereira, E., Rosado, A., & Martins, R. (2021). How has science highlighted sports tourism in recent investigation on sports' environmental sustainability? A systematic review. *Journal of Sport & Tourism, 25*(1), 42–65.

Melo, R., & Gomes, R. (2016). Nature sports and sustainable local development: Practitioners and organizations managers' perspectives in Portugal. In C. Sobry (Ed.), *Sport tourism and local sustainable development* (pp. 75–100). L'Harmattan.

Melo, R., Sobry, C., & Van Rheenen, D. (Eds.). (2021). *Small-scale sport tourism events and local sustainable development: A cross-national comparative perspective*. Springer Nature AG.

Naria, O., Van Rheenen, D., Melo, R., & Sobry, C. (2022). Sport tourism, island territories and sustainable development. In B. Mazza & C. Sobry (Eds.), *Sport tourism and its territorial development and opportunities* (pp. 19–40). Cambridge Scholars Publishing.

Orams, M. (2005). Introduction to sport tourism impacts and environments. In J. Higham (Ed.), *Sport tourism destinations: Issues, opportunities and analysis* (pp. 248–259). Routledge.

Pigeassou, C., Bui-Xuan, G., & Gleyse, J. (2003). Epistemological issues on sport tourism: Challenge for a new scientific field. *Journal of Sport & Tourism, 8*(1), 27–34.

Rockström, J., & Sukhdev, P. (2016, June 14). *How food connects all the SDGs*. Stockholm Resilience Centre.

Saavedra, M. (2018). The sexual and gender politics of sport mega-events: Roving colonialism. *Sociology of Sport Journal, 35*(3), 291–292.

Sachs, J., Lafortune, G., Kroll, C., Fuller, G., & Woelm, F. (2022). *Sustainable development report 2022 – From crisis to sustainable development: The SDGs as roadmap to 2030 and beyond*. Cambridge University Press.

Sobry, C., Liu, X., & Li, J. (2016). Sport tourism: Contribution to a definition and categorization. *Acta Touristica, 28*(1): 7–26.

Sykes, H. (2017). *The sexual and gender politics of sport mega-events: Roving colonialism*. Routledge.

United Nations. Department of Economic and Social Affairs Sustainable Development. (2015). *Transforming our world: The 2030 agenda for sustainable development*. https://sdgs.un.org/2030agenda. Accessed December 2022.

United Nations. Department of Economic and Social Affairs Sustainable Development. (2021). https://sdgs.un.org/goals. Accessed December 2022.

United Nations Conference_on_Enviroment_and Development (UNCED). (1987). *Our common future*. UN. https://sustainabledevelopment.un.org/content/documents/5987our-common-future.pdf. Accessed December 2022.

United Nations World Tourism Association (UNWTO). (2023). *Tourism in the 2030 agenda*. https://www.unwto.org/tourism-in-2030-agenda. Accessed February 2023.

Van Rheenen, D., & Melo, R. (2021). Nature sports: Prospects for sustainability. *Sustainability, 13*, 8732.

Van Rheenen, D., & Roberson, R. (2023). Roving consumers of pleasure: At the conceptual intersection of sport tourism and sex tourism. *Journal of Sport & Tourism, 27*(2), 111–122. https://doi.org/10.1080/14775085.2023.2186926

Van Rheenen, D., Cernaianu, S., & Sobry, C. (2017). Defining sport tourism: A content analysis of an evolving epistemology. *Journal of Sport & Tourism, 21*(2), 75–93. AERES, STAPS, 04/2012.

Weed, M., & Bull, C. (2012). *Sports tourism: Participants, policy and providers*. Routledge.

Acknowledgment

As co-editors of this book, we wish to thank all of the contributors who have participated in this ambitious project. They have been patient and good natured as we have sought to contribute to the literature in a meaningful way, envisioning an aspirational future for these beautiful islands and the diverse peoples and communities who live and thrive in these magnificent places. We would like to thank the Université de la Réunion for its cartographic support.

Contents

Chapter 1
Introduction: A Journey to the Promised Land

Derek Van Rheenen, Olivier Naria, Ricardo Melo, and Claude Sobry

Sometimes I wake up not knowing where I am. The world is full of islands, said Cruso once. His words ring truer every day.

Foe, p. 65 (Coetzee, 1986)

The idea for this book was born from a chance encounter between IRNIST founder and President, Dr. Claude Sobry, and Dr. Olivier Naria, a scholar from La Réunion Island, a French territory in the Indian Ocean. This scholar from the Indian Ocean reached out to Sobry in his capacity as the director of the International Research Network on Sport Tourism (IRNIST) and Professor Emeritus of Economics at the Université of Lille (France). Dr. Olivier Naria, geographer and Associate Professor of Management at the Université de La Réunion (France), asked Sobry what might seem like a relatively simple question, practically speaking: *how can sport tourism contribute to the sustainable development of the islands of the Indian Ocean?*

D. Van Rheenen
Cultural Studies of Sport in Education, University of California Berkeley, Berkeley, CA, USA
e-mail: dvr@berkeley.edu

O. Naria
Department STAPS, Université de La Réunion, Le Tampon, ïle de la Réunion, Réunion, France
e-mail: olivier.naria@univ-reunion.fr

R. Melo
Coimbra Education School, Polytechnic of Coimbra, Coimbra, Portugal

C. Sobry (✉)
Faculty of Sport and Physical Education, Université of Lille, Lille, Nord, France
e-mail: Claude.sobry@univ-lille.frxt

D. Van Rheenen et al. (eds.), *Sport Tourism, Island Territories and Sustainable Development*, Sports Economics, Management and Policy 24,
https://doi.org/10.1007/978-3-031-51705-1_1

Naria's question coincided with the completion of a comparative research project sponsored by IRNIST, a project that analyzed nine half marathon events across three continents and nine countries (Melo et al., 2021). It quickly became apparent through this initial conversation (and the extant literature) that the impact of sport tourism on the islands of the Indian Ocean had been limited to date. Not surprisingly, there was very little scholarship on the subject. The idea of a research collaboration blossomed, whereby island scholars from around the world might weigh in on this topic, merging relevant research with practical application.

Sobry, Van Rheenen, and Melo thought that this proposed project might utilize the comparative methodology they had recently introduced and adopted for their road races project (Van Rheenen et al., 2021). This comparative methodology sought to illuminate conceptual and concrete ways (e.g., best practices) of implementing sport tourism activities and events that intentionally fostered sustainable development principles and practices. While the methodology of the book compared nine half marathon events organized across the globe, the proposed project would seek to reverse the logic and shift the comparative unit of analysis.

The comparative unit of analysis would no longer be the *sport tourism activity or event*, but rather, a *similar site or territory*. The context or site of study would be small island territories, analyzing the different ways in which various sporting activities or events were being offered (or proposed to be offered) as niche and/or diversified markets from a sustainable perspective, with specific sustainability goals incorporated into the design and implementation of such development. By focusing on a geographical or spatial unit of analysis from this theoretical framework, IRNIST determined to contrast similar or like territories and their unique relationships to sport tourism and sustainable development. We anticipated significant variation across both territorial site and time, even within the same oceans, seas, and archipelagos.

Guiding Hypotheses

Just as we specified to potential authors the guidelines for empirical data collection and analysis, this international research project posited three critical hypotheses at its inception:

H1—Sport tourism has the potential to support the sustainable development of small island territories (e.g., *ecoterritories*). This development can be measured at multiple levels, inclusive of the intersecting political, economic, social, and environmental dimensions. More specifically, scholars are able to measure the unique sustainable development goals (SDGs), as articulated by multiple organizations and movements, such as the 2015 Paris Climate Agreement (Stracqualursi & Liptsk, 2021), the Somoa Pathways (Montes, 2016); the U.N.'s Global Sustainability Platform (https://www.thegpsc.org/) and 2020 Agenda (Weiland et al., 2021), the 2022 United Nation's global climate summit (Colman & Mathiesen, 2022; Plumer et al., 2022), and the 2022 Kunming-Montreal global biodiversity framework (Paddison, 2022). That is, sport tourism practices intersect the fundamental dimensions of sustainable development, whether locally, nationally, or globally (Bouchet & Sobry, 2019; Naria et al., 2022; Van Rheenen et al., 2021). Scholars have likewise developed sustainable development indices by nation state, assessing a unique score for each of the 17 SDGs and an overall aggregate score for each geographic location (Sachs et al., 2022).

- At the political level (e.g., governance), sport tourism serves as a lever for the development of stakeholder inclusion and relational accountability, providing a novel framework for land use planning and a guide for the preservation and management of natural resources. The political dimension draws on an island's historical context, such as the multiple impacts of colonialization and its aftermath. This historical context likewise impacts the economic, social, and environmental dimensions of island development and the possibility for, and realization of, local to global sustainability development goals.
- At the economic level, sport tourism has the potential to be a robust entrepreneurial sector, with the opportunity for significant growth as an expanding market. On many islands, this tourism market is often seasonal depending on the specific territory and the given activities or events promoted. As an economic sector, sport tourism on island territories fosters employment growth, innovation, financing and the creation of new businesses and services activities. This economic development enhances island revenue, as well as economic stability and independence. Ideally, this innovation can also reduce inequalities by promoting a circular rather than linear economy.
- At the social level, sport tourism has the potential to foster meaningful relationships between peoples, honor local customs and intergenerational communities, including indigenous island peoples. This can be achieved by promoting relational equity, fairness, and social justice, while making a commitment to equal access to opportunities; sport tourism activities and events help to create civic pride and place recognition. Ideally, the sustainable development connected to sport tourism can help reduce poverty, food insecurity, clean water, and enhanced health care.
- At the environmental level, sport tourism has the potential to foster meaningful relationships between peoples and their environment, from the local to the global. This requires a commitment to protecting and preserving natural resources on land and below water, while seeking to reduce our reliance on a linear rather than circular economy. A focus on responsible consumption and production, as well as climate action undergirds these environmental efforts. This commitment has become a call to action for the world, addressing the multiple ways in which human beings have placed a premium on profits and not on our planet.

H2—A comparative approach to ethical and just local development through sport tourism must address the articulated sustainable development goals as critical to the successful adoption of a viable model for planetary health and well-being. Recent global agreements, such as the Kunming-Montreal global biodiversity framework, suggest that the time is now to act locally in preservation of the planet to avoid what has been characterized as the sixth mass extinction, in which three-quarters of Earth's species could vanish within the next 300 years (Barnosky et al., 2011; Naggs, 2017). Global warming and its deleterious impact have only increased, as the last 8 years have been the hottest on record (Fountain & Rojanasakul, 2023).

H3—A study focusing on the development of sport tourism across several islands within numerous seas and oceans make it possible to better understand the phenomena and model for successful planning and implementation (e.g., the articulation of best practices and policies). We anticipate that such empirical and comparative analyses will foster greater communication and collaboration between key stakeholders, such as the United Nations, the International Olympic Committee and the Fédération Internationale de Football Association (FIFA), NGOs, scholars/scientists, practitioners, and local communities impacted but all too often excluded from the decision-making and governance process. The primary goal of these collaborations, then, should be to promote sustainable development while including local and indigenous communities into these sustainable development goals (SDGs).

Objectives of the Book

With this book, we have sought to realize three objectives. Drawing on the work of our esteemed colleagues in the field since its academic inception, we seek to view sport tourism from an international, inclusive, and interdisciplinary perspective. This approach acknowledges that the field of sport tourism represents a diverse collection of social practices experienced within a fixed continuum of time and space. Sport tourism activities and events exist within a unique logic of production and consumption, socially and historically constructed by peoples in relation to one another and to a given context.

As such, this international, comparative study has fostered a greater understanding of how sport tourism can contribute to sustainable development. In this instance, we explore the sustainable development of small islands, research that has not been undertaken previously at this global scale. The fulfilment of this second objective has led to comparative findings which will assist key stakeholders and decision-makers to more responsibly develop island territories in a collaborative effort to realize specific sustainable development goals. A total of 21 research teams across multiple seas, oceans, and island archipelagos have taken part in this comparative research project.

The third and final objective of this book has been the refinement of a comparative methodology at the intersection of sport tourism and sustainable development. This refined methodology has sought to collect and analyze empirical data across numerous geographical sites worldwide. In comparison to an earlier international project (Melo et al., 2021; see also Naria et al., 2022), island territories became the unit of analysis rather than a specific sport tourism activity or event. This research paves the way for further studies to advance our knowledge of the role of sport tourism and its potential positive impact on local sustainable development. It is our hope that the results of this project will inspire future stakeholders to work together for a more just and sustainable future.

The Journey Begins: Outline of the Book

We begin this journey by constructing a conceptual framework to better refine a comparative methodology previously utilized by the International Research Network in Sport Tourism (IRNIST). Utilizing this conceptual framework, the book offers a global map, an aspirational canvas, of several small island states and their efforts at supporting sustainable development through a diverse array of sport tourism events and activities. This map is also a potential treasure trove to evaluate a set of best practices. We adopt a comparative approach, where place or site become the unit of analysis, a similar but diverse geography—small islands—with varying colonial histories, hybrid cultural identities, and emergent or emerging economies. The methodology measures the dynamic intersections of sport and development, with a specific focus on sport tourism and sustainable development. It seeks to do so through a critical eco-territorial lens, recognizing the possibility of global change and a commitment to ethical and sustained (and also restrained) growth. This conceptual and methodological articulation comprises Part I of the book.

The first chapter of this section more fully defines our unit of comparative analysis, small island states or territories, a common set of geographical, topographical, and socio-historical characteristics spread across multiple oceans and seas. It describes a common history of colonialism, followed by an often turbulent process of cultural and social (in)dependence that continues today through postcolonial policies of capital accumulation and development. Finally, this first chapter acknowledges a present reality and describes a foreseeable and aspirational future, one cognizant of global markets, the desire for economic growth and a corresponding mandate for genuine sustainable development. Such development requires a commitment to social and environmental justice, a moral and political call to action from the local to the global.

Professor Wilfrid Bertile pens Chapter 3, analyzing the intersection of the geomorphology of small tropical islands as an important aspect of island geography and their historical relationship to sport tourism. Adopting a global perspective, the author argues that landforms and morphological processes can be impacted by both natural factors and human actions. As such, island geomorphology has an impact on sport tourism (and tourism broadly), while sport tourism activities and infrastructures impact landforms and geomorphological processes.

Chapter 4 describes the historical relationship of tourism, and in particular sport tourism, with small island states or territories. It draws on existing literature over the past 50 years, describing a niche and diverse market that has evolved over time. This development has been quite uneven, based on island context, as has been the relative attention paid to economic, social, and environmental sustainability. There has been a revitalized focus on sport tourism in the twenty-first century and its potential scalability, recognizing the complimentary contribution of the sport tourism sector to the larger tourism industry. This revitalization has come at a time when there is a global call for sustainability, some of the loudest voices coming from small island stakeholders who have pointed out the inequalities in the global market and the corresponding negative environmental and social impacts on small island states.

Chapter 5 articulates the conceptual framework utilized in this international research project, analyzing small islands as eco-territories. We define eco-territories as unique geographic ecosystems with diverse political economies and socio-cultural histories. These varying island characteristics often structure the dynamic contour of a territory's developmental trajectory. At the intersection of small island eco-territories, sport tourism innovation, and sustainable development, the chapter calls for a moral mandate and the political will to adopt a critical consciousness or eco-pedagogy—a local to global commitment to social and environmental justice.

Finally, Chapter 6 describes the comparative research methodology utilized for this research project. By focusing on small island eco-territories, a geographical or spatial unit of analysis, this book has sought to contrast similar or like territories and their unique relationships to sport tourism and sustainable development. We anticipated significant variation across both territorial site and time, even within the same oceans, seas, and archipelagos.

Part II of the book offers 21 case studies, utilizing the research study's comparative methodology. This section is organized by sea and ocean, with distinct chapters devoted to each of these unique bodies of water. As the global map (Fig. 1.1) illustrates below, these island territories span the globe. We begin in the Mediterranean (including the Aegean) and North (including the Baltic) Seas, depicted as central to the map longitudinally. We then move west from Central Europe across the Atlantic Ocean, culminating this leg of the journey in the Caribbean Sea. We proceed across the vast Pacific Ocean, spanning nearly 20,000 km, ending our circumference of the world in the Indian Ocean archipelago. Along the journey, we seek a new world, a better world, envisioning a healthier and more sustainable planet.

Table 1.1 documents the geographic coordinates of all of these islands, noting their latitude and longitude. Latitudes and longitudes represent the metrics of the geographic coordinate system. Latitudes are horizontal lines that measure the distance north or south of the equator. Lines of longitude are vertical lines, also called meridians, that measure east or west from the prime meridian (0 degrees) in Greenwich, England. These geographic coordinates accompany the map above (see Fig. 1.1) so as to assist the reader in getting his or her bearings. Once within the islands and following the foundational chapters (Part I), this book can be read as an

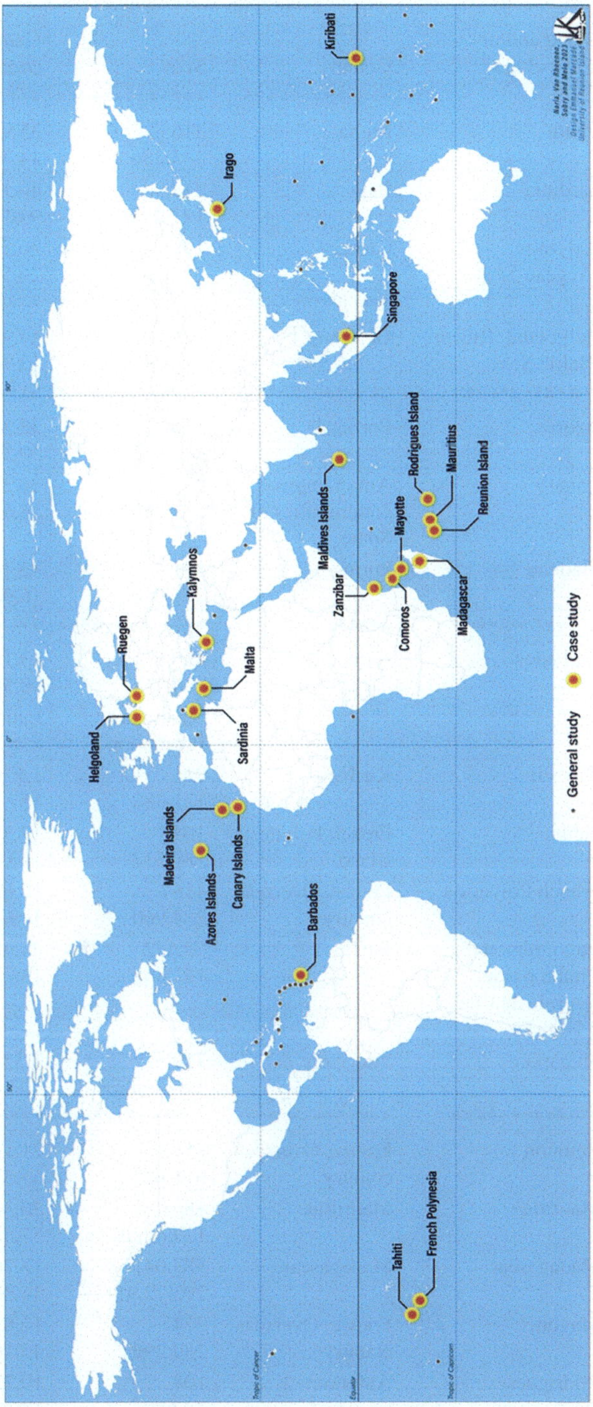

Fig. 1.1 Map of small island case studies

Table 1.1 Island territories by country, size and geographical coordinates

Ocean or sea	Island and/or archipelago	Country	Size	Geographic coordinates
Mediterranean	*Sea*	*************	*************	*************
*************	Malta	Malta	316 533,286	35.9375 N, 14.3754 E
*************	Sardinia	Italy	24,090 1.639,500	40.1209 N, 9.0129 E
*************	Kalymos (Aegean Sea)	Greece	109 29,000	36.5924 N, 26.5906 E
North	*Sea*	*************	*************	*************
*************	Helgoland, Rügen (Baltic Sea)	Germany	926 77,000	54.3520 N, 13.3630 E
Atlantic Ocean	*************	*************	*************	*************
*************	Azores	Portugal	2333 243,862	38.305542 N, 30.38410 W
*************	Canary	Autonomous communities of Spain	7447 2.207 million	28.2916 N, 16.6291 W
*************	Madeira	Portugal	801 251,060	32.7607 N, 16.9595 W
Caribbean Sea	*************	*************	*************	*************
*************	Barbados	Barbados	430 288,255	13.1939 N, 59.5432 W
Pacific Ocean	*************	*************	Square km Population	Latitude, longitude
*************	Kiribati	Kiribati	811 121,388	1.8369 N, 157.3768 W
*************	Tahiti	French Polynesia island	1042 189.517	17.375999 S, 149.265999 W
*************	French Polynesia	France, Overseas country	4167 284,960	17.6797 S, 149.4068 W
*************	Irago, région du Chūbu partie centrale de l'île d'Honshū	Japan	72,572 127,368	34.6505 N, 136.9867 E
*************	Singapore	Singapore	719 5.64 million	1.3521 N, 103.8198 E
Indian Ocean	*************	*************	*************	*************
*************	Reunion	France, Overseas country	2512 911,682	21.1151 S, 55.5364 E
*************	Mauritius	Mauritius	1865 1.266 million	20.3484 S, 57.5522 E
*************	Madagascar	Madagascar	587,041 29,472,911	18.7669 S, 46.8691 E
*************	Mayotte	France, Overseas country	374 289,290	12.8275 S, 45.1662 E
*************	Rodrigues	Autonomous region of Mauritius	108 43,650	19.7245 S, 63.4272 E

(continued)

Table 1.1 (continued)

Ocean or sea	Island and/or archipelago	Country	Size	Geographic coordinates
*************	Union des Comoros	France, Overseas country	1862 844,410	11.39713 S, 43.222135 E
*************	Maldives	Republic of Maldives	298 415,592	3.2028 N, 73.2207 E
*************	Zanzibar	Tanzanian archipelago	2461 800,010	6.1357 S, 39.3621 E

open text (Eco, 1989[1]), beginning where you would first like to travel and proceeding in the direction that most piques your curiosity. The book is also open in promoting a freedom of movement by exploring a critical study of eco-territories. This critical exploration analyzes the present while envisioning an aspirational future, one that, together, we may realize.

Part III offers even larger comparative analyses, drawing from the case studies presented in Part II. Olivier Naria et al. provide a comparative and regional analysis of the primary islands in the Indian Ocean, recognizing both similarities and differences at the intersection of sport tourism and sustainable development across this diverse archipelago. These authors analyze a variety of initiatives utilized by stakeholders seeking economic growth and global visibility in this group of islands, focusing in particular on future prospects for sport tourism and sustainable development within these diverse territories. The articulation of these social practices and their socio-cultural and environmental implications showcase to other like destinations how best to develop more sustainable sport tourism activities in accordance with the United Nation's SDGs. The model presented here, based on the book's comparative methodology, offers a conceptual and ethical framework when analyzing the promotion of sport tourism activities or events in these particular eco-territories.

The next chapter of our comparative analyses of islands as eco-territories highlights a first-of-its-kind global study encompassing three vast oceans, numerous seas, and both large and small archipelagos located within these waters. These geographical locations or sites include a wide variety of island types to this international research project. The conclusion of this comparative research project promises the articulation of best practices to consider for future planning and implementation from the local to regional to global perspective. Finally, the culminating chapter

[1] The idea of an "open text" (Eco, 1989) has been the subject of considerable interest within literary criticism since the middle of the twentieth century. Of note here is Julio Cortazar's revolutionary publication of *La Rayuela* (1963, translated into English as *Hopscotch* in 1963), in which he encourages the reader to participate in the production of the novel. Cortazar suggests two possible ways of reading the book. The first is to read in the normal and linear fashion, or what he refers to as the "passive" approach. The second way of reading the novel takes control of the novel's proposed order by reading chapters in alternative sequence. An open text might also be regarded as critical and liberatory, seeking a hopeful, utopian future.

provides several key snapshots from this comprehensive exploration, mapping the challenges and opportunities at the intersection of small island states, sport tourism activities and events, and sustainable development. Lessons learned from this international research project will inform future directions, contributing to the existing literature. We hope that the proposed conceptual model and revised comparative methodology will help both scholars and practitioners working at this dynamic intersection. We also hope that this international project contributes to the literature as a valuable resource guide based on the multiple characteristics articulated within these diverse case studies.

Following these comparative analyses, Giovanni Di Cola provides the reader a global perspective, highlighting the critical role of large, international organizations, such as the United Nations and the International Labour Organisation (ILO). This chapter outlines the recent international discourse on climate change resilience and the promotion of both blue and green economies among Small Island Developing States (SIDS). This discourse includes innovative global developments such as the Multidimensional Vulnerability Index (MVI), which positions Small Island Developing States (SIDS) at the forefront of local to global sustainable development efforts. This book, and the collective work of IRNIST, hopes to contribute to these global conversations at this critical intersection. With the publication of this volume, we will present our findings and recommendations at the United Nations' 4th International Conference on Small Island Developing States (SIDS) to be held in May 2024 in the sovereign island country of Antigua and Barbuda in the West Indies (at the conjuncture of the Caribbean Sea and Atlantic Ocean).

References

Barnosky, A. D., Matzke, N., Tomiya, S., Wogan, G. O. U., Swartz, B., Quental, T. B., Marshall, C., McGuire, J. L., Lindsey, E. L., Maguire, K., Mersey, B., & Ferrer, E. A. (2011). Has the Earth's sixth mass extinction already arrived? *Nature, 471*, 51–57. https://www.nature.com/articles/nature09678

Bouchet, P., & Sobry, C. (2019). Sports tourism: Contemporary issues and new trends on a global market. In M. Desbordes, P. Aymar, & C. Hautbois (Eds.), *The global sport economy* (pp. 295, 330p–314). Routledge.

Coetzee, J. M. (1986). Foe. New York: Penguin.

Colman, Z., & Mathiesen, K. (2022, November 27). Summit yields "historic win" for climate payments. *Politico*. https://www.politico.com/news/2022/11/19/cop27-climate-deal-payments-00069577. Accessed December 22, 2022.

Eco, U. (1989). *The open work*. Harvard University Press.

Fountain, H., & Rojanasakul, M. (2023, January 10). The last eight years were the hottest on record. *New York Times*. https://www.nytimes.com/interactive/2023/climate/earth-hottest-years.html. Accessed January 2023.

Melo, R., Sobry, C., & Van Rheenen, D. (Eds.). (2021). *Small-scale sport tourism events and local sustainable development: A cross-national comparative perspective*. Springer Nature AG.

Montes, M. F. (2016). The right to development, small island developing states and the SAMOA pathway. *South Centre Policy Brief, 25*, 1–12.

Naggs, F. (2017). Saving living diversity in the face of the unstoppable 6th mass extinction: A call for urgent international action. *The Journal of Population and Sustainability, 1*(2), 67–81.

Naria, O., Van Rheenen, D., Melo, R., & Sobry, C. (2022). Sport tourism, island territories and sustainable development. In B. Mazza & C. Sobry (Eds.), *Sport tourism and its territorial development and opportunities* (pp. 19–40). Cambridge Scholars Publishing.

Paddison, L. (2022, December 19). More than 190 countries sign landmark agreement to halt the biodiversity crisis. *CNN*. https://www.cnn.com/2022/12/19/world/cop15-biodiversity-agreement-montreal-climate-scn-intl/index.html. Accessed December 2022.

Plumer, B., Friedman, L., Bearak, M., & Gross, J. (2022, November 19). In a first, rich countries agree to pay for climate damages in poor nations. *New York Times*. https://www.nytimes.com/2022/11/19/climate/un-climate-damage-cop27.html#:~:text=the%20main%20story-,In%20a%20First%2C%20Rich%20Countries%20Agree%20to%20Pay%20for%20Climate,about%20how%20it%20would%20work. Accessed December 22, 2022.

Sachs, J., Lafortune, G., Kroll, C., Fuller, G., & Woelm, F. (2022). *Sustainable development report 2022 – From crisis to sustainable development: The SDGs as roadmap to 2030 and beyond*. Cambridge University Press.

Stracqualursi, V., & Liptsk, K. (2021, February 19). What you need to know about the Pairs climate agreement now that the U.S. has rejoined. *CNN*. https://www.cnn.com/2021/02/19/politics/what-is-paris-climate-agreement/index.html. Accessed December 26, 2022.

Van Rheenen, D., Melo, R., & Sobry, C. (2021). Defining a methodology for a cross-national comparative study. In R. Melo, C. Sobry, & D. Van Rheenen (Eds.), *Small-scale sport tourism events and local sustainable development: A cross-national comparative perspective* (pp. 43–54). Springer Nature AG.

Weiland, S., Hickmann, T., Lederer, M., Marquardt, J., & Schwindenhammer, S. (2021). The 2030 agenda for sustainable development: Transformative change through the sustainable development goals? *Politics and Governance, 9*(1), 90–95.

Part I
Conceptual and Methodological Frameworks: A Comparison of Place from an Eco-territorial Perspective

Chapter 2
Small Island States or Territories

Derek Van Rheenen, Olivier Naria, Ricardo Melo, and Claude Sobry

I am a rock.
I am an island.
And a rock feels no pain
And an island never cries.

Paul Simon (1965)

From a western, post-colonial perspective, islands evoke an exotic destination, an escape to tropical paradise (Bernardie-Tahir, 2011)—a respite from the cold, rational world of modernity, where pale, rheumatoid fingers poke and prod us to produce and consume, evermore. These islands are home to their own peoples with distinct cultural identities, indigenous social histories, and their unique development is often lost on the visiting tourist, a traveling interloper spending their well-earned time and money to relax. These tourists work to consume; they sense they have earned and therefore deserve the needed rest and relaxation, worn and frayed from their daily labors and productive lives. And so, as tourists do, they come to the

D. Van Rheenen
Cultural Studies of Sport in Education, University of California Berkeley, Berkeley, CA, USA
e-mail: dvr@berkeley.edu

O. Naria
Department STAPS, Université de La Réunion, Le Tampon, ïle de la Réunion, Réunion, France
e-mail: olivier.naria@univ-reunion.fr

R. Melo
Coimbra Education School, Polytechnic of Coimbra, Coimbra, Portugal

C. Sobry (✉)
Faculty of Sport and Physical Education, Université of Lille, Lille, Nord, France
e-mail: Claude.sobry@univ-lille.frxt

D. Van Rheenen et al. (eds.), *Sport Tourism, Island Territories and Sustainable Development*, Sports Economics, Management and Policy 24,
https://doi.org/10.1007/978-3-031-51705-1_2

islands, spend time and money, and then go home. They engage in this modern practice of tourism, whether rested or not, and whether respectful of the place and the nature they have just visited.

Like traveling to an island, enjoying sun, sand, and sea, (3S's) often objectifies the environment as a resource to be used or experienced for personal gain rather than being with, and as a part of, nature. As Warden (2019) challenges, "how can you visit something of which you are a part?" (9). By reframing travel as relational and respectful, can we become one with others and with nature? These are important guiding questions as we conceive of sustainable development through sport tourism *in relation to* small island territories and their local peoples.

Islands also conjure a primordial fear of human survival, a lonely, sometimes singular fight to overcome nature, both physical and human. The popular Simon and Garfunkel 1960s song entitled "I am a Rock" begins with the initial stanza

A winter's day
In a deep and dark December
I am alone
Gazing from my window to the street below
On a freshly fallen silent shroud of snow

The song ends with the following lines:

I am a rock
I am an island
And a rock feels no pain
And an island never cries.

This haunting song captures the numbing and poetic sound of silence, of loneliness, a fearful soul self-sequestered from the world in order to avoid feeling, in order to avoid feeling pain. It is a song of a man lost in his own isolation, metaphorically lost at sea.

Western literature is filled with fictional tales of individuals lost at sea only to navigate the wilds of an inclement rock surrounded by breaking waves. These narratives of self-preservation and potential salvation unearth humankind's quest to understand who he is and who he is meant to be. These stories are parables of possibility and limitation. The shipwrecked, once on the island, seeks elaborate ways to get off of the island, an existential quest to return from "savagery" to "civilization," with all of its comforts and trappings. But since that may never come to pass (Robinson Crusoe was shipwrecked on *his* island for over 28 years before freeing himself), the island survivalist must create their own sustainable ecosystem—food, water, shelter, security, even amusement, and methods to confront isolation and loneliness.

Island tourists are the antithesis of these shipwrecked survivalists. They come of their own volition, stay as long as they want, and do whatever they want while on the island destination. These travelers can also leave whenever they want. They want for little, provided a safe harbor and an all-inclusive smorgasbord of food, drink, and orchestrated activities. In general, they arrive and depart in comfort, by cruise ship or by air, a far cry from the image of the wretched soul crawling through the crashing waves to land, a half-clothed body strewn upon a splash of sand.

Islands are also unique topographies, physical and social geographies that witness coming storms well before the mainland. So too is the case with climate change and dislocation, erosion and cultural eradication, and small island states or territories sending the signal to other places housed in urban fortresses that all is not well with the planet. Seas are rising with global warming, storms are more severe than ever, and the weather forecast for the future is bleak. Some small islands risk survival of their own in this deteriorating climate (Duvat & Magnan, 2012), prompting the urgent need for global change and a concerted commitment to sustainability.

Island Typology: Defining Small Island States or Ecoterritories

This research project focuses on a comparison of a similar unit of analysis, small island states or ecoterritories, and unique geographic ecosystems with diverse political economies and sociocultural histories. Travelers, geographers, explorers, and historiographers have sought to define this distinct "piece of land surrounded by water" (OED, 1989). Small island states have been defined within a larger island typology, differentiated from rocks (Charney, 1999: Schofield, 2009), reef islands and atolls (Dobbs, 2005; McNeill, 1954; Watts, 2019), island dependencies, state islands, and island countries (Armstrong & Read, 2003; Baldacchino, 2007, 2020; Baldacchino & Dehoorne, 2014; Bernardie-Tahir & Taglioni, 2005; Petzold & Magnan, 2019; Taglioni, 2003). State islands, such as Hawaii (USA), the Galapagos Islands (Ecuador), Corsica and Belle (France), Okinawa (Japan), and Lakshadweep and Diu (India), to name a few, are often excluded from global and comparative island data due to the paucity of accurate tourism information, such as number of visits, length of stay, and contributions to Gross National Product (GNP). This data is incomplete primarily because of a lack of border controls within their larger nation-states. Island countries, such as the United Kingdom, New Zealand, Iceland, Singapore, and Japan, among others, may be centered on one or two main islands, while others spread across hundreds or thousands of smaller islands, such as Indonesia, the Bahamas, and the Seychelles (Weaver, 2017).

Small islands exist in both warm and cold water contexts and span the earth as evident when one turns a globe. Weaver (2017: 11) acknowledges that "'small island' status is a subjective designation that compromises between population and area." Relative to area, Nunn (1987) has defined a small island as an island of less than 10,000 square kilometers. Similarly, as Bertile asserts in the next chapter of this book, Small Island Developing States (SIDS) are recognized by international organizations as "lands surrounded by water on all sides in a single block, with a surface area of less than 11,000 km^2 and a population of less than 1.5 million inhabitants. Bertile focuses even more specifically on "small tropical island territories" and their geomorphic relationship to sport tourism historically. The author differentiates between high and low islands, whereby the former are generally volcanic and the latter almost exclusively formed from limestone.

Altitude affects temperatures and rainfall. High islands tend to be lower in temperature, which increases precipitation. In fact, high mountainous tropical islands receive the most rainfall than anywhere else in the world. For example, La Réunion island receives 7.5 billion m^3 of rainfall annually, 3.5 times the amount France receives per year (Météo France, 2023). Conversely, the low or flat islands have very modest altitudes, as their limestone foundations are comprised of thick coral sediment.

Finally, as Bertile notes, the relative level of tourism within the economies of these island territories creates its own typology, based on percentage of GDP. Some islands are highly dependent on tourism, such as the Seychelles, the Maldives, Madeira, and the Canary Islands, with 25–35% of their GDP based on tourism. Additionally, based on the distinctions between high and low islands described above, some islands specialize on beach tourism because they are coral or low-lying, respectively, such as the Maldives and Mauritius. Conversely, other islands have a greater variety of landforms, providing the physical spaces for a diversity of sport tourism activities and events, such as Hawaii, La Réunion, and the Seychelles. Nearly all of these islands have significant potential for increased and diversified sport tourism development. They also have an opportunity for the promotion of genuine sustainable development, whether as "blue tourism," connected to the sea and coastline, or "green tourism," connected to sport tourism activities experienced in the interior of the islands.

The United Nations World Tourism Organization (UNWTO) includes in its definition of small island states a composite of island designations (e.g., Guadaloupe), island countries (e.g., Iceland), and the singular city and island nation of Singapore. The UNWTO collects data for these small island states, including population, land area (square kilometers), number of overnight stays, number of cruise arrivals, and inbound revenue as a percentage of GDP.

As evidenced in Table 2.1., there is tremendous variation across these indicators. Note, for example, that several of these "small" islands boast areas well above the 10,000–11,000 square kilometer threshold as described above (e.g., Bahamas, Fiji, Jamaica, and Papua New Guinea). On the other hand, there remain a number of small island states that do not report this basic tourism data to the UNWTO. While recognized as fitting the working definition of a small island state, these territories are excluded from comparative view, as if they are uninhabited or did not exist at all. We believe that such exclusion does a disservice to island research, a reflection of arbitrary definitions of these spaces, and a paucity of data to date. And yet, small island states are perhaps as diverse as they are similar (Doumenge & Decoudras, 2010).

It is therefore critical to recognize that this designation and definition is not monolithic. This is true across oceans and seas as well as within the same body of water. The vast Pacific Island (PI) archipelago, for example, comprises Polynesia, the "many" islands, Micronesia, the "small" islands, and Melanesia, the "Black" islands. The islands of Micronesia, a region that includes approximately 2100 islands and a total ocean area of 7,400,000 square kilometers (Kirch, 1997, 2017) are not the only "small" islands in the PI archipelago. Niue, a Polynesian island

Table 2.1 Island tourism data

Islands territory	Surface area	Population	International arrivals	Average length of stay	Number of rooms	Inbound tourism expenditure as a percentage of GDP
Antigua and Barbuda	442	98,000	1,062,000	9.8	3000	59.8
Bahamas	13,940	393	6,622,000	6.73	17,028	24.8
Bahrain	771	1,702,000	1,117,000	9	18,033	10.2
Barbados	431	287,000	1,356,000	8.9	6657	21.6
Bermuda	53	61,000	766,000	5.86	2404	9.1
Cabo Verde	4033	556,000	710,000		13,187	26.4
Cayman Islands	264	64,000	2,384,000	6.07	6780	15.2
Comoros	2235	870,000	35,000	8	770	6.4
Cuba	109,884	11,327	4,712,000	9	70,204	2.59
Dominica	750	75,000	198,000	7.7	506	22
Dominican Republic	48,671	10,848,000	7,576,000	8.3	80,526	9.3
Federated States of Micronesia	702	115,000	19,000	9.1	500	7.9
Fiji	18,272	896,000	1,058,000	11.15	10,012	25
Grenada	345	112,000	528,000	9.1	2242	45.4
Haiti	27,750	11,402,000	1,333,000	7.6	3495	6.7
Jamaica	1990	2,961,000	4,319,000	8.8	29,635	19.7
Kiribati	726	119,000	9000	13.4	322	2.1
Madagascar	587,295	27,691,000	291,000	21	28,170	7.3
Malta	315	442,000	3,232,000	6.1	19,057	12.7
Maldives	300	540,000	1,484,000	6.4	22,577	57.9
Marshall Islands	181	53,000	7900	5.3	281	9.5
Mauritius	1969	1,272,000	1,431,000	10.4	13,523	15.2
Mayotte	374	256,500	56,000	30	1113	1.9
Palau	459	22,000	106,000	4.9	2372	42.2
Papua New Guinea	462,840	8,847,000	195,000	23	6195	0.1
Philipines	300,000	109,000,000	7,168,000	9.07	102,752	12.7
Reunion	2512	860,000	533,000	17	3558	3.3
Samoa	2842	198,000	172,000	8.2	5000	22.4
São Tomé and Príncipe	964	219,000	33,000		872	17
Seychelles	457	98,000	405,000	11	6059	38.4
Singapore	719	5,850,000	18,508,000	3.3	66,994	5.6
St. Kitts and Nevis	261	57,000	1,278,000	7	1300	35.3

(continued)

Table 2.1 (continued)

Islands territory	Surface area	Population	International arrivals	Average length of stay	Number of rooms	Inbound tourism expenditure as a percentage of GDP
St. Lucia	539	184,000	1,165,000	8.3	5078	52.7
St. Vincent and the Grenadines	389	111,000	356,000	13.5	2231	29.6
Solomon Islands	28,896	687,000	28,000	15	1556	6.5
Sri-Lanka	65,610	21,413,000	2,334,000	10	38,214	6.3
Timor-Leste	14,919	1,318,000	75,000	4	595	3
Tonga	747	106,000	77,000	7	1193	10.9
Trinidad and Tobago	5127	1,400,000	501,000	14	7731	2.3
Vanuatu	12,189	307,000	350,000	10.9	1722	37.1
Zanzibar	2656	1,671,598	394,185	9	20,991	4.3

country in the South Pacific Ocean, is in fact smaller at 261 square kilometers (and a population of approximately 1600) than several of the small islands of Micronesia (Ali, 2022). Similarly, the Micronesian archipelago is comprised of approximately 2100 islands as opposed to the more than 1000 islands of Polynesia, also known as the many islands.

The Black Islands of New Guinea, Fiji, Vanuatu, and the Solomon Islands (among others) refer to the indigenous populations' perceived racial identities based on skin color or purported phenotype. This racial classification of islanders began early in the "discovery" of colonized lands, seizing assets, and resources, at the expense of indigenous peoples and their communities (Charles & Rah, 2019; Miller, 2019). The original inhabitants of these islands ("First Peoples") were likely descendants of the present-day Papuan people.

While there is genetic differentiation between Melanesians and those inhabitants of Micronesia and Polynesia, there is likewise significant genetic variation and cultural diversity among the groups living in the Melanesian islands. Just as there is genetic differentiation and cultural diversity within groups of the same archipelago, there is likewise tremendous variation in topography from one island to the next, even those in relatively close physical proximity to one another (Friedlaender et al., 2007; Hunley et al., 2008).

In 2020, analysis of the number of international tourist arrivals revealed contrasting situations on islands that have benefited from a century of global tourism growth from the early nineteenth century to the present day. While these figures may vary from one year to the next, depending on international tourism trends, the fact remains that there are both island centers and peripheries and island peripheries. As depicted in Fig. 2.1, the dominant islands relative to these statistics are also the most mature tourist destinations. These destinations include Singapore, followed by the

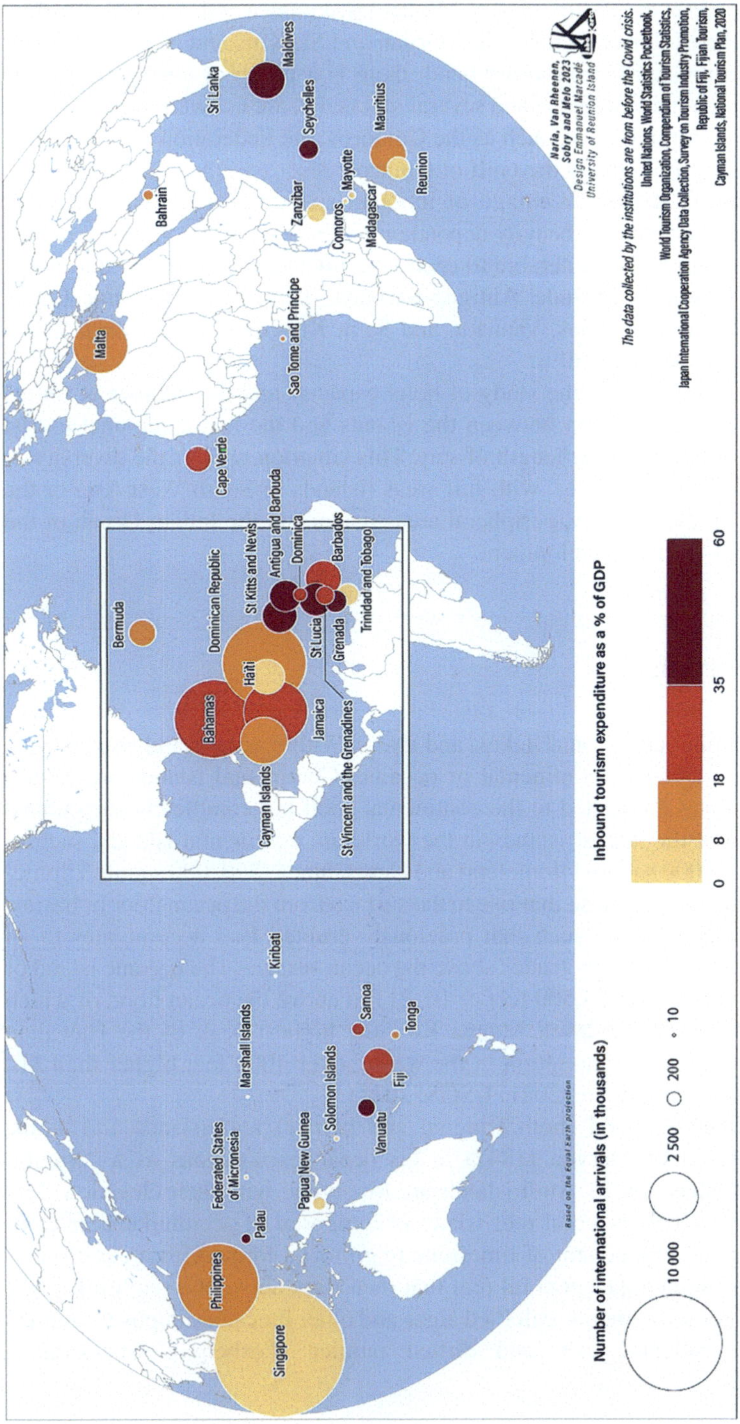

Fig. 2.1 International tourist arrivals, inbound spending, and average length of stays

Dominican Republic, the Philippines, the Bahamas, Jamaica and, to a lesser extent, Mauritius and the Maldives in the Indian Ocean and St. Kitts and Nevis and Saint-Lucia in the Caribbean. On the other hand, there is a group of emerging islands, such as Madagascar and Grenada, and several others that are not well integrated into the international tourism arena, such as the Comoros, the Federation of Micronesia and Sao Tome, and Principe in the Gulf of Guinea.

An examination of the percentage of inbound spending in the islands' GDP shows that several islands are heavily dependent on this sector of commerce, which can make them extremely vulnerable to economic fluctuations or the recent global health crisis. Examples include Antigua-Barbuda, the Maldives, Saint Lucia, Grenada, Palau, the Seychelles, Vanuatu, and Saint Kitts and Nevus, all of which report spending in excess of 30%.

As illustrated in Fig. 2.2, the study of hotel capacity and the number of rooms rented depicts the relationship between the islands and the major regional tourist areas, which can influence the length of stay. This situation reflects the diversity of the level of "tourism influence," with full areas (islands in South-West Asia or the Caribbean) or, on the contrary, peripheral areas (islands in the Indian Ocean or the Pacific) of island tourism development.

Island Topography

Islands exist within oceans, seas, lakes, and rivers. Within oceans and seas, islands can be classified as either continental or oceanic. Continental islands are unsubmerged land masses connected to the continental shelf but completely surrounded by water. Some of the largest islands in the world are continental islands, such as Greenland (2,175,000 square kilometers) and New Guinea (800, 000 square kilometers). Oceanic islands are those that rise to the surface from the ocean floor or basins. These islands are volcanic, such that previously erupted lava accumulates to an enormous thickness until it protrudes above the ocean surface. The oceanic island of Hawaii, for example, rises 33,500 feet or 10, 211 m above the ocean floor, of which 13,796 feet are above sea level. Mauna Kea, the highest peak of the Hawaiian Islands, is also the tallest mountain in the world, over 4000 feet higher than Mt. Everest (National Geographic, 2021; USGS, 2022).

There are likewise coral (both fringing and barrier) reef islands and atolls, famously described by Darwin (1842) in his ocean explorations as a "circular groups of coral islets" (p. 2). Atoll islands are low-lying, with their elevations less than 5 m (Watts, 2019). A coral reef island is composed of accumulated calcium carbonate, the chief component of limestone (e.g., rocks formed from coral skeletons) derived from an adjacent coral reef that then rises above the sea level. Their soils consist of coral fragments, calcified algae and other limestone deposits, humus, seabird guano, volcanic ash, and drifted pumice (Fosberg, 1976; Muller-Parker, 2005).

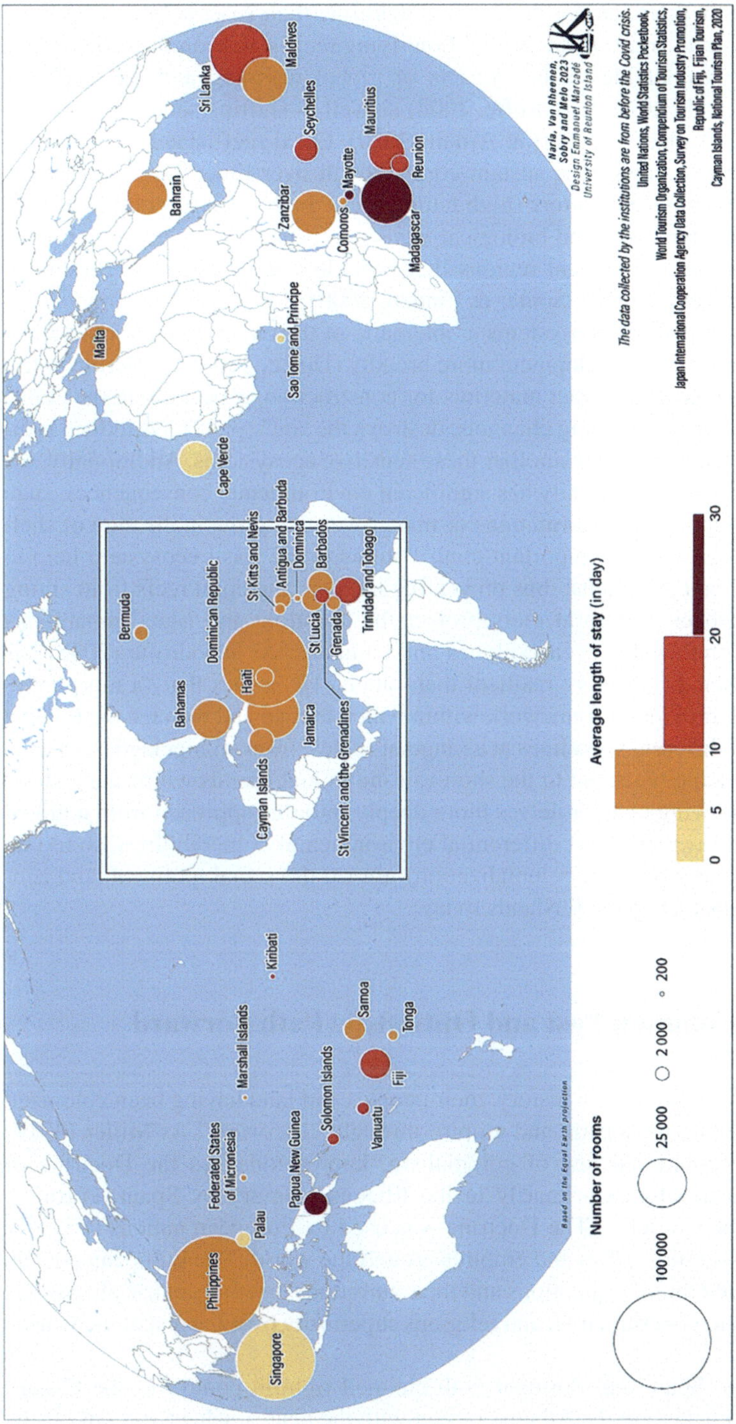

Fig. 2.2 Number of rooms occupied (relative to capacity) and average length of stay (in days)

Islands differ in their geomorphology and therefore their susceptibility to climate change and sea level rise (Duvat, 2015). Low-lying reef islands on the rim of atolls are particularly vulnerable to the impacts of global warming and sea-level rise (Oppenheimer et al., 2019; Woodroffe, 2008) as well as shifting sands and an ever-changing sense of place (Kothari & Arnall, 2020). Coral reef islands can develop only where suitable conditions sustain coral growth over time. These conditions include favorable physical factors (high temperature, high salinity, good light penetration, and low nutrients) and biological factors (especially a supply of coral larvae) in tropical and subtropical regions that provide a firm substrate (Birkeland, 1997; Tyllianakis et al., 2019; Souter & Linden, 2000).

Coral reefs are distinct ecosystems emblematic of the environmental risks associated with tourism and development more broadly (Duvat, 2008). As Bertile notes, "dredging of the reefs to extract materials for construction, especially in the atolls, or to build harbours or shipping channels, destroys the site" (p. 48). Dislodged sediments from such construction smother these sensitive ecosystems. Additionally, the impact of tourism on biodiversity has significant environmental consequences, such as the destruction of woody formations or mangroves that grow in the mud of sheltered bays or inlets. These important contributions to the local ecosystem help to filter sediments and pollutants, thus preventing lagoons and coral reefs from silting up. Both mangroves and coral reefs protect the coastline and island inhabitants against storms and swells, even cyclones and tsunamis. As Woodroffe (2008) has argued, some islands are more resilient than others. He argues that "a morphodynamic approach provides a framework within which to map and manage the relative vulnerability of different shorelines at a range of scales, from archipelagoes, to individual atolls or table reefs and to the shores of individual islands within each atoll" (p. 94). The following chapter delves more deeply into this approach from a global perspective. Just as there exist differential environmental vulnerabilities, so too do the vestiges of the past (e.g., colonialization) inform the social, political, and economic challenges facing small islands today.

An All Too Common Past and Optimistic Path Forward

Most islands share a common history, their peoples and land having been colonized by a nation seeking expansion and empire through "discovery." As Miller (2019) argues, "the international law of colonialism, known today as the Doctrine of Discovery, was developed primarily in the fifteenth century by Spain, Portugal, England, and the Church… The Doctrine was used by European nations to justify their desires to acquire riches and empires around the world. The European powers primarily justified these acquisitions and their ambitions by ethnocentric allegations of cultural, racial, governmental, and religious superiority over the rest of the world" (p. 35).

In the 1430s, Spain and Portugal both claimed rightful control of the Canary Islands, an archipelago in the Atlantic Ocean with the nearest island 108 km off the northwest African mainland. In 1436, Portugal convinced Pope Eugenius IV to issue a papal order granting the mainland country exclusive control of the Islands to civilize and convert the Canary Islanders to the "one true religion" and "for the salvation of the souls of the pagans of the islands" (ibid, 36).

Utilizing the Church to order such decrees of control and colonization, Spain sought papal approval in 1493 for the discoveries Columbus had made in the New World. Soon thereafter, the Pope drew a line of demarcation from the North to the South Poles, 300 miles west of the Azores Islands, granting Spain title to the lands "discovered and to be discovered" west of that line, and granting Portugal the same rights east of the line. Thereafter, Spain and Portugal applied the Doctrine of Discovery in Africa, Asia, and the Americas.

Soon thereafter, England, Holland, and France adopted this international law to claim discovery rights throughout the world and to expand their own empires and fortunes. As Miller states, "the colonial-settler societies that resulted from European colonization of much of the world, including the United States, continue to apply the Doctrine of Discovery" (p. 37). As a result of this colonial history, there is a constellation of subnational island jurisdictions today. According to Baldacchino (2010: xxii), "these constellations share various relevant features of innovative governance. They are not just relics from bygone days but are a response to the opportunities afforded by globalization."

While islands today are often framed within a linear framework of development, moving smoothly through processes of colonialism, decolonialism, and then postcolonialism, Baldacchino (2010: xxii) argues that many "mainland-island" relations can be better understood on the continuum of "shared rule" versus "self-rule." (xxiii). Similarly, in her *Postcolonial Nations, Islands, and Tourism*, Kapstein (2017) asserts that there are a variety of ways in which the center-periphery or mainland-island dichotomy is disrupted. She notes that islands are "socially constructed, historically relative, and creatively fluid" (pp. xvi–xvii) and central to the postcolonial national project. More specifically, the author argues that "islands are the building blocks of nations, and nation-building relies on the formation of island spaces before, during, and after their establishment as spaces of pleasure, punishment, containment, exclusion, labor and leisure" (xvii).

As modern social constructions, island territories have come to rely on tourism as an important staple of their economies. This historical development has created its own form of dependency, often relying heavily on populations from wealthy, industrialized nations, including their former colonizer. According to Kapstein, "tourism reflects the value system of the colonial nation-state, particularly values of surveillance, control, and consumption…as soon as package tourism was extended beyond Europe in the nineteenth century, tourism became 'an agency for the consolidation of Empire'" (p. xviii, as quoted in Carrington, 2009).

This development, particularly in the late twentieth and early twenty-first century, has highlighted the changing relationship between nature and human culture, as well as a prescient debate between economic (high destiny) and ecological (low density) criteria of development (Baldacchino, 2010). Baldacchino differentiates between these two opposing development paradigms. The economic-development approach has poor and degraded local natural environments with typically high population destinies, large pools of immigrant labor, and extensive financing (overseas) investments. Countries such as Japan and city-states such as Hong Kong, Malta, and Singapore are examples of this paradigm. The ecological-development approach, according to the author, represents island contexts with pristine, serene natural environments with low population densities, often accompanied by distinctive cultural practices associated with indigenous communities. Samoa, Seychelles, Tobago, and Dominica are examples of this type of development paradigm.

This debate between opposing paradigms has also led to what Kapstein refers to as "the constructed nature of nature, and the imbrication of local and global economies" (p. xxxiii). She writes "a focus on island tourism ushers in necessary conversations about the myths of untouched nature and about ecotourism, sustainability, and game parks and nature reserves. Donna Haraway (2013) reminds us that these issues cannot be divorced from economies, that the postcolonial nation is as invested in extracting value from its natural resources as the colonial empire ever was" (xxii).

In the pursuit of sustainable development, however, there are multiple dimensions to consider, including the ethical, ecological, social, economic, and political (e.g., governance) (Villeneuve et al., 2017). Based on these multiple, and at times conflicting, dimensions of development, scholars have argued that it will take an integrated and concerted effort to create a sustainable future. Specifically, Beddoe et al. (2009: 2488) surmise, "it will require a whole systems approach at multiple scales in space and time. It will require integrated, systems-level redesign of our entire socio-ecological regime, focused explicitly and directly on the goal of sustainable quality of life rather than the proxy of unlimited material growth." This transformative systemic approach targets focused change and offers strategic responses so as to guide policy interventions. Scoones et al. (2020) recognize that this requires the acknowledgment of three practical and cross-cutting principles underpinning the transformation to sustainability: "taking seriously diverse knowledges, plural pathways and the inherently political nature of transformations" (p. 71).

As more fully developed in Chapter 5 of this book, this systemic redesign simultaneously requires intentionally fostering relational equity among key stakeholders. Also known as relational equity management (Geissdoerfer et al., 2017; Koulouri & Mouraviev, 2019; Sawhney & Zabib, 2020), this governance framework seeks to innovatively congeal the multiple and often competing priorities of a given sector or combination of several sectors (e.g. nexus). This framework fosters systemic sustainability by promoting effective resource management, such as balancing trade-offs and enhancing new and existing synergies (Al-Saidi et al., 2017). Relational equity, then, is a means toward achieving systemic sustainability (Bragdon, 2009; Scoones et al., 2020; Van Rheenen & Melo, 2021). At the intersection of small island territories, sport tourism, and systemic sustainable development, we learn from previous iterations of these historical linkages and envision an optimistic and intentional future.

Small Island Developing States (SIDS)

Small Island Developing States (SIDS) represent an association and intergovernmental organization of 58 low-lying coastal small island countries, 38 of which are United Nations U.N members while 20 are nonmember states. These SIDS are grouped into three broad regions: the Caribbean, Pacific, as well as African, Indian Ocean and South China Sea (AIS). The low-lying coastal countries share many similar characteristics.

The aggregate population of all of these SIDS is 65 million inhabitants, less than one percent of the world's total population (UN, 2023). The majority of the Small Island Developing States are considered Middle-Income Countries (MICs), with a per capita Gross Domestic Product (GDP) of $11,200 USD, weighted by population and ranging significantly from approximately $64.600 USD in Singapore to $800 USD in Guinea-Bissau. The total GDP of the 58 SIDS in 2020 was $733 Billion USD, for which Singapore accounted for about half of that sum (DiCola, Chap. 30, this volume). The Association of Small Island States' (AOSIS) mission focuses on three broad areas: climate change (Allen, 2018; Cook et al., 2016); sustainable development (Clark & Harley, 2020; Purvis et al., 2019), and marine or ocean conservation (Carleton Ray & McCormick-Ray, 2014; Zacharias, 2014). The main purpose of the alliance is to consolidate the voices of Small Island Developing States (SIDS) to address global warming (Armstrong McKay et al., 2022; Chirico, 2018; Fountain & Rojanasakul, 2023). These states advocate for international policy and mechanisms for addressing the inequity of health and climate impacts and the disparate social and economic repercussions of these negative impacts. While the COVID pandemic has exposed the fragility of small island states relying too heavily on the tourism industry, initiatives, and recovery plans offer possible paradigm shifts toward a greener, more sustainable economy. These efforts include the development of a stronger agricultural sector and the diversification of local economies. Sport tourism activities and events have also been the focus of these diversification efforts.

Due to their comparative size and jurisdictions, SIDS are well positioned to serve as catalysts or incubators for innovation, testing, and operationalizing newly designed sustainability policies and practices (Baldacchino, 2010; De Comarmond & Payet, 2010; DiNapoli & Leppard, 2018; Taitingfong, 2020; Whittaker et al., 2017). As Gillespie et al. (2008: p. 45) proffered, "islands can be considered 'nature's test tubes'—simple systems with multiple replicates." This translates into developing both actionable and scalable methods for implementation. The scaling of sustainability studies acknowledges the dialectical relationship between the local and the global (e.g., glocal), where, for example, climate change impacts local communities, while local communities, industries, and practices may well contribute to a larger carbon footprint. For the purposes of this research project, the following case studies support these innovative initiatives focused on the diversification of sport tourism activities and events coupled with the intentional sustainable development of the islands. And the time has never been so pressing, as global attention focused on this critical issue has risen to greater importance.

For example, small island states face the increasing repercussions of global warming with the associated rise of ocean levels (Fountain & Rojanasakul, 2023; IPCC, 2002, Climate Change, 2022, Mitigation of Climate Change, Working Group III Contribution to the Sixth Assessment Report of the Intergovernmental Panel on Climate Change) (Magnan et al., 2022; Terorotua et al., 2020). These islands represent dynamic spaces in flux, characterized by both natural resources and protected ecosystems, unique ecoterritories in which we can test potential trajectories of enhanced sustainability. Numerous international conferences over the last three decades (the Rio Earth Summit in 1992, the Barbados Conference in 1994, the World Summit on Sustainable Development in Johannesburg in 2002, the Mauritius Conference in 2005, the conference on the sustainable development of tourism in the islands in La Réunion in 2013; the five-day virtual Small Island Developing States (SIDS) Dialogues for Sustainable Development in 2022) have thereby emphasized the importance of developing specific, scalable strategies and action plans to promote the sustainable development of SIDS.

Future Implications: A Call to Action

In November 2022, the International Labor Organization (ILO), in collaboration with the United Nations Development Coordination Office (UN-DCO), the United Nations Office of the High Representative for the Least Developed Countries, Landlocked Developing Countries and Small Island Developing States (UN-OHRLLS), the United Nations Sustainable Development Goals (UN-SDG) Fund and the Investment Climate Reform (ICR) Facility, organized the Small Island Developing States (SIDS) Dialogues for Sustainable Development. This five-day virtual conference sought to reinforce the existing network of small islands to share local, regional, and national best practices of development strategies for a sustainable (blue) economy. In an effort to advance the SOMOA Pathway and 2030 Agenda (Montes, 2016), these strategies sought to foster economic innovation and diversification, thereby "bringing prosperity and decent jobs to island communities while restoring ocean and island ecosystems" (ITCILO, 2022). And yet, the promotion of a blue or green economy requires securing investment, finance, and access to technology while at the same time ensuring gender equality and the inclusion of local, marginalized social groups (UNEP, 2014).

This book, based on a comparative research project organized by the International Network in Sport Tourism (IRNIST), hopes to contribute to this ongoing dialogue. And there may be a glimmer of hope in these dark times, as many wealthier countries have recently agreed to pay for longstanding climate damage disproportionally impacting poorer nations, including many island nations. At the 2022 United Nations global climate summit held at the Red Sea resort of Sharm et Sheikh, Egypt, negotiators from nearly 200 countries agreed to create a fund that would help these poorer, more vulnerable nations address the climate disasters exacerbated by the pollution emitted by wealthy, industrial nations.

This first-ever global agreement calls for a committee comprised 24 countries to meet over the next year and work out the details of this agreement, including what form the fund should take, which countries will contribute, and where the money should go (Plumer et al., 2022). Despite the lack of specific details ironed out within this agreement, Avinash Persaud, Special Climate Envoy to the Prime Minister of Barbados, noted "It's a small victory for humankind" (Colman & Mathiesen, 2022).

And yet, the agreement includes few provisions that would cut current levels of greenhouse emissions and shift away from a reliance on fossil fuels. Without a far more ambitious reduction in carbon emissions, scientists, and skeptics fear that the deal all but promises continued global warming, rising seas, worsening storms, and devastating droughts. Of course, all of these catastrophic events will require additional funds to address these climate crises. Some of the locations most negatively impacted by this environmental degradation are small island states.

The following month, the Kunming-Montreal global biodiversity framework, signed by 190 countries worldwide, represented a sweeping agreement to address the global biodiversity crisis and protect the earth's natural resources. Calling this agreement a "major milestone" and "the Paris moment on biodiversity" in reference to the 2015 Paris Climate Agreement, the framework pledges to protect 30% of land and oceans by 2030. Only 17% of land and 10% of oceans are currently protected. The agreement also explicitly seeks to protect local, indigenous peoples and their communities, who have been excluded from these conversations in the past and even displaced from their lands in the name of conservation (Paddison, 2022).

Like the United Nations global climate agreement ratified the month before, however, critics fear that the Kunming-Montreal global biodiversity framework lacks any safeguards for successful implementation. These skeptics note that the framework is not legally binding and lacks specific, quantifiable metrics to reduce unchecked production and consumption, the root causes of the biodiversity crisis today. Despite these valid reservations, the tide seems to be shifting in our *weltanschauung* or worldview, promoting a collaborative and more urgent effort to address the global climate crisis. Small island states or territories provide a kind of test tube for innovation and implementation, given their unique geomorphological characteristics. These physical, social, and environmental characteristics, as well as their political-economic development historically, comprise a diverse set of dimensions worthy of scientific exploration as it relates to sustainable development. Islands are therefore the bellwether or predictor of what is to come for the planet, a local-to-global process relative to resource allocation, environmental protection, and sustainable development.

The following chapter delves more deeply into the geomorphology of small tropical islands globally. The author recognizes the potential relationship between landforms, morphological processes, and the diversification of sport tourism opportunities within these unique ecosystems. These ecosystems are impacted by both natural forces and human actions and must therefore be understood scientifically to best avail, protect, and preserve these unique island contexts.

References

Al-Saidi, M., Elagib, N.A., Ribbe, L., Schellenberg, T., Roach, E., & Oezhan, D. (2017). Water-energy-food security nexus in the eastern Nile Basin: Assessing the potential of transboundary regional cooperation. In *Water-Energy-Food Nexus: Principles and Practices; Salam*, P.A., Shrestha, S., Pandey, V.S., Anal, A.K., Eds.; American Geophysical Union: Washington, DC, USA, pp. 103–116.

Ali, N. (2022). Investigating the potential for sport tourism and wellness experiences. In *Routledge handbook on tourism and small island states in the Pacific* (p. 343). Routledge, Taylor & Francis Group.

Allen, J. T. (2018). Climate change and severe thunderstorms. In *Oxford research encyclopedia of climate science*. Oxford University Press.

Armstrong, H. W., & Read, R. (2003). Small states, islands and small states that are also islands. *Studies in Regional Science, 33*(1), 237–260.

Armstrong McKay, D., Staal, A., Abrams, J. F., Winkelmann, R., Sakschewski, B. S., Loriani, S., Fetzer, I., Cornell, S. E., Rockström, J., & Lenton, T. M. (2022). Exceeding 1.5°C global warming could trigger multiple climate tipping points. *Science, 377*(6611), eabn7950. https://doi.org/10.1126/science.abn7950.hdl:10871/131584. ISSN 0036-8075. PMID 36074831. S2CID 252161375.

Baldacchino, G. (2007). *A world of islands: An island studies reader*. Agenda Academic (Malta) & Institute of Island Studies (Canada), 640 p.

Baldacchino, G. (2010). *Island Enclaves. Offshoring strategies, creative governance and subnational island jurisdictions*. McGill-Queen's University Press.

Baldacchino, G. (2020). *A world of islands. The Routledge international handbook of island studies*. Routledge, 500 p.

Baldacchino, G., & Dehoorne, O. (2014). Island worlds: Spaces, temporalities, resources. *Caribbean Studies*, 27–28, https://journals.openedition.org/etudescaribeennes/6766

Beddoe, R., Costanza, R., Farley, J., Garza, E., Kent, J., Kubiszewski, I., Martinez, L., McCowen, T., Murphy, K., Myers, N., et al. (2009). Overcoming systemic roadblocks to sustainability: The evolutionary redesign of worldviews, institutions, and technologies. *Proc. Natl. Acad. Sci. USA 106*, 2483–2489.

Bernardie-Tahir, N. (2011). *L'usage de l'île*. Pétra, 510 p.

Bernardie-Tahir, N., & Taglioni, F. (2005). *Les dynamiques contemporaines des petits espaces insulaires: de l'île-relais aux réseaux insulaires*. Karthala, 443 p.

Birkeland, C. R. (1997). Introduction. In C. R. Birkeland (Ed.), *Life and death of coral reefs* (pp. 1–12). Chapman and Hall.

Bragdon, J. H. (2009). Capitalism as a human system: The value of relational equity. *Reflections, 10*(1), 1–8.

Carleton Ray, G., & McCormick-Ray, J. (2014). *Marine conservation: Science, policy, and management*. Wiley-Blackwell. ISBN 9781118714430. OCLC 841199230.

Carrington B (2009) Sport without final guarantees: Cultural studies/marxism/sport. In: Carrington B and Mcdonald I (eds) *Marxism, Cultural Studies and Sport*. New York: Routledge, pp. 15–32.

Charles, M., & Rah, S. C. (2019). *Unsettling truths: The ongoing, dehumanizing legacy of the doctrine of discovery*. InterVarsity Press.

Charney, J. I. (1999). Rocks that cannot sustain human habitation. *American Journal of International Law, 93*(4), 863–878.

Chirico, J. (2018). *Global problems, global solutions: Prospects for a better world*. SAGE Publications.

Clark, W., & Harley, A. (2020). Sustainability science: Toward a synthesis. *Annual Review of Environment and Resources, 45*(1), 331–386. https://doi.org/10.1146/annurev-environ-012420-043621

Climate Change (2022), IPCC, 2022: Climate Change 2022: Impacts, Adaptation and Vulnerability. Contribution of Working Group II to the Sixth Assessment Report of the Intergovernmental

Panel on Climate Change [H.-O. Pörtner, D.C. Roberts, M. Tignor, E.S. Poloczanska, K. Mintenbeck, A. Alegría, M. Craig, S. Langsdorf, S. Löschke, V. Möller, A. Okem, B. Rama (eds.)]. Cambridge University Press. Cambridge University Press, Cambridge, UK and New York, NY, USA, 3056 pp., https://doi.org/10.1017/9781009325844

Colman, Z. & Mathiesen, K. (November 27, 2022). Summit yields 'historic win' for climate payments." Politico. https://www.politico.com/news/2022/11/19/cop27-climate-deal-payments-00069577 [Accessed December 22, 2023].

Cook, J., Oreskas, N., Doran, P., & Anderegg, W. R. L. (2016). Consensus on consensus: A synthesis of consensus estimates on human-caused global warming. *Environmental Research Letters, 11*(4), 048002. https://doi.org/10.1088/1748-9326/11/4/048002

Darwin, C. R. (1842). *The Structure and Distribution of Coral Reefs*. Reprinted 1984 by the University of Arizona Press, Tucson. Google Scholar.

De Comarmond, A., & Payet, R. (2010). Small Island Developing States: Incubators of innovative adaptation and sustainable technologies. In *Coastal zones and climate change* (pp. 55–68). Henry L. Stimson Center.

Dictionary, O. E. (1989). *Oxford english dictionary*. Simpson, Ja & Weiner, Esc, 3.

DiNapoli, R. J., & Leppard, T. P. (2018). Islands as model environments. *The Journal of Island and Coastal Archaeology, 13*(2), 157–160.

Dobbs, D. (2005). *Reef Madness: Charles Darwin, Alexander Agassiz, and the Meaning of Coral*. Pantheon. ISBN 0-375-42161-0.

Doumenge, F., & Decoudras, P. -M. (2010). Les iles, "toutes uniques et si semblables". In "Comme un parfum d'îles", sous la direction de Sevin O., Chaléard J.-L. et Guillaud D. (dir.) (pp. 487–489). Presses de l'université Paris-Sorbonne (PUPS).

Duvat, V. (2008). L'evolution de la recherche sur les systems coralliens (1960–2007). *VerigO- la revue electrinique en sciences de l'environment [Online], 8*(2).

Duvat, V. (2015). Changement climatique et risques cotiers dans les iles tropicales. *Annales de geographie, 705*, 541–566.

Duvat, V., & Magnan, A. (2012). *Ces îles qui pourraient disparaître*. Editions Le Pommier, 191 p.

Fosberg, F. R. (1976). Coral island vegetation. In O. A. Jones & R. Endean (Eds.), *Biology and geology of coral reefs* (Vol. 3, pp. 255–277). Academic Press.

Fountain, H., & Rojanasakul, M. (2023). The last eight years were the hottest on record. *New York Times*. Accessed in January 2023 at https://www.nytimes.com/interactive/2023/climate/earth-hottest-years.html

Friedlaender, J. S., Friedlaender, F. R., Hodgson, J. A., Stoltz, M., Koki, G., Horvat, G., et al. (2007). Melanesian mtDNA complexity. *PLoS One, 2*(2), e248.

Geissdoerfer, M., Savaget, P., Bocken, N. M., & Hultink, E. J. (2017). The Circular Economy—A new sustainability paradigm? *J. Clean. Prod. 143*, 757–768.

Gillespie, R. G., Claridge, E. M., & Roderick, G. K. (2008). Biodiversity dynamics in isolated Island communities: Interaction between natural and human-mediated processes. *Molecular Ecology, 17*(1), 45–57.

Haraway, D. J. (2013). *Primate visions: Gender, race, and nature in the world of modern science*. Routledge.

How Big Are Hawaiian Volcanoes?. U.S. Geological Survey (USGS). Archived from the original on February 5, 2022.

Hunley, K., Dunn, M., Lindström, E., Reesink, G., Terrill, A., Healy, M. E., et al. (2008). Genetic and linguistic coevolution in Northern Island Melanesia. *PLoS Genetics, 4*(10), e1000239.

International Training Centre for the International Labor Organisation (ITCILO). (2022, November 21–25). The Small Island Developing States (SIDS) dialogues for sustainable development. https://www.itcilo.org/courses/small-island-developing-states-sids-dialogues-sustainable-development. Accessed January, 2023.

IPCC (2002), H Gitay, A Suárez, RT.Watson, DJ Dokken (Eds). *Intergovernmental Panel on Climatic Change*, Geneva, Switzerland. pp 85.

Kapstein, H. (2017). *Postcolonial nations, islands, and tourism: Reading real and imagined spaces*. Rowman & Littlefield.

Kirch, P. V. (1997). *The Lapita peoples: Ancestors of the oceanic world.* Wiley. ISBN 978-1-57718-036-4.

Kirch, P. V. (2017). *On the road of the winds: An archaeological history of the Pacific Islands before European contact.* University of California Press.

Kothari, U., & Arnall, A. (2020). Shifting sands: The rhythms and temporalities of island sandscapes. *Geoforum, 108*, 305–314.

Koulouri, A. & Mouraviev, N. (2019). *Policy and Governance in the Water-Energy-Food Nexus: A Relational Equity Approach*; Routledge: New York, NY, USA.

Magnan, A., Anisimov, A., & Duvat, V. (2022). Strengthen climate adaptation research globally, *Science, 376*,1398–1400. https://doi.org/10.1126/science.abq0737

Mauna Kea (PDF). National Geographic. Archived (PDF) from the original on August 12, 2021. Retrieved October 10, 2021.

McNeil, F. S. (1954). Organic reefs and banks and associated detrital sediments. *American Journal of Science, 252*(7), 385–401.

Météo France. (2023). http://pluiesextremes.meteo.fr/lareunion/-Evenements-memorables-.html

Miller, R. J. (2019). THE DOCTRINE OF DISCOVERY: The International Law of Colonialism. *The Indigenous Peoples' Journal of Law, Culture, & Resistance, 5*, 35–42. https://www.jstor.org/stable/48671863

Montes, M. F. (2016). The Right to Development, Small Island Developing States and the Samoa Pathway. *South Centre Policy Brief, 25*, 1–12.

Muller-Parker, G. (2005) Coral Reef Islands. In: Schwartz M.L. (eds) *Encyclopedia of Coastal Science. Encyclopedia of Earth Science Series.* Springer, Dordrecht. https://doi.org/10.1007/1-4020-3880-1_101

Nunn, P. D. (1987). Small islands and geomorphology: review and prospect in the context of historical geomorphology. *Transactions of the Institute of British Geographers*, 227–239.

Oppenheimer, M., Glavovic, B., Hinkel, J., Van de Wal, R., Magnan, A. K., Abd-Elgawad, A., ... & Sebesvari, Z. (2019). Sea level rise and implications for low lying islands, coasts and communities.

Paddison, L. (December19, 2022). More than 190 countries sign landmark agreement to halt the biodiversity crisis. CNN. https://www.cnn.com/2022/12/19/world/cop15-biodiversity-agreement-montreal-climate-scn-intl/index.html [Accessed December 2023].

Petzold, J., & Magnan, A. K. (2019). Climate change: Thinking small islands beyond Small Island Developing States (SIDS). *Climatic Change, 152*(1), 145–165.

Plumer, B., Friedman, L., Bearak, M., & Gross, J. (2022, November 19). In a first, rich countries agree to pay for climate damages in poor nations. New York Times. https://www.nytimes.com/2022/11/19/climate/un-climate-damage-cop27.html#:~:text=the%20main%20story-,In%20a%20First%2C%20Rich%20Countries%20Agree%20to%20Pay%20for%20Climate,about%20how%20it%20would%20work. Accessed 22 Dec 2022.

Purvis, B., Mao, Y., & Robinson, D. (2019). Three pillars of sustainability: In search of conceptual origins. *Sustainability Science, 14*(3), 681–695. https://doi.org/10.1007/s11625-018-0627-5. ISSN 1862-4065.

Schofield, C. (2009). Chapter II. The trouble with islands: The definition and role of islands and rocks in maritime boundary delimitation. In *Maritime boundary disputes, settlement processes, and the Law of the Sea* (pp. 19–37). Brill Nijhoff.

Scoones, I., Stirling, A., Abrol, D., Atela, J., Charli-Joseph, L., Eakin, H., et al. (2020). Transformations to sustainability: Combining structural, systemic and enabling approaches. *Current Opinion in Environmental Sustainability, 42*, 65–75.

Souter, D. W., & Linden, O. (2000). The health and future of coral reef systems. *Ocean & Coastal Management, 43*(8–9), 657–688.

Sawhney, M. & Zabib, J. (2020). Managing and Measuring Relational Equity in the Network Economy. *J. Acad. Mark. Sci. 30*, 313–332.

Taglioni, F. (2003). *Recherches sur les petits espaces insulaires et sur leurs organisations régionales, mémoire d'habilitation à diriger des recherches*, vol. II, 218 p.

Taitingfong, R. I. (2020). Islands as laboratories: indigenous knowledge and gene drives in the Pacific. *Human Biology, 91*(3), 179–188.

Terorotua, H., Duvat, V., Maspataud, A., & Ourigua, J. (2020). Assessing perceptions of climate change by representatives of public authorities and designing coastal climate services: Lessons learnt from Frech Polynesia. *Frontires in Marine Science, 7*(160). https://doi.org/10.3389/fmars.2020.00160

Tyllianakis, E., Grilli, G., Gibson, D., Ferrini, S., Conejo-Watt, H., & Luisetti, T. (2019). Policy options to achieve culturally-aware and environmentally-sustainable tourism in Fiji. *Marine Pollution Bulletin, 148*, 107–115.

United Nations. (2023). *About Small Island Developing States (SIDS)*. Retrieved in January, 2023 at https://sdgs.un.org/smallislands/about-small-island-developing-states#:~:text=The%20aggregate%20population%20of%20all,%2C%20economic%2C%20and%20environmental%20challenges

United Nations Environment Programme (UNEP). (2014). *Emerging issues for Small Island Developing States. Results of the UNEP Foresight Process*, Nairobi, Kenya.

Van Rheenen, D., & Melo, R. (2021). Nature sports: prospects for sustainability. *Sustainability, 13*(16), 8732.

Villeneuve, C., Tremblay, D., Riffon, O., Lanmafankpotin, G. Y., & Bouchard, S. (2017). A systemic tool and process for sustainability assessment. *Sustainability 9*, 1909.

Warden, C. (2019). Nature pedagogy: Education sustainability. *Child.Educ. 95*, 6–13.

Watts, L. (2019). *Energy at the end of the world: An Orkney Islands saga*. MIT Press.

Weaver, D. B. (2017), Core-periphery relationships and the sustainability paradox of small island tourism, *Tourism Recreation Research 42*(1), 11–21.

Whittaker, R. J., Fernández-Palacios, J. M., Matthews, T. J., Borregaard, M. K., & Triantis, K. A. (2017). Island biogeography: Taking the long view of nature's laboratories. *Science, 357*(6354), eaam8326.

Woodroffe, C. D. (2008). Reef-island topography and the vulnerability of atolls to sea-level rise. *Global and Planetary Change, 62*(1–2), 77–96.

Zacharias, M. (2014). Marine Environmental Protection Policy: International efforts to address threats to marine biodiversity. In *Marine Policy: An introduction to governance and international law of the oceans*. Routledge. ISBN 1136212477.

Chapter 3
Geomorphology, Sport Tourism and Sustainable Development in the Tropics

Wilfrid Bertile

Walking around the crater of a volcano, even in the middle of fog, is a popular trip for many walkers. An eruption, even in the cold of high altitude, attracts a large number of visitors. Thus, both the shape of a relief and the process of its evolution, the study of which is the subject of geomorphology, can be a factor in the location and attractiveness of outdoor activities. Practitioners and visitors come under the heading of sport tourism, which finds an attractive setting in the small tropical island territories. However, as these environments are fragile, it is important to preserve and promote them with a view to sustainable development.

Tourism, especially in its seaside form, occupies a major, sometimes excessive, place in the economy of most tropical islands, of which it is one of the development levers. Sport tourism, based on geomorphology, is a form of diversification of this "tourism monoculture", which is a source of vulnerability. The "natural environment" can serve as a support for other forms of tourism, provided that local populations are involved and the environment is safeguarded. This sustainable dimension applies both to tourism in particular and to island development policy in general.

The shape of the terrain (beaches, cliffs, valleys, etc.) determines the location of sporting activities that are practised both on the coast (blue tourism) and on land (green tourism). This wide range of activities gives sport tourism a growing economic, social and cultural importance which ends up being part of the landscape and territorial organisation. But the tropical islands are very diverse, due to their geographical location in relation to the major tourist emitting centres (North America, Europe, East and South-East Asia), the variety of their relief and the intensity of their tourism development, which generates very dissimilar situations. It is the islands combining beaches, coral reefs and a mountainous interior which are best able to diversify their tourist activities.

W. Bertile (✉)
Université de Saint-Denis, Reunion, France
e-mail: wilfrid.bertile@wanadoo.fr

D. Van Rheenen et al. (eds.), *Sport Tourism, Island Territories and Sustainable Development*, Sports Economics, Management and Policy 24,
https://doi.org/10.1007/978-3-031-51705-1_3

The aim here is to show how much the structural origin of the islands influences their location in the ocean basins and their differentiation, which gives them reliefs sculpted in particular by morphological evolution. The shapes of the relief are a resource for sport tourism, which appears to be a different tourist activity from mass tourism, while at the same time being an integral part of it. Despite their apparent immutability, landforms and morphological processes can be impacted by both natural factors and human actions. Due to the high vulnerability of island environments, any development has environmental consequences that can be severe. However, geomorphology is not only used as a support for sporting activities, it is also the major component of the ecosystems that make up the living environments of local populations. Through its heritage, it can also be the source of other forms of tourism in a sustainable development perspective, which is an imperative obligation for the islands.

Island Orogeny as a Factor in the Spatial Distribution and Physical Differentiation of Tropical Islands

The structural origin and morphological evolution of tropical islands determine their uneven distribution on the surface of the globe and the variety of their relief. Their location in the intertropical zone gives them a generally warm and humid climate and a biodiversity marked by endemism, making them globally attractive natural environments.

Uneven Spatial Distribution Linked to Structural Origin

The "geographical" distribution of tropical islands is very uneven. Islands and archipelagos are found in Central America, the eastern Atlantic, the mid-western Indian Ocean and the western Pacific. They are rarer in the South Atlantic, the Eastern Indian Ocean and the Eastern and Northern Pacific Ocean.

Most islands have their origin in volcanism. Despite their great variety, they fall into two main categories, islands arranged in rim arcs around the continents and oceanic islands in the middle of marine immensities.

The islands of the island arcs are found in Central America with the Caribbean arc and in the western Pacific Ocean from the Aleutians in the north to the New Hebrides in the south. They are the result of volcanoes born on the edge of a continental plate containing granitic or sedimentary rocks, under which an oceanic plate is sinking. These subduction zones are the site of explosive eruptions and earthquakes.

The Caribbean arc (Fig. 3.1), which is in fact made up of three eruptive arcs of increasing age from west to east, with their convexity facing the Atlantic, comprises

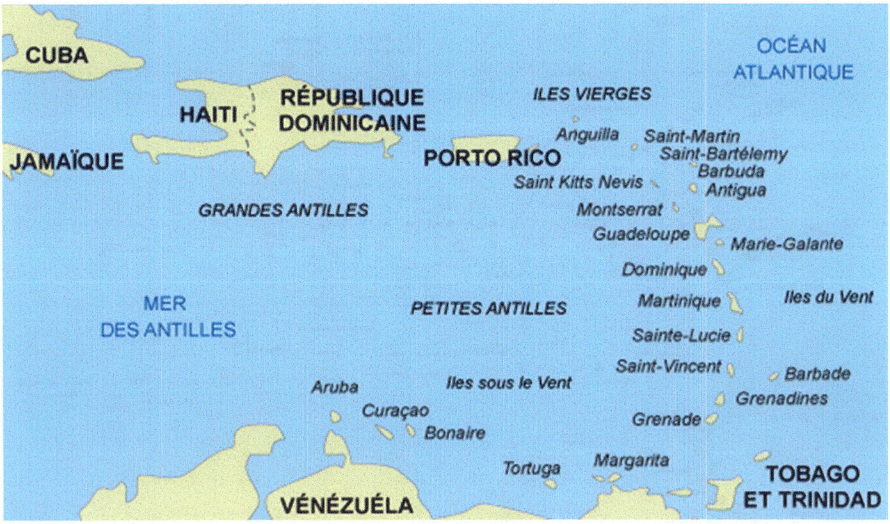

Fig. 3.1 The Caribbean islands

an internal garland of volcanic islands of Pliocene and Quaternary age, whose explosive volcanism, of the Pelean type, is often still active. A second outer arc made up of limestone constructions dating back to the Miocene period and raised a few dozen metres above sea level, leaving small, well-individualised islands (Marie Galante, Désirade, Barbados) or lands adjoining the islands of the aforementioned volcanic arc (Grande Terre of Guadeloupe, south-east of Martinique).

In the inter-tropical West Pacific (Fig. 3.2), the islands also form part of several island chains stretching from the Marianas to the New Hebrides and Fiji. They are the result of explosive volcanism resting on a base that may include fragments of crystalline or sedimentary rocks. This is the case of the Solomon Islands. However, basaltic volcanic rocks are undoubtedly the most represented.

Oceanic islands originate from hot spots resulting from a thinning of the mantle or a fracture of the oceanic crust. They arise from the ocean floor without incorporating continental material. Hawaiian-type effusive eruptions build basaltic stacks "sitting" on abyssal plains standing at around −4000 m. As these volcanoes sometimes exceed 3000 m in altitude, they form edifices that have nothing to envy the highest mountains on the planet. The oceanic islands are usually arranged in alignments linked to the movement of the tectonic plate above the original hot spot.

In the vast Pacific Ocean, the islands emerge along parallel lines oriented east-southeast/north-northwest. In the north, there is the line of Hawaii, in the centre the line of the Carolinas, the Society Islands, the Marquesas Islands, etc. The islands of coral origin (atolls) are oriented in the same way. It is known that the eastern part of the ocean sees only a few islands emerge, whether they are volcanic (Galapagos) or coral (Clipperton).

Fig. 3.2 Islands of the Pacific Ocean

The volcanic islands of the Atlantic are the outcrops of the mid-Atlantic ridge running from Iceland in the north to Gough Island in the south, where the African and American plates meet. The archipelago of the Azores, which is still volcanically active, the extinct strato-volcano of Madeira, the Canary Islands (whose 3718 m high Teide phonolith cone on the island of Tenerife is the highest point in Spain), joins the Cape Verde islands, which are very dislocated, to form Macaronesia.

The islands of the Indian Ocean (Fig. 3.3) range from Reunion Island to the Maldives in relation to the mid-ocean ridge. Born from a hot spot, Reunion Island is formed from the coalescence of two Hawaiian-type volcanoes of different ages, the most recent of which is still active (Piton de la Fournaise). The older volcanic edifices of Mauritius and Rodrigues have been badly eroded. The Maldives are almost entirely coral constructions. At the northern entrance to the Mozambique Channel, the Comoros are the result of the reanimation of a fracture that caused Mayotte to emerge 4.5 million years ago, followed by the other islands of the archipelago, in stages, up to Grande Comore, whose Karthala volcano is still active.

A special place is reserved for the Seychelles, whose main granite islands, with their steep relief, are witnesses to the vast southern continent which, 135 million years ago, began to break up to give birth to the Indian Ocean. According to Pierre Gourou, these are the "only truly oceanic islands that are granitic," with beautiful boulders "grooved by sea spray and rain." The lower islands of the archipelago are of coral origin.

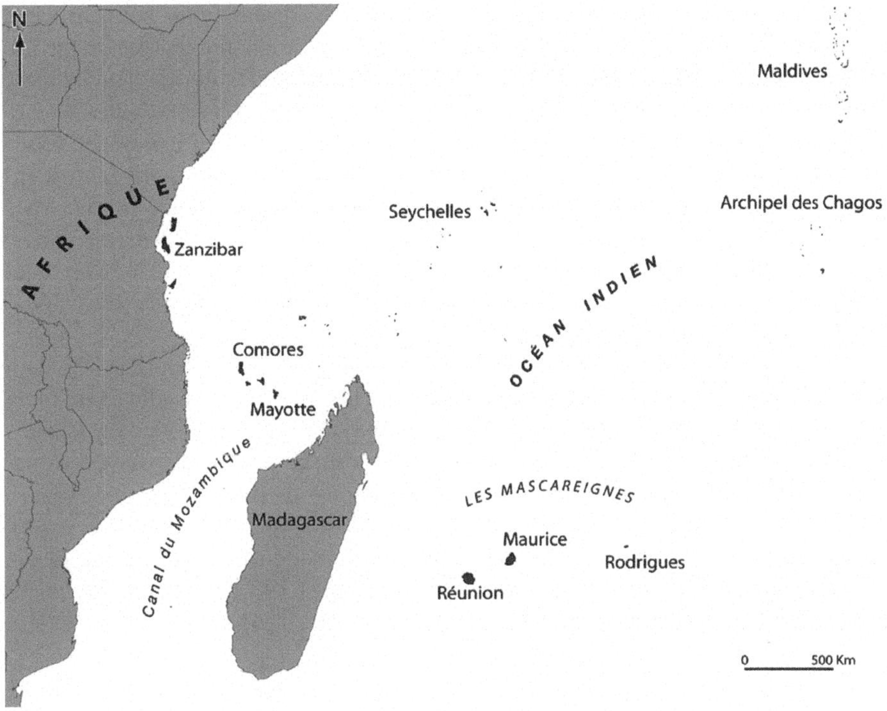

Fig. 3.3 The islands of the Indian Ocean. (Source: COM no. 245-2009)

Although other extra-tropical islands such as those in the Mediterranean offer attractive settings for tourism, what distinguishes tropical islands from islands in other latitudes is the climate. Their landscapes would lose their splendour without their vegetal finery and their generally "restless, humid and warm" atmosphere. Because of their location, the islands of the intertropical belt have a generally warm and humid climate. Average annual temperatures are high, but lower than those at the same latitude on the continents due to the softening influence of the sea and the cooling influence of the prevailing trade winds. The thermal amplitude is small, with average temperatures hardly varying throughout the year. In Papeete (Tahiti), the annual average is 26.3° but the hottest month, March, only reaches 27°6, while the "coolest" month, July, is still 25°6.

Rainfall is generally abundant (2000 mm on average per year in Fort-de-France) but poorly distributed in time and space. Two seasons can be distinguished. The "dry" season, corresponding to the cool season, runs from December to April in the northern hemisphere and from April to October in the southern hemisphere. The wet season, which is also the warm season during which cyclones occur, runs from July to October in the northern hemisphere and from November to March in the southern hemisphere. The distribution of rainfall shows a double imbalance, with areas at high altitude or exposed to the easterly trade winds receiving more rain.

However, these general observations mask quite diverse situations. The Caribbean islands are slightly warmer than the Mascarene Islands in the Indian Ocean, the latter being more open to the cold seas of the high southern latitudes. The Atlantic islands experience contrasting situations: the climate of the Azores, temperate oceanic, is characterised by a mild winter and a dry summer, due to the presence of the anticyclone of the same name. Madeira's climate, oceanic and subtropical, is exceptionally mild. The Canary Islands experience temperatures softened by the trade winds and the eponymous cold current, while the islands closest to Africa are almost desert-like, like the Cape Verde Islands. As for the Oceanic climates, they are constantly hot and humid. Curiously, there is a dry band in the central Pacific and a rather complex trade wind regime in Hawaii.

The isolation of the islands has made them ecological conservatories. The ecosystems have evolved in isolation from the continents. Species have adapted to them and, although their numbers are sometimes low, they are often specific to the islands, which are privileged places for endemism. What is striking is the disparity of situations. If, like Reunion Island, many islands are among the biodiversity "hot spots", others are somewhat poor because of their recent birth or their isolation (Polynesia). For example, in the Mascarene Islands, the rate of endemism for flowering plants (phanerogams) is 60% of all indigenous species, rising to 90% in Hawaii but reaching only 10% in the Azores.

Morphological Evolution: From Strato-Volcano to Atoll, the Variety of Potential of Tropical Islands

The physiognomy of islands is the complex result of a combination of the type of volcanism to which they owe their birth, the length of time during which erosion has taken place, isostatic adjustments linked to the deformation of the oceanic crust under the weight of volcanic buildings and the variation of ocean levels.

The movement of the oceanic crust over the hotspot reflects the succession of eruptive phases, the constructions of which are thus of different ages. The original structural shape of Hawaiian-type volcanoes is generally that of a low-lying dome. However, these stratovolcanoes are soon ripped open by tectonics and erosion. Fractures weaken the volcanic edifices, and craters (calderas) are created by the collapse of the roof of the magma chambers emptied by eruptions. The hot and humid climate alters the basalts and stimulates erosion, which sculpts grandiose reliefs that subsequent volcanic events can further complicate. Valleys, cirques and ridges make up a complex topography. Vesuvian volcanoes, such as those in Vanuatu, are composed of more diverse materials, resulting in particular from products projected during eruptions. In these heterogeneous rocks, differential erosion shapes residual reliefs corresponding to dykes or necks. In any case, the mountainous islands end up being eroded and sometimes, due to subsidence, disappear under the waves, crowned or not by madreporic constructions. This evolution changes the topography from that of the mountainous islands to that of the low islands and the atoll.

The Society Islands (Fig. 3.4) represent a good example of these stages in the evolution of a basaltic shield volcano. They also show how, under the effect of subsidence, coral edifices and, in particular, atolls are built. A fringing reef is built along the shore (Mehitia), then evolves into a barrier reef further offshore (Tahiti), before developing into a ring (Mooréa). The original mountain eventually disappears under the waves, leaving a ring of coral surrounding a lagoon, forming an atoll. As the evolution continues, the atoll itself will be submerged, to become a "guyot", an underwater volcanic edifice crowned with thick limestone constructions.

High and Low Islands

Despite their extraordinary variety, it is usual to distinguish between the high islands, which are generally volcanic, and the low islands, which are almost always limestone. The structuring fact of the high islands is their mountainous relief characterised by altitude and slopes. The altitude can be significant: 4171 m at Mauna Loa (Hawaii), 3718 m at the peak of Teide (Canary Islands), 3069 m at the Piton des

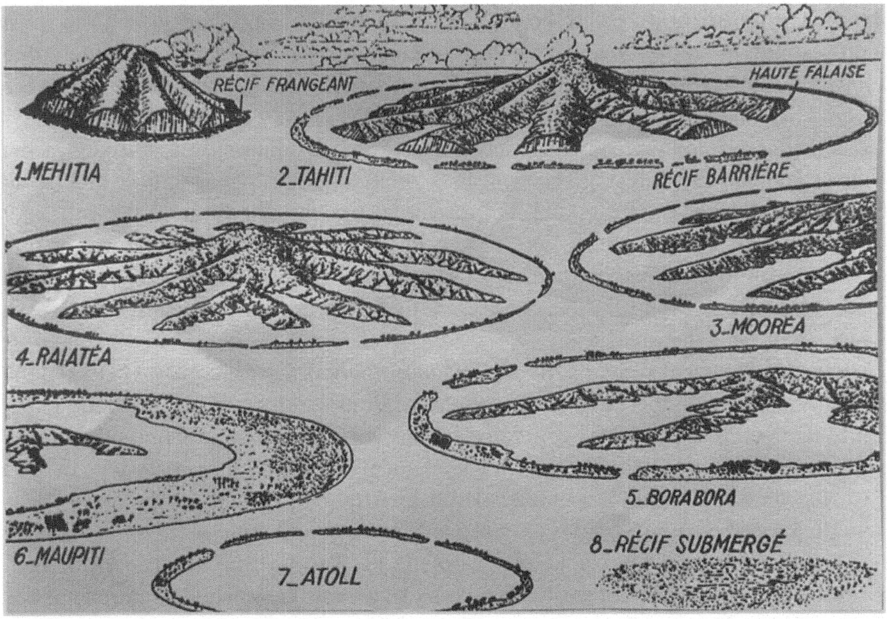

Fig. 3.4 Island types and coral reefs of the Society Islands. *The document shows a succession of shapes, illustrating the transition from high to low island and from fringing reef to atoll*

Neiges (Reunion Island), the highest point in the Indian Ocean and 1467 m at La Soufrière (Guadeloupe), the highest point in the Lesser Antilles. Altitude affects temperatures and precipitation. It lowers the former and increases the latter. In Réunion, the average thermal gradient is 0.65° for every 100 m of difference in altitude. The intensity of rainfall largely follows this thermal gradient, even if above the "sea of clouds" and its rainfall maximum located at around 1600 m, sunshine tends to be closer to that of the coast. The mountainous tropical islands hold many world records for rainfall. Reunion Island receives 7.5 billion m^3 of water per year, i.e., 3.5 times more than France with a comparable surface area.

The two gradients (thermal and pluviometric) cause a staggering of biogeographical environments whose altitudinal limits vary according to the "geographical" location of the islands, their altitude and the orientation of the slopes in relation to the prevailing winds.

Altitudinal zonation of environments according to altitude and rainfall

Altitudinal gradient	Thermal gradient	Rainfall gradient	Ecological areas
High altitudes	Oligotherm	Hygrophilic	Cool and damp
	Mesothermal	Nepheliphile	Cool and very humid
	Megatherm	Ombrophilous	Hot and humid
Low altitudes		Semi-xerophile	Hot and dry

The dense or open tropical forest, wet or dry, is the dominant vegetation formation. It rises to varying altitudes depending on the islands and the orientation of the slopes. The dense rainforest descends to the coast on the windward slopes and stops at around 300 m on the leeward slopes. It often comprises different facies, including a lowland rainforest and a highland rainforest or tropical montane forest with trees covered in mosses and epiphytes. In sheltered areas or on leeward slopes, it is a dry forest, followed downstream by grassy or scrubby formations. At altitude, in the highest islands, the forest gives way to a relatively varied vegetation of ericoid thickets or altimontane grassland. On the highest peaks, the mineral prevails over the vegetation: stunted plants take refuge in the crevices, while lichens and tiny ferns sometimes colonise the bare lava. The coastal vegetation is made up of plant species that creep along the beaches, creating bushy formations within mangrove forests in the marshy areas.

In the intertropical zone, on the high islands, there is a great contrast between the slopes depending on their exposure to the prevailing wind, known as the trade wind. Thus, the windward slopes, generally facing east, experience abundant rainfall, perennial watercourses, a less marked dry season and luxuriant vegetation. In contrast, the leeward slopes are characterised by slightly warmer temperatures, less rainfall, a longer dry season, less vigorous vegetation and coral reefs.

The low islands, or flat islands, have only modest altitudes. Their limestone bedrock comes from coral constructions, some of which are several hundred metres thick and date back to the beginning of the Tertiary period. They have sometimes been raised by a bulge in the oceanic crust (Loyalty Islands). In these limestone tables, karstic erosion sculpts a specific morphology. This is the case on certain islands such as Guadeloupe where it has isolated pitons separated by small poljes with a bottom impermeable by dissolution clay (Grands Fonds). Elsewhere, on the coral islands, erosion is slowed down by the low gradient, the number of cracks and the humidity. It is nevertheless manifested by the presence of sinkholes, dry valleys and underwater caves.

The climate is drier because there is no relief to stop the orographic rains. The dripping peaks of the mountainous islands are contrasted with the thirsty limestone islands. The wells provide water that is often insufficient and brackish. This lack of water is aggravated by the increase in tourist demand. These edaphic and rainfall conditions only allow the existence of relatively poor vegetation, often calciphilous, essentially in the form of savannahs, thorny bushes, succulents, cacti, etc.

However, these low-lying islands do have a choice asset: their white sandy coastlines and sometimes an atoll-like morphology, bringing together with the beach, the blue lagoon, the sun, the coconut palms, etc. all the ingredients of the tourist image of tropical islands.

Geomorphology as a Resource for Sport Tourism in the Tropics?

In the tropical islands, sport tourism has existed for a long time, with swimming, surfing or mountain expeditions. It has recently been renewed and diversified through "new" or reinvented activities such as mountain biking, paragliding, canyoning, rafting, white water swimming and sport climbing all activities for which the shape of the relief (beach, summit, waterfall, cliff, etc..) often serves as a base.

Landforms as a Raw Material for Sport Tourism in Tropical Islands

The landforms of the tropical islands, especially in the mountainous islands (Hawaii, Reunion, Tahiti, etc.), can be very diverse (Fig. 3.5).

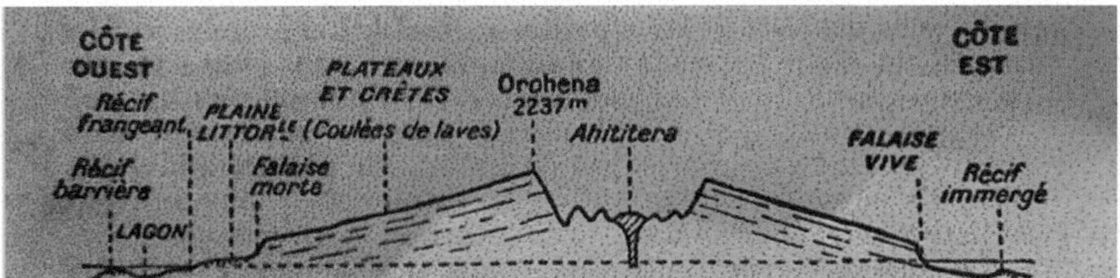

Fig. 3.5 Island landforms. Cross section of Tahiti. The width of the coastal areas has been exaggerated. (Source: A. Huetz de Lemps)

These islands, created by one or more volcanoes, originally had a high and compact relief. In some cases (Tahiti, Reunion, Anjouan, etc.), the tops of the mountains are hollowed out by cirques, depressions of tectonic origin (calderas) and/or hollowed out by erosion. These cirques are limited by very steep abrupts, veritable walls that redeem differences in level of several hundred metres. Their rugged bottoms are made up of tilted or collapsed blocks, residual ridges reinforced by intrusions (dykes), all segmented by a dense network of valleys.

The slopes facing the sea are made up of flows stacked one on top of the other, cut into planezes by a radiating hydrographic network. The main rivers build alluvial fans at their mouths, forming narrow coastal plains. The slopes are either connected to the sea by low, rocky or sandy coasts, or are truncated by cliffs.

Coral reefs are organic structures built mainly by an anthozoan which needs water temperatures above 18 °C, light (which limits the development of corals at depth) and agitated, pure and sufficiently saline water (which explains the absence of reefs opposite the mouths of rivers and off windy slopes that are subject to a lot of water). There are fringing reefs (Seychelles, West Indies, Reunion), barrier reefs (Mayotte), atolls, etc.

This general organisation of the island relief leads to a distinction being made between coastal morphologies and associated practices, the seat of "blue tourism", and the relief of the interior of the islands where sporting activities are more a part of "green tourism". It is the hot spot volcanic islands, which are at the intermediate stage of their demolition with steep relief and barrier reefs, which are the most favourable for tourism.

Whether it concerns continental coasts or islands, coastal morphology displays a wide variety of forms: cliffs, beaches, dunes, etc. The cliffs are very steep slopes ranging from 15° to overhang, located both inland (they are primarily part of the "continental" morphology) and on the coast. In the latter case, they are connected to sea level by an abrasion platform, which does not exist in the case of plunging cliffs. Sculpted by rain, wind and surf, they have only saxicolous vegetation adapted to sea spray and shelter in their crevices for the nesting of sea birds. The highest cliffs in the world are found in the tropical volcanic islands: Kahiwa Falls in Hawaii are on average 660 m high, culminating at Umilehi Point above the Pacific at 1005 m. The cliffs of the coral islands (Palawan, Tonga, etc.) are less high.

Cliffs are a great place for paragliding or extreme jumps. Cliff jumping or dive jumping involves jumping off cliffs to enjoy a free fall before plunging into the water (the world record is 58.8 m). Coastering combines swimming, climbing and rock jumping with the added bonus of visiting some geological curiosity. The via ferrata is a route built into a rock face for an intermediate sport between hiking and climbing. Some sectors are quite intensively developed. Cape Girao in Madeira, for example, is one of the highest cliffs in the world at 580 m high. It can be reached by car or bus from the town of Camara de Lobos. Since 2012, a glassed-in lookout overlooks the ocean, offering a view of the neighbouring towns and even the capital Funchal, as well as the terraced vineyards halfway up or at the foot of the cliff that were only accessible by sea. Since 2003, it has been possible to descend by cable car. The site is also suitable for paragliding and extreme jumps.

Nevertheless, the cliffs are often difficult to access and are generally neglected. They are of great landscape value, with a view of the ocean and the coastline, the spray of the waves on the rocks, the picturesque topography, etc., but generally speaking, the developments are preferable to the beaches. In tropical islands, the beach is always a coveted space due to the presence of the 3 S's (Sand, Sun and Sea). It is made up of light madreporic sand, dark volcanic sand or mixed sand. The sand is picked up by the wind and often forms low-lying dune ridges parallel to the shore. Tourists enjoy the beach for lazing around, as a playground for children, for sports (beach volleyball or beach soccer, sand yachting, etc.). In the triptych of the 3 S's, the presence of sun is as necessary as that of sand, and this is the "added value" of the tropical islands: the rate of sunshine is among the highest in the world (some 3000 hours in the Tuamutu Islands). The sea is even more important in the collective imagination if we remember how much the possibility of a holiday by the sea for workers represented an emblematic social advance. Even if they are sometimes reduced by cold currents (Canary Islands), the surface water temperatures are high. Varying in the Mascarenes throughout the year between 21° and 27°, they are well suited to nautical and underwater activities. The marine waters of the tropical islands are rich in both fauna and flora. The underwater world is remarkable, with coral reefs combining beauty and biological richness (corals, fish and invertebrates).

That is why the sea and the beach provide a wide range of sports and leisure activities. Some of them go back a long way. Others are more recent. The two pioneering activities are sea bathing and surfing. The fashion for sea bathing dates from the end of the seventeenth century in England for therapeutic reasons. It spread with the introduction of rail and then car transport in the early nineteenth century, the introduction of paid holidays and the creation of seaside resorts.

Surfing, which consists of riding a wave that breaks on the beach, either lying down or standing on a board, is a traditional practice in the Pacific. Since the beginning of the twentieth century, it has spread to almost every country in the world with a coastline where it is practised from thousands of "spots": there are 59 in Reunion Island alone. Initially practised in short board and long board, it has taken various forms: body board (lying down or kneeling on the board), body surf and stand-up paddle. In the case of kite surfing, the board is pulled by a kite.

Another popular sport is wind surfing, which was born in the 1960s and has been an Olympic sport since 1984. The "spots" are innumerable. In the islands, there are some famous ones, integrated or not in the course for the world cup: Ho′okipa Paui (Hawaii) with impressive waves, Saint-Leu de La Réunion, world famous or El Medano (Tenerife, Canaries).

There is a long list of leisure activities such as sailing, water-skiing or wake boarding, not forgetting fishing or discovering the sea bed. Scuba diving or snorkelling with mask, tube and fins is a widespread practice (180 sites in Réunion alone!). Marked routes and underwater paths through the white sandy seabed, basaltic substrates, coral reefs or drop-offs allow you to appreciate the variety of landscapes, wrecks and artificial reefs and to take photos and underwater videos.

Other coastal morphologies such as low rocky coasts, alluvial coasts, pebble beaches and mudflats are not very attractive for tourist activities, but they are still popular for hiking and horse riding. The varied morphologies of the interior of the islands, especially in the case of volcanic islands, are conducive to a variety of sporting activities using the mountains, slopes, gorges, etc.

In the world as in the islands, the mountains were for a long time frequented by an elite who went there to study the fauna and flora, to do mountaineering and to discover unknown summits or volcanoes. Since the 1960s, with the evolution of morals, a wider public has taken up "green tourism", which experienced a strong expansion in the 1980s and 1990s. Mountain and nature sports have a playful dimension but are also being promoted as tourist attractions. Hiking is practised almost everywhere, on marked trails, alongside horse riding and mountain biking. With the latter, one descends the slopes on dedicated tracks, unless one prefers different variants such as enduro or mountain biking.

Running is one of the most widespread sports and is most often practised in natural environments. Between 2013 and 2019, the International Trail Running Association counted 25,700 races worldwide in 195 countries, with 1.77 million participants. Cross-country running is run over medium distances (4–12 km). Trails are long-distance foot races on difficult routes, with significant differences in altitude, continuous ascents and descents, changing climatic conditions, etc. Small tropical islands are less represented in the major competitions that emerged in the 1990s in the mountains or deserts. Only Reunion Island, with the Grand Raid, appears in the five most popular trails along with the UMTB (Ultra Trail du Mont Blanc), the WS 100 (Western States) in the Sierra Nevada (California) and the UMTF (Ultra Trail Mount Fuji).

The unevenness of the terrain (cliffs, slopes, etc.) lends itself well to free flight, which consists of piloting an ultra-light glider without motorisation, such as a hang glider or paraglider. The take-off is most often from a relief, the landing on the beach, in a field or a flat area. It is a widespread sport throughout the world. In tropical islands, free flight can be practised all year round. In the Azores, starting from Lagoa do Fogo, you can fly over volcanoes, lakes and grassy slopes in search of updrafts. An annual paragliding festival is held in August.

Tropical islands lend themselves to caving. This is the case of the karst islands with caves that can extend under the sea. But caving is mainly practised in the underground world of lava tunnels. These are created when a flow cools at the surface forming a solid crust below which the lava continues to flow. When the flow of the lava decreases or stops, a gallery-like cavity with unusual shapes remains. Visits are organised in almost all volcanic regions: in Japan, France and the United States, but also in the tropical islands. The Kazamura in the Hawaiian Islands is, at 59.3 km, the longest volcanic land tunnel in the world. There are also famous ones in the Azores, Reunion Island and Easter Island, to name but a few.

The morpho-dynamics of the rivers in the tropical islands determine various whitewater activities. Mountainous islands have a dense network of sub-parallel streams that generally follow the lines of the greatest slope of the relief. Differential erosion reveals rocky escarpments, fronts and edges of flows, which the rivers cross by cascades at the foot of which glisten pools. The flows, linked to precipitation, are abundant especially on the windward slopes; the water is almost always of a pleasant temperature. These young rivers flow through gorges strewn with waterfalls, bordered by steep slopes, often with lush vegetation. These are favourable places for rafting, or descending rivers in an inflatable raft; for canyoning, which consists of progressing on foot or swimming in the bed of a river, passing waterfalls with the help of abseiling ropes; for canoeing and kayaking, when one descends a river in a boat following the direction of the current. These sports can be found on many tropical islands, including the Azores (canyoning in Failal da Terra), Dominica, Reunion, Gran Canaria and Polynesia.

Thus, geomorphology offers a potential for the development of outdoor sports activities, which would give some credit to the controversial notion of "determinism". But many suitable sites are not exploited and Jean-Christophe Gay invites us *"not to reverse the causal link: it is society that chooses the places and not those that attract it*". Moreover, we are witnessing a distension of the links between morphology and sport tourism: For example, cliff jumping can be done from bridges or buildings overlooking a body of water. Sporting practices call upon artificial spaces built in the city or in the countryside (climbing walls; surfing waves; white water stadiums for canoeing and kayaking, etc.) or innovative equipment (accrobranche, i.e., the fact of climbing high up between trees; sports leisure spaces in the forest; zip lines, etc.). Finally, the potential offered by the morphology, which is not fixed, evolves according to demand: for example, the waterfalls, historically visited for their aesthetic character, are today's choice sites for sports activities such as rafting or canyoning.

The landforms used to support these nature sports can also be a constraint. Thus, even if we are in the tropics, the high-altitude environments of the mountainous islands are harsh: the cold, the fog, the uneven topography destabilise hikers, some of whom may lose their lives if they get lost. Phenomena such as the dangers of the sea and floods, which can be brutal, raise the question of the safety of hikers.

Geomorphological Processes, a Limiting Factor but Also Sometimes a Resource for Sport Tourism in Tropical Islands

If the landforms are resources, certain processes, particularly morphoclimatic (cyclones) or geological (landslides, earthquakes, volcanism, etc.), can represent a risk for tourism and the population. Although not a geomorphic process, the tropical cyclone is the most formidable of all weather phenomena. Since the phenomenal energy it releases is contained in the water vapour from the oceans, it occurs most often in the warm season.

The cyclone is a low-pressure system with a diameter of a few dozen to a few hundred kilometres. Over such small distances, the barometric gradient between the centre of the system and its periphery causes very violent swirling winds that cause the most spectacular damage. No water, no electricity, a devastated landscape, roads cut off, flights cancelled and supplies disrupted: the cyclone is a major constraint for tourism in the tropical islands.

The cyclone also exacerbates geomorphological processes. Cyclonic swells accelerate beach erosion, submerge low-lying coasts, demolish coral constructions, etc. Reinforced by the relief, its excessive rainfall causes floods of considerable magnitude that submerge roads and embankments, cause flooding, and lead to landslides that sweep away networks and sometimes houses and their occupants. The drop in atmospheric pressure during the passage of a cyclone raises the sea level, and therefore the base level. As they adjust to this, the rivers spread out, causing flooding at their mouths, which is amplified by the combined effect of floods and storm surges.

Other geomorphological processes affect nature, people and their activities: land movements, earthquakes, tsunamis and volcanism. Ground movements are linked to the topography: steeply sloping planezes, cliffs, steep "ramparts" of erosion cirques, caldera rims or the slopes of deep valleys. The steepness of the slope, due to the vacuum, favours gravity phenomena. Geological elements add to the instability of the slopes: fracturing of volcanic rocks, alternating flows and slag in down-dip structures, faster chemical decomposition of heterogeneous rock banks which become impermeable and direct the underground circulation of water. Water saturation of materials during intense rainfall multiplies land movements such as mudflows, upheavals and landslides which modify the morphology and threaten people and property. According to BRGM, in Reunion, *"in February 1987, during cyclone Clotilda alone, nearly 600 land movements were recorded, 151 of which affected developments"*.

In oceanic islands, the periodic adjustment of the oceanic crust, which bends into a flexure under the weight of the volcanic edifices, causes barely perceptible earthquakes. Sometimes, as in Hawaii, the crust moves under the influence of rising magma, causing microearthquakes. On the other hand, in island arcs located in subduction zones, the accumulated pressure generates telluric tremors that can be destructive and deadly: in the West Indies, for example, an earthquake in 2010 caused 200,000 deaths. The same is true of the South West Pacific, where 30% of the world's earthquakes of magnitude 5 and above occur. Tonga, Solomon Islands and Vanuatu have experienced the most powerful earthquakes (magnitude 8 and above). The damage caused by earthquakes is impressive: buildings collapse, ground cracks, land subsidence and landslides, communication lines are cut, and submarine cables are destroyed.

Tsunamis are caused by earthquakes or submarine landslides, which cause oscillations that break on coasts even thousands of kilometres away in giant waves. In 1960, a terrible earthquake in Chile caused a tsunami that flooded Tahiti, damaged the Marquesas Islands and killed about 50 people in Hawaii. The 26 December 2004 earthquake in Sumatra caused a tsunami that was felt throughout the Indian Ocean.

Volcanism is an ambivalent phenomenon with regard to tourist activities, being both a risk and an attraction. The danger of volcanism varies according to the type of eruption: the explosive type of volcanism, the Pelée type, with burning clouds, a mixture of gas and solid particles that rush down the slopes at high speed, is deadly. In 1902, for example, Mount Pelée (Martinique) destroyed the town of Saint-Pierre, killing 28,000 people and hampering the region's development for a long time. Conversely, Hawaiian volcanoes, of the effusive type, are an asset for tourism. Relatively safe, they make a spectacle of themselves, spouting fountains of lava from their eruptive cones, letting their incandescent lava flows trail for kilometres, exhaling their hot breath and attracting thousands of visitors (Reunion, Hawaii).

The Economic, Social and Territorial Importance of Sport Tourism in the Tropical Islands

Sporting practices or events attract many visitors who lead to the equipping of sites and the creation of services, which structures the area and contributes to local development. Tourism is one of the most dynamic activities in small island economies. It spread to the islands in the wake of its development in the nineteenth century. However, due to their remoteness and landlocked nature, before the 1960s, their use was limited and depended on the development of means of transport. Close to the United States, the Bahamas, as early as the 1850s, welcomed visitors by boat. In the Pacific, Hawaii (Waikiki) also began to attract visitors, but it was not until the inter-war period that the Solomon Islands and the New Hebrides began to receive cruise passengers, as did Fiji and French Polynesia. The development of tourism was even later in the Indian Ocean, far from the major tourist centres. Progress in air transport allowed tourism to develop in the 1970s, as in the Seychelles with the opening of an international airport in 1971.

Globally, tourism is a major economic activity. According to the UNWTO, in 2019, before the health crisis, there were 1460 million tourists. Tourism provides 10% of jobs and 9.8% of GDP worldwide, more than the automotive or food industries.

The situation of the tropical islands (Fig. 3.6) is contrasted. They have little weight in world tourism, since none of them appears in the "top 20" either in terms of the number of tourists received or in terms of revenue from the sector. The islands of Thailand, Indonesia and the Philippines take the lion's share. Tourists are numerous in the West Indies, Hawaii (10 million), the Canary Islands (15.5 million), less so in the Indian Ocean and most of Oceania. Most islands receive less than 100,000 tourists, but as they are narrow and sparsely populated, the tolerance threshold retained by the UNWTO for sustainable tourism (one tourist for one inhabitant) is quickly exceeded.

A typology of islands can be established according to the place of tourism in their economy. Some islands are closely dependent on it, such as the Seychelles, Maldives, Madeira and the Canaries, where it represents between 25% and 35% of the GDP. In others, it constitutes an important sector of a more diversified activity (Mauritius, Reunion and Antilles). Almost everywhere, it is an avenue for development that is being explored. Similarly, a distinction can be made between islands

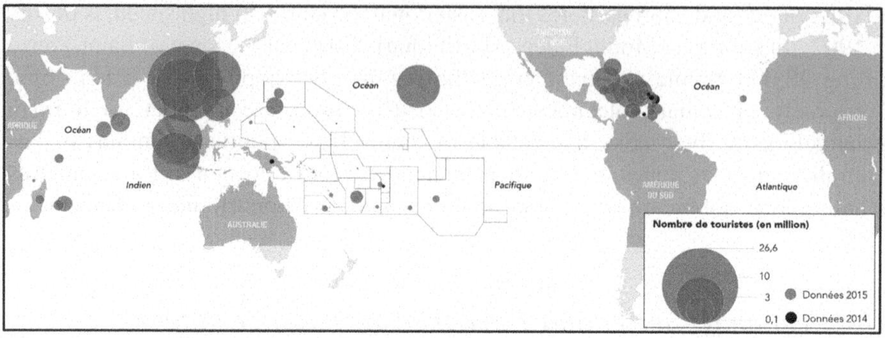

Fig. 3.6 Tourist numbers in the tropical islands. (Source: Caroline Blondy)

specialised in beach tourism because they are coral (Maldives) or low-lying (Mauritius) and other islands (Reunion, Seychelles and Hawaii), which have a variety of landforms in a limited space, serving as physical support for a wide range of sporting activities.

Sport tourism is a factor in local development and in the structuring and recomposition of territories. Its economic, social and cultural effects are added to those of tourism in general, even if the motivations of travellers differ. From an economic point of view, tourism is a vector for the integration of hitherto isolated territories into the world space. It requires investments in infrastructure, equipment and services, which are often financed by foreign capital. But these investments are not without return, since it has been established by the World Bank that 55% of the tourist income of developing countries is taken from outside. In any case, local added value is created which increases the GDP of the countries. The induced activities are diversified, giving pride of place to the large groups, manufacturers and distributors of sports materials and equipment, while the many day-to-day service providers are struggling, with small businesses that are often not very stable, as was noted during the Covid health crisis (Fig. 3.7).

Tourism also has contrasting consequences in the socio-cultural field. On the one hand, it creates "modern" jobs, often occupied by women, which positively raises the question of "gender", distributes income that reduces poverty and emigration and spreads the Western way of life. On the other hand, it increases pressure on the natural environment; it increases social inequalities since the benefits of tourism go primarily to the more privileged; it brings local populations and external visitors into contact, sometimes in a conflictual manner, and instils a consumer society.

Not all islands are equal when it comes to tourism development. There are those that are connected to the outside world by long-haul aircraft and others. In the case of an archipelago, the international airport is often located in the main island, which sets the tourist activities (Antigua) leaving the secondary islands aside (Barbuda), causing great contrasts in development between islands of the same group. The spatial distribution of the islands also acts their proximity favouring cruises as in the West Indies.

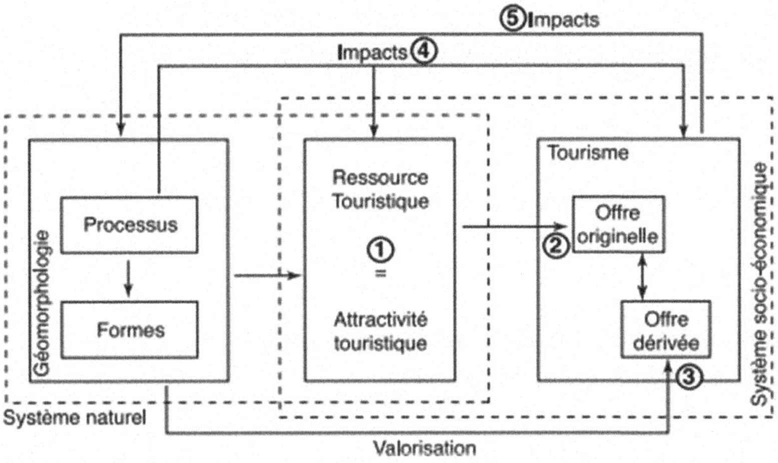

Fig. 3.7 The relationship between geomorphology and tourism. (After Cendreo & Panizza). *Geomorphology is an essential component of landscapes and a tourist resource (1). It represents a potential for tourism development for a place [original offer (2)]. The derived offer (3) includes the tools (structures, paths) and services likely to enhance the original offer. Geomorphology has an impact on tourism (4), and tourism activities and infrastructures have an impact on landforms and geomorphological processes (5)*

The spatial consequences of sport tourism are of two kinds. On the one hand, there is its hold on the environment, and its structuring of space with the development of numerous natural sites. Roads, paths and trails, sometimes steep and dangerous, open up the places where sports are practised. As these practices cover a wide range, the spatial coverage of the sites and specialised itineraries is very variable: take-off and landing areas for hang-gliding, climbing sites, mountain bike trails, canyoning routes, hiking or horse-riding trails and innovative sports facilities (climbing walls, surfing stations, white water stadiums, multi-leisure bases, etc.).

On the other hand, tourism affects the functioning of the areas. The location of tourist activities initially led to a dichotomy in the small tropical islands between a privileged coastline and a neglected interior. Seaside tourism, which is often mass tourism, has developed on the coasts, thus giving a ring-shaped character to its spatial distribution, with the interior remaining empty. For rural and mountain areas, sport tourism has been a factor in local development. It has given them a second wind, even if it is a diffuse and sparse tourism.

Finally, the development of green tourism has reinforced the division of the mountainous islands according to their exposure to the prevailing winds between a windward and a leeward slope. The first, more watery, is suitable for canoeing and canyoning activities, while the second, drier and sunnier, is suitable for aerial, oceanic and aqua-terrestrial activities.

A Vulnerable Geomorphology to be Integrated into Sustainable Tourism and Development in the Tropics

The ecological consequences of sport tourism are added to those of tourism in general: it is often difficult to determine the share of one and the other. They can be observed during the development it requires and during the course of tourist activities.

Impacts of (Sport) Tourism on Island Landforms and Geomorphological Processes

Tourism has repercussions not only on landforms but also on morphological processes, particularly on the coastline, on biodiversity and on particularly fragile environments such as coral reefs and mangroves. The development of tourism requires the construction of infrastructures and facilities such as ports or airports, roads, marinas, hotels, itineraries, development of sites and reception areas. The morphology of the places concerned is impacted. In addition to the clearing of natural areas, which is harmful to biodiversity, the fillings in of wetlands and plains make the pre-existing relief disappear. For example, airports have been built in whole or in part on the flats in the Seychelles, Tahiti or Mayotte or on lagoons (Saint-Martin). Infrastructures, equipment and, more generally, the artificialisation of the coastline leading to protective structures (walls, groins and riprap) can also modify geomorphological processes by disturbing the currentology and therefore the sedimentary transit, which causes the fattening of certain sectors and the thinning of others, thus modifying the coastline.

Another morphological impact is the modification of beach lines and profiles as a result of the overexploitation of materials such as sand for construction, the creation of artificial beaches or the recharging of existing beaches, as well as coral for lime used in the sugar industry or for the building and public works industry.

Dredging of the reefs to extract materials for construction, especially in the atolls, or to build harbours or shipping channels, destroys the site. The reworked sediments in suspension are carried away by the currents and end up smothering the corals in the surrounding areas.

In addition to the impact of the physical facilities, there is also the impact of the tourist activities themselves. The extension and democratisation of these activities result in significant flows with considerable environmental impacts. Thus, nautical and underwater activities can degrade the reefs, such as pleasure boats or diving boats dropping anchor on their outer slopes, or divers' flippers, trampling, kayaking or pedal boats breaking the corals on the plains.

In mountain areas, development and use of sites cause earthworks, clearing, trampling and ruts in place of tyre tracks, which can accelerate erosion and therefore change landscapes.

The impact of tourism activities on biodiversity is even more marked. The clearing of natural areas or mangroves and the filling in of wetlands to build accommodation facilities cause biodiversity losses. Biodiversity is also damaged by over-visitation, which increases pollution from wastewater (only ¼ of hotel wastewater is treated in good conditions) and plastics, not to mention the untimely harvesting of resources (corals, overfishing on the reef flats and outer slopes, etc.), which impoverishes the ecosystems.

The impact of tourism and more generally of human activities is particularly felt on sensitive environments such as mangroves and coral reefs. Mangroves are woody formations established on the mud of sheltered bays. They can include several dozen species of trees, the best known being the mangrove. They benefit coastal ecosystems by filtering mud and sediments, thus preventing the lagoon and coral reefs from silting up, by decomposing waste and by absorbing some of the pollutants, thus helping to clean up the surrounding aquatic environments. They also protect the coasts against erosion, by slowing down the progression of the swell thanks to the very dense root network that anchors the trees. According to the FAO in a 2008 study, a wave that passes through 200 m of a mangrove loses 75% of its power. The economic and social importance of mangroves is as great as their environmental role. They serve as nurseries for fish, allowing the renewal of fisheries resources; they are fishing grounds (oysters, crabs, fish and shrimps); they provide firewood and construction wood. However, mangroves are being destroyed by the construction of infrastructures, deforestation and aquaculture. Its disappearance, beyond its ecological, economic and cultural value, leaves the coastal ecosystems, as well as the populations, unprotected against the natural elements.

Coral reefs are emblematic ecosystems of tropical islands where they are an essential asset for tourism. They are as useful, exposed or threatened as mangroves. They protect coastlines against swell, cyclones and even tsunamis, absorbing 70–90% of the power of the waves. Their disappearance therefore increases coastal erosion. They are also known for their rich biodiversity. They are a breeding ground for many species of fish. They represent only 0.2% of the ocean's surface, but host 30% of marine biodiversity. The economic and social importance of reefs is recognised. According to a 2008 UNESCO publication, globally 500 million people depend on reefs for their livelihoods, and 30 million of the poorest depend on them for food. 1 km^2 of reef can provide 10–15 tons of fish. Yet coral reefs are increasingly under threat. They are under attack from a large starfish, the Acanthaster planci, which feeds on them, and are subject to bacterial diseases (West Indies). But as already mentioned, swells are a major agent in the evolution of coral reefs. They break up the corals and carry organic debris that tends to suffocate them. Coral bleaching is also a degradation factor: when the temperature rises, corals expel their zooxanthellae (small algae present in their surface tissues that give them colour), turn white and then die. When it occurs on a large scale, it is in connection with the El Nino phenomenon (ENSO), which causes a warming of surface waters. Global warming is a major concern in this area. Anthropogenic pressures on coral reefs are well known: soil erosion, urbanisation of slopes and coastlines, agricultural, domestic and industrial pollution cause terrigenous sedimentation and a proliferation of algae that suffocate the corals (eutrophication).

Other Vulnerabilities of Anthropogenic or Natural Origin of the Island's Physical Environments

Island coastlines are a major environmental and societal challenge. While they are often densely populated, they are exposed to anthropogenic and natural threats, as well as to global warming.

The vulnerabilities of anthropogenic and natural origin have already been discussed at length. The impact of human activities on the environment is superimposed on the natural risk to a physical, historical or "humanised" environment (Fig. 3.8).

Tropical islands are also sensitive to global changes. Thus, among the natural causes of beach retreat is basically a deficit in sediment supply, notably a shortage of sand since the end of the Holocene transgression (Cazes-Duvat & Paskoff, 2004). But the most threatening global phenomenon is climate change. Since the last glacial maximum 20,000 years ago, the natural warming of the climate has caused the ice sheets in the northern hemisphere to melt and the sea level to rise by some 120 m. Since the beginning of the industrial era, due to the emission of greenhouse gases, the climate has become unstable, with an accelerated rise in temperature and a corresponding rise in sea level. On average, this rise went from 0.3 mm per year in the eighteenth century to 0.4 mm per year in the nineteenth century, then to 1.7 mm per year in the nineteenth century and to 3.5 mm per year between 2004 and 2015. In 2020, an article in the international journal *Nature Research* predicted a likely rise in sea levels of between 0.63 and 1.32 m by 2100. Global warming is also causing an intensification of cyclones due to an increase in the temperature of ocean waters.

Rising sea levels and more violent cyclones exacerbate beach erosion, modify marine ecosystems with consequences for fauna and flora and, above all, threaten low-lying islands (Tuvalu, Maldives, etc.) with submersion and the risk of disappearing. They also have socio-economic repercussions by impacting infrastructures, agriculture and tourism, with, in the latter case, a loss of attractiveness for tropical islands.

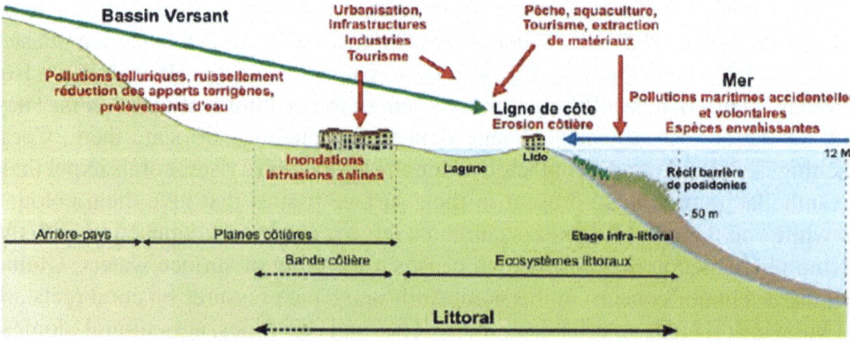

Fig. 3.8 Natural and anthropogenic pressures on the coastline

Taking Geomorphology into Account in a Global Policy for Sustainable Tourism in Tropical Islands

Geomorphology, a component of the living environments of the populations, has a heritage value that must be taken into account in a sustainable tourism policy, which would be a component of a development project for each of the tropical islands.

Geomorphology is more than a resource for sport tourism. It is a major component of ecosystems. Landforms, resulting from geological and morphoclimatic phenomena, are witnesses of the earth's history. They also form the backbone of ecosystems, together with biogeography and climate, and are a constituent element of landscapes. These ecosystems are the environments in which man has settled, which he has organised according to his social project. This initially takes the form of a traditional system comprising a subsistence economy (hunting, gathering, fishing, subsistence farming), a social organisation, values, a culture and an identity. It is increasingly taking the form of an "industrial" system that takes advantage of all the elements of the environment and develops a consumer society. The landforms, which are part of the landscape, culture and identity of the regions, are loaded with values and therefore constitute natural, cultural and economic resources. This approach has given rise to the concept of geomorphosite (Fig. 3.9).

As mass tourism has many disadvantages, other forms of tourism, including sport tourism, are sought after. The geomorphosites constitute a wealth of heritage that can be used for other tourism niches than sport tourism. One of them is geotourism, which aims to enhance the geographical personality of a portion of the earth's relief based on its scientific or socio-cultural values, with the objectives of making them better understood by visitors and preserving the environment. In the same spirit, ecotourism, also based on the resources of a place, was defined in 1992 by the TIES (International Ecotourism Society) as *"a form of responsible travel to natural sites which contributes to the protection of the environment and the well-being of local populations"*.

Ecotourism can thus be seen as a form of sustainable tourism with the objectives of reducing the negative impact on the environment and local culture, while contributing to economic growth, job creation and ecosystem protection. Reducing the negative impact on the environment is an imperative if we want to preserve this resource, which is threatened by infrastructures, over-frequentation, biological invasions, over-exploitation of resources, pollution, etc. Without a preserved environment, there would be little tourism in the tropical islands, which would have lost their main attraction.

All of these threats require regulations that must be drawn up and applied, but the countries concerned do not always have the means to do so. It also requires a change in behaviour, which is sometimes initiated by the practitioners of these new forms of tourism, who are often aware of the need to preserve the environment and local populations.

Sustainable tourism also implies managing overcrowding, which leads to an increased demand for energy and water, resources that are already scarce in some

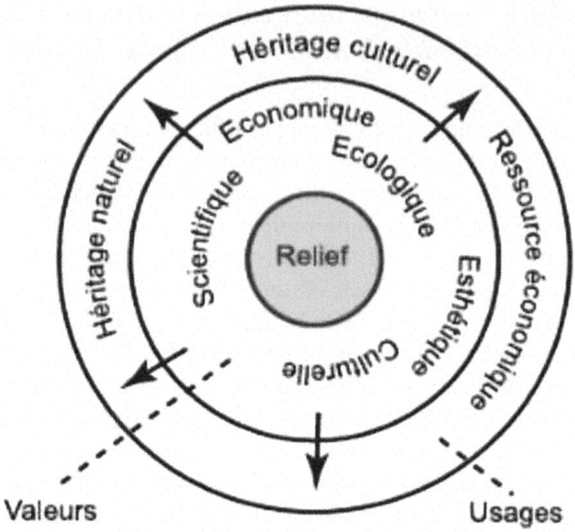

Fig. 3.9 The values of geomorphology. (Source: E. Reynard). M. Panizza and S. Piacente define *geomorphosites* as *"landforms that have acquired scientific, cultural and historical, aesthetic and/or socio-economic value because of their perception and exploitation by man"*

islands, as well as safety problems; cleaning and maintaining sites; managing waste removal and wastewater treatment (essential for the healthiness of beaches, otherwise the water will be declared unfit for bathing); creating protected natural areas (nature reserves and national parks, etc.); implementing integrated coastal zone management. Sustainable tourism is also concerned with integrating local populations, in order to improve their living conditions and to enhance their culture and identity, a factor of authenticity of the islands. It can also be a lever for economic development and job creation. In order to meet all these development goals, it must be framed by public policies in the economic, social and cultural fields, which are concerned with the consequences and the repercussions of tourism for the country.

This means that sustainable tourism must be integrated into a sustainable development project for tropical islands. The Small Island Developing States present specific problems recognised by international bodies such as the UN and recalled by Chapter 17 of Agenda 21 (Rio de Janeiro, 1992). Their level of development is disparate but is generally within the world average. While some islands such as Haiti, Comoros and Solomon Islands are classified as LDCs (less developed countries), most of the Small Independent Island States are classified as MICs (middle-income countries), which is rather encouraging.

A structural approach to the economy of SIDS reveals the coexistence of a traditional subsistence sector (food crops, fishing), which is dominant in certain islands with pre-colonial settlements (Comoros, Polynesia), and a plantation economy, particularly in the "sugar islands" (West Indies, Mascarene Islands, Hawaii). Everywhere, but to varying degrees, there is a diversification of agriculture, often in connection with urbanisation and the tertiarisation of the economy, while the entire economy tends to be absorbed by the sphere of capitalist development. These countries are thus moving from a traditional way of life to a consumer society that is putting pressure on their environment and way of life. Their limited resources and economic structure lead them to rely on imports, resulting in high economic, financial and socio-cultural dependence. This dependence is ultimately political, regardless of the status of the islands.

More than any other type of country, tropical islands thus have the obligation to implement sustainable development policies. They are all “finite” spaces, with permanent constraints such as the narrowness of their territories, their remoteness and insularity, which increase costs, an often difficult terrain and a poorly diversified economic structure that limits growth and job creation.

On the road to development, each island must have its own territorial trajectory, but general principles can be laid down. Development must be based on identity (guaranteeing the uniqueness of each country), endogenous, based on solidarity, sustainable and open to the regional environment.

The development project must start from the specificity of the island territories and the identity of their populations. Just as each human being has his own DNA, no territory is identical to another and no people has exactly the same culture as another. In order to exist in the globalised world, it is necessary to be known and recognised, and each island must play to its own strengths, and its comparative and competitive advantages. Second, since island economies are service based and remain dependent on the outside world, in order to reduce vulnerability and increase resilience, the emphasis should be on more endogenous productive development. Endogenous development starts from the country’s own resources. Even if it is assimilated to an export, tourism can be a pivotal part of a more productive economy, as can the sea-based professions, the “blue economy”, which should be self-evident for islands. Moreover, left behind by tourism and tertiary activities, large sections of the local population have become impoverished: development must be inclusive, supportive and in their favour. Moreover, the small size of the islands limits resources, which leads to the recommendation of a development that is sparing of natural resources, a guarantee of sustainability. Finally, the SIDS engaged in dominant North-South relations must turn more towards the countries of their geographical environment and develop South-South links. Only regional cooperation and integration will make it possible to enlarge markets, bring complementarities into play in the location of value chains, reduce import circuits for a smaller carbon footprint and thus slow down global warming.

Conclusions

The geomorphology and, more generally, the physical environment are a resource for sport tourism in the tropical islands. Insularity allows for a great variety of relief in a small area and the climate offers mild temperatures all year round. Geomorphology is a factor in the location and attractiveness of sport tourism, which finds favourable conditions for its development in tropical islands.

This has resulted in an increase in the number of participants, the organisation of events that are sometimes national or international in scope, a diversification of practices, the equipping and development of new sites, the development of tourism and the structuring of territories.

However, just as outdoor sports activities can be carried out in artificial spaces such as whitewater stadiums, undermining a kind of original determinism, so geomorphology, which often forms the backbone of landscapes, is not just a support for sports activities. It is an aesthetic, scientific and identity-based heritage that can give rise to other forms of tourism than sport tourism. It is important to know it, to safeguard it and to integrate it into a sustainable development policy of which sustainable tourism is a component.

With sport tourism, the non-tropical islands are also finding a new tourist impetus based on nature sports and more globally on sustainable tourism. This evolution is not specific to the islands. For all countries, the need for sustainable tourism is essential and the same approach should be adopted, with, if necessary, adaptations to take into account the specificities.

There remain the questions of awareness, political will and above all means. In the implementation of a sustainable development policy, the islands still integrated into the metropolises of the "North" such as Hawaii (United States), the European or British Overseas Territories, often have the necessary legal, financial and human means. This is not always the case for Small Island Developing States, which remain economically dependent on developed or emerging countries and must be able to rely on international solidarity.

Reference

Cazes-Duvat, V., & Paskoff, R. (2004). *Les littoraux des Mascareignes entre nature et aménagement*. Université de La Réunion/L'Harmattan, 186 p.

Chapter 4
Sport Tourism, Island Territories, and Sustainable Development: A Literature Review

Derek Van Rheenen, Olivier Naria, Ricardo Melo, and Claude Sobry

Introduction

Small island states or territories have vastly different historical relationships with sport and tourism, both in terms of its historical production and consumption of these practices. Often, the sports practices are an amalgam of indigenous and imported games and sports (MacLean, 2021; Singaravelou & Sorez, 2010) often creating unique dynamics of cultural identity and immersion, adaptation, appropriation, and interpretation (James, 2013; Uperesa & Mountjoy, 2014; Walker, 2011; Wheaton et al., 2020). More recently, sport tourism has become a vector for territorial branding of these sites. For the purposes of this volume, sport tourism represents a diverse collection of social practices experienced within a fixed continuum of time and space. Sport tourism activities and events, as well as the corresponding practice facilities and other sport sites, exist within a unique logic of production and consumption, socially and historically constructed by peoples in relation to one another and to a given context. The definition of "islands," "small states," or "island

D. Van Rheenen
Cultural Studies of Sport in Education, University of California Berkeley, Berkeley, CA, USA
e-mail: dvr@berkeley.edu

O. Naria
Department STAPS, Université de La Réunion, Le Tampon, ile de la Reunion, Reunion, Réunion, France
e-mail: olivier.naria@univ-reunion.fr

R. Melo
Coimbra Education School, Polytechnic of Coimbra, Coimbra, Portugal

C. Sobry (✉)
Faculty of Sport and Physical Education, Université of Lille, Lille, Nord, France
e-mail: Claude.sobry@univ-lille.frxt

D. Van Rheenen et al. (eds.), *Sport Tourism, Island Territories and Sustainable Development*, Sports Economics, Management and Policy 24,
https://doi.org/10.1007/978-3-031-51705-1_4

territories" is understood as a common set of geographical and topographical features spread over several oceans and seas. Islands are also characterized by a common history of colonialism, cultural and social independence, and codependence framed by neoliberal policies of capital accumulation and linear models of development. This project aims to challenge these models and provide innovative practices that are more inclusive and culturally sensitive, with a focus on promoting authentic and intentional sustainable development.

Sport tourism and islands share a modern relationship, most recently as a potential vehicle for sustainable development within these unique territories. It was not until the middle to late twentieth century when tourism became a more common practice at these sites, bolstered by necessary island infrastructure, such as modern seaports and air connectivity (Gay, 2000; Lockhardt & Drakas-Smith, 1997; McElroy, 2006). Tourism also relied on potential visitors' disposable income and the emergence of leisure in juxtaposition to work.

Many islands followed the plantation tourism model (Weaver, 1988), based on the historical relationship between agriculture and tourism. Utilizing the Caribbean Island of Antigua as a case study, Weaver describes the sugar-based agricultural economy first established in the mid-1600s. The plantation economy of the sugar industry was the mainstay of most of the islands during this period, with coastal development and port activity (Huetz de Lemps, 2011). It was not until the mid-twentieth century that agriculture and tourism temporally co-exist. By 1980, according to Weaver, tourism had completely supplanted export agriculture as the island's dominant economic sector. This shift in economic activity led to the development of tourism along the coast, initially reserved for colonists and local notables, before being extended to the indigenous population.

This historical shift highlighted the fragile economic dependence that many small island nations had on tourism (Schubert et al., 2011; Weaver, 2017), as well as the undesirable social costs of mass tourism on island communities (MacNaught, 1982; Stephenson & Timothy, 2022; Wong, 2015). And yet, as Sofield and Sivan (2003) demonstrated in their discussion of the Hong Kong Dragon Boat Races, the development of an international festival provided a merging of cultural and sport tourism and a positive force for the revival and maintenance of traditional culture.

In the initial phase of island tourism, then, islands were branded as exotic destinations (Bernardie-Tahir & Taglioni, 2016) based on a pristine, untouched coastal image (Cazes, 1989), focused on the promotion of sun, sea, and sand (3Ss). In these instances, sport served as a secondary or tertiary motivation for travelers, an informal activity in which to engage while enjoying the beach and tropical seascape. The promotion of sport tourism, then, developed by extension to tourism as a modern form of potential revenue generation, a reason for travel in own right (Bouchet & Bouhaouala, 2009).

This historical process led to a myopic focus on specific, niche markets connected to the coastline, such as surfing, diving, and boating. Many island territories continue to promote these particular sport tourism activities, much as some of these island economies relied on a singular product or crop to bolster their livelihood.

Islands also opted to host sport events as a vehicle to brand their identities and draw tourists to these sites, focusing on a particular sport, such as rugby, golf, or surfing, as part of the island's evolving identity in the global market (Dolnicar & Fluker, 2003; Lim & Patterson, 2008; Palmer, 2017).

The second phase of this tourism development has focused on a diversification of offerings, moving away from a singular and often seasonal market. Many islands and island archipelagos are at this stage currently, some more advanced than others in diversifying their tourism products. For sport tourism in particular, this has meant moving away from the beach and utilizing the rest of the island, a playful movement from the horizontal to the vertical plane.

This has led to the promotion of inland activities, such as mountain biking, trail running, and hiking. A further process of diversification on the islands has come to include use of the airspace as a resource for sport tourism activities, such as hanggliding, paragliding, and skydiving. Mauritius is a case in point, having evolved from a mono-crop economy to a well-known island destination offering a panoply of tourism activities (see Chapter 21, this volume; See also Baldacchino and Caruana's, Chapter 7 entitled, "Sport Tourism in Malta: Getting off the Beach.)" There is some evidence that island residents have a positive attitude toward sport tourism diversification, seeing this as a process to combat the ills of mass tourism (Bartolomé et al., 2009). This will of course vary based on the economic differences among islands, even within the same archipelago.

The final phase of this development has been the focus of islands in becoming responsible destinations, connecting their diversified supply with sustainability efforts, such as the implementation of particular sustainable development goals. Activities and events at the sea have become potential contexts for the development of a blue economy, while those activities on land might focus intentionally on the promotion of a green economy. Such branding has turned out to be profitable, appealing to tourists who seek, and are willing to pay extra, to travel to these responsible places.

Given the attractiveness of the paradisiacal image of the islands, the globalization of sport tourism activities has accelerated, spreading far and wide to the most attractive tourist spots on small tropical islands worldwide. Unfortunately, several islands have set up hotel structures directly in and on top of the turquoise waters, endangering nearby coral reefs and fish stocks. Similarly, constructions in the plains and mountains overlooking biodiversity areas have had devastating impacts on the environment.

The stakes involved in developing sport tourism must now include pressing issues of sustainable development (Furt & Tafani, 2014). Awareness of sustainability issues has emerged in varying degrees relative to the promotion of tourism and sporting activities. While it is true that planning documents, charters, and marine and terrestrial protection areas have reflected this desire, stakeholders have been confronted with the powerful engine of the global tourism industry, which has left its mark on the coastlines and inland territories of islands. The development of sport tourism activities and events today must confront this movement and address these potential deleterious impacts on the environment in the islands and archipelagos (Duvat, 2009).

The following sections of this chapter develop these three phases of island tourism more fully, recognizing a shift from being branded as an exotic, niche destination focused on the coastline to a more diversified market, encompassing an assortment of island opportunities. Finally, this development has led to a local to global emphasis on sustainability and the recognition of islands as responsible destinations.

Tourism in Small Island States or Territories: An Exotic Destination of Sun, Sea, and Sand (3Ss)

Tourism is an essential part of the economic activity of small island states or territories, leading to a location's relative competitiveness (WTO, 2014). This is because tourism represents the main market for service industries, with the potential to contribute to their economy and the generation of income for the general population (Craigwell, 2007). For example, islands of the Mediterranean, Caribbean, and South Pacific are important tourist destinations (Butler, 2008); tourism is, for these islands, an important engine for their economies. However, the economic contribution of tourism to small island states or territories also encounters many challenges, namely, monothematic mass tourism, seasonality, and overcrowding (Ziakas & Boukas, 2016).

Small island states or territories are attractive destinations due to the unique natural and cultural elements they offer to tourists (Carlsen & Butler, 2011), particularly their climate and landscapes, but also their tangible and intangible cultural heritage, associated to their rich and vibrant cultures and friendly and welcoming environments (Lewis-Cameron & Roberts, 2010).

Although the islands are places with a rich and diverse flora and fauna, at the same time, they face a lack of resources, segregation, and dependence on the rest of the world (Niles & Baldacchino, 2011). In reality, small island states or territories seek to balance dual purposes that are at times antagonistic, trying to balance economic growth with ecological integrity, such as the quality of life of the community and the need to preserve the unique social and territorial characteristics of these places (Niles & Baldacchino, 2011).

Small island states or territories face problems such as climate change and the scarcity of resources and have to deal with dependence on the global economy and tourism (Bouchard et al., 2010; Klöck & Fink, 2019; Ziakas & Boukas, 2016), which, for the most part, depends on the existence of foreign capital. In general, the local population and island residents are at the bottom of the economic pyramid in terms of ownership and employment (Lewis-Cameron & Roberts, 2010); that is, most local residents do not have the financial power to support large-scale projects, such as hotels and resorts, for example (Ziakas & Boukas, 2016).

In addition to dependence on external forces, these territories face other challenges, such as a reduction in traditional economic activities and vulnerability to various crises and isolation (Carlsen & Butler, 2011). In addition, they suffer from a lack of economies of scale, high transport and communication costs, as well as costly public administration and infrastructure creation (Lim & Patterson, 2008). In this sense, if appropriate policies and strategies are not implemented, the development of tourism in small island states or territories threatens the sustainable management of their limited natural resources and their cultural authenticity (Boukas & Ziakas, 2013). Therefore, the excessive dependence of small island states or territories economies on tourism and foreign capital, together with the occurrence of unforeseen disasters, exogenous failures of the global financial system, political instability, social unrest, environmental or health crises, such as a global pandemic, raise questions about how these territories can become competitive destinations without losing their natural character and damaging their sociocultural characteristics in the long term (Ziakas & Boukas, 2016).

The Emergence of Sport Tourism: Diversification and Territorial Branding

To mitigate some of the challenges of tourism, namely, the fight against seasonality, some island states or territories have promoted sport tourism as an important tourist segment to diversify their tourism products. Sport tourism has been recognized as a means of generating economic growth through internal investment, expansion of the service sector, and the development of special interest tourism (Lim & Patterson, 2008). This process has been documented in the literature, at the intersection of place and sport tourism. Examples, among many, include Greece (Costa & Glinia, 2004; Hudson & Beedie, 2007), Malta (Bull & Weed, 1999), Cyprus (Boukas & Ziakas, 2013; Kartakoullis & Karlis, 2002), the Seychelles (Naria & Sherwin, 2011), the Maldives (Ponting, 2014), Martinique (Desse, 2000), and Barbados (Elcock, 2005). There is a wide range of sporting activities developed on land, water, and air that can be undertaken by small island states or territories to attract tourist visitation and obtain benefits for the community. Among them, road races (marathons) and cycling tours have become particularly popular, perhaps because the nature of these sports allows participants to travel around the main attractions of the island, thus promoting the identity and image of the islands as a touristic destination (Lee & Huang, 2014; Lee et al., 2014; Ziakas & Boukas, 2016). Additionally, these activities take place in outdoor public spaces, or in natural spaces (Naria, 2013), and convey social messages that can be easily linked to community building or charitable purposes (Ziakas & Boukas, 2016). Kayaking (Hudson & Beedie, 2007), sailing and boating activities (Desse, 2000; Rejón-Guardia et al., 2019), scuba diving (Lew, 2013), and surfing (Buckley, 2002; Naria & Sherwin, 2011; Palmer, 2017; Ponting, 2014) are other important nature sport tourism activities that have been developed in some island states or territories as differentiating (sport) tourism products.

Sport tourism is increasingly associated with the construction of a strong tourism image, with the development of the destination. The work of Dolingting et al. (2015) in the Sipadan Islands shows that the construction and dissemination of an attractive image of place and the "branding" of sport tourism represents a development lever to enhance the attractiveness of the islands. These authors also suggest that this theoretical framework is applicable to other activities. Cepernic (2016), on the other hand, analyzes the revitalization of sport tourism on the Croatian islands with "smart destinations" or "marketing strategies" through the effect on the destination's awareness and image. In addition, the effects on sporting events with a potential for attracting tourists have been the subject of several analyses. Sport tourism has been the key sector for revitalizing a strategy for a destination in crisis or resilience, for instance, in Hong Kong (Sofield & Sivan, 2003). Martinique (Séraphin & Gowreesunkar, 2020), Norway (Jaeger, 2020), or in the Balearic Islands (Rejón-Guardia et al., 2019).

One mechanism of sport tourism promotion on the islands has been the decision to host particular sporting events, a strategy of territorial branding (Kapferer, 2011; Naria & Jean-Pierre, 2023). This serves as a lever for global visibility and economic growth. As Fig. 4.1 illustrates, this has become a popular practice among islands around the world. In view of the potential offered by the diversity of their natural environments, sport tourism events have become a viable vector for the economic, cultural, and sustainable development of islands and archipelagos. They enhance the communication policy and image of island destinations, with a view to generating economic spin-offs. They place the venues in which they take place at the heart of visibility with institutional and economic partners likely to generate benefits.

In the Pacific, the Cook Islands organize the "Vaka Eiva," the world's leading sporting event and largest canoe and paddle competition. In Vanuatu, the Vanuatu's Leimalo Surf Festival (2017) attracts sport tourists from all over the world. In Papua New Guinea, the Longboard Championship Tour is organized on the Tupira spot (2017), as well as Kumul PNG World Longboard Championships TUPIRA. On land, the BSP pursuit in Paradise Aitutaki Marathon and the Rarotonga Amateur Golf Open are world renowned.

In the Indian Ocean, Mauritius is particularly active in its hosting of sporting events, such as the Lux Mauritius Marathon and the Bank Mauritius Golf Open. On the ocean, the Kiteival Mauritius AfraSia is a highlight of the annual calendar. In the Maldives, ocean competitions are part of the archipelago's tourism strategy, with the O'Neil Deep Bleu Contest, the Maldives Four Season Championship Surf Trophy, and, more recently, the Coca-Cola Long Run Kandima at Hulhumalé.

In the Caribbean, at Half Moon, the Jamaica Open Golf Championship, the final of the Dunhill Cup of the Americas, and the Red Stripe Pro-Am have international appeal. There is also a marathon event, the Reggae Marathon Event. In Barbados, sport tourism activities range from the Barbados International Fishing Tournament to the Sir Garry Sobers Golf Tournament. In Antigua and Barbuda, ocean activities are at the heart of sport tourism events: the annual Antigua and Barbuda Sport

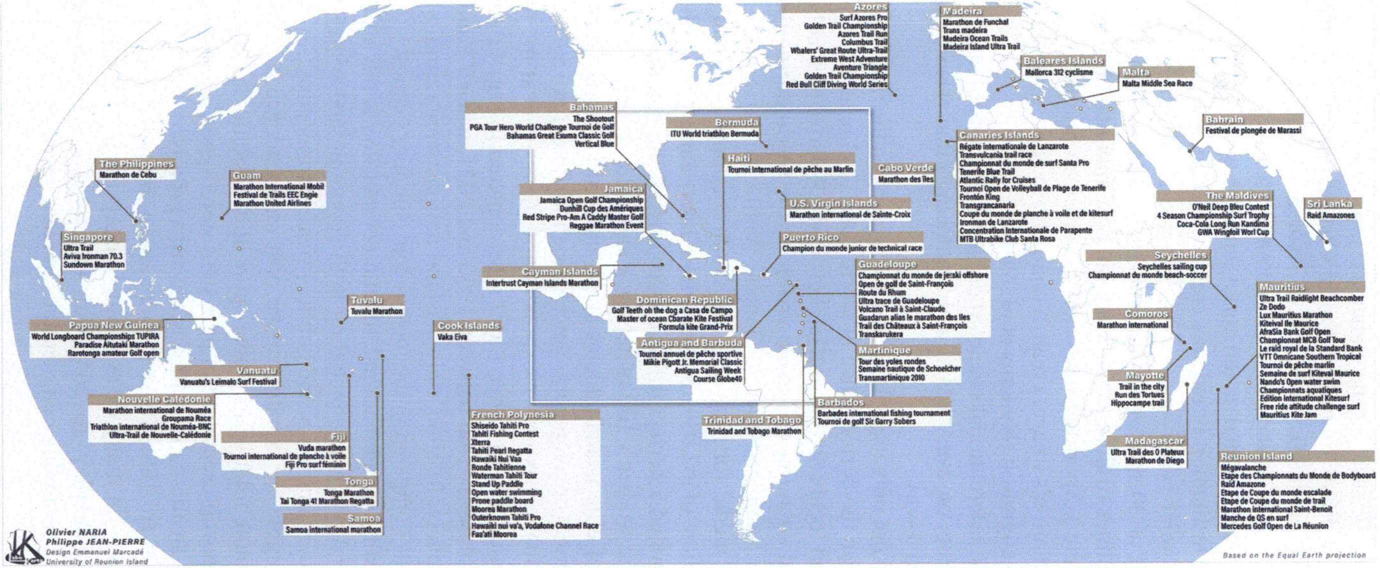

Fig. 4.1 Organizing an international sport tourism event, a territorial marketing lever

Fishing Tournament, the Mikie Pigott Jr. Memorial Classic, Antigua Sailing Week, and even the Globe40 race. In the Bahamas, the Shootout, marlin fishing in the Abacos Islands, the PGA Tour Hero World Challenge Golf Tournament in New Providence (2015), or the Bahamas Great Exuma Classic Golf Tournament are examples of this trend in island destination branding.

Other islands have created a territorial brand based on sport tourism. The European Sports Destination brand was created by Turismo Lanzarote to promote the island as a premier year-round sports destination (SPEL, 2021). The brand is defined by a panel of 30 sport tourism events, including 4 world championships in 13 different disciplines (trail running, triathlon, golf, mountain biking, surfing, sailing, windsurfing, etc.). Sport tourism offers islands the opportunity to enhance and maintain the destination's image, create a competitive edge, and, most importantly, gain worldwide recognition (Nyikana et al., 2014).

Alongside these international events, there are thousands of small and medium-sized sporting events on every island and archipelago that raise the profile of the host territories and towns.

Sport events do not only metaphorically put a place on the map in terms of global visibility; so too can the performance of world class or elite athletes reflect on their home countries and local contexts and backgrounds. Relative to small island territories, this increased attention was made visible to the world during the 2020 (actually, 2021) Tokyo Summer Olympic Games. Within the same week, Flora Duffy won the first gold medal for Bermuda, the smallest country to win a gold, with an estimated population of 62,034. Several days later, Jamaica, another Caribbean Island state, celebrated a perfect trifecta, sweeping all three medals—Gold, Silver, and Bronze—in the women's 100-meter track and field event.

The presence of elite island athletes at the modern Olympic Games has undergone its own historical development. Since its establishment in 1896, representation of small island athletes has grown exponentially, from none at the first Games in Athens to 443 participants in Tokyo (2021). As depicted in Fig. 4.2, the number of athletes has multiplied six-fold between 1960 and 2021, when 77 and 443 athletes competed, respectively. In 1960, only 7 island nations were represented, while 45 small island nations competed in the most recent Summer Games.

This dramatic shift in international athletic competition is the result of contemporary transnational processes, including both economic and social development, including government initiatives and public–private partnerships. Sporting victories symbolize the struggle for equality in the post-independence era. As Upersa and Mountjoy (2014) note relative to global sport in the Pacific, sporting prowess has served as "a counterrepresentation of persistent stereotypes, as a proud representation of a nation, as an alternative pathway toward a promised future, and as a site for cultural resurgence" (p. 265). These authors also point out, however, that the local to global economies "in many cases compel engagement with sport beyond mere forms of recreation and play" (p. 264), with an emphasis on modern, masculinized, hyper-competitive, and highly commercialized team sports.

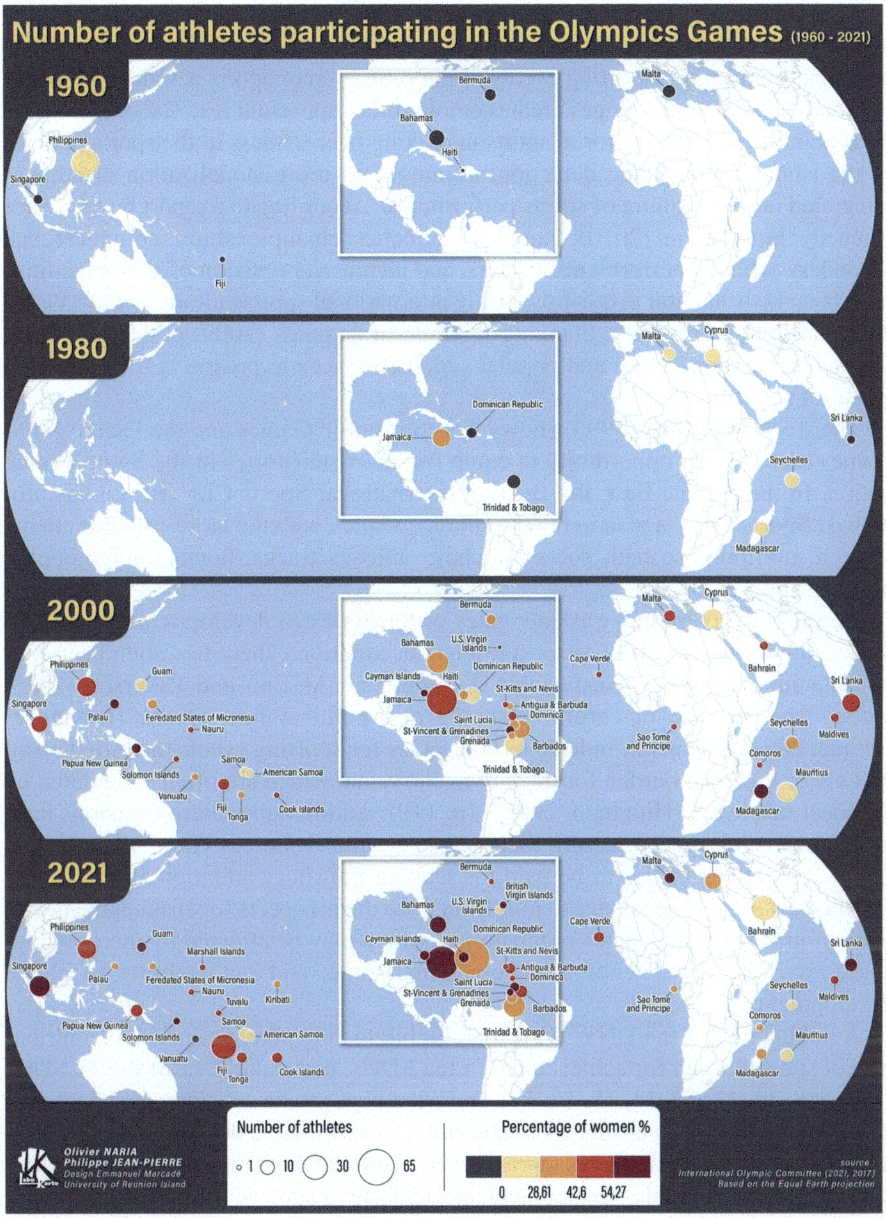

Fig. 4.2 Participating in the Olympic Games puts SIDS on the world map

And yet, modern sport is a tool for promoting integration and social cohesion within societies. It provides development opportunities to combat poverty. The organization of major sporting events or the setting up of sports activities in neighborhoods, districts, or villages creates employment opportunities. The objectives of players in the field of territorial sports marketing bear witness to the sports policies of the islands through the detection, training, and professionalization of athletes integrated into the culture of sports performance. According to a report by the Grace Kennedy Foundation (2010), Jamaica's athletics championships, co-educational secondary schools, performance centers, and increased provision of sports facilities have been instrumental in creating a truly international sports culture. In the "Vision 30" plan drawn up by the Planning Institute of Jamaica (2009, p. 44), increased private-sector sponsorship and popular support are seen as prospects for improving top-level success.

The work of Leloup (1995) shows that the Pacific Games and the Pacific Mini-Games represent an opportunity to equip the sports territories of the South Pacific region. In the Middle East, the Bahrain International Sports City (BISC) has promoted "Sports City," a plan to build a sports complex with the largest stadium in the state, a multipurpose hall, soccer pitches, athletics tracks, hotels, and shopping malls (Boniface, 2021).

Despite efforts by many island states or territories to develop world-class athletes and brand themselves as sport tourism destinations, there has often been limited planning relative to sustainable development. As Lim and Patterson (2008) argue, "the absence of careful planning and political interference (Devine & Devine, 2004) may result in long-term legacies for sporting events that often result in a major financial burden for the city to carry, and which are often shouldered by the local taxpayers (Hingham, 2005)" (p. 119). Additionally, these decisions have often been remiss in protecting their scarce natural resources (Boukas & Ziakas, 2013), a significant price to be paid for seeking transitory attention and visibility. For this reason, it is necessary to jointly analyze the prospects for sustainable development in small island states or territories, which may be affected by their remoteness, excessive dependence on international trade, and vulnerability to global developments.

A publication of the UNWTO (2019) highlights the potential positive contribution of sport tourism for achievement of the SDGs, particularly 11 of the 17 SDGs (1, 3, 5, 8, 11, 12, 13, 14, 15, 16, 17), and what needs to be considered when developing sport tourism to ensure its contribution to sustainable development. The following section illustrates how sport tourism in small island states or territories can contribute to sustainable development, particularly to achieve the SDGs if properly developed and managed. This global call to action has likewise led island territories to become responsible destinations for sport tourism.

Sport Tourism, Island Territories, and the Promotion of Sustainable Development Goals: Building a Responsible Destination

According to UNWTO (2019), the strength of sport tourism to achieve sustainable development, and the 17 SDGs particularly, lies in the engagement in physical activities, the opportunities for interactions, and the development potential of a diversity of small events or activities. As a result, sport tourism can play an important role in achieving several SDGs if developed in a meaningful way (UNWTO, 2019).

The United Nations' SDGs serve as an integrated, multidimensional effort to simultaneously promote the economy, social structure, and environment in a holistic manner. Ideally, a positive economic impact can help maintain ecosystem protection, while a healthy ecosystem provides the venue for sustainable sport tourism market supply and growth (Boley & Green, 2016; Melo et al., 2020) and greater social equity (Melo et al., 2021). In this sense, sport tourism has the potential to generate area-based conservation and the promotion of sustainable development goals (Dudley et al., 2023; Maxwell et al., 2020). According to Dudley et al. (2023), this will require four critical steps: (i) recognition of wider SDG targets addressed by protected and conserved areas; (ii) integration of ecosystem services into site-level policies and national SDG strategies; (iii) enhancement of the relevant values through management approaches; and (iv) consistent reporting of these as a contribution to the SDGs. These efforts are often locally or regionally focused, specific to an island's contemporary context. Hawaii, for example, has sought to emerge as a pioneer in the practical application of sustainability efforts, seeking to develop a sustainable tourism system with improved measurement and analytical tools (Sheldon et al., 2005). And yet, still today, the global sport tourism industry suffers from a lack of reliable data on which to base strategic decision making (Dolnicar & Fluker, 2003; Melo et al., 2021; Van Rheenen, 2017).

In small island states or territories, the economic effects of sport tourism (Table 4.1) can be seen through economic development and diversification (SDG 8) over sustainable production and consumption patterns (SDG 12), which can lead to the eradication of poverty (SDG 1) and reducing social inequalities (SDG 10). In this respect, the work of Elcock (2005) shows that the development and rehabilitation of the sport tourism offer are important assets in Barbados. On the other hand, Kartakoulis and Karlis (2002) question the complementarity between public and commercial services, which diversify the sector in Cyprus. Other effects on the creation of wealth, employment, and the dynamization of a "niche" market have been identified, particularly, in the analysis of the case studies developed by Bull and Weed (1999) in Malta that underlines the direct economic spin-offs of sport tourism in the island. Several authors argue for economic dynamization at other scales. For example, Scol (2006) mentions the role of indirect economic benefits in Kalymnos which complement the sport tourism market, while Pieter and Gevers

Table 4.1 Sport tourism and the 17 Sustainable Development Goals: The economic dimension

Economic dimension	
SDG 1: END POVERTY IN ALL ITS FORMS EVERYWHERE **SDG: 8 PROMOTE SUSTAINED, INCLUSIVE AND SUSTAINABLE ECONOMIC GROWTH, FULL AND PRODUCTIVE EMPLOYMENT AND DECENT WORK FOR ALL**	**Studies on islands about the economic dimension of sustainability**
How sport tourism can contribute to SDGs 1 and 8	Achilleos et al. (2021) Ahmadi et al. (2013) Bartolome et al. (2019) Bessy and Naria (2005) Boukas & Ziakas (2013) Buckley et al. (2017) Cepernic (2016) Dolinting et al. (2015) Elcock (2007) Fadda (2020) González-García et al. (2022) Ito and Higham (2020) Kartakoullis and Karlis (2002) Lee et al. (2014) Lee and Huang (2014) Lim and Patterson (2008) Pieters and Gevers (1995) Scol (2011) Séraphin and Gowreesunkar (2020) Sofield and Atara (2003) Xia et al. (2012)
Jobs and income opportunities: Sport tourism, as other segments of tourism, provides jobs and income opportunities including those for less favored groups. Depending on the nature of sport tourism products and experiences developed, local people can work as instructors and guides who are likely to be paid more than average for their special skills. New businesses catering to sport tourists can be also expected (e.g. equipment rentals, transport services) (UNWTO, 2019, p. 4).	
Visitor spending: Sport tourists tend to spend more than average tourists (for renting equipment, hiring instructors, participation fees, etc.). Sport events, even small ones, attract participants and spectators, whose spending at the destination directly brings in economic benefits. Sport tourists tend to enjoy the other attractions in the destinations, which results in more spending (UNWTO, 2019, p. 4).	
Competitiveness: Sport tourism can enhance a destination's attractiveness and competitiveness by expanding its tourism offer. This helps increase a length of stay and spending of visitors. Sport tourism products and sport events can contribute to dispersal of tourists spatially and temporally when they are organized in a less-visited area of the destination or in off-peak season (UNWTO, 2019, p. 5).	
What needs to be considered when developing sport tourism and advancing the SDGs 1 and 8	
Support of the locals: Hiring local people, encouraging local entrepreneurs, and utilizing local products help minimize economic leakage and maximize local benefits; and businesses can profit from special local knowledge. Employees in sport tourism are often required to have special skills, especially as instructors and guides. Therefore, training opportunities need to be provided for local people. It is preferable that new or expanded business opportunities brought in by sport tourism are materialized by local small and medium-sized enterprises (SMEs) and it is necessary to provide them with business advice and microfinance (UNWTO, 2019, p. 4).	

(continued)

Table 4.1 (continued)

Economic dimension	
Sport events: While planning a sport event, type and size needs to be appropriate to the capacity and characteristics of the destination. Timing and venue of a sport event should be considered with the objectives of the event taken into account. In order to distinguish the event from others, it is important to add "local flavours": for example, combining sport activities with local culture and gastronomy and offering an opportunity to interact with local communities. Offer incentives to participants and tourists to spend more at local shops, restaurants, and other services, for instance in form of coupons. Additionally, promotion and public relations in all types of media are crucial since sport events are a good opportunity for advertising the destination. If the event is held regularly, it is possible to achieve long-term positive impacts (UNWTO, 2019, p. 4).	
Sport events for a cause: Sport events can be held for a cause, where part of participation fees goes to projects or charities to help local communities; moreover, this not only helps communities but also enhances visitor satisfaction and visitors' mental attachment to the destination, which stimulates repeated visits and good word of mouth (UNWTO, 2019, p. 4).	
New sport event: A destination can even create a new sport or sport event taking advantage of its unique natural and cultural heritage. It may start small, but it can be a major attraction if it becomes popular to visitors and residents alike (UNWTO, 2019, p. 5).	
Technology: Sport tourism should take advantage of the latest technology and innovation such as artificial intelligence (AI), augmented reality (AR) or virtual reality (VR), which can make sport tourism more exciting and entertaining. Stadia and arenas have begun smartization utilizing applications and wearable devices to improve audience's convenience. Sport events and museums can enhance visitor experiences with the help of technology. It is also expected that innovation creates new gears for people with disabilities to enjoy sport activities (UNWTO, 2019, p. 5).	
Combining sport tourism with other tourism resources: Sport tourists are likely to want to experience other attractions during their stay at a destination. Participants of a sport event, noncompetitive ones in particular, but also other sport tourist who do not attend sport events, tend to take advantage of the event opportunity to discover other tourist attractions at a destination. Therefore, it is important to combine sport tourism with other tourism resources in a destination and promote them together. This helps distinguish a destination from others and brings additional benefits to a destination as a whole (UNWTO, 2019, p. 5).	

(1995) and Fadda (2020) consider the complementary contribution of the sport tourism segment to the tourism industry in Bonair and Sardinia, respectively.

The relationship between the development of innovation and sport tourism is perceptible with the effects on the synergy of these activities with other tourism sectors. Different approaches have been observed on an island scale. In Japan, Ito and Highman (2020) have shown the added value of the creation of tourism products at the interface of sport, culture, and heritage, while Lim and Patterson (2008) have highlighted that the "business" on the Jeju islands was a sector allied to sport tourism in the search for mechanisms of socioeconomic innovation. In the same vein, the structuring of governance in Gran Canaria (González-García et al., 2022) and partnerships between destinations to enhance the value of the Balearic Islands (Bartolome et al., 2019) present challenges in the development of island territories. Finally, other effects on the marketing of services related to sport tourism activities have been identified to improve management, marketing, and communication related to the experiential aspect of the activity offered (Xia et al., 2012; Achilleos et al., 2021).

The previous discussion draws attention to the complementarity around the development of social cohesion, which has effects on the engagement of indigenous people and island communities. For example, the work of Buckley et al. (2017) has demonstrated that surfing combats poverty in the Maldives, and Ahmadi et al. (2013) interrogate social mobilization for the development of unifying, integrative, and intercultural actions through sport tourism activities on the Kish Islands.

In the sociocultural dimension (Table 4.2), sport tourism in small island states or territories provides residents of a destination with sport opportunities which can have positive impacts on the promotion of leading an active and healthy lifestyle (SDG 3), quality education for all (SDG 4), gender equality (SDG 5), and accessibility (SDG 11). Sport tourism in these contexts is also able to promote mutual understanding (SDG 16) and strengthen the partnerships for sustainable development goals (SDG 17). Indeed, O'Brien and Ponting (2013) outline the creation of social sustainable benefits for host communities in Papua New Guinea. In Reunion Island, sport tourism practices are vectors for fair play, discipline, and respect for young people from disadvantaged neighborhoods by immersing them in mountain areas and islets (Naria, 2010).

Educational objectives have been addressed through the stimulation of didactic and personal development objectives in Hainan (Ai et al., 2021). Other authors refer to the implementation of training of tourism industry actors through sport tourism activities in the Caribbean (Sinclair, 2005; Towner, 2016). The development of the place of women in societies in terms of the effects of women's empowerment can also be supported by sport tourism in island territories, such as on Siargao Island, Philippines (Hansen, 2021).

Considering the environmental dimension (Table 4.3), the originality of the islands, visible through the wealth of their natural heritage and biodiversity, their ecological vulnerabilities, and their spatial imbalances require the various stakeholders to be aware of the role that sport tourism can play in protecting and enhancing the environment but also in enhancing awareness and encouraging environmental

Table 4.2 Sport tourism and the 17 Sustainable Development Goals: The Sociocultural Dimension

Sociocultural dimension	
SDG 3: ENSURE HEALTHY LIVES AND PROMOTE WELL-BEING FOR ALL AT ALL AGES **SDG 5: ACHIEVE GENDER EQUALITY AND EMPOWER ALL WOMEN AND GIRLS** **SDG 11: MAKE CITIES AND HUMAN SETTLEMENTS INCLUSIVE, SAFE, RESILIENT AND SUSTAINABLE** **SDG 16: PROMOTE PEACEFUL AND INCLUSIVE SOCIETIES, PROVIDE ACCESS TO JUSTICE FOR ALL AND BUILD INCLUSIVE INSTITUTIONS** **SDG 17: STRENGTHEN THE MEANS OF IMPLEMENTATION AND REVITALIZE THE GLOBAL PARTNERSHIP FOR SUSTAINABLE DEVELOPMENT**	**Studies on islands about the sociocultural dimension of sustainability**
How sport tourism can contribute to SDG 3, 5, 11, 16 and 17	Ai et al. (2021) Ahmadi et al. (2013) Bartolomé et al. (2009) González et al. (2010) Hansen (2021) Naria (2010) O'Brien and Ponting (2013) Sinclair (2005) Towner (2016) Ziakas & Boukas, (2016)
Well-being of tourists: Sport tourism enhances physical and mental well-being of participants; it helps maintain and/or improve their health. Participants of sport tourism may have positive emotions associated to their experience such as fun and excitement, a sense of fellowship with other travelers, and feelings of accomplishment and self-esteem. Sport tourism helps especially young people achieving both educational and personal development while having fun with friends (UNWTO, 2019, p. 6).	
Welfare of the locals: Sport tourism provides residents of a destination with sport opportunities that can have positive impacts on residents and tourists alike, leading to a healthy and active lifestyle. Sport activities can bring about self-fulfillment and enhance self-confidence among local people, through which the locals are empowered (UNWTO, 2019, p. 6).	
Jobs and income opportunities: The number of women making sport holidays and joining women-only travel groups are on the rise, which may expand opportunities for local women to cater to these groups and expand women's employment in sport tourism (UNWTO, 2019, p. 7).	
Athletes as a role model: Female athletes participating in sport tourism or events can be an inspiring role model for the local women.	
Accessibility: Accessible sport tourism products (e.g. hand bikes) provide people with disabilities, both residents and visitors alike, with opportunities for sport. Parasport activities can be enjoyed by everybody including tourists and residents with and without disabilities. Development of accessible sport tourism can improve general accessibility in the destination by provision of not only such tourism products but also accessible accommodation and transport. Furthermore, accessible sport tourism helps increase the understanding for people with disabilities in society and engage the disabled in social activities (UNWTO, 2019, p. 8).	
Promoting mutual understanding: Sport tourism provides encounters between people of diverse cultural backgrounds from all over the world and let them interact with each other through sport activities. Sport tourism can foster friendship and mutual understanding, laying the foundation of more peaceful societies. It also helps the locals and visitors having friendly relationship through sport and other activities (UNWTO, 2019, p. 10).	

(continued)

Table 4.2 (continued)

Sociocultural dimension	
What needs to be considered when developing sport tourism and advancing the SDG 3, 5, 11, 16 and 17	
Partnerships and community involvement: Successful development of sport tourism requires close cooperative relationship among stakeholders in sport (including sport commissions) and tourism sectors, as well as private and public sectors. Involvement of the communities is also crucial for sport tourism to benefit the destination as a whole and enhance welfare of the locals—the disadvantaged in particular. Engaging local people is crucial in various ways since it enables them to act for tourism development and improvement of quality of life of their own (UNWTO, 2019, p. 4).	
Participation of residents: It is preferable to promote sport activities not only to visitors but also local residents for health and well-being of their own. If visitors and residents participate in the same sport programs, it also brings about good opportunities for them to interact, which enhance visitor satisfaction and local pride. In this sense, sport opportunities for the locals and sport tourism can be developed hand in hand (UNWTO, 2019, p. 6).	
Consideration of women: It is important to provide training and micro finance for women and support local businesses owned by women, sometimes intentionally, to enhance their empowerment. This includes creating and promoting opportunities for female athletes to get inspired and to engage the destination with their female sports representatives (UNWTO, 2019, p. 7).	
Safety: Whatever sport tourism is developed, safety is of utmost importance. Sport activities must be programmed with minimum risk. Participants need to be informed of the risk and safety measures and act responsibly. Instructors and guides need to have a good knowledge of how to act in emergency through trainings. Visitors, employees and businesses should be appropriately insured (UNWTO, 2019, p. 5).	
Enhancement of accessibility: Accessible sport tourism should be considered as an accelerator for social inclusiveness with local needs taken into consideration (UNWTO, 2019, p. 8).	
Encouraging interactions: It is important to ensure opportunities where the locals and visitors can interact; it can be a sport activity itself, a specially created space such as fan zones or programs that both residents and visitors can participate in (UNWTO, 2019, p. 10).	

protection on the islands (SDGs 13, 14, 15). Studies have highlighted the effects of sport tourism on the rebalancing of sites and on the revitalization of spaces. They highlight some of the effects that have been demonstrated with the decompartmentalization of the actors in charge of the organization of the sector (Markwick, 2000) or the structuring of the territory for a better coherence of organization (Guiltat, 2011).

The effects of sport tourism on the development of resources have also been studied. In this regard, Gaspard (2003) highlighted the advantages of geomorphology, landscapes, and the environment for the practice of these activities in Guadeloupe. Bessy and Naria (2004) have qualified the contribution of sport

Table 4.3 Sport tourism and the 17 Sustainable Development Goals: The environmental dimension

Environmental dimension	
SDG 12: ENSURE SUSTAINABLE CONSUMPTION AND PRODUCTION PATTERNS **SDG 13: TAKE URGENT ACTION TO COMBAT CLIMATE CHANGE AND ITS IMPACTS** **SDG 14: CONSERVE AND SUSTAINABLY USE THE OCEANS, SEAS AND MARINE RESOURCES FOR SUSTAINABLE DEVELOPMENT** **SDG 15: PROTECT, RESTORE AND PROMOTE SUSTAINABLE USE OF TERRESTRIAL ECOSYSTEM AND HALT BIODIVERSITY LOSS**	**Studies on islands about the environmental dimension of sustainability**
How sport tourism can contribute to SDGs 12, 13, 14 and 15	Alonso and Medina (2019) Bouchet and Naria (2023) Boukas et al. (2012) Khelifi (2021) Markwick (2000) Martel (2008) Naria and Sherwin (2011) Ramallal et al. (2010)
Enhancing awareness and encouraging environmental measures: Nature-based sport tourism provides opportunities to directly experience and enjoy the nature. As such, it relies on the very presence of natural resources. Global warming has negative impacts on snow sport tourism; it accelerates the loss of biodiversity on land and under water; and the loss of fauna and flora degrades enjoyment of hiking and trekking. Because its close relationship with nature, sport tourism plays an important role to enhance awareness of nature's value and importance of its protection, and to encourage measures for environmental preservation among travellers, the locals and the tourism industry (UNWTO, 2019, p. 9).	
What needs to be considered when developing sport tourism and advancing the SDGs 12, 13, 14 and 15	
Sustainable use of natural resources: Sport tourism must be developed taking into account sustainability of natural resources and biodiversity. Nature-based sport tourism must champion the preservation of natural environment without which it cannot sustain. Preservation of landscape and avoidance of any kind of pollution should also be considered. Tourism fees can be used for preservation of natural resources (UNWTO, 2019, p. 9).	
Environmental measures: Sport tourism operators should make the best efforts in taking environmental measures: for example, to use motors and other equipment with less—preferably no—emissions of CO2, noise, and pollutants, to reuse/recycle sport equipment and to restrict use of hazardous materials (e.g. pesticides). Sport resorts and events should be managed in an environmentally friendly way (efficient water and energy use, waste management, recycling, etc.). Sport tourism operators and sport event organizers (large ones in particular) should consider sustainability while choosing suppliers and products (UNWTO, 2019, p. 9).	
Educating tourists and the locals: Tourism is an agent for change. It is crucial to educate tourists and the locals about the importance of the destination's environment and its protection through signage, leaflets, workshops and classes at school, etc. It should also be considered to communicate the message to the world together with tourists and locals (UNWTO, 2019, p. 9).	

tourism by the improvement of architectural, urban, and landscape quality with the environmental costs generated.

Sport tourism can also play a critical role in the development of environmental protection on islands. In this regard, Martel (2008) examines the effects of these activities on the structuring of the territories of the practice sites and spaces, preventing anarchic uses and delimiting the practice sites in Corsica. Naria and Sherwin (2011) show the awareness and appropriation of the environment in the Seychelles by the local population. Second, Ramallal et al. (2010) relate the effects on the regulation of the use of sites to the orientation of the need to conserve heritage in Tenerife. The writings of Alonso and Medina (2019) in the Canary Islands state that the development of sport tourism sites allows the management of cohabitation and conflicts of use of the space to be regulated.

Concluding Remarks

This chapter has sought to examine the dynamic relationship between sport tourism and sustainability within small island territories. We have attempted to historicize this evolving relationship and demonstrate its future potential, particularly as it relates to the planning and implementation of sustainable development goals. Many islands in the past have been utilized as agricultural export markets, serving the interests of the mainland while offering relatively little to indigenous island communities. Beginning in the twentieth century, island territories began to emerge as attractive tourist destinations, requiring the development of an island infrastructure to allow for such growth. In some instances, these islands drew dependent on the emerging tourism market, having focused singularly on their identity as an exotic paradise destination, offering sun, sea, and sand. The risk of economic dependence and the fragility and vulnerability associated with such a seasonal, singular focus have led to the need to diversify their tourism products. In particular, sport tourism has been heralded as a mechanism to combat mass tourism and overcrowding and as a potential vehicle for sustainable economic growth.

To date, research in the field has been narrowly confined to assessing the economic dimension of sport tourism, generally focusing on specific events, activities, and single case studies. The conceptualization of sport tourism contributed to the territorial branding of islands as "niche" markets. Given the deleterious effects of climate change, however, these island states or territories are particularly vulnerable to global warming and environmental constraints (Hoarau, 2016). This relationship between economic growth and social and environmental health and well-being has created management dilemmas at varying scales (Hall, 2010). Different levels of effects can therefore be considered, and they offer new and strategic approaches for the islands while being weighed against the costs generated. Indeed, sport tourism has the potential to bring about adverse socioeconomic and environmental changes while providing island communities with certain benefits of social progress. Islands are therefore a privileged laboratory (Baldacchino, 2010; Melay, 2019) for examining the challenges of sport tourism and the achievement of the SDGs in island states or territories.

In this regard, we hope that the growing interest in the sustainable development of sport tourism in small island or territories will continue. In order to inform this future direction, robust, comparative research is particularly needed. The following two chapters offer both a conceptual framework and a comparative methodology to address the urgent call to action at these vulnerable, but also promising, sites. This multidimensional and integrated approach hopes to provide a useful set of tools for future island studies at the intersection of place, sport tourism, and sustainability.

References

Achilleos A., Makrominas M., Markides C., Alexandrou R., Konstantinidis A., Papacosta E., Constantinides P., Zikouli E. & Tselepos L. (2021) Promouvoir le tourisme sportif actif grâce à la technologie et évaluer son impact économique : expériences de Chypre. *Journal of Sport & Tourism, 25*(4), 297–315, https://doi.org/10.1080/14775085.2021.1965009

Ahmadi, H., Moeenfard, M. R., & Tabaeeban, S. A. (2013). The impacts of sport tourism development in Kish Island. *The Journal of American Science, 9*(11s), 23–29. http://www.jofamericanscience.org

Ai, K., Liu, S., & Lin, J. (2021). Analysis of Hainan marine folk sports culture and tourism, development approach based on computer. *Journal of Physics: Conference Series, 1744*(2021), 032135. https://doi.org/10.1088/1742-6596/1744/3/032135

Alonso, V., & Medina, X. F. (2019). Impactos, contrapesos y la promoción del turismo deportivo en Canarias. El ejemplo de los deportes de litoral en la isla de Fuerteventura. *Journal of Tourism and Heritage Research, 2*(2), 125–139.

Baldacchino G. (2010) Editorial: The island lure, in *Int. J. Entrepreneurship and Small Business*, Vol. 9, No. 4.

Bartolomé, A., Ramos, V., & Rey-Maquieira, J. (2009). Attitudes des résidents envers la diversification du tourisme sportif dans les Baléares. *Tourism Recreation Research, 34*(1), 55–65. https://doi.org/10.1080/02508281.2009.11081575

Bernardie-Tahir, N. & Taglioni, F. (2016). Les territoires insulaires : des lieux d'exception ?. Cultures et sociétés, 2016, « Île, état du lieu », 40, p.44–49. (hal-01389910).

Bessy, O., & Naria, O. (2004). L'origine géographique des participants au Grand-Raid de La Réunion. *M@ppemonde, 74.*

Bessy, O., & Naria, O. (2005). Les enjeux des loisirs et du tourisme sportif de nature dans le développement durable de l'île de la Réunion. In P. Bouchet, & C. Sobry (dir.), *Management et marketing du sport: du local au global.* Presses Universitaires du Septentrion, collection Sports et Sciences Sociales, Lille, pp. 307–339.

Boley, B. & Green, . (2016). Ecotourism and natural resource conservation: The 'potential' for a sustainable symbiotic relationship. *Journal of Ecotourism.* https://doi.org/10.1080/14724049.2015.1094080

Boniface, P. (2021). *Géopolitique du sport.* Ekho, 228 p.

Bouchard, C., Marrou, L., Plante, S., Payet, R., & Duchemin, E. (2010). Les petits États et territoires insulaires face aux changements climatiques: vulnérabilité, adaptation et développement. *VertigO – la revue électronique en sciences de l'environnement [En ligne], 10*(3).

Bouchet, P., & Bouhaouala, M. (2009). *Tourisme sportif*, Téoros [Online], 28–2 I 2009, Online since 28 April 2016, connection on 16 October 2022. URL: http://journals.openedition.org/teoros/438

Bouchet, P., & Naria, O. (2023, June 5–7). Proposal of a model to analyze the residents' perceptions and attitudes toward impacts of sport tourism on the sustainable development of their islands. In *9th IRNIST Conference, sport tourism development: Local to global perspectives*, Berlin Germany.

Boukas, N., & Ziakas, V. (2013). Exploring perceptions for Cyprus as a sustainable golf destination: Motivational and attitudinal orientations of golf tourists. *International Journal of Sport Management and Marketing, 14*(1/4), 39–70.

Buckley, R. (2002). Surf tourism and sustainable development in Indo-Pacific Islands. The industry and the islands. *Journal of Sustainable Tourism, 10*(5), 405–424. https://doi.org/10.1080/09669580208667176

Buckley, R. C., Guitart, D., & Shakeela, A. (2017). Contested surf tourism resources in the Maldives. *Annals of Tourism Research, 64*, 185–199. https://doi.org/10.1016/j.annals.2017.03.005. ISSN 0160-7383.

Bull, C., & Weed, M. (1999). Niche markets and small island tourism: The development of sports tourism. *Malta, Managing Leisure, 4*(3), 142–155. https://doi.org/10.1080/136067199375814

Butle, R. (2008). The concept of a tourist area cycle of evolution: implications for management of resources. *The Canadian Geographer/Le Géographe canadien. 24.* 5–12. https://doi.org/10.1111/j.1541-0064.1980.tb00970.x

Carlsen, J. & Butler, R (2011). *Island tourism: sustainable perspectives*/edited by Jack Carlsen and Richard Butler. Wallingford, Oxfordshire; Cambridge, MA: CABI.

Cazes, G. (1989). L'île tropicale, figure emblématique du tourisme international. In *dans Îles et tourisme en milieux tropical et subtropical* (pp. 37–53). Talence, CRET – CEGET.

Cepernic, V. (2016). Positionnement global de la marque destination insulaire dans le développement du tourisme sportif. In *Actes 2nd Sports Tourism Congress*, Makarska, Croatie, 11 et 12 novembre 2016, pp. 33–41.

Costa, G., & Glinia, E. (2004). Sport tourism in Greece. *Journal of Sport & Tourism, 9*(3), 283–286. https://doi.org/10.1080/1477508042000320197

Craigwell, R. (2007). *Tourism competitiveness in small island developing states* (No. 19). WIDER Research Paper.

Desse, M. (2000). L'impact économique de la plaisance à la Martinique//Economic impact of yachting in Martinique. *Annales de Géographie, 109*(613), 306–316.

Devine, A., & Devine, F. (2004). The politics of sports tourism in Northern Ireland. *Journal of Sport & Tourism, 9*(2), 171–182. https://doi.org/10.1080/1477508041000l732550

Dolinting, E., Yusof, A., & Chee Chen, S. (2015). Application of push & pull theory in Island sport tourism: A study of Sipadan Island, Sabah. *Journal of Physical Education and Sport, 15*(2), Art 45, 295–304.

Dolnicar, S. and Fluker, M. (2003). Who's Riding the Wave? An Investigation Into Demographic and Psychographic Characteristics of Surf Tourists 2003. *Research Online*. https://ro.uow.edu.au/commpapers/248

Dudley, N., Kettunen, M., Gorricho, J., Krueger, L., MacKinnon, K., Oglethorpe, J., et al. (2023). Area-based conservation and the Sustainable Development Goals: A review. *Biodiversity, 23*, 146.

Duvat, V. (2009). Mondialisation touristique et environnement dans les petites îles tropicales. *Les Cahiers d'Outre-Mer [Online]*, 236 | October–December 2006, online October 01, 2009, accessed December 10, 2020. URL: http://journals.openedition.org/com/616. https://doi.org/10.4000/com.616

Elcock, Y. J. (2005). Sports tourism in Barbados: The development of sports facilities and special events. *Journal of Sport & Tourism, 10*(2), 129–134. https://doi.org/10.1080/14775080500223280

Fadda, N. (2020). Entrepreneurial behaviours and managerial approach of lifestyle entrepreneurs in surf tourism: An exploratory study. *Journal of Sport & Tourism*. https://doi.org/10.1080/14775085.2020.1726801

Furt, J. M., & Tafani, C. (dir.). (2014). Tourisme et insularité: la littoralité en question(s), Karthala et de UMR Lisa, Îles et pays d'outre-mer, 285 p.

Gaspard, G. (2003). Articulations entre potentiels touristiques et réalités géomorphologiques: l'exemple d'un Département français d'Outre-Mer. In E. Reynard, C. Holzmann, D. Guex, & N. Summermatter (dir.), *Actes Géomorphologie et Tourisme*, Finhaut, pp. 67–78.
Gay, J.-C. (2000). La mise en tourisme des îles intertropicales. *Mappemonde, 58*, 17–22.
González, R., Rodríguez, J., & Turégano, M. (2010). Sport and tourism: A potentially conflictual relationship. The case of Marinas in Tenerife. *PASOS Revista de Turismo y Patrimonio Cultural, 8*(2), 265–276.
González-García, R. J., Mártínez-Rico, G., Bañuls-Lapuerta, F., & Calabuig, F. (2022). Residents' perception of the impact of sports tourism on sustainable social development. *Sustainability, 14*(3), 1232. https://doi.org/10.3390/su14031232
Grace Kennedy Foundation (2010), *Sport in Jamaica : a local and international perspective*, Series: Grace, Kennedy Foundation, 134 p.
Guiltat, S. (2011). Le rôle des sports côtiers dans le développement territorial de l'Ouest de La Réunion: entre nouvelle maritimité et jeux d'acteurs, Thèse de doctorat en STAPS.
Hall, M. (2010). C, Island Destinations: A natural laboratory for tourism: Introduction. *Asia Pacific Journal of Tourism Research, 15*(3), 245–249. https://doi.org/10.1080/10941665.2010.503613
Hansen, K. A. (2021). Surfing, masculinity and resistance at Cloud 9: Filipino men who surf negotiating tourism spaces and social hierarchies on Siargao Island, Philippines. *The Australian Journal of Anthropology, 32*(3), 356–371.
Higham, J. (2005) Introduction to Sport Tourism Destination Analysis. In: Higham, J., Ed., *Sport Tourism Destinations*, Elsevier, Oxford, 17–24. https://doi.org/10.1016/B978-0-7506-5937-6.50009-2
Hoarau, J.-F. (2016). *Spécialisation touristique et vulnérabilité: réalités et enjeux pour le développement soutenable des petits territoires insulaires*. L'Harmattan, 378 p.
Hudson, S., & Beedie, P. (2007). Kayaking in. In G. Jennings (dir.), *Water-based tourism, sport, leisure, and recreation experiences* (pp. 171–185). Taylor and Francis Group.
Huetz de Lemps, A. (2011). Grandeur et décadence des "îles à sucre". In "*Comme un parfum d'îles*", sous la direction de Sevin O., Chaléard J.-L. et Guillaud D. (dir.), Presses de l'Université Paris- Sorbonne (PUPS), Paris, pp. 163–173.
Ito, E., & Higham, J. (2020). Supplemental tourism activities: A conceptual framework to maximise sport tourism benefits and opportunities. *Journal of Sport & Tourism, 24*(4), 269–284. https://doi.org/10.1080/14775085.2020.1850322
Jæger, K. (2020). Event start-ups as catalysts for place, sport and tourism development: Momentscapes and geographical considerations. *Sport in Society*. https://doi.org/10.1080/17430437.2018.1555218
James, C. L. R. (2013). *Beyond a boundary*. Duke University Press.
Kartakoullis, N., & Karlis, G. (2002). Developing Cyprus as a sport tourism destination: The results of a SWOT analysis. *Journal of Sport & Tourism, 7*(4), 3–17. https://doi.org/10.1080/10295390208718740
Kapferer, J. (2011). Quelle stratégie pour la marque France, demain ?. Revue française de gestion, 218-219, 139-153. https://www.cairn.info/revue-2011-9-page-139.htm
Khelifi, L. (2021). Une mise en valeur microlocale du littoral par les développements du surf à Madagascar: la construction de nouvelles territorialités et identités sociales. Étude comparative entre Mahambo et Fort-Dauphin. *Annales de géographie, 741*(5), 84–103.
Klöck, C., & Fink, M. (Eds.). (2019). Dealing with climate change on small islands: Towards effective and sustainable adaptation. https://doi.org/10.17875/gup2019-1208
Lee, C.-F., & Huang, H.-I. (2014). The attractiveness of Taiwan as a bicycle tourism destination: A supply-side approach. *Asia Pacific Journal of Tourism Research, 19*(3), 273–299. https://doi.org/10.1080/10941665.2012.739190
Lee, C. F., Chen, P. T., & Huang, H. I. (2014). Attributes of destination attractiveness in Taiwanese bicycle tourism: The perspective of active experienced bicycle tourists. *International Journal of Hospitality & Tourism Administration, 15*(3), 275–297. https://doi.org/10.1080/15256480.2014.925726

Leloup, Y. (1995). Les Jeux du Pacifique de Papeetee en 1995: quand olympisme et enjeux autonomistes se conjuguent in Stadion. In Bolz et Carpentier (dir.), *Olympism and international relation* (Vol. 38–39, pp. 251–272).

Lewis-Cameron, A., & Roberts, S. (Eds.). (2010). *Marketing Island Destinations* (1st ed.). Routledge. https://doi.org/10.4324/9780123849106

Lew, A. A. (2013). A world geography of recreational SCUBA diving. In G. Musa & K. Dimmock (Eds.), *Scuba diving tourism* (pp. 29–51). Routledge.

Lim, C., & Patterson, I. (2008). Sport tourism on the islands: The impact of an international mega golf event. *Journal of Sport & Tourism, 13*(2), 115–133. https://doi.org/10.1080/14775080802170346

Lockhardt, D. G., & Drakas-Smith, D. (Eds.). (1997). *Island tourism. Trends and prospects*. Pinter.

MacLean, M. (2021). Bats, balls and boards: Islands, beaches and decolonising Pacific sport. In *Critical reflections on physical culture at the edges of empire* (p. 175). SUN PReSS.

Macnaught, T. J. (1982). Mass tourism and the dilemmas of modernization in Pacific island communities. *Annals of Tourism Research, 9*(3), 359–381.

Markwick, M. (2000). Golf tourism development, stakeholders, differing discourses and alternative agendas: The case of Malta. *Tourism Management, 21*, 515–524.

Martel, L. (2008). La politique de mise en tourisme des sports de nature et le développement durable en Corse: disjonction entre discours et action publique. In O. Bessy (dir.), *Sport, loisir, tourisme et développement durable des territoires* (pp. 134–139). PUS.

Maxwell, S. L., Cazalis, V., Dudley, N., Hoffmann, M., Rodrigues, A. S., Stolton, S., et al. (2020). Area-based conservation in the twenty-first century. *Nature, 586*(7828), 217–227.

McElroy, J. L. (2006). Small Island tourist economies across the life cycle. *Asia Pacific Viewpoint, 47*, 61–77.

Melay, A. (2019). "Au-delà de l'utopie: l'île comme révélateur de l'état socio-géopolitique du monde actuel", Carnets [Online], Deuxième série – 17 | 2019, Online since 30 November 2019, connection on 24 June 2023. URL: http://journals.openedition.org/carnets/10266. https://doi.org/10.4000/carnets.10266

Melo, R., Sobry, C., & Van Rheenen. (2021). *Small scale sport tourism events and local sustainable development; a cross-national comparative perspective*. Springer.

Melo, R., Van Rheenen, D., & Gammon, S. J. (2020). Part I: Nature sports: A unifying concept. *Annals of Leisure Research, 23*(1), 1–18.

Naria, O. (2010). Les activités de loisirs et de tourisme sportif de nature comme outil de développement local: le cas de l'île de La Réunion (France). In "*Comme un parfum d'îles*", sous la direction de Sevin O., Chaléard J.-L. et Guillaud D. (dir.), Presses de l'université Paris-Sorbonne (PUPS), Paris, pp. 251–259.

Naria, O. (2013). Between cultural marking and sustainable development issues: A geo-historical approach to forms of development in Reunion Upland areas hosting outdoor sport tourism. In J.-M. Jauze (dir.), *The Uplands of Reunion Island, a fabled and fabulous expanse* (pp. 231–256). Océan Editions, Université de La Réunion.

Naria, O., & Sherwin, K. (2011). Tourisme, sports de nature et développement durable aux Seychelles. In O. Dehoorne (dir.), *Etudes Caribéennes, The changing world of coastal, island and tropical tourism* (Vol. 19, pp. 46–67).

Naria O., & Jean-Pierre P., (2023), Tourism and sports, a lever for territorial marketing and sustainable development for small island states ? in 9th IRNIST Conference, Sport Tourism Development: Local to Global Perspectives, 05th – 07th June 2023, Berlin (Germany).

Niles, D., & Baldacchino, G. (2011). *Island futures : conservation and development across the Asia-Pacific region*. Godfrey Baldacchino, Daniel Niles, editors. Springer.

Nyikana, S., Tichaawa, T. & Swart, K. (2014). Sport, tourism and mega-event impacts on host cities: A case study of the 2010 FIFA World Cup in Port Elizabeth. *African Journal for Physical, Health Education, Recreation and Dance, 20*(2): 548–556.

O'Brien, D., & Ponting, J. (2013). Tourisme de surf durable: une approche centrée sur la communauté en Papouasie-Nouvelle-Guinée. *Journal of Sport Management, Tome, 27*(2), 158–172.

Palmer, C. (2017). The Brazilian Hawaii: Surf culture, tourism, and the construction of place. *Global Ethnographic, 4*, 1–10.

Pieters, R., & Gevers, D. (1995). A framework for tourism development of fragile island destinations: The case of Bonaire. Island tourism: Management principles and practice in Conlin M., Baum V, pp. 123–132.

Planning Institute of Jamaica. (2009). Sports sector plan. Jamaica Vision 2030, National Development Plan, 83 p.

Ponting, J. (2014). Comparing modes of surf tourism delivery in the Maldives. *Annals of Tourism Research, 46*. https://doi.org/10.1016/j.annals.2014.01.001)

Ramallal, Manuel Eduardo González, Josefa Rosa Marrero Rodríguez, and Manuel Ángel Santana Turégano. (2010) Sport and Tourism : a potentially conflictual relationship. The case of Marinas in Tenerife. *PASOS Revista de Turismo y Patrimonio Cultural* 8.2 (2010): 265–276.

Rejón-Guardia, F., García-Sastre, M. A., & Alemany-Hormaeche, M. (2019). Sailboat race events: Exploring the effects of a regatta in the Balearic Islands. Trends in tourist behavior: New products and experiences from Europe, pp. 35–51.

Scol J., (2011) « Kalymnos : un paradis de l'escalade au pays des pêcheurs d'éponges », Territoire en mouvement Revue de géographie et aménagement [En ligne], 3 | 2006, https://journals.openedition.org/tem/364; https://doi.org/10.4000/tem.364

Schubert, S. F., Brida, J. G., & Risso, W. A. (2011). The impacts of international tourism demand on economic growth of small economies dependent on tourism. *Tourism Management, 32*(2), 377–385.

Séraphin, H., & Gowreesunkar, V. G. B. (2020). Unlocking changes for sport tourism products using the Blakeley-Visser Model: An application to sports events in Small Islands: The case of Martinique. In *Post-disaster and post-conflict tourism*. Apple Academic Press.

Sheldon, P., Knox, J. M., & Lowry, K. E. M. (2005). Sustainability in a mature mass tourism destination: The case of Hawaii. *Tourism Review International, 9*(1), 47–59.

Sinclair, D. (2005). Sports education – A priority for Caribbean sports tourism. *International Journal of Contemporary Hospitality Management, 17*(6), 536–548. https://doi.org/10.1108/09596110510612176

Singaravelou, P., & Sorez, J. (dir.). (2010). *L'empire des sports. Une histoire de la mondialisation culturelle*. collection Histoire et sociétés, Belin, 231 p.

SPEL (2021): Sociedad de Promoción Exterior de Lanzarote (SPEL) (2021) "Fases 2 y 3 del Plan Estratégico de Lanzarote European Sports Destination" Informe Diagnóstico Octubre de 2021, 237 p.

Sofield, T. H. B., & Sivan, A. (2003). From cultural festival to international sport – The Hong Kong dragon boat races. *Journal of Sport & Tourism, 8*(1), 9–20. https://doi.org/10.1080/14775080306242

Stephenson, M. L., & Timothy, D. J. (2022). Pacific Island tourism. In *Routledge handbook on tourism and small island states in the Pacific* (p. 381). Routledge, Taylor & Francis Group.

Towner, N. (2016). Community participation and emerging surfing tourism destinations: A case study of the Mentawai Islands. *Journal of Sport & Tourism*. https://doi.org/10.1080/14775085.2016.1151819

Uperesa, F. A. L., & Mountjoy, T. (2014). Global sport in the Pacific: A brief overview. *The Contemporary Pacific*, pp. 263–279.

UNWTO (2019): World Tourism Organization (2019), Sport Tourism and the Sustainable Development Goals (SDGs), UNWTO, Madrid, https://doi.org/10.18111/9789284419661

Van Rheenen, D. (2017). Promoting responsible sustainability in sport tourism: A logic model approach. In *Routledge handbook of international sport business* (pp. 317–332). Routledge.

Walker, I. H. (2011). *Waves of resistance: Surfing and history in twentieth-century Hawai'i*. University of Hawai'i Press.

Weaver, D. B. (1988). The evolution of a 'plantation' tourism landscape on the Caribbean island of Antigua. *Tijdschrift voor Economische en Sociale Geografie, 79*(5), 319–313.

Weaver, D. B. (2017). Core–periphery relationships and the sustainability paradox of small island tourism. *Tourism Recreation Research, 42*(1), 11–21.

Weed, M. (2020). The role of the interface of sport and tourism in the response to the COVID-19 pandemic. *Journal of Sport & Tourism*. https://doi.org/10.1080/14775085.2020.1794351

Wheaton, B., Waiti, J., Cosgriff, M., & Burrows, L. (2020). Coastal blue space and wellbeing research: Looking beyond western tides. *Leisure Studies, 39*(1), 83–95.

Wong, A. (2015). Caribbean island tourism: Pathway to continued colonial servitude. *Études caribéennes*, 31–32.

World Tourism Organization. (2014). *Tourism in Small Island Developing States (SIDS) – Building a more sustainable future for the people of islands*. UNWTO. https://doi.org/10.18111/9789284416257

Xia, M., Wang, H., Wang, X., & Su, Y. (2012). A research on the sports tourism special project in Hainan International Tourism Island. *Open Journal of Applied Sciences, 2*, 143–147. https://doi.org/10.4236/ojapps.2012.24B034

Ziakas, V. & Boukas, N. (2013). Extracting meanings of event tourist experiences: A phenomenological exploration of Limassol carnival. *Journal of Destination Marketing & Management. 2*. 94–107. https://doi.org/10.1016/j.jdmm.2013.02.002

Ziakas, V. & Boukas, N. (2016). The Emergence of "Small-Scale" Sport Events in "Small Island" Developing States: Towards Creating Sustainable Outcomes for Island Communities. *Event Management. 20*. 537–563. https://doi.org/10.3727/152599516X14745497664479

Chapter 5
Conceiving of Islands as Ecoterritories

Olivier Naria, Derek Van Rheenen, Ricardo Melo, and Claude Sobry

The critical role and ethic of place has been widely discussed within the sustainability literature (Greunewald, 2003; Orr, 2013; Smith, 2002), a potentially abstract concept with very real social and environmental consequences. One concrete yet dynamic and diverse place of interest within the literature has been the small island destination, a unique geographic location discussed both in the sustainability literature broadly (Apostolopoulos et al., 2002; Briguglio & Briguglio, 1996; Lim & Cooper, 2009; Tisdell, 1993; Weaver, 2017) and within the sport tourism literature specifically (Buckley, 2002; Bull & Weed, 1999; Elcock, 1994; Lim & Patterson, 2008; Silva & Alemedia, 2015). As the preceding and forthcoming chapters illustrate, this comparative research project focuses on the growing and diversified market of sport tourism and the pursuit of sustainable development goals within these physical and sociocultural contexts.

Within this growing and diversified market, the decision to pursue sustainable development goals while seeking economic growth requires a social,

O. Naria
Department STAPS, Université de La Réunion, Le Tampon, île de la Réunion, Réunion, France
e-mail: olivier.naria@univ-reunion.fr

D. Van Rheenen
Cultural Studies of Sport in Education, University of California Berkeley, Berkeley, CA, USA
e-mail: dvr@berkeley.edu

R. Melo
Coimbra Education School, Polytechnic of Coimbra, Coimbra, Portugal

C. Sobry (✉)
Faculty of Sport and Physical Education, Université of Lille, Lille, Nord, France
e-mail: Claude.sobry@univ-lille.frxt

D. Van Rheenen et al. (eds.), *Sport Tourism, Island Territories and Sustainable Development*, Sports Economics, Management and Policy 24,
https://doi.org/10.1007/978-3-031-51705-1_5

environmental, and political commitment. The historical development of a land evaluation and spatial planning framework designed for the purpose of guiding conservation action began in the middle to latter half of the twentieth century (Bailey, 1976; Crowley, 1967; Tansley, 1935; Ducroux, 2002; Purvis & Grainger, 2004).

The concept of region can be defined as a portion of the earth's space of extremely variable scale, ranging from a part of a continent to a small unit of landscape (Brunet, 1972; Fremont, 1976; Claval, 2006). This concept has made it possible to analyze the management of resources for human activities, essential for the future sustainability of our planet. As an extension of this concept, the geographical notion of territory identifies an appropriated space, in which populations have chosen to live. In this territory, people develop and construct their living spaces (Brunet et al., 1992; Di Méo, 2018) and a corresponding identity with this given territory. A population's daily decisions and corresponding actions have begun to be scrutinized within the context of sustainable development, measuring efforts at a territory's social and environmental protection and preservation. This scrutiny also relates to tourists who visit these places. These measurement efforts have sought to utilize the concept of ecosystems (Bailey, 1996, 1998; Rowe, 1961) and ecoregions (Bailey, 1998; 1996; Jepson & Whittaker, 2002; Loveland & Merchant, 2004) as empirical approaches to resource and land management.

Rowe (1961) defined an ecosystem as "a topographical unit, a volume of land and air plus organic contents extended over a particular part of the earth's surface for a certain time" (p. 422). Similarly, Bailey (1996) defines "ecosystems as geographic units of the landscape that include all natural phenomena… identified and surrounded by boundaries" (p. 4). An ecosystem approach recognizes that the earth exists as a series of interconnected systems within which all components are linked, analogous to a spider web. This nested systems approach is the product of the natural and human evolution of a territory and therefore imposes a spatial scale within a systemic hierarchy (Di-Méo, 1991). When a single thread of the web is disturbed, other areas of the web are impacted because of their linkages (Bailey, 1996).

Ecoregions represent areas with relative homogeneity in ecosystems, units of analysis that have become increasingly important in environmental assessment and management (Bailey, 1996, 1998; Jepson & Whittaker, 2002; Omernik & Bailey, 1997). Ecoregion mapping is founded in the disciplines of geography and ecology, although the framework relies on other disciplines as well. Both ecosystems and ecoregions offer a holistic, interdisciplinary framework for flexible, and comparative analysis of complex environmental problems (Loveland & Merchant, 2004).

Climate largely determines natural ecosystem boundaries of all scales (Bailey, 1998). Optimal management and maintenance of ecosystem processes and functionality ensures that all land use consistently sustains resource productivity, otherwise known as ecosystem capability. This relationship is depicted as (Fig. 5.1):

Making predictions about ecosystem behavior requires information about the nature of these unique territories and how they vary geographically (Bailey, 7). This information or data helps to mobilize potential indicators (inputs) of adaptive

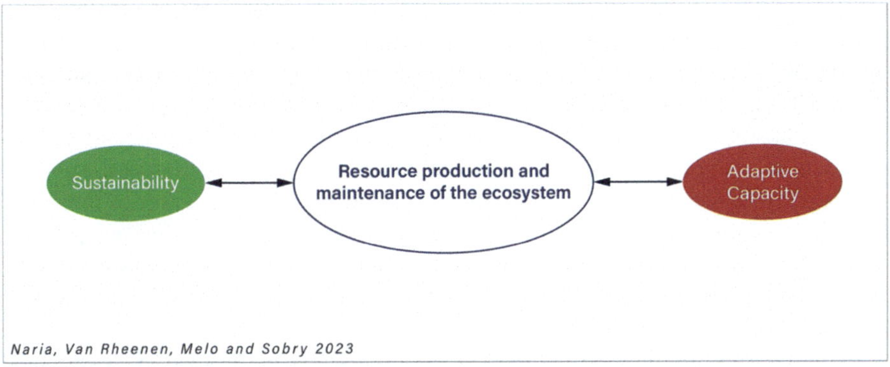

Fig. 5.1 Eco-systemic capacity. (Adapted from Bailey, 1998)

capacity (Jean-Pierre & Perrain, 2021; Simmie & Martin, 2010) in an effort to improve the policy design dedicated to island resource preservation, protection, and even renewal or regeneration. According to Jean-Pierre et Perrain (2022), adaptability or adaptive capacity is an aspirational attribute of the system, an ecoterritorial component of island vulnerability and economic resilience (Dehoorne, 2014; Perrain & Jean-Pierre, 2019; Simmie & Martin, 2010; Tran & Caldicott, 2022).

The notion of vulnerability, mainly of an economic, social, and environmental nature, has become central to the analysis and management of island ecoterritories. This vulnerability influences the relationship between people, tourism, and the environment (Mimura et al., 2007). Because of their remoteness, many islands are particularly dependent on marine resources and blue economy activities (Pauli, 2010, 2019). They must therefore define ways of adapting to change that promote sustainable growth and innovation, such as climate solutions, better management of coastal areas, or more just human development, including restraint. Proactive or adaptive capacity, then, refers to a territory's ability and willingness to change when confronted with an ecosystemic challenge. It is part of a longer-term structural adaptation that moderates potential damage while taking advantage of opportunities. For example, renewal or regenerative tourism moves beyond mitigation efforts by seeking to restore the damages already done to the natural environment. It is a bolder, more ambitious approach, seeking to create the conditions for life to flourish (Wise & Harris, 2017; Wise et al., 2021).

Islands, although similar geographic units of comparative analysis, are extremely heterogeneous. They present multiple assets with varied physical environments, exceptional cultural heritages, and even resilient island societies. At the same time, they are constrained by their vulnerabilities, be it their small size, geographical isolation, limited resources, and human capacity. Additionally, the high cost of living, prohibitive housing, and the effects of global warming require these territories to embark on a path of sustainable development. For this reason, we use the concept of ecoterritories as unique geographic ecosystems with diverse political economies, sociocultural histories, and plural stakeholder strategies. The concept of ecoterritories provides a more flexible tool for analysis at the micro- or meso-level, between islands and islets, bi-island, multi-island, or archipelagos, of very diverse topographies, with the common characteristic of being surrounded by water on all sides.

The concept of ecoterritories is effective in revealing the complexity of islands and their socioeconomic and socioenvironmental challenges and assets. Through the magnifying glass effect, the conceptual framework offers analyses of the existing relationships between humans and the natural world. From this perspective, islands are a shared territory and represent a contested terrain, highlighting evolutions of the living environment and its myriad resources (e.g., patrimonial, natural, and cultural).

Ecoterritories are likewise political in nature, boundaries defined by governance structures with varied but often similar social histories, including colonialization and its turbulent aftermath. These historical to current structures determine the decision-making process—who is and who is not at the proverbial table. Island ecoterritories rely on a moral mandate for change, a focus on a critical consciousness or ecopedagogy. This approach is likewise dialogic, inclusive rather than hierarchical and exclusive. A moral mandate requires political will and a local-to-global commitment to social and environmental justice.

Kahn (2008) draws a distinction between education for sustainable development and eco-pedagogy, an ethical epiphany and fully liberating approach to sustaining life, in all of its forms. This latter form of pedagogy adopts Freire's notion of *conscientizacão*, the awakening of a critical environmental consciousness, a shared, dialogical commitment to social justice, with a genuine ethic of ecological care (Freire, 2018, 2021; Van Rheenen & Melo, 2021).

From this perspective, critical pedagogy promises a moral and political practice rather than a mere method or technique for educational work. As Giroux (2020) reminds us, it should not be conceived as "an a priori set of prescriptions or as a commodity to be exchanged in any context" (p. (91). Instead, he argues, pedagogy "is always implicated in power relations and must be understood as a cultural politics that offers both a particular version and vision of civic life, the future, and how we might construct representations of ourselves, others, and our physical and social environment" (p. 81).

In the same vein, nature itself is not simply a resource, a commodity to be used and exploited for the sole benefit of humankind. Rather, nature is a complex, situated, and relational context or site (Booth, 2020; Melo et al., 2020; Warden, 2019). Nature is also a place, or perhaps more accurately a collection of places and series of sites. The study of place, according to Orr (2013), helps to reeducate people in the art of living by exploring more deeply those places in which we live and work. He writes (p.187), "knowledge of a place—where you are and where you come from—is intertwined with knowledge of who you are. Landscape, in other words, shapes mindscape."

In their *The Geography of Childhood*, Nabhan and Trimble (1995: 106–107, authors' italics) have eloquently argued,

> We need to turn to learning about the land by being *on* the land, or better by being *in* the thick of it. That is the best way we can stay in touch with the fates of its creatures, its indigenous cultures, its earthbound wisdom. That is the best way we can be in touch with ourselves.

Thus, place-based education or ecopedagogy grounds learning in local sites or eco-territories and people's lived experiences within these territories. As such, this critical framework draws attention to not only the economic life of a community but also to its decision-making processes (Smith, 2002), a focus on people in relationship to one another and in relation to all of the natural world. Such decisions—who is and who is not included and whose values prevail as a result—are political acts. This approach seeks to include an indigenous research paradigm with an underlying focus on fostering relational equity and accountability. As Wilson (2020) argues, "the term Indigenous has important implications politically, as in the face of colonialization we assert our collective rights as self-determining peoples at an international level" (p. 34).

Thus, as we interrogate the specific cultural practices, activities, and industries produced at a similar or like site, it is critical to analyze how relational equity and social and environmental justice can be achieved within these historical and contested contexts. It requires critical dialogue and the development of constructive and mutually respectful alliances. The creation of these critical alliances serves as a means to build an ecological movement centered on an environmental politics and justice (Bowers, 2002; Giroux, 2020; Kahn, 2008). As interconnected and relational parts of a larger living system, the economy represents but a subsystem of the biosphere. For equity to be obtained within this biosphere, stakeholders must place a premium on fostering respect and nurturing a harmonious relationship among diverse peoples and between these peoples and places, encompassing the natural world (Van Rheenen & Melo, 2021; Windchief & San Pedro, 2019).

This critical approach also necessitates the importance of understanding the local context (periphery) in *relation to* larger global and transnational forces (center). Islands provide such a place, a set of unique yet diverse contexts. Their economies are often underdeveloped and thus undervalued by international trade organizations (United-Nation, 2010); to date, these islands' economic growth is stymied by high transportation costs (CNUCED, 2014) and the difficulty of creating economies of scale within these small territories (Blancard & Hoarau, 2016). The diverse characteristics of these islands, such as their relative size and geographical isolation, present common obstacles to economic development. Their consistent vulnerability to environmental and economic upheavals requires resilience and innovation in a volatile market. Faced with the growing stakes and challenges facing the islands, only an integrated and multisectoral approach (see Fig. 5.2) will allow the populations to achieve the SDGs, notably through the capacity to transform constraints into opportunities, through the preservation of traditional ways of life, through the collaboration of actors around territorial sustainable projects, and finally through an economic development respectful of the natural, cultural, and social heritage of these island territories.

As Fig. 5.2 illustrates, an island ecoterritory promises a different way of imagining solutions to existing and future problems. It is a question of thinking on a coherent and functional scale; such broad and aspirational thinking requires cooperation and coordination at different levels of governance (e.g., from local to global) in order to aim and achieve sustainable development objectives. It also advocates for approaching projects in their multi-sectoral dimensions, particularly in terms of the intersection of environmental, economic, and social impacts. It also assumes a strategic vision of key stakeholders by articulating the scalability of these territories. Ecoterritories represent a new model, a new way of living and managing resources. It represents sustainable alternatives and resilience to provide strategic responses to the functions of island territories. It is also a vector of integration of sustainable development at the economic level. In this sense, it helps manage economic transitions and upheavals, involving a new relationship between environment and development; at the social level, it initiates the will to fight against poverty and inequality; at the environmental level, it supports the planning and integrated management of heritage protection, preservation, restraint, and regeneration.

In this regard, conceiving of a desired ecoterritory represents one of the best ways to transform targeted islands into dynamic sustainable spaces. A focus on ecoterritories and a corresponding eco-pedagogy critiques widespread appropriations of meaningful words like "sustainable development" as "nothing other than a seductive pedagogical 'greenwash' developed by and for big business-as-usual in the name of combating social and ecological disasters" (Kahn, 2008, p. 7). To combat this ruse of real sustainable development, this approach promotes a global commitment as an "attempt to mobilize diverse peoples to engage with culturally appropriate forms of ecological politics and to engage in movement building on these issues through critical dialogue and constructive alliances" (p. 11).

Since the 1960s, the awareness of environmental protection has grown continuously. Numerous associations, organizations, and events have reinforced this global and local consciousness, the emergence of a critical ecopedgaogy. The various international conferences and programs of SIDS have allowed the United Nations (UN) to provide potential remedies to the major challenges facing these vulnerable territories. These conferences and programs have engendered a local-to-global ecopedagogy. The dialogues seek to address the problems of SIDS in order to work toward genuine sustainability and achieve the UN articulated SDGs (see Table 5.1 below). These efforts focus in part on the reduction of vulnerabilities and the enhancement of assets for positive development. The global crisis of COVID-19 has only accelerated these plans for sustainable post-pandemic recovery so as to strengthen the resilience of SIDS to future shocks.

Table 5.1 highlights the chronological development of a dialogic and global community committed to elevating the local and regional challenges facing SIDs, including economic, social, and environmental vulnerabilities. These vulnerabilities have been exacerbated over the past several decades due to increased globalization

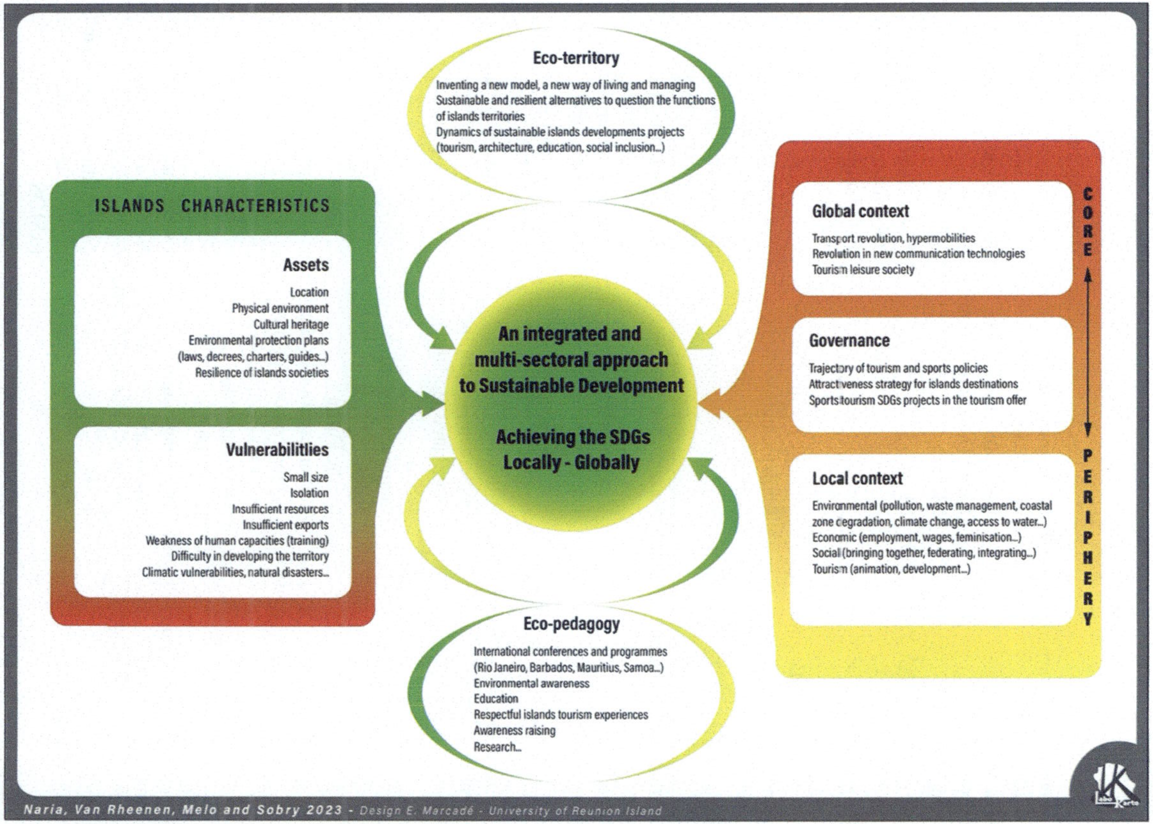

Fig. 5.2 An integrated, multisectoral approach to island eco-territorial development

Table 5.1 International SIDS Conferences and Programs

Conferences	Location	Year	Programs	Objectives
United Nations Conference on Environment and Development	Rio Janeiro	1992	Agenda 21	Commitment of States to study the problems of sustainable development of SIDS and adoption of plans and programs to facilitate sustainable development
Global Conference on the Sustainable Development of Small Island Developing States	Barbados	1994	Programme of Action for the Sustainable Development of Small Island Developing States (BPoA)	Transformation of Agenda 21 into specific policies, actions, and measures for SIDS
22nd Special Session of the UN General Assembly	New York	1999	BPoA +5	Assessment of progress in implementing the BPoA and identification of six priority areas
Mauritius International High-Level Meeting	Mauritius	2005	Strategy for the Further Implementation of the Barbados Programme	Identification of 19 priority areas for action due to limited financial resources and reduced official development assistance
Third International Conference on Small Island Developing States	Samoa	2014	Samoa's Directions	New development trajectory including economic development, food security, risk reduction, and ocean management
74th Session of the UN General Assembly	New York	2019	Follow-up and implementation of the accelerated modalities of action (Samoa Guidelines) and the Mauritius Strategy for the Further Implementation of the implementation of the Programme of Action for the Sustainable Development	Assessment of progress, gaps, and challenges
Fourth International Conference on Small Island Developing States	Antigua and Barbuda	2024	Assessing emerging sustainable development priorities after the COVID crisis	Establishment of the National Focal Points Network (NFP) for better coordination of sustainable development issues related to SIDS

Source: United Nations, https://www.un.org/fr/conferences/small-islands/index

and climate change, impacting off-shore manufacturing, infrastructure investing, and capital flow. These conferences or special sessions have fostered guidelines, pathways, and agendas for the future, perceived at the time as aspirational and realistic. And yet, as target dates for the realization of action items and corresponding results come and go or quickly approach, more urgent calls for action and political will supersede past agendas.

Progress is being made, albeit slowly. At a minimum, we have frameworks and measurement tools to assess such progress, such as global SDG indices. An environmental consciousness or ecopedagogy has developed over time, and SIDS have gained significant political representation within these global conversations. Scholars and practitioners, working in tandem with these organizations, have begun to work together in an effort to mobilize an integrated approach. Policies enacted by SIDS are regularly evaluated and tentative progress is being made in several areas. The transformation of island societies remains a long-term process toward adaptability and the achievement of sustainable development goals.

As demonstrated with this international research project, we seek to construct a conceptual model and comparative methodology at the intersection of sport tourism, small island ecoterritories, and sustainable development. The following chapter outlines this comparative methodology more fully, offering a blueprint for progress and the articulation of varied and best practices at this dynamic intersection.

References

Apostolopoulos, Y., Apostolopoulos, Y., & Gayle, D. J. (Eds.). (2002). *Island tourism and sustainable development: Caribbean, Pacific, and Mediterranean experiences*. Greenwood Publishing Group.

Bailey, R. G. (1976). *Ecoregions of the United States (map)*. US Department of Agriculture, US Forest Service, Intermountain Region.

Bailey, R. G. (1996). *Ecosystem geography*. Springer Verlag.

Bailey, R. G. (1998). *Ecoregions*. Springer-Verlag.

Blancard, S., & Hoarau, J.-F. (2016). Les petites économies insulaires en développement. Des territoires particulièrement vulnérables sur le plan économique ? *Revue économique, 67*(1), 117–142.

Booth, D. (2020). Nature sports: Ontology, embodied being politics. *Annals of leisure research, 23*, 19–33.

Bowers, C. A. (2002). Toward an eco-justice pedagogy. *Environmental Education Research, 8*, 21–34.

Briguglio, L., & Briguglio, M. (1996). Sustainable tourism in the Maltese islands. In *Sustainable tourism in islands and small states: Case studies* (pp. 162–179).

Brunet, R. (1972). Pour une théorie de la géographie régionale. In *La pensée géographique contemporaine* (pp. 649–662). Presses Universitaires de Bretagne. https://doi.org/10.3406/tigr.1972.949

Brunet, R., Ferras, R., & Théry, H. (1992). *Les mots de la géographie, dictionnaire critique* (p. 470). Montpellier/Paris/Reclus/La Documentation Française.

Buckley, R. (2002). Surf tourism and sustainable development in Indo-Pacific Islands. I. The industry and the islands. *Journal of Sustainable Tourism, 10*(5), 405–424.

Bull, C., & Weed, M. (1999). Niche markets and small Island tourism: The development of sports tourism in Malta. *Managing Leisure, 4*(3), 142–155.

Claval, P. (2006). *Géographie régionale*. Collection U, Armand Colin.

CNUCED. (2014). Petits États insulaires en développement: problemes rencontrés en matière de transport et de logistique commerciale, Genève, 24–26 novembre 2014, https://unctad.org/system/files/official-document/cimem7d8_fr.pdf [Consulté le 22 février 2023].

Crowley, J. M. (1967). Biogeography. *Canadian Geographer/Le Géographe canadien, 11*(4), 312–326.

Dehoorne, O. (2014). Les petits territoires insulaires: positionnement et stratégies de développement. *Études caribéennes*. https://doi.org/10.4000/etudescaribeennes.7250

Di Méo, G. (2018). Chapitre 6. Territoires, États, nations, régions et aménagement. Dans. In A. Ciattoni (Ed.), *Les fondamentaux de la géographie* (pp. 145–173). Armand Colin. https://doi.org/10.3917/arco.delle.2018.01.0145

Di-Meo, G. (1991). *L'homme, la société et l'espace*. Anthropos.

Ducroux, A. -M. (dir.). (2002). *Les nouveaux utopistes du développement durable*. Les Éditions Autrement.

Elcock, Y. J. (1994). Sports tourism in Barbados the development of sports facilities and special events. *Journal of Sport Tourism, 1*(3), 18–24.

Freire, P. (2018). *Pedagogy of the oppressed* (4th ed.). Bloomsbury Academic.

Freire, P. (2021). *Pedagogy of Hope: Reliving pedagogy of the oppressed*. Bloomsbury Academic.

Frémont, A. (1976). *La région, espace vécu* (p. 233). Presses Universitaires de France.

Giroux, H. A. (2020). The promise of critical pedagogy in the age of globalization. Towards a pedagogy of democratization. In H. A. Giroux (Ed.), *On critical pedagogy* (2nd ed., pp. 79–98). Bloomsbury Academic.

Gruenewald, D. A. (2003). Foundations of place: A multidisciplinary framework for place-conscious education. *American Educational Research Journal, 40*(3), 619–654.

Jean-Pierre, P., & Perrain, D. (2021). Tourism and economic growth in small island economies: evidence from non-linear threshold models in Tourisme au XXIe siècle: relation entre croissance et tourisme, https://doi.org/10.13140/RG.2.2.35094.50244.

Jean-Pierre, P., & Perrain, D. (2022). *Capacité d'adaptation et résilience de la destination: analyse et mesure avec un indice de capacité d'adaptation du tourisme (TACI)*. The Eighth Conference of the International Association for Tourism Economics, Perpignan, France.

Jepson, P., & Whittaker, R. J. (2002). Ecoregions in context: A critique with special reference to Indonesia. *Conservation Biology, 16*(1), 42–57.

Kahn, R. (2008). From education for sustainable development to ecopedagogy: Sustaining capitalism or sustaining life. *Green Theory & Praxis Journal Ecoped, 4*, 1–14.

Lim, C. C., & Cooper, C. (2009). Beyond sustainability: Optimising Island tourism development. *International Journal of Tourism Research, 11*(1), 89–103.

Lim, C. C., & Patterson, I. (2008). Sport tourism on the islands: The impact of an international mega golf event. *Journal of Sport & Tourism, 13*(2), 115–133.

Loveland, T. R., & Merchant, J. M. (2004). Ecoregions and ecoregionalization: Geographical and ecological perspectives. *Environmental Management, 34*, S1–S13.

Melo, R., Van Rheenen, D., & Gammon, S. (2020). Part I: Nature sports: A unifying concept. *Annals of Leisure Research, 23*, 1–18.

Mimura, N., Nurse, L., McLean, R. F., Agard, J., Briguglio, L., Lefale, P., Payet, R., & Sem, G. (2007). Small islands. Climate change 2007: Impacts, adaptation and vulnerability. In M. L. Parry, O. F. Canziani, J. P. Palutikof, P. J. van der Linden, & C. E. Hanson (Eds.), *Contribution of Working Group II to the fourth assessment report of the intergovernmental panel on climate change* (pp. 687–716). Cambridge University Press.

Nabhan, G., & Trimble, S. (1995). *The geography of childhood: Why children need wild places*. Beacon Press.

Omernik, J. M., & Bailey, R. G. (1997). Distinguishing between watersheds and ecoregions 1. *JAWRA Journal of the American Water Resources Association, 33*(5), 935–949.
Orr, D. (2013). Place and pedagogy. *NAMTA Journal, 38*(1), 183–188.
Pauli, G. A. (2010). *The blue economy: 10 years, 100 innovations, 100 million jobs* (p. 226). Paradigm Publications.
Pauli, G. A. (2019). L'Economie Bleue 3.0, L'Observatoire 2019, 479 p.
Perrain, D., & Jean-Pierre, P. (2019). The smart destination strategy, a key factor for changes in vulnerable tourist destinations? 6 ème Conférence de l'Association Francophone de Management du Tourisme (AFMAT) IAE de Perpignan, 23 et 24 mai 2019.
Purvis, M., & Grainger, A. (2004). *Exploring sustainable development: Geographical perspectives* (1st ed.). Routledge. https://doi.org/10.4324/9781849771290
Rowe, J. S. (1961). The level-of-integration concept and ecology. *Ecology, 42*(2), 420–427.
Secretaire Générale des Nations-Unies. (2010). Examen quinquennal de la Stratégie de Maurice pour la poursuite de la mise en oeuvre du Programme d'action pour le développement durable des petits États insulaires en développement, Assemblée générale des Nations unies, [En ligne], 6 juillet 2010.
Silva, F., & do Céu Almeida, M. (2015). *Nature sport tourism in small islands: Azores Pico Mountain Climb case study. Sport tourism: New challenges in a globalized world.* p. 166.
Simmie, J., & Martin, R. (2010). The economic resilience of regions: Towards an evolutionary approach. *Cambridge Journal of Regions, Economy and Society, 3*(1), 27–43.
Smith, G. A. (2002). Place-based education: Learning to be where we are. *Phi delta kappan, 83*(8), 584–594.
Tansley, A. G. (1935). The use and misuse of vegetation terms and concepts. *Ecology, 16*, 284–307.
Tisdell, C. (1993). Project appraisal, the environment and sustainability for small islands. *World Development, 21*(2), 213–219.
Tran, H. T. G., & Caldicott, R. W. (2022). Developing resilience for small Island tourism planning: A qualitative design infusing the sustainability trilogy with three streams of resilience thinking. *Journal of Marine and Island Cultures*. https://jmic.online/issues/v11n1/10/10.21463/jmic.2022.11.1.10
Van Rheenen, D., & Melo, R. (2021). Nature sports: Prospects for sustainability. *Sustainability, 13*(16), 8732.
Warden, C. (2019). Nature pedagogy: Education sustainability. *Childhood Education, 95*, 6–13.
Weaver, D. B. (2017). Core–periphery relationships and the sustainability paradox of small Island tourism. *Tourism Recreation Research, 42*(1), 11–21.
Wilson, S. (2020). *Research is ceremony: Indigenous research methods*. Fernwood publishing.
Windchief, S., & San Pedro, T. (2019). *Applying Indigenous research methods*. Routledge.
Wise, N., & Harris, J. (2017). Introduction: Framing sport, events, tourism and regeneration. In *Sport, events, tourism and regeneration* (pp. 1–8). Routledge.
Wise, N., Harris, J., & Whittam, G. (2021). Regeneration, communities and contemporary issues in sport and leisure. *Managing Sport and Leisure, 26*(5), 341–345.

Chapter 6
A Comparative Methodology Re(de)fined: Place or Space

Derek Van Rheenen, Olivier Naria, Ricardo Melo, and Claude Sobry

In an international project at the intersection of small-scale sport tourism and local sustainable development, Melo et al. (2021) examined half marathon races across nine countries and three continents. The book project, sponsored by the International Research Network in Sport Tourism (IRNIST), proposed a comparative methodology that utilized a sport tourism activity or event as the unit of analysis. The purpose of the study was to examine and compare the varied experiences and impacts of a sport tourism event on local sustainable development while elucidating a more empirical method to better understand similarities and differences across a like event at roughly the same time (e.g., the same year).

This kind of study promises the articulation of potential best practices to consider for future planning and implementation. While the authors selected the half marathon, one of a diverse collection of road and trail races or events comprising a growing global running tourism industry (Anderson, 2019; Van Rheenen et al., 2021), scholars and practitioners alike have been encouraged to replicate this study and compare other activities or events within different cultural contexts (Gammon,

D. Van Rheenen
Cultural Studies of Sport in Education, University of California Berkeley, Berkeley, CA, USA
e-mail: dvr@berkeley.edu

O. Naria
Department STAPS, Université de La Réunion, Le Tampon, île de la Réunion, Réunion, France
e-mail: olivier.naria@univ-reunion.fr

R. Melo
Coimbra Education School, Polytechnic of Coimbra, Coimbra, Portugal

C. Sobry (✉)
Faculty of Sport and Physical Education, Université of Lille, Lille, Nord, France
e-mail: Claude.sobry@univ-lille.frxt

D. Van Rheenen et al. (eds.), *Sport Tourism, Island Territories and Sustainable Development*, Sports Economics, Management and Policy 24,
https://doi.org/10.1007/978-3-031-51705-1_6

2021; Melo et al., 2021; Naria et al., 2022). The current international research project adopts a similar comparative methodology but shifts the unit of analysis from the activity or event to a given place or space, a geographic location, or socioeconomic "territory" at the intersection of sport tourism and sustainability.

By focusing on a geographical or spatial unit of analysis, this international research project has sought to contrast similar or like territories and their unique relationships to sport tourism and sustainable development. Specifically, this book documents the development of these eco-territories by embracing multiple dimensions of sustainability (and sustainable development goals) at these unique and diverse sites. As a number of limitations emerged during our first comparative research project, it was necessary to be more specific in the international call or request made to potential contributors. We also drew from an exploratory use of the methodology applied to the Indian Ocean islands to better inform our research agenda (Naria et al., 2022). The call for papers read as follows:

> The aim of this research project is to outline the contribution of sport tourism for sustainable local development in several small island territories or island archipelagos worldwide. For this purpose, we invite contributors to participate in this comparative study. The main purpose of comparison is to learn from fellow scholars and practitioners and to understand the cultural and political circumstances of each social geography (Hallman & Petry, 2013; Naria et al., 2022; Melo et al., 2021). In this regard, this project's intention is to adopt a comparative approach across multiple island sites, utilising a similar method of analysis.
>
> For the purpose of this research project, the comparative unit of analysis is location, site or place rather than a given sport tourism activity or event. That is, the social geographical unit of analysis will be a single small island territory or an island archipelago. The definition of 'islands', 'small states' or 'island territories' in this project will be understood as a common set of geographical and topographical features spread over several oceans and seas. Islands are also characterized by a common history of colonialism, a cultural and social independence and codependence framed by neoliberal policies of capital accumulation and linear models of development. This project seeks to challenge these models and provide innovative and indigenous practices that are more inclusive, culturally sensitive, with a focus on fostering authentic and intentional sustainable development.
>
> The geographical site or place accounts for different modalities of action by multiple actors or stakeholders in the organization at the sport-tourism nexus. The articulation of these local, national, regional and international stakeholders determines the contextual meaning of governance, types of collaboration, planning and implementation, particularly as it relates to fostering genuine sustainable development. As such, this process of articulation within the sport tourism sector draws on the spatial, historical, socio-economic and political dimensions of sport tourism and sustainable development.

This international call for book chapters also provided specific guidelines for contributors in an effort to analyze both existing and potential linkages between sport tourism and island sustainability. We identified 10 objectives for potential authors to follow as they embarked upon their island research projects. These included the desire for contributors to:

1. Describe the characteristics of the island, especially their typology, geographical size, population, GDP, annual net earning average, main professional sectors, political system, colonial past and future trajectories.

2. Describe the geographic and natural indicators, such as location, climate, type of geomorphology, accessibility, etc.
3. Introduce more precisely the sport tourism sector on the given island, identifying the most important sport tourism categories.
4. Present the ways in which the sport tourism sector has changed during the new millennium (e.g., over the past two decades) on the island and future trajectories for sustainable development.
5. Analyze the relative importance of the indigenous population within the organization of the sport tourism sector.
6. Articulate the main sport tourism (natural) sites and spaces of the island.
7. Identify key stakeholders operating in the sport tourism sector on the island.
8. Examine the demographic and participation profiles of sport tourists visiting the island (if data exists).
9. Analyze the sustainable development policies on the island, especially those related to sport tourism and how they are and have been implemented (if data exists).
10. Assess the political-economic, environmental, and sociocultural impacts of the sport tourism activities on this particular island or island archipelago.

The culmination of these 10 broad research areas of focus allowed for the design of island studies from a comparative perspective. As such, this research project and book seek to illuminate the unique qualities and best global practices at the intersection of small island territories, sport tourism, and sustainable development. We sought to rely on the indigenous island experience of our Reunionese co-editor and project manager, Dr. Olivier Naria. By extension, we hoped to elicit contributors who were from, and ideally lived on, the islands under study. We hoped that this would promote a more inclusive community of scholars and practitioners by highlighting the lived experiences and indigenous epistemologies or ways of knowing to develop a more critical and sustainable research project and book. Of the 21 island case studies and 35 contributors, 28 scholars identify as indigenous.

This comparative project situates itself within the academic subdiscipline of island studies. In one of his seminal works on the subject, Baldacchino (2010) writes,

> The core of "island studies" is the constitution of "islandness" and its possible or plausible influence and impact on ecology, human/species behavior, and any of the areas handled by the traditional subject disciplines (such as archeology, economics, or literature), subject multi-disciplines (such as political economy or biogeography), or policy foci/issues (such as governance, social capital, waste disposal, language extinction, or sustainable tourism). And this is not to mention the aspect of small islands as somewhat closed (read: manageable) systems, amenable to study: most scholars—who are not necessarily islanders—enter into the study of small islands precisely in order to test and explore conceptual schemes and specific hypotheses emerging from academic and policy debates at a mainland, regional or global level (p. xxx).

These academic and policy debates have traditionally excluded islanders within these discussions, although there has been a growing effort to be more inclusive of local and indigenous populations, as reflected in the United Nations' sustainable development goals (SDGs). These calls for inclusion have emphasized the need to incorporate indigenous epistemologies or ways of knowing and a circular, systems approach that respects the larger ecosystem in an attempt to live in harmony with nature.

Incorporating Indigenous Epistemologies and Approaches to Sustainability

For the purposes of this research project, then, small island states may be understood as ecoterritories, unique geographic ecosystems with diverse political economies and sociocultural histories. The economic diversification of an ecoterritory, particularly around sport tourism, is most often based on partnerships between the public, private, and voluntary sectors. Sports and economic issues related to the promotion of a territory are increasingly at the service of regional or global efforts to integrate the political, sociocultural, and environmental (Augustin & Gillon, 2021). Many indigenous peoples have long witnessed how tourism is both developed and controlled by either foreign or central government interests. Locals are utilized for low-paying, often seasonal positions; additionally, their ways of being and living (e.g., village life) are often appropriated and displayed for the "immersive" experience of island tourists. Locals may feel that this immersive experience for tourists is likewise intrusive and disrespectful of the local and indigenous island communities, even if economically rewarding (Butler & Hinch, 2007; Carr et al., 2016; Steinberg, 1994). That is, these celebrated economic rewards often bear a price for the land and its people, such as the deleterious effects of overtourism, including congestion, air, and noise pollution, damage to the ecosystem, and threats to local culture and heritage (Andersson et al., 2015; Fernandez & Sanchez Riveo, 2009; Fredline, 2005; Gibson et al., 2012; Kozic & Mikulic, 2014).

Thus, if local and indigenous actors do not cooperate with the proposed touristic enterprise, the potential for the genuine sustainability of this territory remains unlikely, particularly regarding the social and environmental dimensions of sustainability. As Koulouri and Mouraviev (2019) argue, "developing and implementing policies with sectoral silos without the engagement of local communities typically faces a multiplicity of barriers that drastically reduce the efficacy of resource management" (p. 2). With the inclusion of local partners and an intentional effort to more fully understand the multiple dimensions of sustainability, innovative opportunities increase.

Adopting Weaver's (2017) nuanced understanding of small island tourism and sustainability, this international project seeks to envision small islands as "sustainable sites of impressive resilience and innovation capable of providing peak experiences that foster positive tourism centrality within contexts of optimal autonomy and retained cultural and environmental distinctiveness" (p. 19). This optimistic rendering of small islands runs counter to the common, pessimistic narrative that presents these geographic locations as unsustainable, marginal entities, economically dependent, and environmentally vulnerable.

Weaver acknowledges that there is no "instruction manual" that advises local stakeholders on how to achieve positive sustainable tourism within small islands; he calls on the research community to compile a collection of examples or case studies to better understand how this ideal can be realized in practice. This book project begins from such a premise and proposes to address this call to action, with a particular focus on intentional localized planning and decision making, where the researchers and collaborators are inclusive of local and indigenous members of the island community. Such inclusion promises the possibility of being open to nontraditional ways of thinking and viewing the world. This worldview may be more sensitive to the dynamic and respectful relationships between and among people, place, and history. This reflects indigenous ways of knowing and caring for the environment that sustains life and community.

As scholars have argued, indigenous, diverse cultures often maintain manifestly different epistemological relationships to nature and long-standing practices and traditions of sustainability (Bowers, 2002, 2007; Kahn, 2008). Bowers (1993), for example, critiques the ways first-world cultural values—a "modernizing and Westernizing mode of consciousness" (p. 34)—universalize, and, in turn, institutionalize, complex understandings of ecological and historical knowledge.

In order to intentionally address these cultural biases toward linear, material development, and globalization, this book attempts to recognize and honor indigenous ways of knowing, being, and doing in a relational, circular, and evolving context. It seeks to incorporate indigenous research methods comparatively (Kinloch & San Pedro, 2014; Kovach, 2021; Tuck & Mckenzie, 2014; Wilson, 2020; Windchief & San Pedro, 2019). As Wilson (2020) has articulated, "within an Indigenous research epistemology and ontology is the recognition that research and thinking need to be (and are) culturally based...Once we recognize the importance of the relational quality of knowledge and knowing, then we recognize that all knowledge is cultural knowledge. The foundations of this cultural knowledge guide the way that our societies come to be formed" (91). This knowledge might also guide how sport tourism is implemented, thereby increasing the social and economic empowerment of indigenous groups.

Indigenous tourism provides a means for facilitating these forms of empowerment globally. As Carr et al. (2016) have asserted, "Like all forms of tourism, the development, implementation and management of Indigenous tourism should arguably be underpinned by the principles of sustainable development and natural resource management" (1068). For the purposes of our research project, we recognize that such forms of tourism may complement ongoing efforts to use sport tourism as a powerful vehicle for the realization of sustainable development goals.

Using Sport as a Vehicle for Sustainable Development

The concept of using sport as a mechanism for sustainable development, promoting a triple-bottom-line strategy of beneficial economic, environmental, and sociocultural outcomes has been discussed at some length to date (Gibson et al., 2012; Jagemann, 2004; Lindsey & Darby, 2019; Loland, 2006; Melo et al., 2021). But the world and the literature on the subject have shifted beyond a tripartite understanding of sustainability. Within these broad frameworks or categories of an emerging sustainability ontology, as noted in the introduction of its book, the United Nations (UN) has articulated seventeen sustainable development goals that map onto the economic, environmental, and social spheres of sustainable development originally articulated in the Brundtland Report (UNCED, 1987). In planning for activities, events, and capital projects, island stakeholders must question how these SDGs contribute to the economic, sociocultural, and territorial challenges for the sustainable development of small islands and the potential empowerment of indigenous communities. These questions must be at the heart of proposals and implementation plans for the promotion of sustainable development at the small island—sport tourism nexus.

Based on a more thorough understanding of sustainable development, then, the proposed and revised comparative methodology highlights the complex and dynamic relationship of island sport tourism and local sustainable development. Specifically, we sought to better understand the complex political history, geography, and social and ecological vulnerability of small island territories. We also sought to develop a model or heuristic that incorporated the 17 sustainable development goals, as articulated by the WTO (Naria et al., 2022). This model is illustrated in Fig. 6.1.

Figure 6.1 illustrates a model or heuristic to analyze the relationship of sport tourism and sustainable development globally. The model offers a methodological approach that encompasses the United Nation's seventeen articulated sustainable development goals. The tool can be used comparatively, such that small islands become the unit of analysis by highlighting clearly articulated social, environmental, and political-economic indicators. The comparison of place or island site(s) can occur within an island archipelago, across a particular ocean or sea, ultimately transcending these geographic divisions. In comparing the different island experiences, we are better able to understand the varied contexts and contours of sustainable development. As such, this research methodology identifies a broad range of sport tourism opportunities (and corresponding demand) within these island contexts.

Geography, population, relative vulnerability, adaptability, and sociohistorical legacy constitute the comparative variables under study. Geographically, the location of an island, its distance, and relative isolation from mainland locations (e.g., core-periphery) will determine the size and types of sport tourism opportunities in the islands. This is most often the result of transportation costs and the amount of

Fig. 6.1 A model of sustainable sport tourism development. (Revised from Naria et al., 2022)

time required to travel to these sites. In the field of geomorphology, particular landforms, and vulnerabilities constrain opportunity; they also constitute landscape assets, such as abrupt slopes, pitons, ramparts, coral reefs, and atolls, which offer new or revitalized eco-territories for the practice of these sporting activities. Finally, any comparative analysis of sport tourism must be put into perspective with regard to the sociohistorical legacy of an island and its corresponding limitations and future potential.

The comparative model acknowledges critical factors, such as socioeconomic influences, local cultural contexts, and the relative level of isolation and vulnerability that characterize these unique and highly diverse islands. The history of the introduction and diffusion of sport tourism onto these islands often illustrates the initial role of exogenous stakeholders importing many of these sporting activities as "civilizing," and thereby ideologically laden, games, and pastimes (Featherstone et al., 2018; James, 2013). On many of these islands, there has been a progressive appropriation of these activities by the local populations, at times articulated as resistive practices vis-a-vis the mainland of former colonizers (MacLean, 2021; Walker, 2011; Wheaton et al., 2020). Of course, indigenous peoples of these islands had their own traditional practices, which remain intact and remain cultural fodder for island tourism. For example, Bachimon and Tchekemian's (see Chap. 16, this volume) discussion of *ma' ohi* sports and games, such as traditional wrestling, stone lifting, and fruit-carrying races in French Polynesia, offers a case in point. Dominant cultures have likewise appropriated these sports, such as surfing, to become a global phenomenon and recent induction into the Olympic Games.

Relative to the socioeconomic growth sector, policies enacted by the islands since their independence have been based on several factors, including governance structure and level of investment. Finally, health vulnerabilities (e.g., Covid-19, zika viruses), ecosystemic occurrences (e.g., hurricanes, volcanic eruptions), and social upheavals (e.g. political instability, social protests) influence the development of sport tourism on island territories. As a result, sport tourism has become an attractive means of developing these territories while intentionally focusing on the multiple dimensions of sustainable development. Global SDGs, envisaged for small islands, are well suited for sport tourism development. Sustainable development in these unique territories (e.g., eco-territories) has the potential of more efficiently managing and protecting natural resources while including the local and indigenous populations within the governance and management structures.

A series of tables (please see Appendix 6.1) provided a template of indicators and data that helped inform and map onto the comparative model depicted above. These indicators included physical (geomorphological) and demographic data, island typology and particular vulnerabilities, as well as economic variables associated with this niche form of tourism (e.g., contribution to GDP per capita and Gini Index). These comparative tables provided a more detailed accounting of the sport tourism market on a given island, its relative breadth and depth based on sun, sea, and sand, and/or interior areas of the land. These tables were provided to the contributing authors as a means to clarify the desired data collection for these island territories (where such data exists).

These tables sought to collect geographic and demographic data, including the size (in square kilometers) of the island or island archipelago, its location (which sea or ocean), size of the population, and the political status of the island or archipelago (island dependencies, state islands, and island countries). We also asked contributors to note the geomorphological type of island(s) under study (continental island, oceanic island, reef island, etc.) and the main means of transportation to and from this location. The tables also included requested economic data, such as gross domestic product (GDP) per capita and the unemployment rate. More specifically to the economy of the island(s), contributors were asked to provide general tourism indicators, as well as data relative to sport tourism. This included tourism revenue as a direct contribution (%) to total GDP, number of overnight stays, number of beds, etc. In order to determine the relative impact of sport tourism supply, we asked for the participation rates of tourists who engage in sport tourism activities and total sector revenue of sport tourism activities and events within these island territories.

Thus, this model and set of tables make it possible to define a common methodology to research how sport tourism as a local, national, and regional enterprise has the potential to integrate the multiple dimensions of sustainable development within a small island context. Part II provides case studies of the use of this comparative methodology focused on a given place, such as individual islands or island archipelagos. The following chapters present case studies from a number of these sites and provide a critical discussion of the complex relationship between sport tourism and sustainable development globally. As noted in the Introduction of this book, we begin our journey in the Mediterranean and North Seas. We then move west from Europe across the Atlantic Ocean, culminating in the Caribbean Sea. We proceed across the vast Pacific Ocean, ending our circumference of the world in the Indian Ocean archipelago. Thus, the 21 case studies begin in Malta, a chapter written by Professor Godfrey Baldacchino, pioneer in the field of island studies. We are honored to begin Part II of this book with Dr. Baldacchino, Ambassador-at-large for Islands and Sustainability, as well as current President of the International Small Islands Studies Association (ISISA).

The final section (Part III) of this book includes regional and global analyses of these sites, spanning numerous oceans and seas across the world. A key objective of this international research project has been the refinement of a comparative methodology previously introduced by members of IRNIST (Melo et al., 2021; Naria et al., 2022). This refined methodology has sought to collect and analyze empirical data across numerous islands as unique and diverse ecoterritories. Findings from this comparative analysis will assist key stakeholders and decision makers to more responsibly develop island territories in a collaborative effort to realize specific sustainable development goals.

Appendix 6.1

Notes to complete the tables: If the unit of analysis is a singular island, please enter data pertaining to this island; if the unit of analysis is an island archipelago, please enter data about pertaining to the entire archipelago
Table 1
Island or archipelago under analysis (facts in 2019)
Island or archipelago under analysis (name)
Capital (city name)

Geographical size of the island [if an island archipelago indicates the size of the island where the capital is situated/if about an island indicates the size of the island] (number in km^2)
Population (number in millions)
Share of population under 20 years old (%)
Share of population above 65 years old (%)
Unemployment rate (%)
Unemployment rate of population under 20 years old (%)
Average annual income (US Dollar, per capita)
Sport participation index (%)
Table 2
Characteristics of the archipelago (if more than one island) or Island (facts in 2019)
Island or archipelago typology (mono-island, multi-islands, archipelago)
Number of islands [if an archipelago] (number)
Geographical size of the archipelago [if an archipelago] (number in km^2)
Political status of the island or archipelago (island dependencies, state islands, island countries, etc.)
Main colonizing country (name)
Table 3
Indicators of island or archipelago type (facts in 2019)
Location [ocean or sea] (name)
Type of geomorphology of islands (continental island, oceanic island, reef island, etc.)
Inhabited islands (number)
Main means of transportation to and from the island or archipelago (number and type).
Seasonality (number of seasons)
Conservation areas [national parks, natural parks, regional parks, protected areas, etc.] (name, number, and type)
Marine conservation area(s) (number and names)
World heritage site(s) (number and names)
Table 4
Global indicators (facts in 2019)
Gross domestic product (GDP) per capita (US Dollars, in millions)
Gini Index (0.0 to 1.0)
Principal vulnerabilities (environmental threats, natural hazards, health challenges, security risks, etc.)
Table 5
Island tourism statistics (facts in 2019)
Tourists (number)
Number of cruise passengers (number)
Average length of stay (number of days)
Total revenues (US Dollars, in millions)
Tourism revenue as direct contribution (% of total GDP)
Domestic travel agencies (number)
Tourist beds (number)

Table 6
Sport tourism in islands (facts in 2019)
Local sport tourism companies (number)
Jobs in the sport tourism sector (number)
Total sector revenue (US Dollars, in millions)
Sites and spaces of sport tourism activities and events (number)
Planning documents of sport tourism sites and locations [2010-2020] (number, name and year)
Participation rates of tourists who engage in sport tourism activities [in national or local tourism survey] (%)
Main sport tourism activity # 1 (name)
Sites and spaces of sport tourism activity # 1 (number)
Sport tourism companies of activity # 1 (number)
Revenue in sport tourism activity # 1 (US Dollar, in millions)
Sport tourists in activity # 1 (number)
Sport tourism activity # 1 project (name, location and cost)
Events organized in activity # 1 (number)
Main sport tourism activity # 2 (name)
Sites and spaces of sport tourism activity # 2 (number)
Sport tourism companies of activity # 2 (number)
Revenue in sport tourism activity # 2 (US Dollar, in millions)
Sport tourists in activity # 2 (number)
Sport tourism activity # 2 project (name, location, and cost)
Events organized in activity # 2 (number)

References

Andersen, J. (2019). *The state of running 2019*. Retrieved on March 2, 2020, from https://runrepeat.com/state-of-running

Andersson, T. D., Armbrecht, J., & Lundberg, E. (2015). Triple impact assessments of the 2013 European athletics indoor championship in Gothenburg. *Scandinavian Journal of Hospitality and Tourism*. https://doi.org/10.1080/15022250.2015.1108863

Augustin, J. P., & Gillon, P. (2021). *Les jeux du monde, géopolitique de la flamme olympique* (p. 224). Armand Colin.

Baldacchino, G. (2010). *Island enclaves: Offshoring strategies, creative governance, and subnational Island jurisdictions* (Vol. 14). McGill-Queen's Press-MQUP.

Bowers, C. A. (1993). *Education, cultural myths, and the ecological crisis: Toward deep changes*. SUNY Press.

Bowers, C. A. (2002). Toward an eco-justice pedagogy. *Environmental Education Research, 8*(1), 21–34.

Bowers, C. A. (2007). *Critical essays on the enclosure of the commons: The conceptual foundations of Today's mis-education*. Ecojustice Press.

Butler, R., & Hinch, T. (Eds.). (2007). *Tourism and indigenous peoples: Issues and implications*. Routledge.

Carr, A., Ruhanen, L., & Whitford, M. (2016). Indigenous peoples and tourism: The challenges and opportunities for sustainable tourism. *Journal of Sustainable Tourism, 24*(8–9), 1067–1079.

Featherstone, D., Gair, C., Høgsbjerg, C., & Smith, A. (Eds.). (2018). *Marxism, colonialism, and cricket: CLR James's beyond a boundary*. Duke University Press.

Fernández, J. I. P., & Sánchez Rivero, M. (2009). Measuring tourism sustainability: Proposal for a composite index. *Tourism Economics, 15*, 277–296.

Fredline, E. (2005). Host and guest relations and sport tourism. *Sport in Society, 8*(2), 263–279. https://doi.org/10.1080/17430430500087328

Gammon, S. (2021). *Preface*. In R. Melo, C. Sobry, & D. Van Rheenen (Eds.), *Small-scale sport tourism events and local sustainable development: A cross-national comparative perspective* (pp. v–viii). Springer.

Gibson, H. J., Kaplanidou, K., & Kang, S. J. (2012). Small-scale event sport tourism: A case study in sustainable tourism. *Sport Management Review, 15*(2), 160–170.

Hallman, K., & Petry, K. (Eds.). (2013). *Comparative sport development: Systems, participation and public policy*. Springer.

Jagemann, H. (2004). *Sports and the environment: ways towards achieving the sustainable development of sport.*

James, C. L. R. (2013). *Beyond a boundary*. Duke University Press.

Kahn, R. (2008). From education for sustainable development to ecopedagogy: Sustaining capitalism or sustaining life. *Green Theory & Praxis: The Journal of Ecopedagogy, 4*, 1–14.

Kinloch, V., & San Pedro, T. (2014). The space between listening and storying: Foundations for projects in humanization. *Humanizing Research: Decolonizing Qualitative Inquiry with Youth and Communities, 21*, 42.

Koulouri, A., & Mouraviev, N. (2019). Introduction: The water-energy-food nexus through the lens of relational equity management. In *Policy and Governance in the Water-Energy-Food Nexus* (pp. 1–8). Routledge.

Kovach, M. (2021). *Indigenous methodologies: Characteristics, conversations, and contexts*. University of Toronto Press.

Kozic, I. & Mikulic, J. (2014). Research Note: Measuring Tourism Sustainability: An Empirical Comparison of Different Weighting Procedures Used in Modelling Composite Indicators. *Tourism Economics. 20*. 429–438. https://doi.org/10.5367/te.2013.0283

Lindsey, I., & Darby, P. (2019). Sport and the Sustainable Development Goals: Where is the policy coherence? *International Review for the Sociology of Sport, 54*(7), 793–812.

Loland, S. (2006). Olympic sport and the ideal of sustainable development. *Journal of the Philosophy of Sport, 33*(2), 144–156.

MacLean, M. (2021). Bats, balls and boards: Islands, beaches and decolonising Pacific sport1. *Critical Reflections on Physical Culture at the Edges of Empire*, 175.

Melo, R., Van Rheenen, D., & Sobry, C. (2021). Sport tourism events and local sustainable development: An overview. In *Small scale sport tourism events and local sustainable development* (pp. 19–42).

Naria, O., Van Rheenen, D., Melo, R., & Sobry, C. (2022). Sport tourism, Island territories and sustainable development. In B. Mazza & C. Sobry (Eds.), *Sport tourism and its territorial development and opportunities* (pp. 19–40). Cambridge Scholars Publishing.

Steinberg, M. K. (1994). Tourism development and indigenous people: The Maya experience in Southern Belize. *Focus on Geography, 44*(2), 17.

Tuck, E., & McKenzie, M. (2014). *Place in research: Theory, methodology, and methods*. Routledge.

United Nations Conference_on_Enviroment_and Development (UNCED). (1987). *Our common future*. UN. https://sustainabledevelopment.un.org/content/documents/5987our-common-future.pdf. Accessed Dec 2022.

Van Rheenen, D., Sobry, C., & Melo, R. (2021). Running tourism and the global rise of small scale sport tourism events. In *Small scale sport tourism events and local sustainable development* (pp. 1–17). Springer.

Walker, I. H. (2011). *Waves of resistance: Surfing and history in twentieth-century Hawai 'i.* University of Hawaii Press.
Weaver, D. B. (2017). Core-periphery relationships and the sustainability paradox of small Island tourism. *Tourism Recreation Research, 42*(1), 11–21.
Wheaton, B., Waiti, J., Cosgriff, M., & Burrows, L. (2020). Coastal blue space and wellbeing research: Looking beyond western tides. *Leisure Studies, 39*(1), 83–95.
Wilson, S. (2020). *Research is ceremony: Indigenous research methods.* Fernwood publishing.
Windchief, S., & San Pedro, T. (2019). *Applying Indigenous research methods.* Routledge.

Part II
From Islands to Archipelagos: Sport Tourism in a Globalized World

Chapter 7
Sport Tourism in Malta: Getting off the Beach

Godfrey Baldacchino and Simon Caruana

Introduction

Sport tourism involves sport as an integral part of a general holiday, whether pre-planned or serendipitous. There is growing evidence to show that sport tourism is being seen as a more important part of the general holiday, with websites, travel brochures and even hotels specifically promoting sport events and opportunities (Bull & Weed, 1999).

Most of the world's small island states became independent in the latter part of the twentieth century, at around the same time that tourism took off as one of the world's dominant service industries. Thus, such locales quickly became targets of international tourism, as the fledgling industry took on the status of an easy and beguiling 'passport to development' (Conlin & Baum, 1995; De Kadt, 1984; Fletcher et al., 2017). Meanwhile, many small island territories also became sought-after refuges for domestic tourists, keen to experience somewhere calm(er) and different, but with a familiar language, culture and currency (Archer, 1978). A gripping island paradise narrative quickly took hold, with small island jurisdictions, and their people, becoming incorporated—often with the connivance of their own policy makers and tourism marketing agencies—into a discourse of exotic otherness and servility (Ateljevic & Doorne, 2005; Connell, 2003; Cohen, 2010).

Such a neo-colonial psychological 'host-guest' syndrome could easily get stuck in and restricted to the tight geographical confines of beach tourism. Here, on the

G. Baldacchino (✉)
University of Malta, Msida, Malta
e-mail: godfrey.baldacchino@um.edu.mt

S. Caruana
Institute of Tourism Studies, Luqa, Malta
e-mail: simon.caruana@its.edu.mt

D. Van Rheenen et al. (eds.), *Sport Tourism, Island Territories and Sustainable Development*, Sports Economics, Management and Policy 24,
https://doi.org/10.1007/978-3-031-51705-1_7

narrow stretch between land and sea, opportunities for local value added have been heavily restricted: dealing in umbrellas, sun beds, beach towels, ice cream and perhaps a conch shell as souvenir? Coastlines and their natural habitats have also been subject to erosion and degradation, especially on small islands (Lithgow et al., 2019). Sea level rise risks can also swamp these low-lying habitats and render them unsuitable for any beach infrastructure (Perch-Nielsen, 2010; Seetanah & Fauzel, 2018).

Yet, the 'sun, sea and sand' paradigm has proved hard to dislodge. The beach has become the ultimate site for performative warm water tourism (Kluwick & Richter, 2020). Cruise ship tourism has added to this dynamic (Wood, 2000). Getting tourists 'off the beach' and doing other things that could prove much more profitable to the small island economy has thus become a key development and marketing challenge. It is now an accepted mantra that tourist impact should be diversified and so visitors should be offered cultural- and natural-related experiences that require them to venture 'inland'. For example, Binkhorst (2007) proposes an 'off the beach' agenda that comprises: (1) cultural heritage (passive sightseeing to museums, art galleries, monasteries, markets and so on, and routes passing modernist buildings); (2) observing, spectating and participating in festivals and events; (3) wine trails, cheese trails and other food- and drink-related settings; (4) passive and/or active activities in national parks and reserves, appreciating nature; and (5) active or adventure sightseeing such as hot-air ballooning, helicopter tours, sailing, yachting, 4 × 4 excursions and other activities such as carting and skydiving.

Of course, not all small island jurisdictions have pleasant climates to warrant beach tourism, even some of the time (Baldacchino, 2006). Iceland's swimming 'beaches and bays' are restricted to those few that are heated by geothermal energy (Nautholsvik, 2022). But then, Iceland is the world's only small island state that does not lie in a tropical or temperate zone. All other SISs, and many small island territories, offer some opportunities for a beach experience. Indeed, various small island jurisdictions—such as those in the Caribbean Sea and Indian Ocean—have become small island tourist economies (SITEs), with tourism contributing a considerable proportion of their net foreign exchange receipts (McElroy, 2006; Shareef & McAleer, 2005).

Malta and the UN's Sustainable Development Goals (SDGs)

Malta scores highly on the UN's Sustainable Development Goals (SDGs). With an SDG Index score of 76.82, it ranked 33rd out of 174 countries (and the 22nd out of the EU-27) in 2022. The SDG Index has been rising steadily from a value of 68.82 in 2000, indicating a general positive disposition towards the achievement of most SDGs. Indeed, a situation of practically full employment, free public schooling (including undergraduate education), a strong welfare state and a stubbornly resilient extended family structure means that Malta has been acknowledged as having achieved SDG 1 (No Poverty) and SDG 4 (Quality Education), while no downward trends are reported in connection with any other SDG.

Significant or severe challenges, however, remain in relation to eleven SDGs: 2, 3, 5–9, 12–14 and 17. Most critical of all are initiatives addressing zero hunger (SDG 2), clean water and sanitation (SDG 6), responsible consumption and production (SDG 12) and life below water (SDG 14), although the situation on these four fronts is reported as moderately improving. The symptoms are typical of small, densely populated island states that are transitioning towards a circular economy and cleaner energy. Better waste collection and separation strategies and a bottle and can reversing scheme, both recently introduced, will help address some of these challenges. Malta's relatively low score in SDG 2 is due mainly to high rates of obesity across the native population, including children—the result of a lack of exercise and excessive carbohydrate-rich food consumption (SD Report: Malta, 2022; Sachs et al., 2022, pp. 14, 298–9).

Malta and Tourism

Malta is a small island state, a parliamentary democracy that secured its independence from the United Kingdom in 1964 and became a republic in 1974. It is located between Sicily (Italy) and Libya in the central Mediterranean Sea. The country boasts two official languages: English and Maltese, the latter derived from the Siculo-Arabic spoken in the Middle Ages. Its land area of 316 km^2 makes it the world's tenth smallest country; although that comes with a very high population density, since it supports some 515,000 residents. Having spent millennia as a colony, there is no Indigenous population to speak of: the population has mixed origins and a recent heavy wave of immigration is responsible for around 22% of the current resident population not being born in Malta (even after Covid-19 hit). In spite of its small size, the country's economy is fairly diversified. As a result, and unlike most small island states or territories, tourism contributes directly and indirectly to 'only' around 17% of Malta's GDP (Cassar et al., 2017). Other sectors—mainly export-led manufacturing, real estate and construction, electronic gaming, banking and finance, aircraft maintenance, aquaculture, trans-shipment, wholesale and retail trade—are responsible for the rest and ensure a diversified economic portfolio.

The central Mediterranean remains the preferred destination of around 30% of the world's tourists (Drius et al., 2019), and Malta benefits handsomely from this location. Some 2.8 million tourists have visited the Maltese islands annually during the period immediately before Covid-19 struck, and that number is forecast to be reached once again by 2024. This is a huge difference from tourism statistics drawn from the 1960s, when the annual number of tourists was still below 100,000, with most of these being British (Vella, 2019, p. 290). And yet, mercifully, Malta is *not* a 'small island tourist economy' (Armstrong & Read, 2021; McElroy, 2006). Airline maintenance, aquaculture (including tuna penning and fattening), container transhipment, high-end manufacturing, online gam(bl)ing, construction and real estate, and financial services count amongst its various economic sectors, apart from tourism (Debono & Baldacchino, 2021).

It is ironic that one of the long-term *advantages* of this small archipelago state is that it has a very *small* supply of sandy beaches, along a limited, national coastline of just 270 km, and along most of which—particularly on Malta's rugged south, south-west and western fringes—there are cliffs and no beaches. This means that, unlike many other warm-water island states and territories, the dominant island tourism model in Malta could never have been driven exclusively by 'sun, sea and sand' (Vella, 2019). Of course, beaches remain popular, also with the locals, especially in the peak summer months of July and August, when they afford some welcome spaces for recreational swimming, snorkelling or just cooling off. Meanwhile, beach-related sport activities now include beach volleyball, paragliding, kayaking, canoeing and (wind) surfing.

Malta's Mediterranean climate is suitable for much of this sporting activity throughout the whole year, unlike most of continental Europe, which is Malta's main source of tourists. But options are limited by the sheer, finite and limited size of the island and its sandy beaches, as well as competing pressures for land use. Some projects have focused on beach enrichment: Pretty Bay and St George's Bay, Birżebbuġa, in Malta's southeast, is a good example of one such successful initiative; the locality now has an extensive beach sports area (Zammit Pace et al., 2019). Here, beach replenishment has been the result of the deposition of large amounts of fine sediment that had been dredged from a nearby site during the construction of the nearby Malta Freeport container terminal (Borg et al., 2006).

Marine (and Non-beach) Sport Tourism in Malta

A number of marine-related tourism services and events have been assiduously nurtured in Malta, exploiting the availability of the sea while not depending on the availability of beaches. Notable is the construction of yacht marinas in practically all available bays and creeks that are sheltered enough to permit all-year-round berths. Sailing, canoeing and kayaking continue to increase in popularity, with lots of little creeks, coves and caves to explore along the long shoreline of the archipelago. Bi-annual boat races in the Grand Harbour—called *regattas*—attract both a domestic and international audience. The Malta Middle Sea Race (2022), with yachts racing from Malta to Sicily and back, is a notable event in the annual calendar.

Vessels of various sizes can be chartered for group tours, including the circumnavigation of any of the three main islands (e.g. Barbarossa Excursions, 2022; Hera Cruises, 2022). Finally, travelling between Malta and Gozo, part of a typical tourist experience, is undertaken by ferry; Gozo has no air services. The choice falls either on a 25-min crossing across the Gozo Channel, the narrowest distance between the two islands; or a 45-min crossing via fast ferry from the harbour in Valletta, the capital of Malta, to Mġarr harbour in Gozo. Travel to Comino is by small boats only; the turquoise waters of the Blue Lagoon, in Comino, draw in crowds of local and international visitors daily (Ergin et al., 2008).

Significant tracts of the territorial sea have been declared marine protected areas, with reduced fishing efforts. Malta's marine Natura 2000 network encompasses 18 sites and covers some 4100 km^2. Established over a decade ago, it comprises Special Areas of Conservation (SACs), designated for the protection of marine habitats and species pursuant to the EU Habitats Directive, and Special Protected Areas (SPAs) designated for the protection of seabirds under the EU Birds Directive (ERA, 2022).

Inland Sport Tourism in Malta

Meanwhile, various inland, non-beach-based sport and leisure activities have developed over time. These mainly concern walking, jogging, running, cycling, abseiling, geocaching, etc., across the many trails and relatively rural areas along Malta's North, West and South-West regions, as well as on the (much less developed and less heavily populated) sister islands of Gozo and Comino. They come on stream in the autumn, at around the same time that beach and marine tourism declines, and thus help to support a thriving non-summer tourism economy. These initiatives have also been fuelled by Malta's involvement in international films and TV series that have highlighted the unique topography of the islands: such as the first season of *Game of Thrones* (2011), *The Count of Monte Cristo* (2002) and, more recently, *Jurassic World Domination* (2022) (Wikipedia, 2022). And they tend to attract environmentally friendly tourists who are more respectful of the natural environment and are better able and willing to spend more to enhance their holiday experience (Hudson, 2011).

Malta's small geographical size, plus its dense population and concomitant transport and road network, makes inland tourism feasible, safe and affordable. As we know from personal experience, everywhere is within easy reach: you can cross the main island of Malta at its widest point (from west to east) *on foot* in less than 3 h. For those who are adventurous and would not mind using public transport—buses are safe, efficient and not expensive—one can take a bus to one destination and then walk—along a planned route or serendipitously—and catch a bus back from somewhere else. In spite of the smallness of the islands, their semi-arid climate, and the absence of dangerous animals—the largest mammal is a reclusive weasel—there is considerable diversity of flora and fauna, best appreciated in the early spring and in the various 'river valleys', hills and garigue patches (Dillon, 2016; Haslam & Borg, 1998). The archipelago's geology, including the erosion of the local limestone, makes for interesting ridges, caves, natural depressions and other features.

Spiteri and Stevens (2019, p. 360) describe the following as the most relevant natural landscape elements in Malta: (1) cliffs and boulder screes, known as *rdum*; (2) a system of valleys known as *wied* (singular) or *widien* (plural) which form an intricate network and are sometimes referred to as river valleys; (3) garigues, including phrygana, often collectively known as *xagħri* in Maltese, are the most characteristic natural landscape, typically consisting of low shrubs (such as thyme and Mediterranean heather) growing on coralline limestone and karst; and grasslands common on limestone and blue clay, particularly with esparto grass and sulla (clover). One natural landscape feature was the 28-metre tall 'Azure Window', a limestone arch on the western coast of Gozo and that island's most photographed natural formation, which sadly collapsed and crashed into the sea during a storm in March 2017 (Caruana et al., 2022). There are 25 other classified natural arches in the Maltese archipelago; and some locals and tourists have asked for the Azure Window to be 'rebuilt' (Mordà, 2020).

There are also a few nature parks and reserves. A rising swell of environmental consciousness has resisted, and at times successfully blocked, construction projects, with some of the land—and including coastal areas and waters—which had been originally earmarked for 'development' preserved instead as public open spaces (Boissevain & Selwyn, 2004). Examples include the Majjistral History and Nature Park in the north-west, declared in 2007 and previously targeted as a site for a golf course (Majjistral, 2022; Rolé, 2019); and the Inwadar National Nature Park in the south-east, declared in 2016, saved from being handed over to build the campus of an international university (Inwadar, 2022; Spiteri & Stevens, 2019). Għadira Reserve is a rare permanent brackish lake and saltmarsh habitat in Malta's north and is a fully protected bird sanctuary (BirdLife Malta, 2022).

Being a small island state with a very high population density, however, means that the enjoyment of such rural sport tourism can encroach on, and interfere with, private interests and practices: hunting and trapping are the most notorious examples of this in Malta (Camilleri et al., 2004). They remain passionate pastimes for a well-organized and politically savvy segment of the population. Areas of the countryside are likely to include areas marked as 'reserved' for such practices. Public footpaths and trails occasionally get blocked by gates and other impediments to discourage ramblers and casual hikers (Falzon, 2008).

Sport Tourism in Malta: Scuba Diving Stands Out

Statistics collated by the Malta Tourism Authority (MTA) for 2019—the last full year before Covid-19 struck—suggest that around 2.8% of the total number of tourists visiting Malta identified practising some form of sports as their *main* reason or travel (MTA, 2019a). This translates to 75,600 sports tourists out of a total of 2.7 million visitors. This segment is still deemed too small to warrant its own analysis. The MTA however has been taking note of tourists who declare that SCUBA diving—'scuba' is actually the acronym for 'self-contained underwater breathing apparatus'—is their main reason for visiting Malta: in 2019, 131,380 visitors—4.8% of the total—declared this to be the case. These figures exclude visitors who decide to take part in a sport-related event—for example, go for a 'try-dive' in shallow water—while already in Malta, without having made plans to do so prior to their actual visit. Taking this into account for scuba diving, the percentage of scuba diving tourists increases from 4.8% to 6.5%, which translates to 177,640. In other words, 46,260 individuals decided to try out scuba diving only while already on holiday in the Maltese Islands in 2019. This particular activity is attractive; it has developed an international reputation (e.g. Press Reader, 2019); and it is perhaps less elitist and reserved for professional users than is often assumed.

Thanks to advances in technology and in safety measures, the marine environment is no longer seen as a hostile and unsafe one but rather a 'relaxing' place where suitably trained and certified scuba divers can interact with the marine environment (Musa & Dimmock, 2013). Ironically enough, almost half of these scuba divers acknowledge that the quality of sandy beaches in Malta is not exceptional; while that of rocky bays, accommodation, nature and the quality of the air and sea in general, are all 'good' or 'very good' (MTA, 2020).

The MTA collates other statistics that distinguish between the main reason for coming to Malta and activities engaged in while in the archipelago. Sport tourism of some kind, including scuba diving, is practised by around an impressive one-third of all visitors to Malta; and confirms the primacy of non-beach tourism, which is by definition more sedentary. This statistic is consistent across the years 2017–2019, with an upward trend. In 2019, around 637,000 tourists declared that they had taken the opportunity to go for a guided tour, walk or hike; some 21,000 opted for cycling; another 24,000 decided to try out rock climbing. Around 162,000 declared having attended sports events, whether as spectators or as active participants (see Table 7.1).

These figures provide a rather generic picture; this would benefit from further analysis. *Visit Malta* is the country's official tourism website: it includes a link to 'Activities and Sports' (Visit Malta, 2022). This page offers some details as to what types of sport tourism are in vogue. Other than diving, outdoor sport activities available to tourists in Malta referred to on the MTA webpage include horse riding, kayaking, rock climbing, cave exploration by sea or land, abseiling, hiking, biking, archery, stand up paddle boarding, zip lining, trekking, sea level traversing,

Table 7.1 Sport-related tourism statistics in Malta (2017–2019)

	2019	2018	2017
Main motivation for visiting Malta			
Scuba	4.8%	4.9%	5.2%
Other sports	2.8%	2.8%	2.9%
Activities engaged in whilst in Malta			
Scuba diving	6.5%	6.5%	6.8%
Trekking/hiking	23.6%	22.7%	19.4%
Cycling	0.8%	1.2%	0.9%
Rock climbing	0.9%	1.1%	0.9%
Attend/participate in sport event	6.0%	4.3%	5.8%
Total engaging in sport-related activities	*37.8%*	*35.8%*	*33.8%*

Source: MTA (2019a, b)

coasteering, paragliding, trail running, windsurfing and free fall. There is also a list of various sports clubs and facilities, as well as hotels, where one can go to bowl, swim, play golf, play tennis, soccer/football, basketball, exercise in a gym under supervision, as well as practice clay pigeon shooting.

Separated from the main island by a five-kilometre channel, Gozo is the second largest island of the Maltese archipelago, with a resident population of around 25,000. It has its own tourism-related website, *Visit Gozo*, developed jointly by the Gozo Tourism Association (GTA) and the Ministry for Gozo (Visit Gozo, 2022a). This island, more hilly and agricultural, and less densely populated than Malta, and therefore greener, is marketed as a distinct holiday destination; it lacks its own airport and a fixed link in the form of an undersea tunnel between the two main islands is being considered (Chaperon & Theuma, 2015). The *Visit Gozo* website has its own dedicated page for sports and outdoor activities (Visit Gozo, 2022b). Compared to the Visit Malta site, the Visit Gozo page has more detailed information, including a series of proposed walking tour itineraries. There are also specific categories related to sports and outdoor activity tourism, namely, abseiling, rock climbing, bouldering, day adventure tours, eco tours, Segway tour, horse riding, cycling and bicycle tours, kayaking and other water sports. Each section has a detailed description of the type of activities available and a list of service providers to choose from. These details reflect how sport tourism in Gozo plays a proportionately more significant role than on the main island of Malta. In 2019, 10.4% of all visitors to Gozo came with the intent to practise 'other sport' and 16.5% to practise scuba diving; much more than the respective 2.8% and 4.8% indicated for Malta as a whole (MTA, 2019b).

Scuba Diving in the Maltese Islands and Its Sustainability

In this discussion, scuba diving is being treated apart from the rest of the other activities as it is a distinct niche that has been progressively developed since the first dive centres opened in the Maltese Islands in the late 1960s (Caruana & Sultana, 2021). Indeed, the evolution of scuba diving as an international tourism niche in and for the Maltese Islands may suggest ways for 'other' sport tourism segments to be identified and nurtured. To date, the local association representing the majority of dive operators, the Professional Diving Schools Association has 44 dive centres within its ranks (PDSA, 2022). If one includes the independent dive centres, the number increases to around 70. The economic sustainability of this relatively large concentration of scuba diving centres in Malta is possible only due to the influx of tourists who visit the Maltese Islands with the intention to carry out scuba diving.

The Professional Association of Diving Instructors (PADI) is by far the largest diving certification agency, accounting for almost 80% of all diving qualifications issued worldwide. PADI describes Malta as an island with a great diversity of underwater plants and wildlife, thanks to the warm Mediterranean Sea that surrounds the archipelago. The *Visit Malta* website (2022) states that the islands' various reefs, caves and wrecks make diving alluring to first-timers, novice, advanced, as well as technical divers. Divers can easily engage in a recreational dive while appreciating the marine flora and fauna since the Maltese Islands provide excellent underwater visibility, which makes the archipelago's waters exceptional also for photography (PADI, 2022).

PADI promotes some of the Maltese Islands' best dive sites. These include various wrecks, scuttled purposely for diving off Malta and Gozo, such as the tanker *Um el Faroud*, the tugboat *Rożi*, the Gozo-Malta car ferry *Cominoland* and the triple-deck tourist ferry *Karwela*. Such sites encourage divers to better appreciate the marine environment and spread awareness for its further protection and conservation (Rich, 2006). There are various other historical wrecks ranging from a 2700-year-old Phoenician wreck, Roman and Early Modern shipwrecks, to dozens of aircraft crash sites, submarines and battleships from the First and Second World Wars. Some of these wrecks are designated war graves (Dive Magazine, 2022). The national heritage agency, Heritage Malta, set up a Underwater Heritage Unit in 2019 to better monitor and manage these dive sites and has issued a series of best practices to be adopted by divers and dive operators when visiting these sites (Heritage Malta, 2022). The effective enforcement of existing legislation would help to improve the sustainability of the industry (Caruana & Sultana, 2020).

In 2021, the Malta Tourism Authority and the non-governmental organization Nature Trust signed a memorandum of understanding leading to the establishment of the first marine park that comprises three of the most popular shore dive sites at Cirkewwa, Malta's north-west tip (Arena, 2021). Nature Trust has been tasked with the management of the park and to protect against illegal fishing practices therein (Nature Trust Malta, 2021). Such partnerships presage a blueprint for 'other' sports to develop: identifying and involving key stakeholders, and locking in government support so as to push for the necessary legislative framework. Such 'best practices' help to future-proof these sites and assure their sustainability.

Conclusion

Sport (non-beach) tourism helps to diversify the tourism sector and attract visitors with a propensity to spend more. It could therefore prove to be an important niche industry and a partial solution to the problems associated with mass tourism in small islands, although its development comes with its own slate of challenges (Briguglio et al., 1996).

Malta offers an interesting case study because, on one hand, it has been successful in nurturing a tourism product that is not dependent on 'sea, sun and sand' summer visitations; but, on the other hand, the country is still developing a keen sensibility for the sport potential that the country offers, for locals and visitors alike. Government support, stakeholder consultation, site and activity management via 'public-private' partnerships, and stringent accreditation and licensing, will help lift this niche industry to the respect and significance that it deserves. Scuba diving in Malta (and Gozo) shows how it can be done.

References

Archer, B. (1978). Domestic tourism as a development factor. *Annals of Tourism Research, 5*(1), 126–141.

Arena, J. (2021, June 10). Malta to get first-ever marine park in nature trust, MTA collaboration. *Times of Malta*. https://timesofmalta.com/articles/view/malta-to-get-first-ever-marine-park-in-nature-trust-mta-collaboration.878348

Armstrong, H. W., & Read, R. (2021). Tourism and sustainable growth in small (Island) economies. In *Shaping the future of small islands* (pp. 93–108). Palgrave Macmillan.

Ateljevic, I., & Doorne, S. (2005). Dialectics of authentication: Performing 'exotic otherness' in a backpacker enclave of Dali, China. *Journal of Tourism and Cultural Change, 3*(1), 1–17.

Baldacchino, G. (Ed.). (2006). *Extreme tourism: Lessons from the world's cold water islands*. Routledge.

Barbarossa Excursions. (2022). *Barbarossa excuarsions: Gozo, Malta & Comino*. https://www.barbarossaexcursions.com/

Binkhorst, E. (2007). Creativity in tourism experiences: The case of Sitges. In G. Richards & J. Wilson (Eds.), *Tourism creativity and development* (pp. 125–144). Routledge.

BirdLifeMalta. (2022). *Għadira nature reserve*. https://birdlifemalta.org/nature-reserves/ghadira/

Boissevain, J., & Selwyn, T. (Eds.). (2004). *Contesting the foreshore: Tourism, society and politics on the coast*. Amsterdam University Press.

Borg, J. A., Gauci, M. J., Magro, M., & Micallef, M. A. (2006). Environmental monitoring at St. George's bay (Malta) in connection with beach replenishment works. In A. Micallef, A. Vassallo, & M. Cassar (Eds.), *Proceedings of the second international conference on the management of coastal recreational resources: Beaches, yachting and coastal ecotourism* (pp. 25–33). Euro-Mediterranean Centre on Insular Coastal Dynamics; Foundation for International Studies, University of Malta.

Briguglio, L., Butler, R., Harrison, D., & Leal Filho, W. (Eds.). (1996). *Sustainable tourism in islands and small states: Vol. 2: Case studies*. Pinter.

Bull, C., & Weed, M. (1999). Niche markets and small Island tourism: The development of sports tourism in Malta. *Managing Leisure, 4*(3), 142–155. https://doi.org/10.1080/136067199375814

Camilleri, M., Hili, M., Magro Conti, J., Attard, R., Stevens, D., Gambin, M. T., & Galea, R. (2004). Malta. In G. Baldacchino (Ed.), *Extreme heritage management: The practices and policies of densely populated islands* (pp. 47–74). Berghahn.

Caruana, S., & Sultana, T. (2021). Exploring the scuba diving tourism sector in Malta and its sustainable impact on the Island. In V. Katsoni & C. Van Zyl (Eds.), *Culture and tourism in a smart, globalized and sustainable world* (pp. 85–103). Springer.

Caruana, J., Wood, J., Nocerino, E., Menna, F., Micallef, A., & Gambin, T. (2022). Reconstruction of the collapse of the 'Azure Window' natural arch via photogrammetry. *Geomorphology*, 108250. https://doi.org/10.1016/j.geomorph.2022.108250

Cassar, I. P., Vella, K., & Buttigieg, S. (2017). *Understanding the economic contribution of tourism in Malta: A literature review* (Working paper series). Economic Policy Department, Ministry for Finance, Malta. https://finance.gov.mt/en/epd/Documents/Working_Papers/Tourism_Working_Paper.pdf

Chaperon, S., & Theuma, N. (2015). The Malta-Gozo-Comino story: Implications of a Malta-Gozo fixed link on tourist activity. In G. Baldacchino (Ed.), *Archipelago tourism: Policies and practices* (pp. 51–65). Ashgate.

Cohen, C. B. (2010). *Take me to my paradise*. Rutgers University Press.

Conlin, M. V., & Baum, T. (Eds.). (1995). *Island tourism: Management principles and practice*. Wiley.
Connell, J. (2003). Island dreaming: The contemplation of Polynesian paradise. *Journal of Historical Geography, 29*(4), 554–581.
De Kadt, E. B. (1984). *Tourism: Passport to development? Perspectives on the social and cultural effects of tourism in developing countries*. World Bank & UNESCO.
Debono, M., & Baldacchino, G. (Eds.). (2021). *Working life: The transformation of Malta, 1960–2020*. Malta University Press.
Dillon, P. (2016). *Walking on Malta: 33 walks on the Mediterranean islands of Malta*. Cicerone Press.
Dive Magazine. (2022). *Malta's deep wrecks*. https://divemagazine.com/scuba-diving-travel/malta-deep-wrecks-fairey-swordfish-junkers-ju88
Drius, M., Bongiorni, L., Depellegrin, D., Menegon, S., Pugnetti, A., & Stifter, S. (2019). Tackling challenges for Mediterranean sustainable coastal tourism: An ecosystem service perspective. *Science of the Total Environment, 652*, 1302–1317.
ERA. (2022). *Marine protected areas*. Environmental Resources Authority. https://era.org.mt/topic/marine-protected-areas-2/
Ergin, A., Micallef, A., & Williams, A. T. (2008). Coastal scenic evaluation of Gozo/Comino, Malta, as a tourism product. In *Proceedings of the international Pluridisciplinary conference: The littoral: Challenge, dialogue, action. Lille, France*. https://meshs.fr/page/datas/files/docs/publi/2008/littoral/ergin.pdf
Falzon, M. A. (2008). Flights of passion: Hunting, ecology and politics in Malta and the Mediterranean. *Anthropology Today, 24*(1), 15–20.
Fletcher, J., Fyall, A., Gilbert, D., & Wanhill, S. (2017). *Tourism: Principles and practice*. Pearson UK.
Haslam, S. M., & Borg, J. (1998). *The river valleys of the Maltese islands: Environment and human impact*. Islands and Small States Institute, University of Malta.
Hera Cruises. (2022). Hera Cruises. https://heracruises.com/
Heritage Malta. (2022). *Underwater Cultural Heritage Unit*. https://heritagemalta.mt/departments/underwater-cultural-heritage-unit/
Hudson, S. (2011). Working together to leverage film tourism: Collaboration between the film and tourism industries. *Worldwide Hospitality and Tourism Themes, 3*(2), 165–172. https://doi.org/10.1108/17554211111123023
Inwadar. (2022). *About the park*. http://inwadar.com
Kluwick, U. M., & Richter, V. (2020). Of tourists and refugees: The global beach in the twenty-first century. In S. Ferdinand, I. Souch, & D. Wesselman (Eds.), *Heterotopia and globalisation in the twenty-first century* (pp. 116–130). Routledge.
Lithgow, D., Martínez, M. L., Gallego-Fernández, J. B., Silva, R., & Ramírez-Vargas, D. L. (2019). Exploring the co-occurrence between coastal squeeze and coastal tourism in a changing climate and its consequences. *Tourism Management, 74*(1), 43–54.
Majjistral. (2022). *Welcome to Majjistral nature and history park*. https://www.majjistral.org
McElroy, J. L. (2006). Small Island tourist economies across the life cycle. *Asia Pacific Viewpoint, 47*(1), 61–77.
Middle Sea Race. (2022). https://www.rolexmiddlesearace.com/
Mordà, B. (2020, August 16). The azure window: To rebuild or not to rebuild? *The Heritage Call*. https://theheritagecall.com/the-azure-window-to-rebuild-or-not-to-rebuild/
MTA. (2019a). *Tourism in Malta: Facts and figures 2019*. Malta Tourism Authority Research Unit. https://www.mta.com.mt/en/file.aspx?f=34248
MTA. (2019b). *Tourism in Gozo*. Malta Tourism Authority Research Unit. https://www.mta.com.mt/en/file.aspx?f=34519
MTA. (2020). *The profile of diving travellers in Malta*. Malta Tourism Authority Research Unit. https://www.mta.com.mt/en/file.aspx?f=34514
Musa, G., & Dimmock, K. (Eds.). (2013). *Scuba diving tourism*. Routledge.

Nature Trust Malta. (2021). *Cirkewwa Marine Park.* https://naturetrustmalta.org/environmental-education/biodiversity/habitats/marine/cirkewwa-marine-park/

Nautholsvik. (2022). *Nauthólsvík Bay: Our own Ibiza.* https://nautholsvik.is/en/

PADI. (2022). *Diving in Malta.* https://www.padi.com/diving-in/malta/

PDSA. (2022). Professional Diving Schools Association. http://pdsa.org.mt/

Perch-Nielsen, S. L. (2010). The vulnerability of beach tourism to climate change: An index approach. *Climatic Change, 100*(3), 579–606.

Press Reader. (2019, November 7). *Malta voted second best dive destination of the year at Diver Awards.* https://www.pressreader.com/malta/the-malta-business-weekly/20191107/281801400773874

Rich, S. A. (2006). Scuba diving as Mediterranean culture: Preservation and presentation of Gozo's maritime heritage. *Omertaa: Journal for Applied Anthropology.* http://www.omertaa.org/archive/omertaa0004.pdf

Rolé, A. (2019). Landforms and processes at Il-Majjistral Park and its environs. In R. Gauci & J. A. Schembri (Eds.), *Landscapes and landforms of the Maltese islands* (pp. 305–316). Springer.

Sachs, J. D., Lafortune, G., Kroll, C., Fuller, G., & Woelm, F. (2022). *From crisis to sustainable development: The SDGs as roadmap to 2030 and beyond* (Sustainable development report 2022). Cambridge University Press.

SD Report: Malta. (2022). Malta. https://dashboards.sdgindex.org/profiles/malta

Seetanah, B., & Fauzel, S. (2018). Investigating the impact of climate change on the tourism sector: Evidence from a sample of Island economies. *Tourism Review, 74*(2), 194–203.

Shareef, R., & McAleer, M. (2005). Modelling international tourism demand and volatility in small Island tourism economies. *International Journal of Tourism Research, 7*(6), 313–333.

Spiteri, L., & Stevens, D. T. (2019). Landscape diversity and protection in Malta. In R. Gauci & J. A. Schembri (Eds.), *Landscapes and landforms of the Maltese islands* (pp. 359–372). Springer.

Vella, L. (2019). Tourism and hospitality: Come visit our islands. In G. Baldacchino, V. Cassar, & J. G. Azzopardi (Eds.), *Malta and its human resources: Management and development perspectives* (pp. 287–316). Malta University Press.

Visit Gozo. (2022a). *Experience Gozo: The best of Gozo.* https://www.visitgozo.com/

Visit Gozo. (2022b). *Adventure and sports.* https://www.visitgozo.com/experience-gozo/adventure-sports/

Visit Malta (2022). *Land sports.* https://www.visitmalta.com/en/a/land-sports/

Wikipedia. (2022). *List of films shot in Malta.* https://en.wikipedia.org/wiki/List_of_films_shot_in_Malta.

Wood, R. E. (2000). Caribbean cruise tourism: Globalization at sea. *Annals of Tourism Research, 27*(2), 345–370.

Zammit Pace, M. L., Bray, M., Potts, J., & Baily, B. (2019). The beaches of the Maltese islands: A valuable but threatened resource? In R. Gauci & J. A. Schembri (Eds.), *Landscapes and landforms of the Maltese Islands* (pp. 213–227). Springer.

Chapter 8
The Emerging Sport Tourism Destination of Sardinia

Nicoletta Fadda

A Sketch of Sardinia

Geographic and Natural Indicators

Sardinia island, located in the western region of the Mediterranean basin in the Tyrrhenian Sea, west of the Italian peninsula and after Sicily, but before Cyprus, is the second-largest island in the Mediterranean Sea. Sardinia is the island furthest from any mainland. The nearest lands are the French island of Corsica, the Italian Peninsula, Sicily, Tunisia and the Spanish archipelago of the Balearic Islands. Sardinia has an area of 24,100 km^2 (9305 square metres), 1897 km of coasts and 1,628,384 inhabitants.

Coasts are generally high and rocky; however, there are long stretches of wide sandy beaches. Indeed, along the shores, many outstanding headlands, deep bays and inlets can be observed, as well as various smaller islands off the coast. Sardinia is an ancient land not prone to earthquakes, different from Sicily and the Italian peninsula. Geomorphologically, the landscape varies; coasts, valleys, hills and mountains can be observed in various places and locations. Generally, mountains are located in the inland regions; the major mountain chain is Gennargentu in the central-eastern region, the highest peak being Punta La Marmora (1834 m). It is worth noting the existence of several valleys and flatlands that separate hills and plateaux; the most well-known are the Campidano in the southwest and the Nurra in the northwest, where most agricultural firms are located.

The island's climate is Mediterranean, temperate and variable from area to area. Along the coasts, the summer heat is mitigated by the fresh sea breeze, and the

N. Fadda (✉)
Department of Economics and Business, University of Sassari, Sassari, Italy
e-mail: nfadda@uniss.it

D. Van Rheenen et al. (eds.), *Sport Tourism, Island Territories and Sustainable Development*, Sports Economics, Management and Policy 24,
https://doi.org/10.1007/978-3-031-51705-1_8

temperature does not go below zero during the winter season. In the inner areas of the island, where mountains are located, temperatures are cooler by some degrees, with snowfall occurring during the winter. Rainfall is generally scarce throughout the year, particularly during summer and is mainly concentrated in the autumn and winter months. The mistral from the northwest is the dominant wind on and off throughout the year, especially during winter and spring. It can also be frequent during the summer season and can blow rather robustly.

Sardinia and Its Inhabitants

Sardinia belongs to the Italian territory, and it is one of the 20 Italian regions. It benefits from a special degree of autonomy, together with four other regions due to its insularity, which renders Sardinia a unique case in the Italian State. Sardinia is governed at the local level by six provinces (Nord-Est, Nuoro, Ogliastra, Oristano, Sulcis-Iglesiente, Medio Campidano) and two metropolitan cities (Sassari and Cagliari), as shown in Fig. 8.1. It counts 377 municipalities, which vary from cities, towns and villages.

The last population census in 2020, registered 1,590,044 inhabitants and showed a continuing decrease in the total population across all provinces and areas; however, this decrease was less in the main urban zones and the metropolitan city of

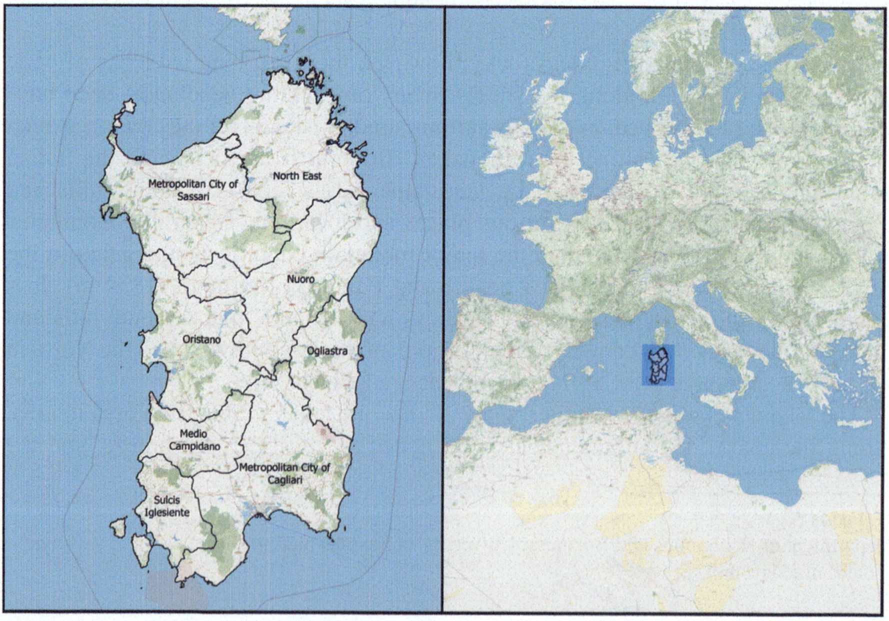

Fig. 8.1 Map of the island of Sardinia and its provinces

Cagliari, which has the Regional County Seats. Of the total population, 56.5% live in the areas encompassing the two metropolitan cities, Sassari and Cagliari, which cover 37.1% of the territory. In the area of Cagliari, the density of inhabitants per km^2 is 377.6, compared to the regional average of 66 km^2. By comparison, in the provinces of Nuoro and Oristano, the more rural areas, which cover 35.8% of the regional surface, the density is 35.7 and 51 km^2, respectively (Istat, 2022).

The structural downtrend of the population, due to the decline in births and migration from Sardinia to other Italian regions and abroad, has also been influenced by the increase in the number of deaths, caused by the Covid pandemic.

Sardinians are older by comparison with Italian inhabitants. The structure of the population per age group is slightly higher, with the average age in 2020 being 47.3 years compared with the average Italian age of 45.4 years; the old age index (people over 65 on population 0–14 years) has continuously increased. The younger inhabitants live in the metropolitan areas of Sassari and Cagliari, where a lower average age has been registered, while the older population structure has been observed in the provinces of Oristano, South Sardinia and Nuoro (Istat, 2022). This latter province, particularly the area known as Barbagia, located in the central-eastern region on the slopes of the mountain chain, Gennargentu, has become known for the longevity of its inhabitants. A significant concentration of centenarians lives in the villages in this part of the island, which several studies have identified as one of the five worldwide 'Blue Zones'. In the Sardinian blue zone, the cultural and anthropological characteristics are respectful of ancient traditions. Indeed, it has been observed that nutrition and other lifestyle habits are well-preserved and health-giving (Poulain et al., 2004). Isolation and the strong attachment to local heritage characterize the culture and traditions of the area, which remains a rather remote location, due to the poor state of its roads and infrastructures, while its economy is still mainly based on agro-pastoral activities (Hospers, 2003).

Economic Fabric and Hallmarks

Sardinia's economy is one of the poorest compared to the other Italian regions, although it is in the best position among those located in the country's south. GDP in 2020 was €32,121 million, while the 2020 GDP per capita was €20,066. The per capita income in Sardinia is the highest in the southern half of Italy (Banca d'Italia, 2022).

The economy is mainly focused on the tertiary sector (80.8% of value-added) even if the primary sector (4.4% of value-added) and farming are most related to traditions and the economic base of the ancient civilization and current inland populations (Banca d'Italia, 2022).[1] Among the tertiary sector, the main activities are commerce, services, information technology, public administration and tourism.

[1] Added value by sector of economic activity and GDP in 2020. Banca D'Italia (2022, p. 67).

The latter has been a driving industry since its expansion in the second half of the last century.

The labour market in Sardinia is mainly focused on the macro area of services. Around 8 out of 10 employees work in these sectors, and one-third in the commerce, accommodation and transport segments. Compared to the national average, Sardinia constantly registers a higher percentage of employees in the agricultural sector (6% against 4.1%) compared to previous years (Crenos, 2022). When considering the percentage of firms operating in the different sectors, agriculture in Sardinia—including traditional activities, such as sheep breeding, olive growing and winemaking—remains a relatively essential industry, accounting for 24.1% of the total activities in 2021 (Mezzogiorno: 19.4%; Italy: 14%), a value only lower than commerce (25.1%) but higher than other services.

The last economic crisis negatively impacted the labour market in Sardinia. Although there were weak signs of recovery in 2021, unemployment, especially among young people, is still a current problem, with a total unemployment rate of 13.8% in 2021, higher than the national average rate (9.7%) (Istat, 2022). This value is also related to the emigration of young people, who leave Sardinia to seek a job after studying for a career. Although public incentives and regional programmes prevent this brain drain, the structure of the labour market, which is directly affected by the economic crisis and the isolation of Sardinia does not help limit this phenomenon.

However, it is worth noting how the latest trends in economic development, also supported by public policies, have contributed to tourism sector development, with several incentives for new entrepreneurship initiatives and new businesses. These are often related to innovative types of tourism, such as community-based tourism, active tourism, sport tourism, slow tourism and ecotourism that exploit the peculiarities of a specific context (geographical, cultural and geomorphological), the general trends of the external environment (i.e. the push towards sustainability, combatting climate change and a slow pace of life) and new requirements being demanded.

Tourism in Sardinia: Development and Statistics

Several visitors have defined Sardinia as a combination of a sunny holiday destination and an 'outdoor museum' (Hospers, 2003). The development of tourism in Sardinia began during the second half of the nineteenth century by Sardinians who initiated leisure activities along the coasts and beaches, with the first beach resorts developed in Cagliari (1862) and Alghero (1863). However, the significant flourishing of tourism and Sardinia's growth as a tourist destination began along the north-eastern coast in the 1960s, when certain tycoons from the Middle East discovered the land there and established the Costa Smeralda Consortium in 1962. In particular, the main supporter of this project, who made huge investments, was the prince, Karim Aga Khan IV; he bought more than 50 km^2 of coasts in the municipalities of Olbia and Arzachena (Price, 1983). Within a few years, one of the most important Italian tourism systems took off with international connotations, favoured by the presence of kings, heads of state, businessmen and artists, an exclusive jet-set that would start to visit and spend leisure time in this destination, created from nothing and concentrated in three main settlements, Porto Cervo, Cala di Volpe and Razza di Juncu (Battino, 2014).

The island is currently a renowned destination for sea-sun-sand tourism segments. As shown in Table 8.1, arrivals and overnight stays have gradually increased from 2007 until 2019, with a slight reduction in total values in 2012 and 2013.

The arrivals' concentration during the high season creates seasonal problems, such as significant pressure on tourism resources in certain coastal destinations for a short period of the year, unemployment outside summer months and a general territorial overload. To tackle seasonality, other forms of tourism that exploit the territorial characteristics of the island and that can be enjoyed all year round have been promoted, such as climbing, cycling, trekking, cannoning, kayaking, diving and wine and food experiences. Most of these activities can be carried out in the central region of the island and they especially attract guests from abroad.

The international component has transpired to be essential in terms of pursuing de-seasonality policies, although the latest unofficial data regarding the international flow of visitors have not completely recovered to pre-Covid figures. Seaside tourism aside, over the last three decades, different segments of tourism have been developed, mainly related to outdoors activities (excursions linked to trekking, rock climbing, cycling, motorbiking, etc.), cultural events and sport segments, primarily associated with watersports along the coast.

Outdoor activities can primarily be practised in the inland region of the islands, thanks to the natural and morphological characteristics of the territory.

Cultural events are mainly related to countryside festivals, associated with cultural, archaeological and traditional heritage, which are spread throughout the island, mainly during the spring, summer and autumn months. The most important in terms of the villages and municipalities involved (32 in the 2022 edition) is 'Autumn in Barbagia', a successful multilevel tourism network, recognized at an international level; this is essentially a cultural event that takes place in Barbagia during the autumn months (Rotondo & Fadda, 2017).

Thanks to its geology and various territories, Sardinia is suitable for practising different types of sports and consequently hosts sporting events that often require a unique setting and unique climate conditions, such as those related to watersports. The emerging sport segment can include both sport events and sport participation. The former can be related to each type of sport, while the second is primarily associated with watersports.

Table 8.1 Arrivals, overnight stays and average duration of stay in Sardinia (2007–2020)

	Domestic			International			Total		
Year	Arrivals	Overnight stays	Average length of stay	Arrivals	Overnight stays	Average length of stay	Arrivals	Overnight stays	Average length of stay
2007	1,528,445	8,238,807	5.4	802,554	3,920,022	4.9	2,330,999	12,158,829	5.2
2008	1,575,538	8,452,914	5.4	803,537	3,894,419	4.8	2,379,075	12,347,333	5.2
2009	1,564,219	8,243,836	5.3	883,178	4,066,854	4.6	2,447,397	12,310,690	5.0
2010	1,544,211	8,149,164	5.3	840,212	4,023,759	4.8	2,384,423	12,172,923	5.1
2011	1,355,554	6,975,572	5.1	884,739	4,467,864	5.0	2,240,293	11,443,436	5.1
2011	1,357,215	6,979,435	5.1	885,492	4,469,248	5.0	2,242,707	11,448,683	5.1
2012	1,247,003	6,442,528	5.2	872,115	4,400,649	5.0	2,119,118	10,843,177	5.1
2013	1,172,186	5,772,427	4.9	1,002,046	4,908,201	4.9	2,174,232	10,680,628	4.9
2014	1,295,919	6,046,540	4.7	1,095,489	5,316,299	4.9	2,391,408	11,362,839	4.8
2015	1,400,968	6,554,291	4.7	1,208,724	5,838,536	4.8	2,609,692	12,392,827	4.7
2016	1,554,087	7,057,088	4.5	1,325,408	6,428,656	4.9	2,879,495	13,485,744	4.7
2017	1,603,118	7,124,396	4.4	1,494,248	7,097,936	4.8	3,097,366	14,222,332	4.6
2018	1,629,624	7,237,520	4.4	1,651,270	7,702,591	4.7	3,280,894	14,940,111	4.6
2019	1,705,190	7,418,767	4.4	1,738,868	7,727,118	4.4	3,444,058	15,145,885	4.4
2020	1,099,327	4,739,031	4.3	376,193	1,582,080	4	1,475,520	6,321,111	4.3

Source: Our elaboration—data from Sardinia Statistics on tourism—http://sardegnastatistiche.it

Sport Tourism in Sardinia: Sport Events and Watersport Practice

Among the various motivations behind, travel is the desire to practise sport in attractive surroundings. Sport practice and events are able to attract many people to specific destinations, where the conditions are optimal for practising such sports (Weed, 2009).

In recent times, Sardinia has exploited this new segment. Initially, it sprung from the initiative of isolated actors, but later became the result of a targeted strategy, supported by the regional government to counteract seasonal trends in tourism demand, aimed at attracting tourists throughout the year (Fadda, 2019).

This strategy can also be defined as seaside destinations' product diversifications and is in line with those implemented by other islands in the Mediterranean zone (Karyopouli & Koutra, 2013; Parra-López & Martínez-González, 2018). Over the last two decades, Sardinia has hosted several international sport events, such as the World Rally Championship, the America's Cup, the World Padel Tour and Giro d'Italia (cycling race). Furthermore, the island has been the venue of many national sport events, such as horse races and golf opens, as well as surfing and windsurfing championship races. In addition, it is worth noting the presence of a football and a basketball team in the top Italian leagues.

Table 8.2 shows the latest main events hosted in Sardinia in recent years. Thanks to the breath-taking landscapes, the coastal roads and dirt roads in the inland regions, the island has been a perfect setting for cycling, motorcycle and rally car races for years. Among others, the most important annual event is the international World Rally Championship, first hosted in 2003 and being held for the nineteenth occasion this year (Pischedda et al., 2020). In the wake of the success of this event, enduro races have taken hold along the same dirt roads among unique natural beauty, mainly in the mountains inland and in coastal areas.

In recent years, racket sports have also been welcomed in Sardinia with important events, such as the ATP 250 Sardinia Open and the Women's Coppa Devis event for tennis being held there, as well as the World Padel Tour. However, the most important segment of sport tourism, conditioned by the suitability of the Sardinian Sea and shores is related to watersports, particularly water boardsports, i.e. windsurfing, kitesurfing and surfing.

These sports are very closely associated with equipment, weather conditions and the sea or lake where they are practised. Many of these sports are fairly new, and the tourism related to them has only emerged recently (Buckley, 2010). During a typical day along the shores of Sardinia, it is common to observe skies crowded with giant kites, the strong sails of windsurfers and lines of tiny figures on the horizon formed by surfers seated on their boards waiting for the right wave. Activities related to water boardsports are a recent addition to the island, even though boardsports have been practised for some time. The first pioneers were surfers in the late 1970s, who mainly visited the north and west coasts. Windsurfing became very popular during the 1980s and 1990s. Meanwhile, kitesurfing appeared in the late 1990s and has

Table 8.2 Sport events in Sardinia

Sport	Event	Relevance	Last edition	Location
Rally car	World Rally Championship	International	June 2022	Alghero-Olbia
Enduro (off-road motorcycle sport)	Swank Rally	International	September–October 2022	Olbia-Orosei
	Italian Championship	National	May 2022	Olbia
Cycling	Giro d'Italia Men	National	May 2017	Alghero-Olbia-Tortolì-Cagliari
	Giro d'Italia Women	National	June–July 2022	Cagliari-Villasimius-Dorgali-Olbia
Padel	World Padel Tour	International	September 2021	Cagliari
Tennis	Women Coppa Devis	International	April 2022	Alghero
	ATP 250 Sardegna Open	International	April 2021	Pula-Cagliari
Sailing	America's Cup World Series	International	October 2019	Cagliari
Kitesurfing	Formula Kite Youth World Championship	International	May 2022	Oristano
	Kite World Championships	International	October 2022	Poetto (Cagliari)
Windsurfing	Italian Championship Windsurfing	National	April–June-September 2022	Oristano-Calasetta-Cagliari
Surfing	Italian Championship Shortboarding	National	October–November 2022	Capo Mannu (Oristano)

become widespread all around Sardinia's coasts. Nowadays, due to the weather conditions, it is commonplace to observe riders in suitable places, while, in recent years, many schools teaching surfing, windsurfing and kitesurfing have been established.

Over the years, the island has become a privileged field for international and national contests, professional training and, in general, for practising these sports. In this regard, the city of Cagliari, also known as 'The City of Wind', hosts the Luna Rossa Team (an Italian sailboat racing syndicate, first created to compete for the 2000 America's Cup) and is the main hub for training and developing new sailing solutions. In relation to water boardsports, i.e. windsurfing, kitesurfing and surfing, national and international contests have taken place for many years on the principal shores suitable for practising these sports. At the same time, the diffusion of these sports has led to a significant rise in practitioners along the 1897 km of Sardinian coasts due to the suitable conditions: wind in the case of windsurfing and kitesurfing, and waves in the case of surfing. Furthermore, there has been a notable increase in the number of schools located in prime locations that attract both beginners and experienced practitioners all year around (Fadda, 2019, 2020).

Some Empirical Evidence from Water Boardsport Tourism in Sardinia

Methodological Issues

To understand tourism implications in terms of the ability of these sports to attract tourists to a renowned sea-sun-sand tourism destination and to highlight governance practices, which impact on sustainable development policies in terms of managing deliveries for such tourists, some empirical evidence will be presented relating to research carried out during 2016 and 2017. The exploratory study has dealt with the supply and demand of water boardsport activities in the tourist destination of Sardinia.

The supply analysis focused on schools for boardsports, as these form part of the specialized services offered to tourists in surf destinations (Ponting & McDonald, 2013; Brochado et al., 2018). Surfing, windsurfing and kitesurfing schools, located in Sardinia, were first mapped using primarily Internet sources and dedicated maps: 55 schools were identified. Figure 8.2 shows the location of the schools that took part in the study.

All 55 schools were contacted by email and provided with an overview of the research objective; their participation in filling out a web-based questionnaire and taking part in a qualitative investigation through semi-structured interviews was

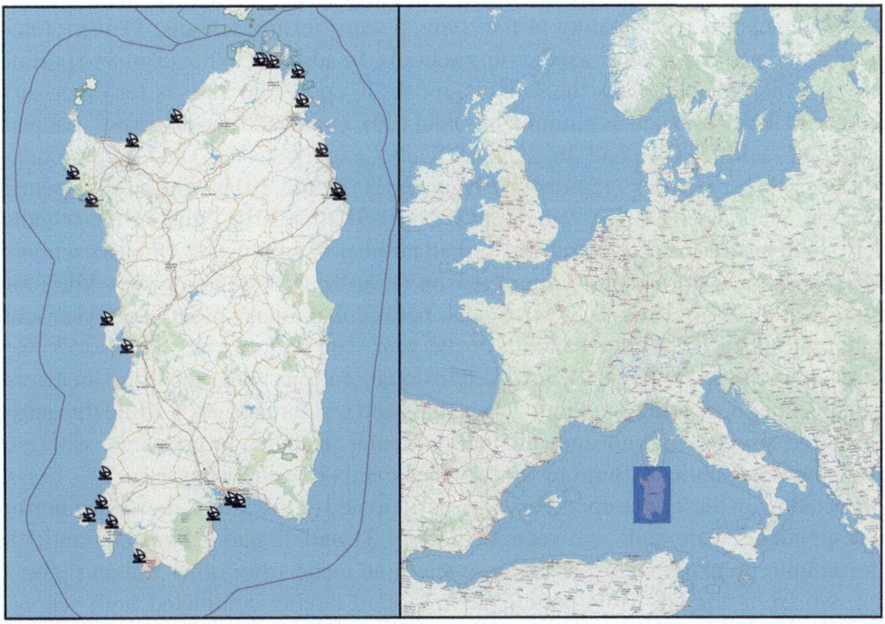

Fig. 8.2 Map of the island of Sardinia—schools' location. (Source: Own elaboration using the software Qgis)

requested. Twenty-two entrepreneurs, the owners of the schools, agreed to take part in the qualitative analysis (response rate of 40%) (Fadda, 2020), while 21 filled out the questionnaire with a response rate of 38.18% (Fadda, 2019).

The demand analysis focused on boardsport practitioners (residents and non-residents) to better understand travelling behaviours, preferences and sustainable inclinations. Thus, at the same time, a survey was carried out, addressed to boardsport riders who have practised on Sardinia's shores.

The questionnaire was created using the Google Forms platform, which was shared via social networks (Facebook and WhatsApp). A snowball sampling technique was used to identify survey participants (Patton, 2002). This consisted of asking a preliminary list of participants to fill in the form and then to share it with their contacts. The questionnaire was subsequently published on dedicated websites (4surf.com, 4windsurf.com). During this process, 311 completed questionnaires were collected.

The main results of the study will be reported below, describing, on the one hand, the characteristics of the supply (the profile of schools and entrepreneurs) and, on the other, the demographic and participation profiles of sport practitioners (demand), with a focus on the tourists visiting the island.

Results

Schools and Entrepreneurs' Profiles

Of the 22 schools participating in the study, 5 were surfing schools, 11 were kitesurfing schools, and 6 were windsurfing schools. In addition to the primary sport of surfing which characterizes the schools, most (20) provided classes for other surf sports and services, such as equipment rental (20), yoga and fitness (seven), bar and restaurant facilities (five and three, respectively), excursions (six) and accommodation (six). Most schools (12) opened between 2000 and 2010, while four opened between 1980 and 2000; the remaining six opened after 2010. Most of the schools were small businesses that employed a small number of employees. Only six schools were operational during the summer season (from May to September), while four were open all year round, four were open between six and 9 months per year and eight were open for all but 1 month of the year. The majority of schools (52.3%) were happy or very happy to be involved in collaborations with other tourism actors (establishments offering accommodation, bars and restaurants, etc.), while the questionnaire responses did not indicate any particular tendency to cooperate with the community or public institutions (only 38% agreed or strongly agreed).

Focusing on the 22 entrepreneurs who participated in this study, the vast majority of respondents were male, between the ages of 31 and 40 and between 41 and 50. Eleven entrepreneurs were Sardinian, while nine came from other Italian regions and two from abroad. Most participants reported having completed high school, while four stated that they held a university degree. The majority of the respondents stated that they spoke at least one foreign language (English). Half of the participants reported having more than 10 years' experience of running a business.

Furthermore, the majority of participants possessed many years' experience in surf sports that had affected their life habits in terms of how much time they spent practising and the extent to which they travelled (Sotomayor & Barbieri, 2016). Moreover, the results revealed that the entrepreneurs studied in this research were likely to have travelled extensively and visited the main surfing destinations, which could have fueled their ambition to become entrepreneurs. Hence, the backgrounds of the entrepreneurs seem to have influenced their motivation and decision to open a school and to migrate to a different destination. The union of different personal motivations, mainly related to the necessity to practise sports and enjoy a good lifestyle, led them to establish a school and a new type of business within the empirical context analyzed.

The Market

Analyzing the customers of these schools from the entrepreneur's point of view, more than 70% of the schools declared that almost 20% of their customers are Sardinian. Meanwhile, 52% of respondents stated that 30–60% of their customers are from other Italian regions and 56% indicated that between 40% and 100% of their customers are foreign, mainly from Germany, England and France. Moreover, customers from other countries, such as Scandinavia and the United States were also noted, although these visited less frequently. This evidence is meaningful in light of the higher willingness to pay a tax earmarked for environmental protection, prevalent among surf tourists from DACH countries (Germany, Austria and Switzerland) in destinations in the Algarve (Frank et al., 2015). Hence, this tourism sub-segment attracts those more inclined to support environmental sustainability. The majority of schools (11) reported that the average age of their customers is between 31 and 40 years, while seven indicated that their customers are between 21 and 30 years, and three stated that their customers are in their early 20 s. The majority of respondents (around 80%) perceived that the market was growing, while a minority (three) reported a declining status, with only one respondent stating that the market was sluggish.

Practitioners of Boardsports

Socioeconomic Profile

A large proportion of practitioners are surfers (126), while 90 are kitesurfers and 95 are windsurfers. Most respondents (212) are Sardinian, while 31.83% are non-residents, of whom 21.86% are riders, who often visit the island expressly to practise their boardsport (Table 8.3).

As shown in Table 8.4, the majority of respondents are male (85%). Female participation is highest (around 21%) in kitesurfing, compared to the other sports. A substantial proportion of respondents (44%) are aged between 31 and 40 years. The

Table 8.3 Respondents by sport and origin

	Surfing	Kitesurfing	Windsurfing	Total	%
$N = 311$	126	90	95	311	
Percentage	40.51	28.94	30.55	100	
Sardinians	92	68	52	212	68.17
Non-residents	34	22	43	99	31.83
Surfing is the main motivation to visit Sardinia	21	12	35	68	21.86
Other motivations to visit Sardinia	13	10	8	31	9.97

Table 8.4 Gender, age, educational level and income indicators of participants by boardsport practised

Gender	Surfing	Kitesurfing	Windsurfing	Total	%
Male	107	71	87	265	85.21
Female	19	19	8	46	14.79
Age					
20 or less	6	0	7	13	4.18
21–30	31	19	20	70	22.51
31–40	59	51	27	137	44.05
41–50	24	15	27	66	21.22
50 or more	6	5	14	25	8.04
Educational level					
Compulsory education	6	10	4	20	6.43
High school degree	63	43	46	152	48.87
University degree	35	24	29	88	28.30
Master's degree or higher	22	13	16	51	16.40
Marital status					
Single	40	37	27	104	33.44
Married	29	16	38	83	26.69
Other	57	37	30	124	39.87
Household income					
I do not earn any money yet	7	1	11	19	6.11
I barely make enough money to live	16	10	5	31	9.97
I can afford my basic needs	33	33	27	93	29.90
I live with some comfort	40	20	29	89	28.62
I am able to save some money every month	13	6	11	30	9.65
Income is not a problem for me	17	20	12	49	15.76

other most frequently observed age groups were 21–30 and 41–50 years. Around 75% of riders have a high level of education, holding at least a high school degree, a university-level degree or a higher degree. It should be noted that a quarter of respondents (33.4%) are single, 84% of whom have a stable financial status. Thus—and consistent with their high education levels—a significant proportion indicated that they could afford their basic needs (29.9%), while 28.6% reported that their standard of living comes with at least some comfort and 25.4% stated they are able to save money or have no economic concerns. Results regarding male predominance, high educational levels and the good financial status of riders are in line with previous analyses of the profile of surfers, conducted in various countries across different continents (Portugal et al., 2015; Frank et al., 2015; Barbieri & Sotomayor, 2013; Sotomayor & Barbieri, 2016; Porter & Usher, 2019).

Travel Behaviour of Tourists (Non-resident Riders)

As previously discussed, the objective of this study was to investigate the tourism implications derived from the practice of boardsports. Thus, it has been observed that the travel behaviour of riders, who occasionally or frequently visit Sardinia to practise their boardsport, is determined by favourable weather conditions.

Among the total responses, 99 were from non-resident riders, of which 68 (21.86% of the overall total) reported that practising their boardsport was the main motivation for visiting Sardinia (see Table 8.3). Regarding the average length of the trip, a significant proportion (35.4%) took trips lasting one to 2 weeks, 25.3% less than 1 week and 22.2% 2–4 weeks, as shown in Table 8.5. Among the surfers, the most frequent option was a trip of a duration of 1 week, while the riders practising the other two sports reported spending between one and 2 weeks. Looking at the overall percentage data, the results confirm previous evidence among surf tourists of a common length of stay of 1–2 weeks (Portugal et al., 2015; Porter & Usher, 2019). It should be noted that riders travel with family (33.3%), a partner (28.3%) or friends (22.2%), although 'lone wolves' can also be observed (14.1%). These figures are consistent with percentages found in previous studies regarding choices made to travel alone or with a partner (Portugal et al., 2015; Porter & Usher, 2019), while in this study, a higher percentage chose to travel with family and a lower percentage travelled with friends. The choice of travel companions clearly seems to be associated with marital status.

Regarding accommodations, 28.3% of the riders who responded stay with family or at a friend's house, 22.2% rent an apartment and 15.2% choose to sleep in a van, which is the typical means of transport/accommodation for riders. A surf house was chosen by a few respondents, probably as it is a recent accommodation solution on the island. Figures related to spending per day are consistent with accommodation solutions. The demand for low-cost accommodation is reflected in the choice of apartment type (own, rented or friend's/family member's house). At the same time, other data emphasize a low-cost tourism segment, which is likely to be focused on sport practices (see Table 8.5). This latter piece of evidence is also supported by findings regarding the cost of equipment, given that 'expert riders' typically bring their own equipment. While daily expenditure was low for accommodation and transport—the latter due to the short travelling distances on Sardinia—participants spent relatively more in restaurants and bars (the majority less than €50/day) and at supermarkets (between €50 and €100/day). These figures are high by comparison with the surf destination of Mentawai, Indonesia, for example, where the average

Table 8.5 Travel behaviour of non-resident riders who responded to the survey ($N = 99$)

Average length of trip to Sardinia	Surfing	Kitesurfing	Windsurfing	Total	%
2–3 days	2	0	4	6	6.1
Less than 1 week	14	2	9	25	25.3
1–2 weeks	9	10	16	35	35.4
2–4 weeks	6	7	9	22	22.2
1–2 months	2	1	0	3	3.0
> 2 months	1	2	5	8	8.1
Travelling companionship					
Alone	5	4	5	14	14.1
With family (including children)	11	5	17	33	33.3
Partner	10	11	7	28	28.3
Friends	7	2	13	22	22.2
Group	1	0	1	2	2.0
Accommodation solutions					
Hotel/residence	1	1	2	4	4.0
B&B	3	1	6	10	10.1
Rental apartment	8	4	10	22	22.2
Family or friends' house	9	9	10	28	28.3
Own home	3	3	4	10	10.1
Surf house	1	0	0	1	1.0
Camping	4	0	4	8	8.1
Van	4	4	7	15	15.2
Anywhere	1	0	0	1	1.0

daily expenditure was found to be $22; the figures relating to Sardinia are more in line with both the broader European context and mature destinations, if we consider comparable values of approximately €100/day or less in Portugal and €120/day in Mundaka, Spain (Portugal et al., 2015; Murphy, 2008).

In terms of the main attractions beyond surf activities while in Sardinia, 92.9% of non-resident participants visit beaches and natural reserves, 45.5% experience agritourism or vineyards and 37.4% go on boat tours. Other activities of significance, not related to the sea, include visits to archaeological sites (32.3%), the main urban centre (33.3%) and inland villages (23.2%) (Table 8.6).

Sardinia as a Boardsport Tourism Destination

To investigate the impact that boardsports can have on tourism, one item included in the questionnaire focused on the best period in which to practise boardsports in relation to weather conditions. Most respondents (residents and non-residents who frequently visit the island to practise their boardsport) indicated the period from October to December (64.6%), followed by April to June (47.9%). Summer was also chosen by a minority of respondents (31.4%), as well as the winter season (25.7%). As reported in Table 8.7, certain surfing spot conditions were analyzed in

Table 8.6 Spending per day

Spending per day	€0	Less than €25	€25–50	€51–100	€100–200	More than €200
Accommodation	34	26	31	6	1	1
Restaurants and bars	3	53	37	5	1	0
Supermarkets	3		13	82	1	0
Transport	11	62	14	10	2	0
Rental equipment	82		3	9	5	0
Other	52	1	1	31	14	0

Table 8.7 Practitioners' perceptions of surfing spot conditions in percentages

Surfing spot conditions	N	Strongly disagree	Disagree	Neutral	Agree	Strongly agree
Cleanliness of beaches	280	5.7	9.6	32.1	38.2	14.3
Respect for the natural environment	280	1.4	5.1	13	29.3	51.2
Localism	280	3.9	11.4	35.4	30.0	19.3
Easy access and dedicated zone	280	25.7	29.6	28.9	11.1	4.6
Local surfing community	280	1.8	10.7	34.3	31.8	21.4

line with the importance of the quality of the natural environment, as requested by sustainable development goals (SDGs), 14 and 15, for surf tourists. The majority of respondents perceived that the cleanliness of the beaches was sufficient (more than 50% reported that they agreed or strongly agreed with this statement), while another interesting aspect noted was the respect for the natural environment shown by surfers, compared to the average beachgoer. Around 80% reported that they agreed or strongly agreed with this statement.

This result is consistent with many riders' respect for nature, which is relatively high compared to typical bathers, due to surfers' intimacy with their environment (Tantamjarik, 2004). Another interesting aspect emerged regarding the lack of dedicated beach zones to set up equipment or provide easy access to the sea: 50% of respondents disagreed or strongly disagreed with this statement. Instead, more than 50% of participants reported the presence of a friendly, local community of practitioners, although they perceived a 'localism' phenomenon (around 49% agreed or strongly agreed with this). Surfers who are attached to a spot perceived this as something of their own, which they sought to protect from use by others (Reineman & Ardoin, 2018). This phenomenon relates to an issue of territoriality such that, when practising boardsports, standard rules have evolved whereby locals have the first opportunity to enjoy the best breaks in the waves (Sotomayor & Barbieri, 2016).

The last section of the questionnaire, relating to boardsport school entrepreneurs and riders, dealt with their perceptions as to whether Sardinia is a destination for boardsport tourism (see Table 8.8). The island's suitability in terms of weather conditions was reported favourably by the vast majority of respondents (more than 90% of practitioners and around 95% of entrepreneurs). The island's insularity and the high transport costs to reach it, however, were perceived as obstacles for the growth of tourism, as were the lack of beach services (toilets, showers, organized beaches, etc.). Localism, however, was not considered to be an obstacle to the development of this tourist segment, either by practitioners or by entrepreneurs. More than 90% of riders and almost all entrepreneurs reported being positive or very positive as regards structured activities initiated by public institutions, to promote the island as a boardsport tourism destination and to facilitate dedicated services. The organization of boardsport contests and events was seen as an appropriate strategy to mainstream these sports around the island and to encourage new practitioners (around 75% of practitioners and 95% of entrepreneurs agreed or strongly agreed with this statement). Overall, all school entrepreneurs expressed strong support for Sardinia becoming a boardsport tourism destination, along with around 87% of practitioners, who agreed or strongly agreed.

Table 8.8 Practitioners'/entrepreneurs' perceptions of Sardinia as a boardsport tourism destination in percentages

Destination items	*N*	*Strongly disagree*	*Disagree*	*Neutral*	*Agree*	*Strongly agree*
Suitability of Sardinia (weather conditions) as a boardsport tourism destination	311	0.0	5.0	7.9	26.8	70.7
Insularity and transport: weakness of the system	311	8.9	7.5	18.2	31.8	44.6
Lack of services	311	13.6	15.0	25.0	27.1	30.4
Localism as an obstacle	311	24.3	34.3	28.2	13.9	10.4
More promotional efforts required by public institutions	311	1.8	3.6	13.6	24.6	67.5
Contest and event organization	311	3.9	6.8	21.8	27.5	51.1
"I would like Sardinia to become a boardsport tourism destination"	311	7.1	1.8	14.3	21.4	66.4
Destination items	*N*	*Strongly disagree*	*Disagree*	*Neutral*	*Agree*	*Strongly agree*
Suitability of Sardinia (weather conditions) as a boardsport tourism destination	21	0.0	0.0	0.0	4.8	95.2
Insularity and transport: weakness of the system	21	4.8	4.8	4.8	23.8	61.9
Lack of services	21	4.8	4.8	4.8	23.8	61.9
Localism as an obstacle	21	23.8	42.9	4.8	19.0	9.5
More promotional efforts required by public institutions	21	0.0	0.0	0.0	9.5	90.5
Contest and event organization	21	0.0	4.8	0.0	14.3	81.0
"I would like Sardinia to become a boardsport tourism destination"	21	0.0	0.0	0.0	0.0	100.0

Boardsports in Sardinia as an Opportunity for Sustainable Tourism Development

The empirical pieces of evidence that emerged from the exploratory study investigated the tourism implications regarding the practice of boardsports in the seaside tourism destinations of Sardinia. According to the supply analysis, schools for boardsports are widespread along the island's coasts, which confirms an increase in the growth of these activities and exposes an emerging niche tourist segment. Schools, which are widespread around Sardinia's shores, were mainly established between 2000 and 2010 by entrepreneurs, many of whom came from other Italian regions to seize business opportunities in this context. The opening periods of these small firms extend beyond the summer season, highlighting their capacity to attract tourists for at least 9 months each year. The market study highlighted a growing trend, with most riders coming from other European countries, especially Germany, France, England and Northern Europe. Entrepreneurs reported high levels of satisfaction with economic performance. At the same time, in terms of networking with other actors, frequent collaborations were observed with other tourism businesses, but fewer with public institutions.

The investigation into demand presented interesting pieces of tourism-related evidence. The study of the travel behaviours of non-resident practitioners, who have visited Sardinia to practise their boardsport, revealed that they prefer a medium-term vacation with family, a partner or friends. However, certain 'lone wolves' were identified among the surfers who chose to stay in their own apartment or with family or friends. The boardsport tourist who was found to be interested in cultural attractions should be defined as an 'active tourist', that is, a traveller who is inclined to discover the culture and traditions of the host country and come into contact with local communities. Furthermore, the average daily expenditure was in line with both the broader European context and mature surf destinations in Spain and Portugal.

These travel features reflect a form of tourism that could belong to 'recreational surf tourism', a previously defined branch of 'surf tourism', which is characterized by the search for an authentic adventure tourism experience (Buckley, 2002). This form of tourism has been overtaken by a commercial offer in which tour operators play a fundamental role in linking demand with supply. Thus, by considering boardsport practices as an emerging tourism segment in the destination of Sardinia, it is not surprising that a recreational tourism typology is identifiable. Moreover, it is noteworthy that riders develop strong ties with nature and display a disposition towards respecting the surrounding environment and natural resources. Thus, the form of tourism that emerged from this research can be considered environmentally sustainable (Tantamjarik, 2004). This element is meaningful in light of sustainable tourism development in a mature seaside destination. Indeed, the sustainable management of coasts and natural resources requires an understanding, not only of the environmental context but also of users, tourists and stakeholders (Reineman & Ardoin, 2018; Machado et al., 2018).

The respect for nature investigated in this study is related to both the 14 and 15 SDGs, respectively, 'life below water' and 'life on land'. Water boardsports are nature-based sports in which practitioners have a close relationship with nature, providing opportunities to directly experience, admire and enjoy nature, below the sea and on the coastal resources. These should enhance awareness of nature's value and the importance of its protection, stimulating measures for environmental preservation among practitioners, locals and travellers (UNWTO, 2019). The empirical evidence from the exploratory study also analyzed Sardinia as a boardsport tourism destination. Favourable weather conditions support this potential, although surfing spot conditions may require structural measures to enhance easy access and create dedicated zones. However, the lack of overall services and facilities was not significant in terms of choosing a surf travel destination (Barbieri & Sotomayor, 2013). Other areas of improvement have been identified in relation to transport and the lack of other services, while localism is perceived as being rather widespread in Sardinia's surfing spots.

Furthermore, it is noteworthy that boardsport tourism constitutes a potential deseasonality segment in a seaside destination, such as Sardinia, which has to cope with high tourist peaks during the summer season. Autumn and spring emerged as the most suitable seasons for practising boardsports. The fact that the opening period of many schools extends beyond the summer season supports the strong potential of these activities to expand local seaside offers. Therefore, boardsports help to diversify the traditional tourism product, facilitating a de-seasonality strategy. Regional and local public policy should adequately consider this tourism niche, which mainly engages higher-spending tourists who, it turns out, are attracted by cultural offers when visiting the island during the off-season period.

Public policies to support this niche segment have not yet been implemented. In recent years, a constant commitment by the regional government to promote sporting events has been observed but no specific actions have been dedicated to this area. A commitment from public institutions to promote Sardinia as a boardsport tourism destination and to support contests and events is expected by practitioners and entrepreneurs. Indeed, as the coastal areas in which boardsports are practised in Sardinia are public property, they require public intervention. In this regard, the intervention of local government authorities is fundamental in planning such activity in the territory and in developing beach supports and infrastructures.

On the other hand, tourism destination managers should ensure that their services for boardsport practitioners match the profile of boardsport tourists and respond to their needs. In this regard, private–public networking at both local and regional levels, as well as collaborations between private actors, could be implemented to boost the supply of services, which already seems to consist of a significant flow of tourists. A shared course of action between private and public actors in terms of policies and effective activities could lead to the implementation of deseasonality strategies, the aim of which is to increase services for boardsport practitioners, improve surfing spot conditions and strengthen an emerging tourism segment in an environmentally sustainable manner in light of the respect for the natural environment shown by surfers and the importance of the quality of the natural environment for surf tourists. Thus, when developing boardsport tourism and advancing the 14 and 15 SDGs, sustainable use of natural resources should be considered, as the avoidance of any kind of pollution. In this sense, environmental measures should be the adoption of means of transportation with no emissions of CO_2, encouraging recycling of sport equipment, and for sport tourism operators and sport event organizers, efficient water and green energy use, waste recycling, choosing sustainable suppliers and products, etc.

References

Banca D'Italia. (2022). *Economie regionali. L'economia della Sardegna*. Giugno 2022. https://www.bancaditalia.it/pubblicazioni/economie-regionali/2022/2022-0020/2220-sardegna.pdf

Barbieri, C., & Sotomayor, S. (2013). Surf travel behavior and destination preferences: An application of the serious leisure inventory and measure. *Tourism Management, 35*, 111–121.

Battino, S. (2014). *Turismo sostenibile in Gallura: prospettiva vincente o modello illusorio? I principali caratteri distintivi del cuore turistico della Sardegna*. Patron editore.

Brochado, A., Stoleriu, O., & Lupu, C. (2018). Surf camp experiences. *Journal of Sport & Tourism, 22*(1), 21–41.

Buckley, R. C. (2002). Surf tourism and sustainable development in Indo-Pacific Islands. I. The industry and the islands. *Journal of Sustainable Tourism, 10*(5), 405–424.

Buckley, R. C. (2010). *Adventure tourism management*. Elsevier Butterworth-Heinemann.

CRENOS Centro Ricerche Nord Sud. (2022). *Economia della Sardegna 29° Rapporto 2022*. Cuec.

Fadda, N. (2019). Tourism from water boardsports: Evidence from the emerging destination of Sardinia. *Journal of the Knowledge Economy, 10*(4), 1720–1737.

Fadda, N. (2020). Entrepreneurial behaviours and managerial approach of lifestyle entrepreneurs in surf tourism: An exploratory study. *Journal of Sport & Tourism, 24*(1), 53–77.

Frank, F., Pintassilgo, P., & Pinto, P. (2015). Environmental awareness of surf tourists: A case study in the Algarve. *Journal of Spatial and Organizational Dynamics, 3*(2), 102–113.

Hospers, G. J. (2003). Localization in Europe's periphery: Tourism development in Sardinia. *European Planning Studies, 11*(6), 629–645.

Istat. (2022). *Il censimento permanente della popolazione in Sardegna. Anno 2020*. https://www.istat.it/it/archivio/268393

Karyopouli, S., & Koutra, C. (2013). Mature coastal Mediterranean destinations: Mitigating seasonality. *Trends in European Tourism Planning and Organisation, 60*, 48.

Machado, V., Carrasco, P., Contreiras, J. P., Duarte, A. P., & Gouveia, D. (2018). Governing locally for sustainability: Public and private organizations' perspective in surf tourism at Aljezur, Costa Vicentina, Portugal. *Tourism Planning & Development, 15*(6), 692–704.

Murphy, M. (2008). *The impact of surfing on the local economy of Mundaka, Spain*. Save The Waves Coalition.

Parra-López, E., & Martínez-González, J. A. (2018). Tourism research on Island destinations: A review. *Tourism Review, 73*(2), 133–155.

Patton, M. Q. (2002). *Qualitative evaluation and research methods* (3rd ed.). SAGE Publications, Inc.

Pischedda, G., Marinò, L., & Del Chiappa, G. (2020). Expenditure behaviour and satisfaction in motor sport events: Insights from the FIA world rally championship in Sardinia. In *Principles and practices of small-scale sport event management* (pp. 233–248). IGI Global.

Ponting, J., & McDonald, M. G. (2013). Performance, agency and change in surfing tourist space. *Annals of Tourism Research, 43*, 415–434.

Porter, B. A., & Usher, L. E. (2019). Sole surfers?: Exploring family status and travel behaviour among surf travellers. *Annals of Leisure Research, 22*(4), 424–443.

Portugal, A., Ramos, A., Campos, F., & Melo, R. (2015). Surf tourism: Understanding the relation between serious leisure characteristics and surf travel behaviour. In *Sport tourism: New challenges in a globalized world* (pp. 66–77). Cambridge Scholars Publishing.

Poulain, M., Pes, G. M., Grasland, C., Carru, C., Ferrucci, L., Baggio, G., …, & Deiana, L. (2004). Identification of a geographic area characterized by extreme longevity in the Sardinia Island: The AKEA study. *Experimental Gerontology, 39*(9), 1423–1429.

Price, R. (1983). *Una geografia del turismo: paesaggio e insediamenti umani sulle coste della Sardegna*. Cagliari.

Reineman, D. R., & Ardoin, N. M. (2018). Sustainable tourism and the management of nearshore coastal places: Place attachment and disruption to surf-spots. *Journal of Sustainable Tourism, 26*(2), 325–340.

Rotondo, R., & Fadda, N. (2017). The governance of multilevel networks in tourism. The case of 'Autunno in Barbagia'. *Review of Business Research, 17*(2), 69–80.

Sotomayor, S., & Barbieri, C. (2016). An exploratory examination of serious surfers: Implications for the surf tourism industry. *International Journal of Tourism Research, 18*(1), 62–73.

Tantamjarik, P. (2004). *Sustainability issues facing the Costa Rica surf tourism industry*. Master's thesis, University of Hawaii, HA.

UNWTO World Tourism Organization. (2019). *Sport Tourism and the Sustainable Development Goals (SDGs)* (pp. 1–9). https://www.e-unwto.org/doi/book/10.18111/9789284419661

Weed, M. (2009). Progress in sports tourism research? A meta-review and exploration of futures. *Tourism Management, 30*(5), 615–628.

Chapter 9
Sport Tourism and Local Sustainable Development in Small Islands: The Case of Kalymnos as a Sustainable Destination

Dimitrios Belias and Athanasios Koustelios

Introduction

Sport tourism has been developing rapidly over the past 20 years. According to Gibson (1998a), sport tourism is defined as a "leisure-based travel that takes individuals temporarily outside of their home communities to participate and/or watch activities or to venerate attractions associated with physical activities." In other words, Gibson (1998b) suggested three behavioral types which are associated with sport tourism: a) active sport tourism (which is referred to the people who travel in order to participate to a sport event), b) event sport tourism (which is referred to the people who travel in order to watch a sport event, and c) nostalgia sport tourism (which is referred to the people who travel in order to visit sports museums and sports venues). These three types of sport tourism were enriched by Pigeassou (2004) and Sobry et al. (2016), with the terms of (i) activism sport tourism (Pigeassou, 2004) and (ii) opportunity sport tourism (Sobry et al., 2016). Activism sport tourism is referred to as the "supply" side of the sport tourism which contains all the efforts that are made by the professionals of the sector in order to propose services, improve their products, and create direct and indirect activity (Pigeassou, 2004). Opportunity sport tourism is referred to all the offers of sport activities that

D. Belias (✉)
Department of Business Administration, University of Thessaly, Larissa, Greece

Department of Business Administration and Tourism, Hellenic Mediterranean University, Heraklion, Greece
e-mail: dbelias@uth.gr

A. Koustelios
Department of Business Administration, University of Thessaly, Larissa, Greece
e-mail: akoustel@uth.gr

D. Van Rheenen et al. (eds.), *Sport Tourism, Island Territories and Sustainable Development*, Sports Economics, Management and Policy 24,
https://doi.org/10.1007/978-3-031-51705-1_9

aim at attracting tourists who visited a location and motivated by other reasons (Sobry et al., 2016).

One of the most important questions that researchers have taken into consideration over the past years is: "How can sport tourism contribute to the sustainable development of an area?" In accordance with the World Tourism Organization (UNWHO, 2019), there are several reasons why sport tourism can contribute to the development of a tourist destination:

- Jobs and income opportunities: Sport tourism can provide jobs and income opportunities, including those for less favored groups (Belias et al., 2018). For example, local people may work as instructors and guides and might be paid more than their average salaries.
- Visitor spending: Sport tourists have the tendency to spend more money than the average tourists as they tend to rent equipment, hire instructors, pay participation fees, as well as enjoy other attractions in the destinations.
- Competitiveness: Sport tourism can increase the attractiveness and competitiveness of a touristic destination by expanding its tourism offer.
- Well-being of tourists and welfare of the locals: Sport tourism can increase the physical and mental well-being of the tourists, as it can improve their health. As far as the locals are concerned, it provides them with sport opportunities, which can lead to a healthy and active lifestyle.
- Athletes as a role model: Athletes can play an inspiring role for the locals in order to persuade them to participate in a healthier way of life.

The main purpose of the International Research Network in Sport Tourism (IRNIST) is to outline the contribution of sport tourism for sustainable local development in several small island territories or island archipelagos worldwide (Melo et al., 2021).

Kalymnos is a small island in Greece, located in the south-eastern Aegean Sea. It belongs to the Dodecanese island chain. Kalymnos is an excellent climbing destination. The island became famous among the mountaineering community as the locals improved the basic infrastructure and promoted the island as an alternative tourist destination for climbing, using mainly the Internet as a means of promotion. According to Zafeiropoulos and Drinia (2021), the rocks of the islands can support many new routes to keep the interest of climbers alive, if necessary. This particular form of tourism remains dynamic with great prospects for the future.

The history of climbing tourism on this island began in earnest in 1996, after hiking and climbing enthusiasts from Italy, when they went to the island and were amazed by the unique environment of cliffs and steep and inaccessible peaks. These natural formations have been attributed to the remnants of volcanic eruptions or even meteorites. In the most recent recording made by geologists, Kalymnos is said to maintain about 52 caves and 210 mountaineering trails; Kalymnos has at least 80 places for climbing (Belitsi, 2020). The Cave of Kefalas and the Virgos are of unique speleological beauty (Higgins & Higgins, 1996), while several speleologists insist on coming every year for the exploration of other areas. From 1996 until today, mainly with the help of the Internet but also with word-of-mouth communication,

Kalymnos has emerged worldwide as a remarkable destination for climbing, popular for the uniqueness of its rocky landscape and scenic beauty.

The aim of this chapter is to examine the case of Kalymnos as a sustainable destination. With the development of its image as a top climbing destination, Kalymnos has been able to maximize its opportunities as a destination. On the one hand, there is an increase in tourists to a destination, which was relatively unknown until a few years ago. This increase in sport tourism has led to financial gains for the local population; at the same time, tourists appear to respect the natural environment of the island as well as the local community. The outcome is that the island is considered to be a role model of sustainability. However, it seems the chapter investigates what other possible types of tourism can be developed on this island that will not disrupt the local sociality and its fragile natural environment. Also, the chapter discusses the fact that there are some who point out that the island, particularly during the climbing festivals, becomes overcrowded. For this reason, this chapter offers some recommendations on how the island and its local population will handle the increase in demand for the island; this can occur by expanding the seasonality of the sport and by focusing on attracting rock climbers during the winter time. As such, Kalymnos can become a year-round destination. Moreover, this chapter makes additional recommendations such as the expansion of Kalymnos airport in order not only to facilitate more international flights but also to develop a new campsite that will accommodate climbers who do not want to stay in hotels. Finally, there is a need to develop new hotels, which will operate in a sustainable way.

General Characteristics of Kalymnos Island

Kalymnos is a Greek island, which is located in the southeastern Aegean Sea. Kalymnos is a part of the Dodecanese island archipelago, being the fourth island in size of this island chain. Its geographical size is about 111 km^2 (110,875 m^2). Thus, the island of Kalymnos holds the 26th place in size among the Greek islands. Kalymnos has its own municipality which consists of the following smaller habitable islands: Telendos (4.6 km^2), Pserimos (14.6 km^2), Kalolimnos (1.9 km^2), and Plati (0.71 km^2) as well as the following very small uninhabitable islands: Sari, Gaidouronisi, Agia Kiriaki, Mauropinakaki, Hteni, Nera, Imia one and two and Kalamvros. The capital of Kalymnos is Pothia, which is also the capital of the Municipality of all the islands of Kalymnos.

As far as the population of Kalymnos is concerned, there are 16,179 residents (Municipality of Kalymnos, 2017). This size of population makes Kalymnos the 15th place (on population) when compared to all Greek islands. Of the total of 16,179 residents, 49.3% are male, whereas 50.7% are female. In addition, the share of population under 20 years old is 25%; on the contrary the share of population above 20 years old is 75% as well as the share of population who are above 65 years old is 15.5%. Regarding the employment and the unemployment rate, the workforce of Kalymnos consists of 58.25% of the total population of the island. The unemployment rate is about 29.8%. Additionally, 71% of the unemployed residents are younger; 31.7% of the unemployed are men, whereas 68.3% are women (Municipality of Kalymnos, 2017).

The average annual income of island inhabitants is 11,379 euros. More specifically, the GDP of Kalymnos Island is approximately 187,086,381 euros. This contributes 5.76% to the total GDP of the Dodecanese island archipelago, as well as a minor percentage of 3.52% of the total GDP of the South Aegean Region. The economic sectors that contribute to the GDP of Kalymnos are as follows: the services sector contributes 64.2% to the total GDP of Kalymnos. The primary productivity sector contributes 4.7%, while the secondary productivity sector contributes 20.8% to the total GDP of Kalymnos. The tourism sector contributes by 10.3% (INSETE Intelligence, 2019).

The main professional sectors in which residents are employed are fishery, construction, commercial stores, tourism and hospitality industry, transfers and storage businesses, as well as public organizations (educational institutions, hospitals, and public administration). In addition, 96.8% of the total population of the residents of Kalymnos have Greek citizenship, 1.1% have Australian citizenship and 1.5% have US citizenship. A very small fraction (0.6% of Kalymnos's population come from Egypt, Albania, and Russia (Municipality of Kalymnos, 2017).

The political system of Kalymnos is a parliamentary republic. Kalymnos is a Greek island so it follows the Hellenic Republic's political system. Specifically, the Greek president is the head of state. He is elected by parliament every 5 years as the prime minister of the head of the government. The ministerial council consists of the prime minister, the ministers, the deputy ministers, and ministers without portfolio. They are the collective decision-making body that constitutes the government of Greece (MFA.GR, 2022). All the above descriptive variables are presented in Table 9.1.

Table 9.1 General characteristics of Kalymnos Island

Island of Kalymnos	
Island or archipelago under analysis (name)	Kalymnos
Capital (city name)	Pothia
Geographical size of the island [if an island archipelago indicates the size of the island where the capital is situated/if about an island indicates the size of the island] (number in km^2)	110.875
Population (number in millions)	0.016179
Share of population under 20 years old (%)	25%
Share of population above 65 years old (%)	15.5%
Unemployment rate (%)	58.25%
Unemployment rate of population under 20 years old (%)	29.8%
Average annual income	11,379.26 euro

Geographic and Natural Characteristics of Kalymnos Island

Kalymnos is a Greek Island and municipality located in the southeastern Aegean Sea. Its geographical location is placed between the islands of Kos (12 km south of Kos) and Leros (2 km north of Leros). The island is rectangular in shape, with a length of 21 km and a width of 13 km, covering an area of 110.875 km^2. Kalymnos's geographical coordinates are 36°59′24″N 26°59′06″E (GeoHack, 2022).

Kalymnos is a mountainous island with a rugged topography. There are three main mountain chains moving from West and North-West to East and South East, as well as a fourth chain, which extends from North-West the length of the peninsula. Stratigraphically, Kalymnos Island is structured by the following units (Mimides et al., 2007):

- Paleozoic unit: The oldest unit, its elements are mostly thick-bedded white and dolomite limestone.
- Tectonic unit of amphibolites, schists and gneisses.
- Tectonic unit of the Ionian Zone: This unit is composed of fine-grained to coarse-grained sandstone, conglomerates, metashales, clay schists, and micritic and breccial limestones; Triassic dolomites and dolomite limestones; Middle limestones which are the typical upper development of Triassic dolomites and dolomite limestone; Malmian-Cretaceous upper and middle cherty limestone.
- Tectonic unit of Gavrovos zone of upper Jurassic-Cretaceous age.
- Neogene formations: These formations occur in several locations throughout the island. They are made up of white marls, yellow-red clay, clay-marls with a gray-yellow tint, white marl limestone, brown-gray and gray thin-middle, and thick layered conglomerates.
- Quaternary formations: These formations consist of Pleistocene materials comprised consolidated slide rock and fan conglomerate, volcanic tuffs, and a volcanic pyroclastic formation with rhyolite as the main component; Holocene formations which comprise talus slope screes and detrital cones, terra-rossa alluvial valley deposits, and coastal sands.

Moreover, Kalymnos has several springs, one of which is thermal, as well as a large number of caves. There are more than 50 small or large caves, all of which are characterized as very important geological monuments. Some of the most famous caves of Kalymnos are Daskalio, Kefala, Choiromantres, Skalia, Stimenia, and Sikati. Kalymnos has two climbing fields: Arginonta and Grande Grotta (Kalymnos-info, 2020).

When a tourist wants to come and visit Kalymnos, there are several means of transportation to reach the island. More specifically, there are ships departing from Piraeus port to Port Kalymnos (Pothia) on a regular basis (more frequently at the summer season); there are daily routes from the Dodecanese islands (as Rhodos and Kos). Additionally, Kalymnos has a small airport, which was renovated and began operations in August 2006. This airport hosts airplanes departing either from Eleftherios Venizelos airport—Athens or from Kos airport and from Turkey

Table 9.2 Geographic characteristics of Kalymnos Island

Kalymnos Island	
Location [ocean or sea] (name)	Aegean Sea
Type of geomorphology of islands (continental island, oceanic island, reef island, etc.)	Continental Island
Inhabited islands (number)	5
Main means of transportation to and from the island or archipelago (number and type)	2 (aiplane, boat)
Seasonality (number of seasons)	2 (spring, summer)
Conservation areas [national parks, natural parks, regional parks, protected areas, etc.] (name, number and type)	Geological Caves and Climbing Fields (52)

(Bodrum airport) (Kalymnos-island, 2022a). When a tourist is on the Kalymnos islands, he/she can use the municipality transportation (buses), utilize a taxi, or rent a car or motorbike.

As a touristic destination, Kalymnos is frequented mainly in the spring and summer seasons. Kalymnos has its greatest number of tourist arrivals from April to September. According to the South Aegean Region Office (INSETE Intelligence, 2019), there were approximately 732,492 visits to Kalymnos in 2018. All the above elements are presented in Table 9.2.

Sport Tourism on Kalymnos Island

Climbing

Over the past 20 years, Kalymnos has been widely recognized as a rock climbing destination, offering climbing and bouldering during the full year (Visiting Greece, 2022). As has been described, rocky Kalymnos Island is considered to be a unique geological and geomorphological destination that includes not only a magnificent geoheritage but also a unique landscape. *"Climb to Seventh Heaven"* is the advertising phrase that Visiting-Greece is using to describe this unique climbing destination. The large number of caves, deep ravines, and mountains are important natural resources for Kalymnos Island. Many researchers call Kalymnos the "island of limestone," as the majority of the island consists of limestone rock (Zafeiropoulos & Drinia, 2021). This magnificent rocky topographical environment is the main element that put Kalymnos Island on the worldwide map as a famous rock-climbing destination. Indeed, the 57-h agency places Kalymnos as one of the world's favorite rock climbing destinations, along with Yosemite National Park in the USA, Dolomites in Italy, and Fontainebleau in France (Levatic, 2022). Thus, Kalymnos has gained a worldwide recognition and reputation as a climbing destination.

The history of rock climbing on this island began in 1995. Giannis Torelli, one of the best Greek climbers, visited the island. Although he did not do any climbing on the rocks, he took some very inspiring photos of the crags, which he then shared with the climbing community. The next summer (summer of 1996), the famous Italian film director and climber Andrea di Bari visited Kalymnos for his summer holidays. Admiring the amazing quality of the rocky environment and the huge climbing potential of the area, Andrea di Bari returned to Kalymnos in May 1997 with friends (climbers) in order to climb several routes. They opened 43 sport routes on Kalymnos (in the areas of Arhi, Odyssey, and Poets). In October 1997, Andrea di Bari returned to Kalymnos with his friend Andrea Gallo, a photographer who worked for Alp magazine. Andrea Gallo's photos were given to Alp and Rotpunkt magazines. Their articles were issued in the spring of 1999. The rock climbing era began for the Kalymnos Island. In order to make Kalymnos a famous rock climbing destination, a multilingual website was created. This website led to the island hosting the first international climbing event on Kalymnos in October 2000, organized by the Acharnes Greek Alpine Club. French alpinist Catherine Destivelle was a special guest at the event. Apart from Catherine Destivelle, 180 climbers from 13 different countries participated in the event. They exchanged points of views and climbed together. During this year, the first Kalymnos Rock Climbing Guide was published by the local council and was distributed for free to the climbers. Hence, sport-climbing tourism on the island of Kalymnos began in earnest (Climb Kalymnos, 2022).

Since 2000, there have been many climbing festivals, organized approximately every 2 years: the second Climbing Festival of Kalymnos took place on 2–8 October 2004, the third Climbing Festival of Kalymnos occurred on 25–30 October 2006, the fourth Kalymnos Climbing Festival happened on 22–26 May 2009), and the fifth Climbing Festival of Kalymnos took place on 18–21 May 2011. All these festivals were organized by the Municipality of Kalymnos (Climb Kalymnos, 2022). Additionally, in 2012, 2013, and 2014, the North Face organized the North Face Kalymnos Climbing Festival, giving a certain prestige to the island of Kalymnos. Since 2015, the Municipality of Kalymnos has organized annual festivals in September or October, promoting Kalymnos as one of the best rock climbing destinations in the world (Climb Kalymnos, 2022).

Normally, festivals organized in October would be ill-advised on a small Greek island due to the fact that this time of year is considered off-season. In most Greek islands, the touristic period generally runs from May to September. Thus, climbing sport tourism offers a significant departure from other forms of tourism. Due to the dry climate of Kalymnos, climbing can take place throughout the year and during all four seasons. According to the Climbing Kalymnos Guidebook (Theodoropoulos & Roussos, 2019), there are many advantages for climbing in every season:

- Autumn: The weather in autumn is ideal for climbing. The temperatures are comfortable for all-day climbing. Generally, it does not often rain in Kalymnos during September and October. Hence, October is usually a peak season for climbing.

- Winter: Mild, sunny, and clear days in Kalymnos are common during the winter months. Some areas, such as Arginonta and Emporios, are ideal for cold-weather climbing.
- Spring: The weather in spring is similar to autumn. There is a greater possibility of rain but this does not seem to pose a major problem, as the rocks in Kalymnos dry quickly. If the winter was rainy, tufas may be seeping. Also, stalactites sometimes drip in the spring and they are softer, so they are more likely to break. For that reason, it is recommended for athletes to climb inside the caves when it rains, but not during a thunderstorm.
- Summer: Many people believe that the summer is too hot for climbing. But this is not the case in Kalymnos. There are many places, such as the Secret Garden or the Argigonta Valley, that are in the shade almost all day. These places usually offer a cool breeze so it is ideal for someone to climb, especially during summer mornings.

As has been mentioned, Kalymnos became a popular rock-climbing destination over a relatively short period of time. It was not long after that when the local authorities realized the development potential of the island. For that reason, they collaborated with the mountain coach Aris Theodoropoulos (2016) for a series of important steps to strengthen sport tourism in Kalymnos:

- Careful maintenance of mountaineering trails.
- Creation of a new field for climbing and additional special constructions with facilities that can host events.
- Creation of a strategic marketing and promotion plan to invite specialized professionals in the field, sports journalists, etc. to the island.

In order to ensure safety and quality standards for climbing, the municipal council of Kalymnos established a New Route Protocol which has become effective since September 2018 (Climb Kalymnos, 2022). More specifically, if a climber wants to take a new route, he/she is subject to approval by a three-person committee (a representative from Kalymnos municipality, a representative of the Kalymnos Rescue Team and mountain guide Aris Theorodopoulos). In a few words, a climber must notify the municipality first and wait for approval before starting a climb over a new route. Otherwise, the climber would not be permitted to climb.

The main goal of the Municipality of Kalymnos is to make climbing routes as safe as possible. In this way, climbers can concentrate on the technical difficulty of the routes without worrying about getting hurt, injured, or even dying (Climb Kalymnos, 2022). In order to make Kalymnos a safe climbing destination, Kalymnos Rescue Team was formed in 2013. Kalymnos Rescue Team is a nonprofit organization that consists of 25 people, all volunteer climbers, and non-climbers, who are trained in rope rescue and first aid. The main purpose of Kalymnos Rescue Team is to aid in any accident concerning climbing or other activities that take place on land and in rough places where medical care cannot easily access (Kalymnos Rescue Team, 2022). Kalymnos Rescue Team is funded by donations.

Today, Kalymnos Island has approximately 52 caves and more than 2500 well-equipped sport climbing routes for all climbing levels, from easy (for beginners) to extremely hard routes (for professionals) (Kalymnos-island, 2022b). Routes are graded according to the French grading system. About half of the routes on Kalymnos are graded up to 6b+, and more than one-third of the total routes are between 6a-6b+. Also, there are more than 700 routes that are graded 7b and up. Finally, there are some routes (such as Los Revolucionarios at sector Odyssey) that are graded as 9a. Thus, one reason for the broad-based attractiveness of Kalymnos is that every climber (new or old) will find interesting climbing routes on the island. Figure 9.1 shows some of the most important caves and climbing fields of Kalymnos.

In March 2015, the first tourist institution of Kalymnos, Systas, decided to establish the Climbing ID Card for guest climbers (offered at the price of 20 euros). The Climbing ID Card offers a packet of vouchers that can redeemed at a list of local businesses, associated to Systas. This is an act of reward toward all the benefits that the locals gain from the existence of climbing in Kalymnos, which hosts approximately 15,000 climbers per year (Kalymnos-island, 2022c).

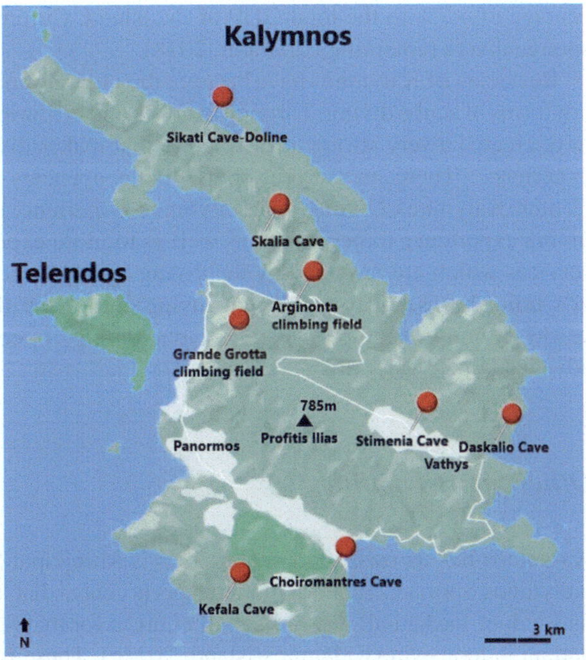

Fig. 9.1 Satellite map of Kalymnos Island indicating the most important caves and climbing fields of the island. (Zafeiropoulos & Drinia, 2021)

Diving/Fishing Sponges

For many decades, Kalymnos has been known as the "Sponge-divers' Island." Sponge diving was the main occupation on Kalymnos for ages. In other words, fishing sponges were the main source of income for the residents of Kalymnos, bringing wealth and reputation to the island community and making it famous worldwide. Sponges can be found in several depths, from a few centimeters depth to 8500 m deep. In Kalymnos, there are 5000 different species of sponges. These species vary in size, quality, and color; some are velvety and smooth and others are rough and hard (Kalymnos-island, 2022d).

In order to honor Kalymnos divers, the University of Gabriele d' Annuzio in Chietti, Italy, suggested in 1984 that a research center for Hyperbaric Underwater Medicine should be established in Kalymnos. For thousands of years, fishermen of Kalymnos have been diving for sponges. From the old methods of holding their breath to more modern methods, diving has become second nature to fishermen. The skills of diving have passed from one generation to the next, making Kalymnos fishermen some of the best divers worldwide (Kalymnos-island, 2022d).

Unfortunately, things changed in October 1986. When "sponge disease" appeared and destroyed many species, even in shallow waters (5–8 m) (Castritsi-Catharios et al., 2016). Due to this destructive disease, Kalymnians were forced to turn to other kinds of fishing, such as tuna and swordfish. For this reason, during the last two decades, the previous primary productivity sector contributes the minor percentage of 4.7% to the total GDP of the island. Currently, there are only 80 active sponge divers remaining (Charami, 2018).

Residents of Kalymnos have brought the old tradition of sponge diving into modern forms of scuba diving. Through decades and centuries of experience, Kalymnians have created a very strong infrastructure that makes diving a safe and unforgettable experience. There are several sport diving agencies (such as Kalymnos Diving Center, Kalymnos Diving, and Kalymnos Experience) (2022) that offer the scuba diving experience to beginners, as well as to more experienced divers. Some of the services which are provided by the diving agencies include traditional sponge diver's trail, the use of rongdifferent diving equipment through the ages, use of the world famous "skafandro" suit, scuba diving, open water diver courses, and advanced open water diver courses.

Other Sport Facilities

Kalymnos has its own sport center. This is Municipal Sports Center of Kalymnos. Kalymnos Sports Center is a modern sports center that has a Natatorium, an outdoor and indoor basketball and volleyball court, a football/soccer field, a running track, and an indoor gym (Kalymnos-island, 2022e). These facilities can be used both by residents and tourists alike. There is also a modern tennis court which can be used by the locals and visitors. Finally, in Kalymnos, a visitor may experience water sports, such as windsurfing (Kalymnos-island, 2022f).

Climbing, Diving, and Other Sport Facilities as Factors for the Development of Sustainable Tourism on Kalymnos Island

The aim of the current chapter is to discover ways that sport tourism can contribute to the sustainable development of the island of Kalymnos. As has been mentioned in the previous section, over the past 20 years Kalymnos has upgraded its tourism "image" from being a small island that only offers summer vacation with its restaurants, rooms to let, and wonderful beaches, into being one of the best climbing destinations in the world.

Although several climbing festivals have been organized since 2000, the majority funded by the Municipality of Kalymnos or the North Face brand, there remains a "window" of opportunity for improvement. Kalymnos Municipality, as well as the local community (businesses and organizations), should do more to promote the island as a world-class climbing destination. There are rock climbing magazines with global recognition by the climbing community (such as *Gripped Climbing Magazine*, *Alpinist Magazine*, *SA Mountain Magazine*, *Climbing Business Journal*, Rock and Ice and Climb) in which Kalymnos can advertise its "image" as a premiere climbing destination.

Apart from the climbing magazines, most of which are marketed to professional climbers or experts, Kalymnos's image as a climbing destination should be promoted in a larger scale to several touristic sites and forums in order to attract more tourists who enjoy climbing at an amateur level. Social media can play an important role in building the image of Kalymnos (Belias et al., 2022; Skagias et al., 2022; Tsiotas et al., 2020). In addition, Visiting-Greece (2022) campaigns should promote this image of Kalymnos on multiple media platforms, such as social media, internet advertising, and traveling television channels.

The image of Kalymnos as a climbing destination can help the sustainable development of the island due to the following opportunities:

- Increased climbing tourism means more visitor spending than the average tourist (UNWTO, 2019). Climbing tourists need special equipment, such as bolts and ropes. Moreover, they usually need to hire instructors and guides in order to help them with new or existing routes. Additionally, some of these tourists will need to take courses for learning basic or more advanced elements concerning their safety and climbing rules.
- Climbing tourism may also create new jobs. Climbing equipment stores, climbing schools, climbing travel agencies, climbing instructors, and rescue—First Aid Instructors are some of the new jobs that have been created on Kalymnos Islands.

- As more climbing tourists arrive, economic opportunities expand. More specifically, climbing tourism on Kalymnos is considered to be an all season type of tourism. This likewise increases the economic opportunities created for the local tourist community (travel agencies, restaurants, and bars, hospitality businesses—hotels, rooms to let, Airbnb, cars and motorbikes rental cars, climbing equipment stores, etc.). The expansion of year-round seasons the spring/summer season lessens the demand and potential overcrowding during the normal high (summer) season (Tsiotas et al., 2019).
- Climbing tourism offers the opportunity for exploration, recording, and promotion of the island's speleological treasures (Zafeiropoulos & Drinia, 2021). These treasures can attract scientists and geologists (amateur or professional) to visit the island and view all the magnificent geological resources that Kalymnos Island offers. In addition, a climbing/geo-museum can be created combined with geoeducation activities (Zafeiropoulos & Drinia, 2021). As a result, new jobs can be created within this area of specialization.
- Climbing tourism can also create new touristic facilities. For example, new restaurants and hotels can be created in order to accommodate climbers. As an alternative facility for those climbers who do not like staying in hotels, a new campsite can be created to appeal to the climber who might prefer to be immersed within the natural environment of Kalymnos.
- In addition, new cultural events and festivals throughout the year can be created in order to attract more tourists who come to Kalymnos for climbing. They may also spend time enjoying the cultural events the island has to offer them (Belias et al., 2020; Vasiliadis & Belias, 2020).
- Another opportunity that climbing tourism creates is the expansion of the airport to host more international flights during the year.
- Creation of tourism institutions (such as Systas) (Kalymnos-island, 2022c) that represent all business sectors and clubs of tourism, involved in the development of a natural environment, culture, and outdoor activities. These collaborations among local businesses help expand the tourism market by creating opportunities and rewards for tourists (coupons, discounts) in order to attract them.
- The Municipality of Kalymnos should carefully examine all the funds that are given by the European Union in order to select and supervise public or private projects concerning the development of climbing tourism in Kalymnos.

As previously noted, Kalymnos has also been considered to be the "island of sponge diving" for ages. So, diving and fishery should be factors that will help Kalymnos be further developed, in order to be a sustainable touristic destination. Similarly to climbing tourism, diving tourism should raise the following development opportunities for Kalymnos Island:

- Diving tourists spend more money than the average tourists as they have to buy/rent extra equipment for diving, attend training courses for diving, and pay travel agencies for diving experience services. This can create new jobs (as diving instructors and travel agents). The residents of Kalymnos are traditionally familiar with the diving techniques so they are the best representatives in order to teach the secrets of diving to both amateurs and experts. For this reason, the municipality should seriously consider applying for funds regarding the reopening of the Public Professional Diving School of Kalymnos (Facebook, 2022).

- For the same reasons as were previously mentioned on climbing tourism, diving tourism can also help to create new tourism facilities (Tsiotas et al., 2020b), new cultural events, and increase economic touristic opportunities.
- Similarly to climbing tourism, the Municipality of Kalymnos should carefully examine all the funds that are given by the European Union in order to select and supervise public or private projects concerning the development of diving tourism in Kalymnos.

Last but not least, sport water facilities, regarding sport water amusements—e.g. water sport games—seem to be in a "primitive" stage of development in Kalymnos. The local community should invest in creating facilities for water sports (such as water amusement parks). Diving, combined with water sports, can offer an amazing experience for tourists who want to be familiar with the marine environment of Kalymnos.

Conclusions

Kalymnos is the fourth largest in size, island of the Dodecanese archipelago, which is located in the South Eastern Aegean Sea (Greece). As a typical Greek Aegean island, the majority of tourists have the tendency to visit the island in the summer season so as to live the experience of "Summer in Greece" (Discover Greece, 2022). "Summer in Greece" for many tourists, both domestic and foreign, is often equated to the sun, sea, and sand.

Today, there is an opportunity to expand and diversify the touristic image of Kalymnos (Rossidis et al., 2019) to be widely known as one of the best destinations in the world for rock climbing and diving. This type of tourism (sport tourism) offers an excellent opportunity for the promotion of genuine sustainable development of the island.

Kalymnos, for ages, was known as the "sponge island." Unfortunately, in the past decades, things have changed and sponge fishing is no longer the primary occupation of the residents of Kalymnos. Services have become the major economic sector that contributes to Kalymnos's GDP.

Thus, a new era has begun for the island of Kalymnos: to search for alternative ways of promoting tourism (Belias et al., 2018b; Belias, 2019; Kyriakou & Belias, 2017; Skagias et al., 2021; Niavis et al., 2020). Local communities should invest more heavily in promoting Kalymnos as a "climbing and diving" destination. The advantages of this strategy include creating new jobs for residents, increasing the touristic opportunities, extending the touristic summer season to a four-season supply, increasing visitors' spending, creating new cultural events and sport festivals, and increasing the well-being of the tourists and welfare of the locals.

Despite the fact that climbing and diving tourism can offer several opportunities for sustainable development, there are several locals who have reacted negatively to the potential degradation of the natural environment of Kalymnos as a result of year-round tourism. For this reason, the municipality of Kalymnos should take all of the precautions (via rules and protocols) in order to protect the environment from human overuse. The Kalymnos municipality should inform the local community about these precautions so as to lessen their doubts and fears about the destruction of the environment due to the development of sport tourism in the island (Belias et al., 2022b). Kalymnos, therefore, hopes to become a role model as a responsible, sustainable island destination due to the growth of sport tourism.

References

Belias, D. (2019). Adventure tourism-examining cases of how it can contribute on rural development in Greece. In A. Kavoura, E. Kefallonitis, & A. Giovanis (Eds.), *Strategic innovative marketing and tourism* (Springer proceedings in business and economics). Springer. https://doi.org/10.1007/978-3-030-12453-3_104

Belias, D., Velissariou, E., Kyriakou, D., Vasiliadis, L., Aspridis, G., Mantas, C., Roditis, A., & Koustelios, A. (2018). Greece as a sports tourism destination. In V. Katsoni & K. Velander (Eds.), *Innovative approaches to tourism and leisure* (Springer proceedings in business and economics). Springer. https://doi.org/10.1007/978-3-319-67603-6_38

Belias, D., Velissariou, E., Kyriakou, D., Varsanis, K., Vasiliadis, L., Mantas, C., Sdrolias, L., & Koustelios, A. (2018b). Tourism consumer behavior and alternative tourism: The case of Agrotourism in Greece. In *Innovative approaches to tourism and leisure*. Springer.

Belias, D., Vasiliadis, L., & Zaftis, E. (2020). The impact of cultural routes on traditional settlements: The case of Greece. In V. Katsoni & T. Spyriadis (Eds.), *Cultural and tourism innovation in the digital era* (Springer proceedings in business and economics). Springer. https://doi.org/10.1007/978-3-030-36342-0_10

Belias, D., Malik, S., Rossidis, I., & Mantas, C. (2022). The impact of social tourism entrepreneurship on tourism sustainability – A systematic review. *World Journal of Entrepreneurship, Management and Sustainable Development, 5*, ISSN:2042-5961.

Belias, D., Rossidis, I., Papdemetriou, C., & Marco, V. (2022b). Destination governance: The role of local authorities in Greek tourism marketing. In *New governance and management in touristic destinations* (pp. 1–18).

Belitsi, M. E. (2020). *Alternative tourism: The case of Kalymnos island* [in Greek]. Master thesis under the supervision of professor Giannoukou I. Open University of Greece. Department of Tourism Management.

Castritsi-Catharios, J., Zouganelis, G. D., Berillis, P. J., Ekonomou, G., & Kefalas, E. (2016). Kalymnos, the Island which made history in the sponge fishery. Data on the identification of Spongia officinalis officinalis & Hippospongia communis, the most common species in Aegean Sea. In *Conference: ISISA Islands of the world XIV conference 2016 Nisiology and utopia: Back to the roots of Island studies at: Mytilini, Lesvos Island, Greece*.

Charami, O. (2018). *Climbers' and divers' paradise: A guide to Kalymnos* [online]. Greece is Travel, Culture, Gastronomy and more. Retrieved from https://www.greece-is.com/climbers-divers-paradise-guide-kalymnos/. Last Accessed 15 June 2022.

Chatziioannou, E., Grasemann, B., Schneider, D., Hubmann, B., & Soukis, K. (2015). Variscan tectonics in Dodecanese, Kalymnos Island, Greece. In *Proceedings of the EGU general assembly conference abstracts, Vienna, Austria, 12–17 April 2015*.

Climb Kalymnos. (2022). *Climbing in Kalymnos* [online]. Retrieved from: https://climbkalymnos.com/climbing/. Last Accessed 15 June 2022.
Discover Greece. (2022). *Summer Holidays in Greece* [online]. Retrieved from https://www.discovergreece.com/travel-ideas/article/summer-holidays-greece. Last Accessed 15 June 2022.
Facebook. (2022). *Professional diving school of Kalymnos* [in Greek] [online]. Retrieved from https://www.facebook.com/esdk2013. Last Accessed 15 June 2022.
GeoHack. (2022). *Geo-Hack Kalymnos Global/Trans-national services*. Retrieved from: https://geohack.toolforge.org/geohack.php?pagename=Kalymnos&params=36_59_24_N_26_59_06_E_type:isle_region:GR-L. Last Accessed 15 June 2022.
Gibson, H. J. (1998a). Sport tourism: A critical analysis of research. *Sport Management Review, 1*, 45–76.
Gibson, H. (1998b). Sport tourism: A critical analysis of research. *Sport Management Review, 1*, 45–76.
Higgins, M. D., & Higgins, R. (1996). A geological companion to Greece and the Aegean. In *The Dodecanese and the Carian shore*. Cornell University Press/Duckworth Publishers.
INSETE Intelligence. (2019). *Annual report of Competiveness on the section of Tourism for South Aegean Region* [in Greek]. Retrieved from: https://insete.gr/studies/%CE%B5%CF%84%CE%AE%CF%83%CE%B9%CE%B1-%CE%AD%CE%BA%CE%B8%CE%B5%CF%83%CE%B7-%CE%B1%CE%BD%CF%84%CE%B1%CE%B3%CF%89%CE%BD%CE%B9%CF%83%CF%84%CE%B9%CE%BA%CF%8C%CF%84%CE%B7%CF%84%CE%B1%CF%82-%CE%BA%CE%B1-9/. Last Accessed 15 June 2022.
Kalymnos Rescue Team. (2022). *Information about Kalymnos Rescue Team*. Retrieved from: https://www.kalymnosrescueteam.org/#!/the-team. Last Accessed 15 June 2022.
Kalymnos-info. (2020). *Kalymnos Island Greece, The Official Travel Guide* [online]. Retrieved from: http://www.kalymnosinfo.com/. Last Accessed 15 June 2022.
Kalymnos-island. (2022a). *Transportation – How to come to Kalymnos* [online]. Retrieved from https://www.kalymnos-isl.gr/en/information/transportation. Last Accessed 15 June 2022.
Kalymnos-island. (2022b). *Climbing activities* [online]. Retrieved from https://www.kalymnos-isl.gr/en/activities/climbing-trekking. Last Accessed 15 June 2022.
Kalymnos-island. (2022c). *What about alternative tourism and climbing in Kalymnos*. Retrieved from https://www.kalymnos-isl.gr/en/information/news-events/24-climbing-news/64-what-s-new-about-alternative-tourism-and-climbing-in-kalymnos. Last Accessed 15 June 2022.
Kalymnos-island. (2022d). *Sponge Diving in Kalymnos*. Retrieved from https://www.kalymnos-isl.gr/en/about-kalymnos/sponge-diving Last Accessed 15 June 2022.
Kalymnos-island. (2022e). *Sports facilities*. Retrieved from https://www.kalymnos-isl.gr/en/activities/sports-facilities. Last Accessed 15 June 2022.
Kalymnos-island. (2022f). *Other activities*. Retrieved from https://www.kalymnos-isl.gr/en/activities/other. Last Accessed 15 June 2022.
Kyriakou, D., & Belias, D. (2017). Is silver economy a new way of tourism potential for Greece? In V. Katsoni, A. Upadhya, & A. Stratigea (Eds.), *Tourism, culture and heritage in a smart economy* (Springer proceedings in business and economics). Springer. https://doi.org/10.1007/978-3-319-47732-9_28
Levatic Jurica. (2022). *Your 2022 guide to finding the best rock climbing in the world* [online] 57 Hours own your weekend. Retrieved from https://57hours.com/best-of/rock-climbing-worldwide/. Last Accessed 15 June 2022.
Melo, R., Sobry, C., & Van Rheenen, D. (Eds.). (2021). *Small scale sport tourism events and local sustainable development. A cross-national comparative perspective*. Springer.
Mimides, T., Psomiadis, E., & Trikili, I. (2007). Monitoring land use changes and soil degradation using spot and temporal aerial photograph data-the case of Vathy basin (Kalymnos Island-Dodecanese Greece). *Bulletin of the Geological Society of Greece, 2007*(40), 1476–1484.
Ministry of Foreign Affairs of Greece (MFA.GR). (2022). *Government and politics*. Retrieved from: https://www.mfa.gr/usa/en/about-greece/government-and-politics/. Last Accessed 15 June 2022.

Municipality of Kalymnos. (2017). *Management plan on the island of Kalymnos* [in Greek]. Retrieved from: https://www.kalymnos.gov.gr/documents/diavoulefseis/olokliromeni_parembasi_leader_sxedio_24-8-2017.pdf. Last Accessed 15 June 2022.

Niavis, S., Belias, D., & Tsiotas, D. (2020). The contribution of wine tourism in the development of rural areas in North Greece. In A. Kavoura, E. Kefallonitis, & P. Theodoridis (Eds.), *Strategic innovative marketing and tourism* (Springer proceedings in business and economics). Springer. https://doi.org/10.1007/978-3-030-36126-6_105

Pigeassou, C. (2004). Contribution to the definition of sport tourism. *Journal of Sport Tourism, 9*(3). https://doi.org/10.1080/1477508042000320205

Rossidis, I., Belias, D., Varsanis, K., Papailias, S., Tsiotas, D., Vasiliadis, L., & Sdrolias, L. (2019). Tourism and destination branding: The case of Greek Islands. In *Strategic innovative marketing and tourism* (pp. 93–100). Springer. https://doi.org/10.1007/978-3-030-12453-3_11

Skagias, K., Vasiliadis, L., Belias, D., & Papademetriou, C. (2021). From mass tourism and mass culture to sustainable tourism in the post-covid19 Era: The case of Mykonos. In *Culture and tourism in a smart, globalized, and sustainable world* (pp. 347–358). Springer.

Skagias, K., Belias, D., Vasiliadis, L., & Papademetriou, C. (2022). Digital tourist marketing: The latest developments and recommendations on how Mykonos can take advantage of digital and influencer marketing. In *Transcending Borders in tourism through innovation and cultural heritage* (pp. 1007–1020). Springer.

Sobry, C., Liu, X., & Li, J. (2016). Sport tourism: Contribution to a definition and categorization. *Acta Touristica, 28*(1), 7–26.

Theodoropoulos, A. (2016). *Rock Climbing Guidebook*. Anavasi.

Theodoropoulos, A., & Roussos, K. (2019). *Kalymnos Climbing Guidebook* (Revised ed.). Vertical Life.

Tsiotas, D., Niavis, S., Belias, D., & Sdrolias, L. (2019). Modeling the international tourism demand as a complex network: The case of the global inbound tourism market. In A. Kavoura, E. Kefallonitis, & A. Giovanis (Eds.), *Strategic innovative marketing and tourism* (Springer proceedings in business and economics). Springer. https://doi.org/10.1007/978-3-030-12453-3_93

Tsiotas, D., Niavis, S., Belias, D., & Sdrolias, L. (2020). What can the TripAdvisor tell us about the complaints management strategies? The case of the Greek hotels. In *Strategic innovative marketing and tourism* (pp. 999–1005). Springer.

Tsiotas, D., Belias, D., & Niavis, S. (2020b). Internal marketing on the tourism sector: Examination of how the staff can develop on Greek hotels. In *Strategic innovative marketing and tourism* (pp. 963–970). Springer.

UNWTO (World Tourism Organization). (2019). *Sport tourism & the sustainable development goals (SDGs)* [online]. Retrieved from: https://www.e-unwto.org/doi/epdf/10.18111/9789284419661. Last Accessed 15 June 2022.

Vasiliadis, L., & Belias, D. (2020). The value of cultural routes in Greece: Examination of the current situation. In A. Kavoura, E. Kefallonitis, & P. Theodoridis (Eds.), *Strategic innovative marketing and tourism* (Springer proceedings in business and economics). Springer. https://doi.org/10.1007/978-3-030-36126-6_119

Visiting Greece. (2022). *Rock Climbing on Kalymnos* [online]. Retrieved from https://www.visitgreece.gr/experiences/activities/land-activities/rock-climbing-on-kalymnos/. Last Accessed 15 June 2022.

Zafeiropoulos, G., & Drinia, H. (2021). Kalymnos Island, SE Aegean Sea: From fishing sponges and rock climbing to Geotourism perspective. *Heritage, 4*(4), 3126–3146.

Chapter 10
Sustainable Sport Tourism on German Islands: Case Studies of Helgoland and Rügen

Alexander Hodek

Introduction

Sustainability has long been an issue for sport (e.g., Taks, 2013), as well as for tourism (e.g., McCool, 2015). It has become apparent that the importance of sustainability has grown in recent years, especially in the tourism industry (Streimikiene et al., 2021). German tourists in particular have been attaching extremely high importance to this topic for some time (FUR, 2023). This development is no different in sports tourism (Taks, 2013; Gibson et al., 2012; Ma et al., 2011). Since the Corona crisis, it has become apparent to German tourists that a holiday in their own country has become increasingly attractive. It is true that in the wake of the Corona restrictions, German tourists' desire to travel plummeted. However, this has not fully recovered to the precrisis level in 2022 compared to 2020. At the same time, travel in Germany itself has gained significantly in importance (Destatis, 2023). The trend toward traveling closer to home is not new and can probably also be linked to the attitude toward more sustainable travel. Above all, this is linked to the desire to produce less CO_2 through long air journeys. Therefore, this chapter deals with two very popular travel destinations in Germany—Helgoland and Rügen. Even if Germany is not the first country in the world to be associated with water tourism, it does have coasts on two seas. On the one hand, in the northwest of Germany, there is the North Sea as a saltwater sea with direct access to the Atlantic Ocean. The island of Helgoland is located in the North Sea, along with other islands that have been developed for tourism. On the other hand, Germany borders the Baltic Sea in the northeast, a freshwater inland sea that only has access to the world's oceans via a channel. Here, among other islands, is the largest of the Baltic Sea islands, the

A. Hodek (✉)
ISM – International School of Management (Berlin), Berlin, Germany
e-mail: Alexander.hodeck@ism.de

D. Van Rheenen et al. (eds.), *Sport Tourism, Island Territories and Sustainable Development*, Sports Economics, Management and Policy 24,
https://doi.org/10.1007/978-3-031-51705-1_10

island of Rügen. This chapter first examines the sports tourism offers on both islands and then their contribution to sustainable development in the respective destinations.

Characterization of the Island

Helgoland

General Data

The German Island Helgoland is a small archipelago in the North Sea. Located 60 kilometers away from the German coast, approximately 1500 people live on the island. As Helgoland is an archipelago, it consists of the main Island "Helgoland" as well as the smaller "Helgoland Dune" (Helgoland Tourismus-Service, n.d.-a). Together, they spread over 2.1 square kilometers, which makes Helgoland the smallest Island in the North Sea and the seventh smallest Island in Germany (Kreulitsch, 2022). Reviewing the Island from an economical perspective, tourism is the major income factor. In 2019, about 400,000 tourists visited the small island (Preuß, 2021). Comparing this figure to the population, the importance of tourism becomes clear. What makes the island special from an economic and political perspective are the regulations related to taxes and duties. Because the Island is neither part of the German tax area nor part of the customs area of the European Union, tourists can buy goods and services duty-free and VAT-free (Zoll, n.d.). These circumstances are mainly caused by historical reasons, as Helgoland was owned by different countries in the past. At the beginning of the eighteenth century, the small Island was an important base for sea pilots and one of the largest trading centers in Europe. In 1714, Helgoland became Danish, before it turned British in 1826. In 1890, the Island became German for the first time. A significant part of Helgoland's history was the Second World War. Being involved in the war as a German base and nearly destroyed by the British, it was not until 1952 that Heligoland was returned to Germany (Helgoland Tourismus-Service, n.d.-b).

Geographical and Nature Factors

Located 60 kilometers away from the German coast, Helgoland can only be reached by boat or plane. The smaller of the two islands, "Helgoland Coast," has an airport that can be reached within 20 min from the German North Sea coast (Helgoland AirService, n.d.). Furthermore, the Island can be reached by boat. For tourists, there is a ferry service from Cuxhaven (Germany) and Büsum (Germany) throughout the year.

From a geological perspective, a flat profile is typical for similar islands in the North Sea. With an area size of 2.1 square kilometers, the island's height of up to 40 meters is one of Helgoland's unique characteristics (Ross, 1990).

To describe Helgoland's climate, the indicators of precipitation, temperature, duration of sunshine, as well as wind speed, are utilized. The data include the years 1991 until 2020. The mean monthly sum of precipitation reaches from 35.8 millimeters in April up to 88.9 millimeters in August. The average temperature (monthly mean) amounts to 3.2 degrees Celsius as minimum in February and increases up to 18 degrees Celsius in August. The average duration of sunshine is at least 2248.6 meters in December and reaches a maximum of 15,051.2 meters in May. The lowest average of wind speed amounts to 23.9 kilometers per hour in July, while the highest wind speed average could be detected in December and January at 35.6 kilometers per hour (Meteostat, 2023a).

Sport Tourism Activities

The island of Helgoland does not have an extensive offer of sporting activities, particularly for their tourists. Possible activities that are advertised by the island itself are minigolf, trampoline, boule as well as open-air chess, which are rather leisure activities for families or group activities without too much action. Furthermore, the island has private tennis courts, a football field, and open-air fitness equipment with free access. The owner of the football field is the club VfL Forsite Helgoland, which is an official member of the Hamburg Football Association. Due to the distance to the German coast, they only play home friendly games and are not part of any league. As an article by the German Football Association demonstrates, at least 10–15 teams traveled to Heligoland per year to have friendly games there. Furthermore, the island's representatives have tried to invite first and second-division football teams to have a training camp on Heligoland, but without success to date (DFB, 2012). Another sporting activity especially offered for tourists is yoga at the beach. In summer, there is an offering of four yoga courses a week (Helgoland Tourismus-Service, n.d.-c).

Looking at the wind speed as mentioned before, the island seems attractive for water sports like wind- or kitesurfing. This type of sport is generally forbidden in "Wattenmeer" national parks. As there are only a few areas with special permission and Helgoland is not one of them, there is no sport tourism related to wind- or kitesurfing on the island (Nationalpark Wattenmeer, 2020).

As such, it can be said that sport tourism is not an important part of Helgoland's tourism. Due to the small area size, environment protection regulations, and long distances to the German coast, that status has not changed over the past 20 years.

Rügen (Germany)

General Data

The German Island of Rügen is the biggest German Island located in the Baltic Sea. Located very close to the mainland, Rügen is officially part of the federal state of Mecklenburg Western Pomerania. The Island has 70,000 inhabitants spread over four cities and 38 municipalities. In total, Rügen has an area size of 926 square kilometers. One of Rügen's special characteristics is its infrastructure. Streets with a total length of 1995 kilometers as well as 80 kilometers of rails and 250 kilometers of cycling paths guarantee mobility as well as the 25 bus lines on the island. For water mobility, there are 51 ports including 22 sport and yachting ports (Tourismusverband Rügen, n.d.-a, -b).

Geographical and Natural Factors

The Island of Rügen is located very close to the German Coast and is linked to the mainland by a bridge, known as the "Rügendamm." Furthermore, there is the option to reach Rügen via Ferry from April until October within 15 min (NDR, 2022). Rügen has 140 kilometers coastline with 56 kilometers of sandy beach as well as 27 kilometers of nature beach. Next to the beaches, there are several tourist magnets like the national park Jasmund with its impressive chalk cliffs (Tourismusverband Rügen, n.d.-a).

Like Helgoland, Ruegen's climate can be described by highlighting its annual precipitation rate, temperature, duration of sunshine, and wind speed. The data include the years 1991 until 2020. The mean monthly sum of precipitation reaches from 31 millimeters in April up to 71 millimeters in August. The average temperature (monthly mean) amounts to 0.7 degrees Celsius as a minimum in January and increases up to 17.9 degrees Celsius in August. The average duration of sunshine is at least 1909.6 meters in December and reaches a maximum of 15,750 meters in May. The lowest average wind speed amounts to 12.7 kilometers per hour in August, while the highest wind speed average could be detected in February at 18.4 kilometers per hour (Meteostat, 2023b).

Sport Tourism Activities

In 2021, nearly one million (988,652) tourists visited the island of Rügen (Statistisches Amt Mecklenburg-Vorpommern, 2022: 7). Due to its long coastline, the main sports activities on Rügen are water sports like surfing, kitesurfing, windsurfing, stand up paddling and canoeing. There are more than 13 kitesurfing schools around the island that offer other sports as well. One of these schools is "ProBoarding Rügen Kitesurfschule" located in Thiessow, which resides in the southwest of

Rügen. Customers can either learn kite surfing through regular courses or take part in camps where the provider also offers accommodations and a more extensive program. Next to offering courses in kitesurfing, windsurfing, wingsurfing, wing foil, and stand-up paddling, tourists also can rent the sports material at this location (ProBoarding Rügen, n.d.).

Next to this water sports area, there is also a water ski and wakeboard facility named "Cable Park Rügen." At 670 meters long, the Cable Park offers different obstacles. Like the kitesurfing school, the Cable Park offers a Camp, including 2 h of water skiing a day for 1 week in total. The target group are children between 10 and 16 years, who are housed in tents during the camp (Cable Park Rügen, n.d.). The climbing park, "Kletterwald Binzprora," offers another sport option (Kletterwald Binzprora, n.d.). Here young and old climb together. Usually, compared to the sport options mentioned before, this activity is more of a day trip.

Several sport events take place throughout the year. These events have two distinct target groups. On the one hand, competitive athletes travel to Rügen to actively take part. On the other hand, sport fans of any age are attracted and travel to Rügen to follow the events as spectators. Reviewing the 2019 event plan, the main events that took place were running, cycling, and swimming events. One of the biggest events is the "Strandräuber Ironman 70,3 Rügen," a triathlon including 1.9 kilometers of swimming, 90 kilometers of cycling, and 21.1 kilometers of running (Tourismusverband Rügen, n.d.-b).

To sum up, it can be said that sport tourism plays a vital role for the island of Rügen. For active athletes, water sports are the most important activities.

Analyzing Sustainable Aspects of Sport Tourism on the Islands

Analyzing the sport tourism activities on both islands, it can be seen that these activities primarily contribute to five of the Sustainable Development Goals (SDGs) formulated by the United Nations. These are "SDG 3—Ensure healthy lives and promote well-being for all at all ages," "SDG 12—Ensure sustainable consumption and production patterns," "SDG 13—Take urgent action to combat climate change and its impacts," "SDG 14—Conserve and sustainably use the oceans, seas and marine resources for sustainable development" and "SDG 15—Protect, restore and promote sustainable use of terrestrial ecosystem and halt biodiversity loss." The following section of the chapter will provide concrete examples of the sport touristic development at both destinations during the last year, which underscores the contribution toward reaching these goals.

SDG 3—Ensure Healthy Lives and Promote Well-Being for All at All Ages

In principle, active sports tourism is associated with a healthy lifestyle. Since both of the German islands are clearly focused on active sports tourism, it is not surprising that the offers created are aimed at promoting a healthy lifestyle for the visitors, but surprisingly also for the inhabitants themselves. The tourism organization of the island of Ruegen (Rügen, 2023a) includes a healthy lifestyle in three of its ten official tips for a conscious Rügen holiday. Two of them refer to walking or cycling rather than traveling by car. The third takes up the conscious consumption of healthy and sustainable food. To ensure that holidaymakers can actually accomplish this recommendation, concrete measures have been launched, such as green4rent (where eBikes can be easily rented all over the island) or the bike bus (where holidaymakers and locals can use the public bus during the high season and transport their bikes in the bus trailer free of charge to cover even longer distances) (Picture 10.1).

The situation is somewhat different on the island of Helgoland. A special paragraph of the German Road Traffic Regulations applies here. §Section 50 of the STVO actually prohibits driving motor vehicles and BICYCLES on the island of Helgoland. The reason for this is the limited space. The only exceptions are police and fire brigade officials, tradesmen, and other special groups. As a tourist, you can either use the electric island train or the very popular pedal scooters. Most holidaymakers, however, walk exclusively on the island. For this purpose, the tourist board

Picture 10.1 Green4rent. (Downloaded from https://www.ruegen.de/nachhaltig)

has created five specific themed trails on the approximately four square kilometer island, which focus on specific topics such as nature, culture, history, or architecture. Tourists who do not enjoy the pure air or observe wildlife have the opportunity to swim, play tennis, or take advantage of regularly offered yoga classes in their free time (Helgoland, 2023). These offers, as well as the ban on driving cars and bicycles, naturally apply to the island's inhabitants. Thus, a healthy lifestyle is also taken into account here. Children up to the age of 14 are allowed to cycle on the island during the low season from October to April. §Section 50 of the STVO provides for this exception in order to enable children to learn how to ride a bicycle so that they are not disadvantaged when attending secondary schools and universities outside of the island.

SDG 12—Ensure Sustainable Consumption and Production Patterns

As mentioned, sustainable tourism on the two islands is always aimed at the inhabitants. It is therefore not surprising that a sustainable tourism concept also includes the production and sustainable consumption of sustainable products and food. As an example, the activities of the island of Rügen are particularly worth mentioning here. In 2023, there will be again a week of sustainability in Rügen, addressing different stakeholders, e.g., inhabitants, owners of hotels, tourists, and others. Within this project, regional products are especially highlighted. A destination-specific trademark standing for sustainability, "Rügen Produkte" (products of Rügen), has been created, signaling these local products to all consumers (Rügen, 2023b) (Picture 10.2).

There are also manufacturers under the trademark, which produce fresh and regional goods, sold in any speciality shops on the island. These products reach markets outside of the tourist destination in Germany, as visitors seek to enjoy these products back home.

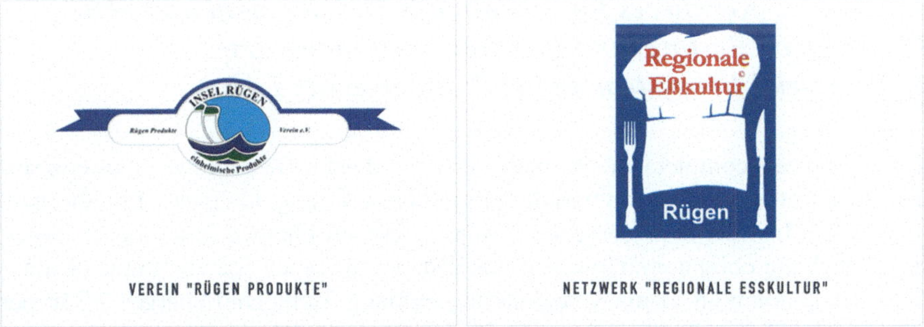

Picture 10.2 Rügen Produkte. (Downloaded from https://www.ruegen.de/ruegen-entdecken/regionale-produkte)

SDG 13—Take Urgent Action to Combat Climate Change and Its Impacts

In German holiday regions, individual transport still plays an important role. Therefore, both islands are worth mentioning in terms of transportation concepts. While Helgoland can only be reached by boat and individual transport is basically restricted, Rügen can be reached by car. Nevertheless, guests are advised of the possibility of traveling by train before they arrive. Rügen also offers a comprehensive local public transport service, especially during the peak holiday periods. The use of electric vehicles is favored by an increasing number of charging stations. On Helgoland, electric minibuses are also used for tourist transport. However, the focus on measures is not limited to the topic of mobility. For hostels in particular, information is offered on how to build sustainably, and which measures can be used in the operation of accommodation facilities to conserve energy and other resources. Corresponding subsidies to promote sustainable projects are offered by the responsible state authorities.

Tourists on both islands are made aware of the effects of climate change at various points and are given information on how they can contribute to a reduction in energy consumption and emissions with their behavior while on holiday. In addition to well-known tips, new and niche tips can also be discovered. On Rügen, for example, it is pointed out that tourists should not bring their own books on holiday, but should use local libraries or exchange offers in the hotels in order to save weight during transport and to use books several times. The same applies to bringing your own bicycles; there are plenty of them available on site. In order to include residents, owners of hotels, holiday apartments, shops, or restaurants in these sustainable education efforts on Rügen, there are camps for sustainability regularly. The last camp focused on regional and sustainable foods was held in January 2023 (Tourismus Mecklenburg-Vorpommern, 2023).

SDG 14—Conserve and Sustainably Use the Oceans, Seas, and Marine Resources for Sustainable Development and SDG 15—Protect, Restore, and Promote Sustainable Use of Terrestrial Ecosystem and Halt Biodiversity Loss

Both islands are completely or partially considered nature reserves and thus contribute per se to the preservation of nature. Strict behavioral guidelines set by the state limit traffic. In order to preserve these areas, guests are charged fees for each day of stay, which are obligatory. However, this visitor's tax is not specific to the destinations but is common in many regions in Germany. Helgoland charges 2,75€ per adult per day during the summer and 1,50€ in winter. On Rügen, the tax differs from the cities, charging between 1,10€ and 3,25€ per adult per day depending on the city and season. On Helgoland in particular, the local government is used to protect the unique nature and birdlife of the island territory. A large number of information boards with rules of conduct and explanations of the consequences of noncompliance are in place. In Helgoland, the number of visitors is also limited by the number of ships arriving on the island, in order to protect the natural habitat of animals on land and in the sea.

Conclusion

In summary, it can be said that sustainability is generally considered important to both island destinations in the German North Sea and Baltic Sea. This applies holistically to both destinations. Various sustainability goals based on the UN definition are promoted and regulated. This has not only recently become part of the strategic orientation but has been an important part of the strategy of the island destinations for decades, even before the articulation of the goals by the United Nations. In this respect, this can be considered credible. It is remarkable that individual measures have been aimed at multiple stakeholder groups. In addition to tourists, these measures also include locals and operators of sport tourism offers. A special feature is certainly the island of Helgoland, for which its own laws were created a long time ago in order to significantly reduce traffic and make it sustainable. In this respect, both destinations can definitely be described as ecologically sustainable. Looking at the development of visitor numbers in these destinations, it is clear that these destinations have also succeeded in making sustainable sport tourism economically sustainable. Through the active involvement of the local population in individual measures, such as public transport, as well as the policies of the federal states, one can also speak of social sustainability. In this respect, the two islands can definitely be seen as providing best-practice solutions with regard to sustainable sport tourism. Nevertheless, there remain negative challenges to overcome in these destinations. These include a shortage of skilled workers in the tourism industry and an increasingly aging society, which is typical for Germany, especially in the rural regions. These two developments could well have a negative impact on the development of tourism on the islands under consideration. Future research should therefore focus on the attitudes of residents and workers, in addition to addressing tourists on these two German islands.

References

Cable Park Rügen. (n.d.). *Wasserskianlage Zirkow*. https://wasserskiruegen.de/cable-park/. Accessed on 15 Jan 2023.

Destatis. (2023). *Tourism in Germany from 2012 to 2022*. https://www-genesis.destatis.de/genesis/online?operation=abruftabelleBearbeiten&levelindex=1&levelid=1678014672130&auswahlo

peration=abruftabelleAuspraegungAuswaehlen&auswahlverzeichnis=ordnungsstruktur&auswahlziel=werteabruf&code=45413-0004&auswahltext=&nummer=4&variable=4&name=RSDAU1&werteabruf=Werteabruf#abreadcrumb. Accessed on 05 Mar 2023.

DFB. (2012). *Fussball auf der Hochinsel: Nur Heimspiele auf Helgoland*. https://www.dfb.de/news/detail/fussball-auf-der-hochseeinsel-nur-heimspiele-auf-helgoland-32094/. Accessed on 08 Jan 2023.

FUR. (2023). *Kurzfassung der deutschen Reiseanalyse 2022*. Kiel.

Gibson, H. J., Kaplanidou, K., & Kang, S. J. (2012). Small-scale event sport tourism: A case study in sustainable tourism. *Sport Management Review, 15*(2), 160–170. https://doi.org/10.1016/j.smr.2011.08.013

Helgoland. (2023). *Helgoland service*. https://www.helgoland.de/service. Accessed on 03 Feb 2023.

Helgoland AirService. (n.d.). *Fiegen nach Helgoland: Sicher, schnell und bequem*. http://www.flughafen-helgoland.de/mitfliegen-nach-helgoland.html. Accessed on 08 Jan 23.

Helgoland Tourismus-Service. (n.d.-a). *Helgoland zum ersten Mal*. https://www.helgoland.de/so-ist-helgoland/helgoland-zum-ersten-mal/. Accessed on 05 Jan 2023.

Helgoland Tourismus-Service. (n.d.-b). *Geschichte*. https://www.helgoland.de/so-ist-helgoland/geschichte/. Accessed on 05 Jan 2023.

Helgoland Tourismus-Service. (n.d.-c). *Sport & spiel*. https://www.helgoland.de/freizeit-gestalten/sport-spiel/. Accessed on 08 Jan 2023.

Kletterwald Binzprora. (n.d.). *Abenteuerliche Freiheit in den Bäumen*. https://kletterwald-binzprora.de. Accessed on 15 Jan 23.

Kreulitsch, J. (2022). *Die 10 kleinsten Inseln Deutschlands—mit großen Geschichten*. https://www.reisereporter.de/artikel/15767-kleine-inseln-grosse-geschichten-deutschlands-kleinste-inseln. Accessed on 05 Jan 23.

Ma, S.-C., Egan, D., Rotherham, I., & Ma, S.-M. (2011). A framework for monitoring during the planning stage for a sports mega-event. *Journal of Sustainable Tourism, 19*(1), 79–96. https://doi.org/10.1080/09669582.2010.502576

McCool, S. (2015). Sustainable tourism: Guiding fiction, social trap or path to resilience. In *Challenges in tourism research* (Vol. 70, pp. 224-234). Bristol: Channel View Publications.

Meteostat. (2023a). *Helgoland*. https://meteostat.net/de/station/10015?t=2022-12-25/2023-01-01. Accessed on 08 Jan 2023.

Meteostat. (2023b). *Bergen auf Rügen*. https://meteostat.net/de/place/de/bergen-auf-rugen?s=10093&t=2022-12-25/2023-01-01. Accessed on 08 Jan 2023.

Nationalpark Wattenmeer. (2020). *Aktuelles zum Kitefurfen im Nationalpark*. https://www.nationalpark-wattenmeer.de/news/aktuelles-zum-kitesurfen-im-nationalpark/. Accessed on 10 Jan 2023.

NDR [Norddeutscher Rundfunk]. (2022). *Anreise nach Rügen*. https://www.ndr.de/ratgeber/reise/inseln/Anreise-nach-Ruegen,ruegen380.html. Accessed on 10 Jan 2023.

Preuß, O. (2021). *Helgoland neu entdecken*. https://www.welt.de/regionales/hamburg/article231438471/Tourismus-Helgoland-neu-entdecken.html. Accessed on 05 Jan 2023.

ProBoarding Rügen. (n.d.). *Kitesurfcamp*. https://www.proboarding.de/kitesurfen-camp/. Accessed on 14 Jan 2023.

Ross, P.-H. (1990). Beitrag zur Ingenieurgeologie Helgolands. *Die Küste, 49*, 71–80.

Ruegen. (2023a). *Insel Ruegen—10 Tipps für einen bewussten Ruegenurlaub*. https://www.ruegen.de/nachhaltig. Accessed on 03 Feb 2023.

Ruegen. (2023b). *Insel Ruegen—Regionale Produkte*. https://www.ruegen.de/ruegen-entdecken/regionale-produkte. Accessed on 05 Mar 2023.

Statistisches Amt Mecklenburg-Vorpommern. (2022). *Tourismus in Mecklenburg-Vorpommern*. https://www.laiv-mv.de/static/LAIV/Statistik/Dateien/Publikationen/G%20IV%20Tourismus%2c%20Gastgewerbe/G%20413/2021/G413%202021%2012.pdf. Accessed on 10 Jan 2023.

Streimikiene, D., Svagzdiene, B., Jasinskas, E., & Simanavicius, A. (2021). Sustainable tourism development and competitiveness: The systematic literature review. *Sustainable Development, 29*(1), 259–271.

Taks, M. (2013). Social sustainability of non-mega sport events in a global world. *EJSS. European Journal for Sport and Society, 10*(2), 121–141.

Tourismus Mecklenburg-Vorpommern. (2023). *Nachhaltigkeitscamp am 19.* Januar 2023 auf Rügen. https://tourismus.mv/artikel/nachhaltigkeitscamp-ruegen. Accessed on 05 Mar 2023.

Tourismusverband Rügen. (n.d.-a). *Zahlen & Fakten*. https://www.ruegen.de/ueber-ruegen/zahlen-fakten. Accessed on 08 Jan 2023.

Tourismusverband Rügen. (n.d.-b). *Termine 2019 im Überblick*. https://www.ruegen.de/sportveranstaltungen/. Accessed on 14 Jan 2023.

Zoll. (n.d.). *Geltungsbereich—Umsatzsteuerrechtliches Inland*. https://www.zoll.de/DE/Fachthemen/Steuern/Einfuhrumsatzsteuer/Geltungsbereich/geltungsbereich_node.html. Accessed on 05 Jan 23.

Chapter 11
Azores Archipelago: Sport Tourism Development Toward Sustainability

Francisco Silva and Ricardo Melo

Introduction

The Azores are an ultraperipheral archipelago located in the North Atlantic Ocean and one of the most economically depressed regions in Europe (European Commission, 2019). This autonomous region, belonging to the Portuguese Republic, has tourism as one of its main sources of income and job creation, along with the agri-food industry and fisheries. Recognizing the strategic importance of tourism for the archipelago, Azorean governmental bodies started to develop strategies and implement measures to enhance sustainable development through tourism. In 2019, Azores became the first archipelago in the world to be recognized as a sustainable tourist destination. Sport tourism, particularly nature sport tourism, has contributed to enhancing the image of the destination and to the growth of demand and, consequently, of an increasingly structured and specialized offer.

Despite all these developments, few studies have analyzed the sustainability of sport tourism in the territory, although some studies researched the demand, the offer, and the impacts of sport tourism products or services. Within the IRNIST research framework produced specifically for this project, outlining the contribution

F. Silva
Centre of Geographical Studies (IGOT), University of Lisbon, Lisbon, Portugal

Centre for Tourism Research, Development and Innovation (CiTUR), Coimbra, Portugal

Escola Superior de Hotelaria e Turismo do Estoril, Vila do Conde, Portugal

R. Melo (✉)
Polytechnic Institute of Coimbra – Coimbra Education School, Coimbra, Portugal

Centre for Research, Development and Innovation in Tourism (CITUR), Coimbra, Portugal

International Research Network In Sport Tourism (IRNIST), Lille, France

D. Van Rheenen et al. (eds.), *Sport Tourism, Island Territories and Sustainable Development*, Sports Economics, Management and Policy 24,
https://doi.org/10.1007/978-3-031-51705-1_11

of sport tourism for local sustainable development in several small island territories, the main goal of this chapter is to present an overview of sport tourism in the Azores archipelago and to analyze two of its emerging products: canyoning and scuba diving. This chapter presents results essentially from the analysis of secondary data, resulting from scientific publications, as well as technical documents prepared with the support of governmental bodies. For the two sport tourism products analyzed, some data was also collected through interviews with the companies that provide those services.

This chapter is structured into five sections, including this introduction. The next section presents a characterization of the Azores destination, including a geographical and sociodemographic analysis of the archipelago, followed by an analysis of the sustainability of tourism in the Azores. In Section "Sport Tourism in the Azores Archipelago", focus is given to sport tourism in the Azores, in particular the offer dimension, especially regarding natural sports spaces, services related to sports tourism activities in nature, and nature sports events on the island. In the final section, the fundamental issues presented in this chapter are discussed and some suggestions are presented for the key stakeholders involved in the tourism sector in the Azores, as well as some recommendations for the development of studies that will allow others to obtain more data on this important tourism segment in the Azores.

Geographic and Sociodemographic Characteristics of the Azores Archipelago

Azores is a Portuguese Autonomous Region located in the Macaronesia region of the North Atlantic Ocean. This small archipelago, with 2322 km^2 and approximately 240,000 inhabitants, is composed of nine volcanic islands (Fig. 11.1) emerging from the Mid-Atlantic Ridge, divided into three groups: (i) Eastern (Santa Maria and São Miguel); (ii) Central (Terceira, Graciosa, São Jorge, Pico, and Faial); and (iii) Western (Flores and Corvo).

This outermost region has a GDP per capita below 75% of the European Union average. This is largely a consequence of geography and economic structural problems arising from the development model, the limitations associated with the Portuguese economy, its small size, dispersion, isolation, limited resources, and accessibility problems (European Commission, 2019; Silva, 2022). The most important sectors of its economy are services (public administration, commerce, transport, and tourism), the agri-food sector (especially milk production and agribusiness), and fisheries (Azevedo, 2017).

Insularity is also reflected in the limitations in terms of accessibility that are increased by territorial dispersion. Except for some visitors who arrive in the territory by sailboat or cruise ship, all the others only have the option of using the plane. The cost of tickets and the reduced number of air connections have important constraints on the region's development, but the liberalization of airspace, as of 2015, has helped to alleviate this problem (Fraga, 2016).

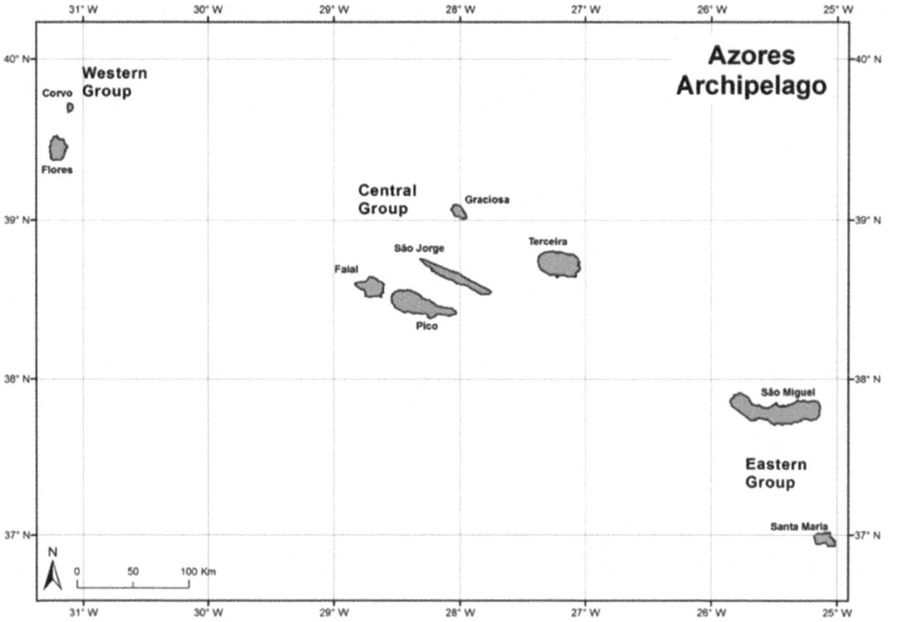

Fig. 11.1 The Azores archipelago

The region is also marked by a large disparity between the islands, with São Miguel and Terceira concentrating around 50% of the territory, 79% of the population, and 80% of regional GDP (Table 11.1).

Located in the middle latitudes, the Azores has a temperate climate, characterized by low thermal amplitudes, high humidity, and regular rainfall (Azevedo et al., 2004). The summer is relatively hot and sunny, and the sea has water temperatures above 20 °C, making it the most favorable period for tourism and nature sports.

In the Azorean landscape, the volcanic reliefs stand out, highlighting the green of the pasture fields and the blue of the omnipresent sea. In its remarkable geodiversity, numerous volcanoes abound, such as the Pico Mountain with 2351 meters of altitude, as well as numerous volcanic caves, basaltic flows, "caldeiras," lagoons, "fajãs," and cliffs with imposing waterfalls and lava fields.

The cultural identity of the Azores is another of the distinguishing features of this archipelago, expressed in the hospitality of its people, multiple religious manifestations, gastronomy, rural architecture, monuments, and landscape (Beier & Kramer, 2018). With this multitude of natural, historical, and cultural resources, initiatives that have led to its protection and preservation have been privileged for more than two decades (Fraga, 2016). The tourism sector has also benefited from these initiatives and a strategic plan has also been followed by the multiple agents involved.

Table 11.1 Area, population (2021) and GDP (2018) in the Azores

	Santa Maria	São Miguel	Terceira	Graciosa	São Jorge	Pico	Faial	Flores	Corvo	Total
Area (sq./km^2)	97,9	744,6	400,2	60,7	243,7	444,8	173,1	141,0	17,1	2322
Population	5414	133,390	53,311	4095	8381	13,895	14,356	3429	386	236,657
%area	4,2%	32,1%	17,2%	2,6%	10,5%	19,2%	7,5%	6,1%	0,7%	100%
%Population	2,29%	56,36%	22,53%	1,73%	3,54%	5,87%	6,07%	1,45%	0,16%	100%
%GDP	2.85%	58.11%	21.95%	1.44%	3.08%	4.85%	6.28%	1.22%	0.21%	100%

Sources: SREA (2023)

Sustainability of Tourism in Azores

Adopting the motto "Azores—towards sustainability," the territory's agents have promoted the continuous development of sustainability policies and initiatives in the region, which have been internationally recognized by organizations such as UNESCO, among others (see Table 11.2). These efforts led to the classification of almost 25% of its terrestrial territory as Protected Areas, integrated into the Natural Parks created in each of the nine islands of the Azores archipelago, and to the creation of 50 Marine Protected Areas (MPA), included in the Natural Parks of the Azores (e.g., coastal parks) and in the "Parque Marinho dos Açores" (e.g., oceanic paks) (Azores DMO, 2021).

The region boasts 36 areas classified under the Community Directives of the Natura 2000 Network and 10 under the Convention for the Protection of the Marine Environment of the Northeast Atlantic (OSPAR), representing about 4% of the Exclusive Economic Zone and Territorial Sea and 7% of the Extended Continental Shelf. Another nine small areas, designated fishing restriction areas, promote the conservation of resources and non-extractive maritime-tourist activities. The five underwater Archaeological Parks and the 82 coastal bathing waters found on all the islands constitute nature refuges with implications for the conservation of existing natural resources, with the objectives of preserving heritage and common enjoyment (Azores DMO, 2021).

In addition to nature conservation and biodiversity protection, sustainability policies have also been developed in terms of energy production and management, noise management, air quality, waste, and water, as well as carrying out actions of environmental promotion and education in appreciation of endogenous products, heritage, and Azorean culture (Azores DMO, 2021).

The worldwide recognition of the significant cultural and natural heritage value of the Azores is not a recent phenomenon, and the territory is one of the only two regions in the world that have the full classifications attributed by UNESCO, namely, "World Heritage," "Biosphere Reserve" and "Geopark," also having an international Ramsar classification, attributed to wetlands and to which UNESCO is associated. In 1983, UNESCO recognized the Historic Center of Angra do Heroísmo on Terceira Island as a World Heritage Site, and in 2004 the Vineyard Culture Landscape on Pico Island. Corvo Island and Graciosa Island (2007), Flores Island (2009), and São Jorge Fajãs (2016) also became part of the UNESCO World Network of Biosphere Reserves. In 2015, the Azores Geopark also became a UNESCO territory, reinforcing its value and role in the implementation of strategies for the conservation and enhancement of the Geological Heritage of the Azores (Azores DMO, 2021).

Marine diversity and the importance of the coast also occupy an important place in the international panorama and constitute a strategic resource, which the Autonomous Region of the Azores has invested in through its dissemination, management, and conservation. As a result, the Azores have obtained several international recognitions, being certified since 2014 with the Platinum Seal of the Quality

Table 11.2 Chronology of Azores sustainability policies and sustainable tourism awards

Year	Sustainability policies and sustainable tourism awards
2022	The Azores achieved Silver Level III certification as a Sustainable Tourist Destination by EarthCheck.
2021	European Best Destination awards the Azores as a safe destination for 2021 Distinguished by Green Destination in the "Communities & Culture" category Considered the "Best Adventure Destination in Europe" in the World Travel Awards
2020	Are once again part of the TOP 100 of Sustainable Destinations National Tourism Award: The Certification of the Azores wins the category of Sustainable Tourism
2019	Involvement in the BLUE Azores project Launch of LIFE Vidalia and LIFE IP Azores Natura Programs Azores with bronze status in EarthCheck tourist destination certification process *Azores become the first archipelago in the world as a sustainable tourist destination*
2018	Creation of Azores DMO EarthCheck benchmarking process begins to certify the Azores as a sustainable tourist destination First place in "Best of the Atlantic" category by Green Destinations
2017	Beginning of the Azores certification as a sustainable tourism destination (stakeholders' workshop) Congress: Azores, Towards Sustainability Creation of the Sustainability Charter of the Azores Top 100 of most Sustainable World Destinations by Green Destinations QualityCoast Platinum Award by QualityCoast—Coastal and Marine Union of the European Union
2016	Recognition of São Jorge Fajãs as a UNESCO Biosphere Reserve Top 100 of most Sustainable World Destinations by Green Destinations
2015	Azores Geopark became a UNESCO territory Implementation of the "Parque Escola" project
2014	Achievement of Coast Platinum Award
2013	Integration into the European and Global Geopark Networks Achievement of Coast Gold Award
2012	Achievement of Quality Coast Top 100 Award European Charter for Sustainable Tourism
2018	Creation of Azores DMO
2010	Creation of Geopark Azores Best of the Best—Nature Award by the European Commission Second Best Islands in the World for Sustainable Tourism by the National Geographic Traveler
2009	Recognition of Flores Island as UNESCO Biosphere Reserve
2008	Creation of Islands Natural Parks
2007	Recognition of Corvo and Graciosa Islands as UNESCO Biosphere Reserve
2004	UNESCO recognizes the Landscape of the Pico Vineyard Culture as a World Heritage Site
1989	Natura 2000 network creation
1988	Implementation of the Blue Flag Program in the Azores
1983	Recognition of Central Zone of the Town of Angra do Heroísmo as World Heritage site
1972	Creation of "Caldeira do Faial" and Pico Mountain Reserves

Sources: Visit Azores (2023) and Sustainable Azores (2023)

Coast program. The Azores were also nominated, by the European Best Destination, as one of the most beautiful destinations for whale watching in Europe. In 2019, the underwater archaeological map of the Azores was also considered by UNESCO as one of the five examples that represent the best practices for the protection of underwater cultural heritage (Azores DMO, 2021).

Other government initiatives have been implemented, of which, for example, the "LIFE IP AZORES NATURA—Active Protection and Integrated Management of the Natura 2000 Network" project stands out, which is the largest conservation project conceived for the Azores and which will be developed over the next several years (Azores DMO, 2021).

The endogenous resources of the region, combined with these policies and initiatives for the conservation and protection of nature, have also generated a tourism sustainability policy, especially from 2017 onwards, through the creation of the Azores Sustainability Charter. In the same year, the process of the Azores certification as a sustainable tourism destination began, which culminated in 2019 in recognition by EarthCheck Sustainable Destination Programme, an entity accredited by the Global Council for Sustainable Tourism, as the first archipelago in the world to achieve international certification of sustainable destination, making Azores one of the most sustainable tourism destinations in the world, currently shared by only 13 regions across the entire globe. As a result, in 2020, the Azores won the National Tourism Prize in the Sustainable Tourism category, for the certification work carried out by the destination. In 2021, already in a pandemic period, the Azores were recognized as the European Best Destination category as one of the safest destinations. The Azores were also awarded the "Best Adventure Destination in Europe" in the World Travel Awards (Azores DMO, 2021).

The Azores Tourist Destination Sustainability Management Structure (Azores DMO), created in 2018 by Government Council Resolution No. 58/2018 of May 15, 2018, and with a mandate renewed by Resolution No. 91/2021, of April 21, 2021, is the entity responsible for managing the sustainability of the destination under the Regional Secretariat with competence in matters of tourism. The Azores DMO is thus constituted as the structure mandated to plan, organize, implement, manage, and supervise the certification of the Azores as a Sustainable Tourist Destination, as well as to monitor the implementation of the strategy for the tourism sector in the Azores (Azores DMO, 2021).

The Azores tourism strategy was designed so that the archipelago asserts itself as a leading and reference tourist destination, which stands out for its sustainable performance, and allows a balance to be found between the environmental, economic, and sociocultural dimensions. Through consultation with the Destination Sustainability Monitoring Groups (consisting of representatives of the Regional Secretariats and Directorates), 10 priority Sustainable Development Goals (SDGs) emerged for the management of the destination, in line with the 2030 Agenda (Fig. 11.2). Thus, the development of tourism in the Azores must necessarily be based on a unique environmental offer, combined with energy sustainability and carbon neutrality, in a concerted strategy to safeguard natural resources, so that

Fig. 11.2 Prioritized sustainable development goals for tourism in Azores (Azores DMO, 2021)

tourist activity in the Azores archipelago is enhanced as a nature destination (Azores DMO, 2022).

Considering the sustainable development policies implemented in the Azores territory, sustainability key performance indicators (KPI) were accessed in 2021, which can be considered of clear importance both for the local population and for the tourism sector as well (Azores DMO, 2022). Examples of important environmental KPI are as follows:

- 35% of the electricity produced in the Azores comes from renewable sources.
- 99% of water is safe for consumption.
- 45 beaches (55%) of coastal bathing had a blue flag.
- Azores generate around 150 m (ton) of urban waste in 2021; 57% of the Azores' urban waste was recovered; 5 islands reach 100% recovery of waste.
- From 2018 to 2021, greenhouse gas emission per person decreased by 12%.

The creation of the sustainability strategy comes in a context in which tourism has increasingly asserted itself as a key sector for the development of the territory, despite the obvious asymmetries between the nine islands of the archipelago (see Table 11.3). The number of tourists in the archipelago has grown on all of the islands, showing in 2019 a record number of visitors (971516), of which 48.7% were international, in a total of 3,008,715 overnight stays, with a growth trend of 192% and 130%, respectively, for the year 2015 (SREA, 2023). The economic contribution of the sector is extremely relevant and growing, considering the evolution of the Gross Value Added (GVA) generated in 2015 (8.2%) and 2019 (12.9%), as well as the Gross Domestic Product (GDP) that has also increased from 15.0% in 2016 to 18.1% in 2018 (Azores DMO, 2022). Despite this growth trend dissipating in the pandemic period caused by COVID-19 (in 2020 and 2021), preliminary results for 2022 indicate a recovery to values similar to the pre-pandemic period: 906,484 tourists and 2,871,794 overnight stays (SREA, 2023).

In 2019, tourist intensity was around 12.4 overnight stays per resident, and tourist density stood at 1282.5 overnight stays per Km^2, values well below those of competing destinations, which shows that the Azores are a destination with tourist

Table 11.3 Azores tourism data in 2019

	Santa Maria	São Miguel	Terceira	Graciosa	São Jorge	Pico	Faial	Flores	Corvo	Total
Guests	16,456	623,909	143,523	7946	23,439	57,704	79,390	17,906	1243	971,516
% of national guests	74,9	48,2	60,3	89,7	49,3	49,3	45,1	60,2	61,6	51,3
Overnight stays	47,894	2,072,511	391,781	19,673	58,842	170,346	192,950	51,521	3197	3,008,715
Average stay	2,9	3,3	2,7	2,5	2,5	3,0	2,4	2,9	2,6	3,1
Tourism intensity[a]	8,5	15,1	7,1	4,7	7,1	12,5	13,3	14,2	6,9	12,4
Tourism density[b]	493,8	2730,6	972,2	317,3	239,2	382,8	1115,3	360,3	188,1	1282,5
Tourist saturation index[c]	293,0	454,4	260,4	188,8	283,2	422,7	546,7	493,6	267,9	400,1
Tourists per area (Km^2)	169,6	822,0	356,1	128,2	95,3	129,7	458,9	125,2	73,1	414,1

[a]Number of overnight stays per population
[b]Number of overnight stays per area (Km^2)
[c]Guests per inhabitants X100

pressure relatively low (SREA, 2023). Also compared to other competing destinations, such as Madeira and the Canary Islands, the weight of national guests in traditional hotels is much more expressive in the Azores (51.3% in 2019), as shown in Table 11.3.

In short, the cultural and natural resources of this region, combined with the security and hospitality of the Azoreans, have created conditions for visitation highly valued by the tourist demand, which has positioned the Azores as one of the most sustainable tourist destinations. Internationally, embodied in a portfolio of Portuguese products (cultural and scenic tourism, wildlife observation, geotourism, sports tourism, etc.), the archipelago has supported sustained and high-quality tourism development (Ponte et al., 2018).

However, excessive demand and particularly very rapid growth, as current data indicate, will likely result in a very significant increase in pressure on some resources and areas, making sustainable management difficult and may also have an impact on the quality of the tourist experience. As will be seen in the next section, of the wide range of tourism products in the Azores, sport tourism has great potential in terms of supply, particularly activities related to nature sports, through the use of the excellence of its natural resources.

Sport Tourism in the Azores Archipelago

Sport tourism, in particular nature sport tourism, is one of the most important segments of tourism in the Azores, although there are no known studies that specifically present their impacts (environmental, economic, and sociocultural) nor the size of their demand. In general, in the Azores, and responding to the different types of nature sport tourists (Melo et al., 2020), the sport tourism offer consists of:

1. Practice spaces where sports tourists can practice their sport independently, such as hiking or diving, for example
2. Services organized by companies, whose activities are guided by specialized technicians, aimed at sport tourists whose main motivation for going to the destination is to practice this activity, or who travel to the Azores for another reason but who take advantage of the offer to carry out these activities, such as canyoning or diving (Silva et al., 2022)
3. Organized events where participants participate to compete, in events such as trail running or cliff diving

With regard to nature sports spaces (Table 11.4), the Azores have a diverse network of spaces, especially canyoning (120), hiking (91), and diving (131). Also worthy of mention are the spaces for practicing water sports such as surfing (56), SUP (21), and windsurfing (7). Spaces for free flight practice (32) are also significant in the context of the region. There are also natural spaces where other activities take place but which are not yet properly referenced, such as trail running, caving, coasteering, kayaking, swimming in open waters, and horse riding, among others.

Table 11.4 Nature sports spaces in Azores

	Santa Maria	São Miguel	Terceira	Graciosa	São Jorge	Pico	Faial	Flores	Corvo	Total
Bike trails[a]	4	21	3	0	0	2	0	0	0	30
Canyoning[b]	8	22	3	0	31	0	13	43	0	120
Climbing zones[c]	1	3	6	0	2	3	0	1	0	16
Diving spots[a]	20	19	17	14	13	14	16	12	6	131
Free flight[d]	2	14	3	1	2	4	4	1	1	32
Hiking trails[a]	7	29	11	4	9	14	10	5	2	91
SUP spots[a]	1	8	3	2	2	1	1	2	1	21
Surfing spots[a]	2	17	8	6	9	5	4	5	0	56
Windsurf[a]	0	5	1	0	0	1	0	0	0	7

Sources:[a]Visit Azores (2023)
[b]Silva et al. (2022)
[c]Climbing Portugal (2022)
[d]Farias (2017)

Regarding the business offer of sport tourism activities, the analysis of the government database (Azores Government, 2023) carried out in 2022 showed that there were 260 companies registered with an offer of at least one of the different types of nature sports activities (Table 11.5). São Miguel is the island with the highest number of companies offering sport tourism (44.6%), followed by Pico (13.8%) and Terceira (13.1%). Regarding the type of activities, 70.4% of the companies offer land activities, 23.5% offer nautical activities and 6.9% offer both land and nautical activities. A more specific analysis of the activities offered by the companies (Table 11.6) shows that hiking and nature observation activities (73.5%) are the activity that most companies present in their offer, followed by diving (15.8%) and nautical activities, such as canoeing and windsurfing (15.8%).

In a broader context of activities related to adventure tourism, but which includes many of the activities considered in our study, Ponte et al. (2018) found that the number of tourist entertainment companies with this offer has been increasing since 2013 to 2018. For this reason, we can say that the business offer of sport tourism, particularly nature sports, has been growing in recent years.

Regarding sport tourism events, there are no statistical data or studies quantifying the number and growth rate of those organized events. However, the number of organized sport tourism events are growing, demonstrating that some of them have a long history. One of the most popular cases is related to trail running, where Azores Trail Run® stands out. This is a nonprofit organization that has been organizing Trail Running events in the Azores Islands since 2012, with the purpose of: (a) developing and spreading the practice of Trail Running in the Azores; (b) improving the local economy of the islands where the event takes place during low tourist seasons; (c) promoting the image of the Azorean Islands around the world as a destination for nature and for practicing outdoor sports of excellence; (d) promoting sports practice and health maintenance of the Azorean people (Azores Trail Run, 2023). As can be analyzed by the data provided on their site (Azores Trail Run, 2023), there are events held with organizers for more than a decade and with a growing number of annual events, on the most diverse islands, namely, the Columbus Trail (Santa Maria), the Whalers' Great Route Ultra-Trail (Faial), the Extreme West Adventure (Flores and Corvo), the Triangle Adventure (Pico, São Jorge and Faial islands); and the Golden Trail Championship (Faial and Pico). A study developed in 2018 about Triangle Adventure Trail (Valente, 2020) showed that this sporting event follows the principles of local sustainable development defined by the Azores regional Government and promotes the Azores as a sustainable tourism destination.

Another important international annual sporting event in Azores is the *Red Bull Cliff Diving World Series, an event organized* on São Miguel Island, more precisely in the cliffs of Vila Franca do Campo, since 2012. A study about the event organized in 2018 revealed that this event has an important economic impact, and most respondents consider it to be important for tourism and economic development, and a large proportion consider the event important for the social and cultural development of the region (Avelar et al., 2021).

Table 11.5 Sport tourism companies and typologies of activities offered global and by island

		Santa Maria	São Miguel	Terceira	Graciosa	São Jorge	Pico	Faial	Flores	Corvo	Total
Total ST companies	*n*	13	116	34	4	18	36	22	11	6	260
	%	5,0	44,6	13,1	1,5	6,9	13,8	8,5	4,2	2,3	100,0
Land ST activities	*n*	6	87	22	1	14	26	14	7	6	183
	%	46,2	75,0	64,7	25,0	77,8	72,2	63,6	63,6	100,0	70,4
Water ST activities	*n*	6	21	10	2	3	9	7	3	0	61
	%	46,2	18,1	29,4	50,0	16,7	25,0	31,8	27,3	0,0	23,5
Land and nautical ST activities	*n*	1	8	2	1	1	3	1	1	0	18
	%	7,7	6,9	5,9	25,0	5,6	8,3	4,5	9,1	0,0	6,9

Sources: Azores Government (2023)

Table 11.6 Main sport tourism activities offered by ST companies global and by island[a]

		Santa Maria	São Miguel	Terceira	Graciosa	São Jorge	Pico	Faial	Flores	Corvo	Total
Hiking and nature observation	*n*	7	90	23	2	15	25	15	8	6	191
	%	53,8	77,6	67,6	50,0	83,3	69,4	68,2	72,7	100,0	73,5
Caving and mountaineering	*n*	0	12	6	0	1	15	1	1	0	36
	%	0,0	10,3	17,6	0,0	5,6	41,7	4,5	9,1	0,0	13,8
Canyoning	*n*	1	9	1	0	2	2	1	4	0	20
	%	7,7	7,8	5,9	0,0	16,7	5,6	0,0	36,4	0,0	8,1
Free-flying and ballooning	*n*	0	1	1	0	1	1	0	0	0	4
	%	0,0	0,9	2,9	0,0	5,6	2,8	0,0	0,0	0,0	1,5
Nautical activities (kayaking, windsurfing, etc.)	*n*	0	22	7	0	1	7	3	1	0	41
	%	0,0	19,0	20,6	0,0	5,6	19,4	13,6	9,1	0,0	15,8
Diving activities	n	7	9	5	3	3	7	4	3	0	41
	%	53,8	7,8	14,7	75,0	16,7	19,4	18,2	27,3	0,0	15,8

Sources: Azores Government (2023)

[a]Companies that declare to present a certain activity in their offer does not mean that they develop it; sometimes the organization of this activity is subcontracted to another company

Surfing events have also increased in the last decade, as the example of the Surf Azores Pro, a stage of the World Surf League European Qualifying Series, as well as other national and regional competitions. Despite the increased number of organized events, especially trail running events, hiking, diving, and canyoning remain the three main sport tourism products in the Azores.

In the case of canyoning, after three international meetings (Canyoning International Meeting of Azores—CIMA) were organized between 2014 and 2016 on the islands of Flores and São Jorge, in 2023, there will be an event again on the island of Flores and continuity is expected in the following years.

Discussion and Conclusions

The Azores archipelago has been considered one of the most sustainable tourist destinations in the world, the result of a number of factors ranging from geography, environment, isolation, culture, history, economy, and human action, in particular in shaping the landscape. At the same time, it is also important to consider the efforts made by agents of the territory that have promoted the continuous development of sustainability policies and initiatives in the region, which culminated in obtaining the award attributed by the EarthCheck Sustainable Destination Program, as the first archipelago in the world to achieve international certification of sustainable destination.

The archipelago has also received other international awards that have catapulted it into the media sphere, namely, they were considered in 2021, at the World Travel Awards, as the "Best Adventure Destination in Europe." Other international reports and awards confirm the high levels of sustainability of the tourist activities carried out on all nine islands of the archipelago (Ponte et al., 2018). Furthermore, compared to other competing destinations, such as Madeira and the Canary Islands, tourism intensity is lower, although it has increased significantly in recent years (Ponte et al., 2018; Silva, 2022; SREA, 2023).

Remoteness, natural beauty, and relatively untouched nature allow the development of very diversified tourism products, which embrace the central concept of sustainable tourism (Couto et al., 2021; Santos et al., 2012; Silva & Almeida, 2018; Vieira et al., 2014). Sport tourism, associated with nature and adventure, is a structural tourism product in the Azores.

However, those who practice sports in nature in the region are particularly external visitors, with some sports in nature having little expression among the local community (Silva et al., 2022). According to these authors, in the years of the COVID-19 pandemic, there was an increase in practitioners from the local community, but there is no information on whether it was a temporary or structural situation.

According to the Azores Tourism Strategic and Marketing Plan (PEMTA) published in 2016, nature tourism is the most consolidated tourism product in the destination. In the same document (IPDT, 2016), it can be seen that this is also the "priority product" for regional tourism development, incorporating many activities identified as nature and adventure sports tourism, such as mountain biking and downhill, kayaking, hiking and trekking, horse riding, paragliding, climbing, and abseiling. In addition, it chooses nautical tourism as one of the fundamental "complementary products," including sport tourism activities in nature such as diving, surfing and bodyboarding, sailing, and windsurfing.

According to the analysis carried out, the Azores archipelago presents unique conditions for developing these activities, due to the commitment to structuring an offer of practice spaces, such as hiking and biking trails, canyoning, diving and surfing spots, take-off spaces, and landing for free flight, among other natural sports spaces (Silva, 2022; Visit Azores, 2022). Also in terms of the offer of specialized services in sports and adventure tourism, there has been an increase in the offer, both in water and on land activities (Ponte et al., 2018). On the other hand, the focus on organizing sporting events throughout the archipelago has also been growing, with many of these events reaching an international dimension, such as the Azores Triangle Adventure, a trail running competition held on the islands of the Triangle of Azores (Faial, Pico, and São Jorge), the "Rassemblement Internationale de Canyon" and the Red Bull Cliff Diving World Series.

Concerning the potential of each island in terms of nature sports tourism offer the following is highlighted: (a) Santa Maria is one of the Azorean islands with the most potential for scuba diving; (b) São Miguel is the main touristic destination of all Azorean archipelago, composed by important sport tourism natural spaces, namely, for MTB, canyoning, surfing, hiking, scuba diving, and free flight; (c) Terceira is the second touristic destination of the archipelago and its potential is mostly for scuba diving, hiking and coasteering; (d) Graciosa is an interesting island for scuba diving and hiking; (e) São Jorge is an island with great potential for canyoning, hiking and surfing; (f) with the highest mountain of Portugal, the trekking to climb Pico mountain is one of the most important sport tourism products of the Pico island (and of the archipelago), along with scuba diving, and hiking; (g) Faial is a very interesting island for scuba diving and hiking; (h) Flores island is the Azores archipelago hotspot for canyoning; and (i) Corvo island stands out as the best European spot for the observation of migratory birds. The nine islands of the Azores archipelago are very different geographically and, because of this, the nature sport tourism offer is also very different. As a whole, the nine islands present an enormous potential to develop nature sport tourism in an integrated way.

Unfortunately, there is no data on the profile and number of practitioners of sports and active tourism activities in nature in the Azores, which can analyze the importance of each of the sports in the region, what is its true dimension and the practice of these sports by the local community. However, as shown in the cases of canyoning and diving, external demand is very strong and has been growing in recent years (except for the pandemic years), contributing to reinforce the image of the destination and the economy of the region. The appreciation of these sports and tourism products results both from the increase in visitors and from the reinforcement in the offer of tourism and sports in nature.

There are several reasons for verifying a significant increase in visitors, among which stand out: the growing notoriety of the destination; the implementation of the new air transport model, open to low coast companies; and the existence of a growing number of tourists looking for exotic destinations, off the main circuits and with strong authenticity (Hall, & Williams, 2019; IPDT, 2016; Silva, 2022).

Despite these positive aspects, the destination needs to reinforce planning and development, particularly with regard to the areas of sport and nature tourism, namely, ensuring a safe, sustainable practice and greater involvement of the local community. The commitment to training, encouraging the practice of sports in nature at school sports level, ensuring the collection of data on demand, involving different stakeholders, defining best practices, and establishing load (carrying) capacities for the most sensitive areas are some of aspects to consider.

Considering the image and tourist resources of the archipelago, the enhancement of tourism and nature sports must be transversal to all islands, but at the same time must be considered a differentiation and a vision that allows the creation of clusters, contributing to tourists staying longer in the destination while visiting various islands. This fact is particularly relevant, as there is a strong concentration of tourist demand on the island of São Miguel (SREA, 2023). Furthermore, there are differentiated resources that can be leveraged so that within a more general brand (nature tourism), the more peripheral islands can become more attractive and capture specific market niches. Examples are the islands of Flores and São Jorge with canyoning and those of Santa Maria and Graciosa with diving.

The absence of studies and data on the dimension of demand for sport tourism activities in the Azores archipelago and that characterize and analyze their impacts (environmental, economic, and sociocultural), the profile of consumers, and the practice of sports in nature by the Azorean population, is a strong limitation to establish a more sustained analysis of the sport tourism sector in the Azores. Therefore, it is suggested that, in the future, studies be carried out on these important areas for the development of sport tourism.

References

Avelar, M., Sousa, Á., & Menezes, A. (2021). The impacts of sport tourism events: The Red Bull Cliff Diving World Series Case in the Azores Islands. In J. V. de Carvalho, Á. Rocha, P. Liberato, & A. Peña (Eds.), *Advances in tourism, technology and systems*. ICOTTS 2020. Smart Innovation, Systems and Technologies (Vol. 208). Springer. https://doi.org/10.1007/978-981-33-4256-9_6

Azevedo, F. (2017). *Estudo para a Comissão REGI - A situação económica, social e territorial dos Açores (Portugal)*. European Commission.

Azevedo, E., Rodrigues, M., & Fernandes, J. (2004). O clima dos Açores. In E. V. H. Forjaz (Ed.), *Atlas básico dos Açores* (pp. 25–58). OVGA - Observatório Vulcanológico e Geotérmico dos Açores.

Azores DMO. (2021). *2019–2030 action plan for the sustainability of the Azores tourist destination*. Azores Government.

Azores DMO. (2022). *Sustainability report 2022 Azores, sustainable tourist destination*. Azores Government.

Azores Government. (2023). *Empreendimentos & Atividades Turísticas*. Available at: http://www.azores.gov.pt/ext/drt-pa/listagem.aspx?ilhas=&atividades=9|11|8

Beier, R., & Kramer, J. (2018). A portrait of the Azores: From natural forces to cultural identity. In U. Kueppers & C. Beier (Eds.), *Volcanoes of the Azores. Active volcanoes of the world*. Springer. https://doi.org/10.1007/978-3-642-32226-6_2

Climbing Portugal. (2022). *Zonas de Escalada nos Açores*. Available at: http://climbingportugal.blogspot.com/search/label/A%C3%A7ores.

Couto, G., Castanho, R. A., Pimentel, P., Carvalho, C. B., & Sousa, Á. (2021). The potential of adventure tourism in the Azores: Focusing on the regional strategic planning. In A. Abreu, D. Liberato, E. A. González, & J. C. Garcia Ojeda (Eds.), *Advances in tourism, technology and systems. ICOTTS 2020. Smart innovation, systems and technologies* (Vol. 209). Springer. https://doi.org/10.1007/978-981-33-4260-6_2

European Commission. (2019). *Country report Portugal 2019. Including an In-Depth Review on the prevention and correction of macroeconomic imbalances*. European Commission.

Farias, P. (2017). *Azores Paragliding Guide*. Associação de Turismo dos Açores.

Fraga, V. (2016). Azores tourism: A development and marketing strategy. *Worldwide Hospitality and Tourism Themes, 8*(6), 705–710. https://doi.org/10.1108/WHATT-09-2016-0045

Hall, C. M., & Williams, A. M. (2019). *Tourism and innovation* (2nd ed.). Routledge.

IPDT. (2016). *Plano Estratégico e de Marketing do Turismo dos Açores*. Instituto de Planeamento e Desenvolvimento do Turismo.

Melo, R., Van Rheenen, D., & Gammon, S. J. (2020). Part I: Nature sports: A unifying concept. *Annals of Leisure Research, 23*(1), 1–18.

Ponte, J., Couto, G., Pimentel, P., & Oliveira, A. (2018). *Tourism & Management Studies, 14*(4), 25–38.

Santos, C., Couto, G., Pimentel, P., & Vieira, J. (2012). Quality of the Azores destination in the perspective of tourists. *Tourism and Hospitality Research, 12*(1), 32–42. https://doi.org/10.1177/1467358411429639

Silva, F. (2022). Inland and Island tourism, contrasting geographies touching each other: Azores, a peripheral archipelago. In G. Fernandes (Ed.), *Challenges and new opportunities for tourism in inland territories: Ecocultural resources and sustainable initiatives* (pp. 48–63). IGI Global. https://doi.org/10.4018/978-1-7998-7339-6.ch004

Silva, F., & Almeida, M. C. (2018). *Dossier produto: Canyoning Açores*. Oficina do Lazer Consultoria Lda. & Turismo dos Açores.

Silva, F., Lopes, T., & Silva, M. (2022). The resilience of tourism recreation companies in a pandemic context: The case of canyoning in the Azores. *Social Sciences, 11*, 558. https://doi.org/10.3390/socsci11120558

SREA. (2023). *Statistics Azores*. Serviço Regional de Estatística dos Açores. Available at: https://srea.azores.gov.pt/default.aspx?lang_id=2.

Sustainable Azores. (2023). *O percurso da sustentabilidade*. Available at: https://sustainable.azores.gov.pt/cronologiaa/

Valente, T. (2020). Azores Trail Run® as a sustainable vehicle for promotion and local development. *Cadernos de Geografia, 41*, 97–103. https://doi.org/10.14195/0871-1623_41_7

Vieira, J., Couto, G., Pimentel, P., Menezes, A., Moniz, A., & Sousa, F. (2014). The satisfaction of the Nordic tourist with the Azores as a destination. *Scandinavian Journal of Hospitality and Tourism, 13*(1), 59–73. https://doi.org/10.1080/15022250.2014.959806

Visit Azores. (2022). *Azores sustainable by nature*. Available at: https://www.visitazores.com/

Visit Azores. (2023). *Azores sustainable by nature*. Available at: https://www.visitazores.com/

Chapter 12
Sport Tourism in the Canary Islands: The Diversification of Mass Tourism

Manuel Ángel Santana-Turégano, Pablo Rodríguez-González, Carlos Calderon-Chouquet, and Ricardo Melo

Introduction

The Canary Islands are a volcanic archipelago of 7492 km^2, made up of eight islands and inhabited by 2.2 million inhabitants, with an average daily population of tourists that can be estimated at around 0.3 million. The archipelago is situated 100 km west of the African coast, at the south of Morocco, almost 1300 km southwest of the closest point of the Spanish mainland. Being one of the 17 "Comunidades Autonomas" (Autonomous Communities), a territorial similar to a Federated State that makes up Spain, they are one of the nine European Union (EU) "Outermost Regions," the largest considering its population and economic importance and enjoying a special status both within Spain and the European Union (EU).

The Canary Islands belong to the Eurozone, sharing currency with more than 340 million citizens from 19 states, and to the Schengen Area that allows almost 420 million citizens to move freely. However, unlike the Portuguese Autonomous regions of Madeira and Azores, and just as the French outermost regions (Reunion, Guyana, Martinique, Guadeloupe, Mayotte, and Saint-Martin), the archipelago does not belong to the common European VAT area. As all outermost regions of the EU, they enjoy some specific measures, such as lower taxes that help them address

M. Á. Santana-Turégano · P. Rodríguez-González · C. Calderon-Chouquet
University of La Laguna, Canary Islands, Spain

R. Melo (✉)
Polytechnic Institute of Coimbra – Coimbra Education School, Coimbra, Portugal

Centre for Research, Development and Innovation in Tourism (CITUR), Coimbra, Portugal

International Research Network In Sport Tourism (IRNIST), Lille, France

D. Van Rheenen et al. (eds.), *Sport Tourism, Island Territories and Sustainable Development*, Sports Economics, Management and Policy 24,
https://doi.org/10.1007/978-3-031-51705-1_12

the major challenges they face due to their remoteness and insularity among other issues (see European Commission, 2021).

After the islands were colonized by the Crown of Castile in the fifteenth century, a plantation economy was developed. First, sugar was the main product, with an important number of slave workers coming from Africa, traders coming from Europe (Flanders and Genoa, mainly) and settlers arriving mainly from Castile, Portugal, and other parts of Europe. In the seventeenth century, wine substituted sugar as the main crop until the late nineteenth century. From the late nineteenth until the mid-twentieth century, the exportation of bananas and tomatoes, combined with the specialization in maritime supplies, mostly to British steamships that needed a base in the mid-Atlantic, was the base of the island's economy, and this constant flux of people and commodities between the Canary Islands and northern Europe (mostly, with Britain) set the basis for the next stage (Macías-Hernández, 2011).

Since the 1970s, and even though bananas, tomatoes, other crops, and maritime supplies continue to play a role, the economy of the Canary Islands has been dependent upon tourism. According to official data (see, for instance, Hernández Martín, 2011), tourism accounts for more than one-third of the island's gross domestic product (GDP) and employment. From the point of view of the tourism industry, the development of tourism in the Canary Islands is a success story. The Canary Islands are the number one region in tourism overnight stays (Kotzeva et al., 2019), surpassing such touristic hotspots as the Parisian region, the Balearic Islands, Catalonia, or the Croatian Adriatic Coast, mostly due to the fact that while in the rest of the European Regions, tourism is a highly seasonal phenomenon, in the Canary Islands, it is quite evenly distributed throughout the year, making it very interesting for European tourism companies from the point of view of the profitability of the investment.

Even though after the impact of the COVID-19 pandemic, figures have needed some time to recover, it can be estimated that more than 15 million tourists visit the islands every year, producing more than 100 million overnights stay in the almost 400,000 beds that make up the accommodation capacity of the islands (ISTAC, 2021b). Tourists come mostly from Britain, Germany, the Nordic countries, and the rest of Europe and Spain, even though there are significant differences between the main touristic islands of Tenerife, Gran Canaria, Lanzarote, and Fuerteventura. While Britons account for almost half of foreign tourists in Tenerife and Lanzarote, Germans play a more significant role in the tourism of Fuerteventura and Gran Canaria, an island where Nordic tourists, especially in the winter season, are very important (Promotur, 2022: 4).

However, if tourism in the Canary Islands can be considered a success from the point of view of the tourism industry, the balance must be more nuanced from the local population's point of view. With an average annual per capita GDP of around 21,000 US$ for 2019 (INE, 2022a), the islands are one of the poorest regions in Spain, and wages, income inequality, and other indicators usually considered to measure socio-economic performance suggest that the high specialization of the Canary Island's economy in tourism has not been so beneficial. Even though counterfactual reasoning is always a complicated issue, other Spanish regions with a high degree of specialization in tourism, such as the Balearics or Catalonia, enjoy higher levels of well-being. And even though it might also be argued that, coming from a plantation economy the islands have not performed poorly, it should always be considered that it is not only the specialization in a certain sector of the economy that determines the level of well-being a society can achieve, but rather the way in which that specialization is done (Santana Turégano, 2006). In that vein, prior to analyzing the development of sport tourism in the Canary Islands, we should consider the intended and unintended goals of this modality of tourism in the islands. This includes the role it plays in the tourism system and in the tourism strategies that are developed by the multiple stakeholders involved in the governance of tourism in the Canary Islands, both private and public.

Tourism Development in the Canary Islands

The beginning of tourism in the Canary Islands dates to the beginning of the twentieth century when a small number of affluent European travelers began to arrive. However, it was in the 1960s when mass tourism really materialized in the islands, first in the northern parts of Tenerife and Gran Canaria (around Puerto de la Cruz and Playa de Las Canteras, respectively). Tourists from the Nordic countries and Germany played an important role in this first wave of development. In the 1970s, tourism development moved to the South of these two islands, and to Fuerteventura and Lanzarote, where sun is much more frequent, following the standard "sea and sun model." By the 1980s, mass tourism was already consolidated and had become a very important sector in the island's economy (see Domínguez-Mujica et al., 2011). Development continued in the 1990s, and while concerns about stagnation, following Butler's "*life-cycle model of tourism destinations*" (Butler, 2006), already started to appear, growth continued, although with some crisis, until the new millennium.

By the 1990s, the tourist development of the smaller western islands La Palma, La Gomera, and El Hierro began, following a model oriented toward nature and rural tourism, inspired by the European Union framework on rural development, synthesized in the term Green Islands, as opposed to the previously developed tourist islands. Because of their geographical situation, air transportation has always been a key issue in the Canary Islands' tourism development. Since the beginning of mass tourism, major European Tour operators, such as TUI, for German-speaking countries, Thomas Cook and Thomson for Great Britain and Ireland, and Ving or Fritids Resor for the Scandinavian markets, among others, have played a major role in the transportation of tourists to the Canary Islands; for decades, their interest in the destination has made it possible for millions of Europeans to reach the islands in a convenient way. Tour Operators also invested directly, in many cases via loans with local hoteliers, in the development of the tourism infrastructure, but the ownership of the accommodation capacity remained for the bigger part in the hands of local small and medium investors, in many cases owners of second homes that were rented to tourists during the winter season (see, for instance, Santana Turégano, 2003, for the case of Gran Canaria).

Due in part to a 1973 law that promoted the acquisition of second homes in the Canary Islands by German citizens, since it was considered "aid development" and therefore entitled to tax reduction, the investment of German private citizens was also important in this phase of the Island's development (see Domínguez-Mujica et al., 2011). It is also worth mentioning that during the first stages of tourism development, the winter season was not only the high season, as it still is, but foreign tourism was mostly a winter phenomenon. Up until the third decade of the twenty-first century, there are areas that are used as vacation spots for locals during summer and as tourism areas, welcoming tourists mainly from the Nordic countries, during winter. And this phenomenon has been facilitated by the social distribution of the ownership of the accommodation capacity, mainly (but not only) of apartments (Simancas Cruz, 2011; Simancas Cruz & Mañoso Valderrama, 2016).

In this first tourism model of the Canary Islands, two subsectors coexisted. On the one hand, a non-hotel sector emerged, comprised of many small owners (one or two apartments, in many cases), many of them locals, that rented apartments, bungalows, or villas, to tourists. These transactions were usually managed via the mediation of tour operators, who usually facilitated the arrival of tourists through their charter airlines, even though it was quite common that a firm, providing cleaning and reception services (sociedad de explotación), mediated these relationships. On the other hand, in the "hotel sector," the booking departments sold their accommodation capacity to tour operators who sold them, either directly or via a travel agent, to the final client, comprised mostly of middle-class Northern Europeans (including Britons) who sought to take a break from their harsh winters (see Domínguez-Mujica et al., 2011).

By the late 1980s and early 1990s, the fact that in many apartment buildings, tourists coexisted with permanent residents and the lack of professionalism of many of these "extra hotel accommodations""was seen as a weakness of the island's tourism model. Therefore, first by the *Ley de Turismo de Canarias de 1995* (Canarian Tourism Law of 1995) and then by the *Leyes de Moratoria Turística* de 2011 (Rodríguez González & Santana Turégano, 2012; Simancas Cruz, 2015), regulations were passed to address those weaknesses. The diagnosis considered that the future of tourism in the Canary Islands should promote the construction of quality hotels and make efforts to turn apartments into hotels.

It is in this context that the idea of sport tourism first appears as a key issue. In the first decade of the twenty-first century, a "moratorium" on building (Bianchi 2010; Hernández-Martín et al., 2015; Inchausti-Sintes & Voltes-Dorta, 2020) only allowed new tourism accommodations in certain cases. One of the accepted ways to build new construction was to include a golf course in the planned development. By then, the development of certain types of sport tourism, mainly golf and sailing tourism, was seen as a tool for the rejuvenation of the destination and to avoid stagnation following Butler's model (Agarwal, 2002).

However, it is very important to consider the major changes that had by then occurred, both in the socio-economic phenomena underpinning tourism behavior in the tourism industry at large and in other tourism destinations. By the mid- to late 1990s, both academic and tourism professionals were speaking of the change from the so-called "fordist tourism" to the "post fordist tourism" (Ioannides & Debbage, 1997). On the one hand, the development of the "low-cost airlines" proved to be both a challenge and an opportunity for tourist destinations such as the Canary Islands, where the practical totality of tourists arrive by airplane. In the previous model, tour operators played a key role in transporting tourists. Following vertical integration strategies, they had either directly bought tourism capacity or established relations with some hotels via special contracts. In the multiple changes that followed European integration, the tourism sector in the EU changed completely between 1990 and 2010. Particularly, some of the dominant players in the field of traditional tour operations in Europe became involved in the development of hotels in the Canary Islands.[1]

If up until then it had traditionally been said that tour operators play a key role in tourism development and that they can divert tourism flows to where they can extract more profit (Britton, 1982), the new situation seemed like a double-edged sword. On the one hand, establishing strategic alliances with the main players in the European Tourism industry could guarantee a constant flow of tourists to the islands, at least in the standard 1-week flight and hotel tourist package. After almost 20 years since the beginning of the model of low-cost air transportation to the Canary Islands, it can be said that this model was quite successful, primarily because of the low seasonality of tourism in the Canary Islands.

On the other hand, considering how the boom of low-cost airlines has changed the sector, if one compares the Canary Islands to other destinations such as *Catalonia, Comunidad Valenciana, Costa del Sol,* or the Balearic Islands, the island's story has not been one of success. The emergence of new means of transportation has allowed new mobilities to thrive, like British or German doctors that commute weekly between their countries and places such as Alicante, Palma de Mallorca, or Marbella, lifestyle migrations (Huete et al., 2013) that contribute to blur the frontiers between tourism and other forms of travel and leisure and that certainly make the case for Urry's mobilities' paradigm. It is in this context that the development of sport tourism must be contemplated if we truly seek to understand its development.

Sport Tourism in the Canary Islands

According to data provided by the Canary Islands' Tourism official board, it can be estimated that approximately 1.3 million people, or 8.0% of the tourists that visit the islands, mention sport practice as a motive to visit the Canary Islands and can

[1] To name just a few, see the complex relationship between Tui, its own hotels brands (they currently operate several resorts under their own brand in the Canaries), and the special relationship they have had with hotel chains such as RIU and or IFA-Lopesan.

therefore be considered "sport tourists."[2] It is much harder, however, to provide an estimate of the number of "sport tourists," following Higham's (2005) terminology, in which the term "sports tourism" is used to refer to those travels in which "sports," either practicing them or assisting as a spectator to main tourism events, is the main motivation for traveling. This fact can be attributed to the multifactorial character of motivation in travel: People decide to choose a certain destination for their holidays for several reasons, and it is only in a small number of cases that a single motivation can be pointed out as the sole reason for traveling to a certain place.

For example, in the currently so-called "VFR (Visiting Friends and Relatives) Tourism," if someone's friend or relative is getting married in one of the Canary Islands, that is the reason behind the choice of destination (Fuerteventura and not Lanzarote, for instance). This can also be the case when attending, both as a participant or as a spectator, to a tourism event: Some people might want to visit La Palma and not Tenerife, for instance, because they want to run the Transvulcania Trail Race (and not Tenerife Blue Trail). But even in those cases, the decision-making process is always a complex one (Bargeman & van der Poel, 2006). In other cases, as, for instance, in ski tourism, motivations related to the sport can be the main driver in choosing a destination, but, even in those cases, other motivations (the *après-ski*) are very important. All these considerations have nothing to do with the importance of "sports" within the tourism sector but with the necessity to reconsider, following the mobilities' paradigm (Sheller & Urry, 2016; Urry, 2012, 2016), the definition of a whole industry on the motivation for traveling (tourism means travel without a labor motivation).

Analysis of the Tourism Sport Offer of the Canary Islands

In this broad concept in which the sport tourism sector caters to a wide range of "sport involvement," for instance, from dedicated windsurfers who choose the Canary Islands mainly to practice their sport to more traditional tourists that decide to give a try at this sport while vacationing at the Canary Islands, the sector of Sport Tourism comprises more than 1400 firms (Gobierno de Canarias, 2022c), which create more than 12,500 jobs and a total revenue that can be estimated, for the year 2017, of 1836,89,200 US$. It is important to bear in mind that both tourism and tourism communication in the Canary Islands are based on the weather, one of the main reasons for visiting the islands. In the last 25 years, a key theme of tourism's communication has been "*come and enjoy outdoor activities at a time of the year where you cannot enjoy them at home.*" It is no wonder that active sport tourism has

[2] It is worth mentioning that this estimation was provided by the tourism board using the responses to the open question "Indicate the three most important aspects for which you decided to choose the Canary Islands as a destination for this trip" in the tourist expenditure survey (ISTAC, 2022), considering sport tourists those who mention at least one sport in these aspects. Unfortunately, the methodology was changed in 2018, precluding this method of estimation (PROMOTUR, 2018).

been highlighted: Visitors to the Canary Islands can enjoy outdoor tennis, football, swimming, running, cycling, or hiking, among other activities, whereas in Europe's winter season, these activities can, at most, be endured.

As previously mentioned, the first development of a sport tourism strategy in the Canary Islands, appeared in the late 1990s to first 2000s, connected to both golf and sailing. As several authors have noted (Martín & Martín, 2000), the construction sector has always played a major role through lobbying in the islands' economy and politics. Therefore, the development of golf courses and marinas can be seen partly as a way to keep the machinery of the building industry running and not only from the suitability of the investment (Domínguez-Mujica et al., 2011; González Ramallal et al., 2010). By the 2010s, it became clear that the islands were ill-equipped to compete against other major tourism destinations, many of them within Spain. The sector of sail cruises has developed significantly in the last 25 years in places such as Croatia and the Balearic Islands. It has failed to do so in the Canary Islands quite simply because, even though weather conditions are good, sea conditions are often quite harsh, and therefore, even sailing from one island to another can become a challenge, best suited for experienced sailors.

Despite this, a small sector of so-called "nautical tourism" has developed in the Canaries too. Currently, there are several marinas, with several thousand moorings in the islands, which cater mostly to tourists that enjoy short 1-day trips (Gobierno de Canarias, 2022b). There are also events, like the Atlantic Rally for Cruises (ARC), which cater to more experienced sailors. As for the case of the golf courses, there are 24 in the islands, their main attraction being the good weather (Gobierno de Canarias, 2022a). However, the land and water supply scarcity in the Canary Islands make these islands lag well behind other Spanish *Comunidades Autónomas*, such as Andalucía as a golf tourism destination (Hudson & Hudson, 2014; Markwick, 2000). Therefore, and spite of some initially ill-conceived expectations about the development of certain types of "sport tourism," the Canary Islands booming sector can be summarized around two main sectors, "sea" and "land" sports. The sports that have been most developed are those that benefit more from the natural conditions of the islands.

Analysis of the Sport Tourism Demand to the Canary Islands

To provide an overview of sport tourism in the Canary Islands from the demand side, we will now present the results of the statistical analysis performed on the Data of the *Encuesta del Gasto Turístico de 2019* (Tourist Expenditure Survey) provided by ISTAC (Canary Islands Statistics Institute), a survey with a significant sample size ($n = 38,761$ tourists over 16 years old) (for methodological notes and microdata downloaded, ISTAC, 2022). According to this survey, 91.1% of tourists interviewed stated that the main purpose of their travel was "holiday, leisure and entertainment," and only 0.6% declared that their main reason for visiting the Canary Islands was "sports training." When asked about why they had chosen these islands, 54.0% of tourists declared "rest," 17.3% "explore the destination," and 14.0% "enjoy family time." Only 1.8% of tourists interviewed declared "practice my hobbies," among them sports practice. However, if we focus not on the reasons for traveling, but on activities tourists enjoy while on vacation, it can be estimated that 13.4% of tourists visiting the Canary Islands engage in sports activities, a figure that is certainly well below the sport participation index at home. Even if we add those who state that they have enjoyed sea (10.0%) and nature activities (10.3%) while enjoying holidays in the Canary Islands, the sport participation index of tourists could then be estimated to be around 25%, still quite low by European standards (European Commission, 2018).

Following Higham's (2005) differentiation between "sport tourism" as the activity pursued by those people whose motivation for traveling is mostly related to sports and "tourism sports" and the sport activities practiced by people who, for some other reason, had decided to travel, the estimation for the Canaries are as follows. Out of the almost 15 million tourists the islands received every year before 2020, around 3.2 (13.0%) practice sports while in the Canaries. Of these, only around 5.7% of them (less than 400,000 people) can be considered truly "sport tourists" since their motivation was mainly sports-related. The rest, around 2.8 million people, can be considered participants in "tourism sports": They came to the islands for other reasons than practicing sports or attending sporting events, but nevertheless, they engaged in sport activities on the islands.

As hypothesized by the "fordist tourism" approach (Ioannides & Debbage, 1997), many tourists suspend their everyday sports activities when they travel. Even though the "sports participation index" in the countries that make the bulk of tourism demand in the Canary Islands is around 40–50% (European Commission, 2018), the sport participation index among tourists in the Canary Islands is much lower, around 25%. These figures should be considered to fully understand the development of sport tourism in island destinations. Even though the effort of many Destination Management Organizations (DMO) is often focused on developing sport tourism as a tool for promoting a more endogenous and sustainable tourism, our data suggest that in mature destinations, the development of sport tourism has more to do with offering a wider range of experiences, spaces, and alternatives for a *soft* sport tourist, for whom the possibility to enjoy some degree of physical activity while on vacation is a trait that makes a tourism destination more attractive, but not the sole reason for choosing it.

According to our data for the Canary Islands, it might be the case that, in the long term, sport tourism can only survive inside an active tourism ecosystem connected to a massive tourism node.[3] Differences in the sports participation index of tourists in the different Canary Islands provide an illustration of this idea. As previously noted, the tourism development of the different islands of the archipelago took place in different eras. Tenerife and Gran Canaria are two islands that can be considered

[3]This seems to be the case also in other cases, like, for instance, the development of Cycling Tourism in Girona-Costa Brava or the Balearic Islands.

as mature, consolidated, tourism destinations, with between 3 and 6 million tourists per year. Lanzarote (2.5 million tourists) and Fuerteventura (1.7 million tourists) started their tourism development later (in the 1980s and 1990s), while tourism in the island of La Palma has never reached mass scale, and the island receives less than 0.3 million tourists per year.[4] Data provided in Table 12.1 suggest that the rate of active tourists in a destination decreases as it grows and becomes mature. In the Canary Islands, Gran Canaria, Tenerife, and Lanzarote, more mature destinations, have lower rates of "active tourists" (tourists that engage in sports activities while on vacation) than Fuerteventura and La Palma, where tourism development is more recent. Since the situation of sport tourism in these two islands has some interesting features, we will elaborate on them in the next section.

Sport Tourism in Fuerteventura and La Palma

In 1990, the island of La Palma had 82,131 inhabitants and received only 40,157 tourists per year, while Fuerteventura, with half that population, already received 380.951 tourists annually (Hernández Luis et al., 2017). Even though some enclaves of traditional, sea and sun mass tourism had been developed in both islands in the late 1960s and the beginning of the 1970s, the scarce development of some key resources, such as airports, roads, water desalination plants, and human capital, have hindered for decades the development of tourism and prevented it to reach mass scale. In the 1990s, after Spain joined in 1986 the EU (by then, European Economic Community), the Canary Islands became a major beneficiary of the European Regional Development Fund. The millions of funds received, both private and public, promoted an unprecedented socio-economic transformation of the archipelago (Macías Hernández, 2011; Sosvilla Rivero et al., 2010).

In the case of Fuerteventura, quantitative indicators suggest that this transformation was a success. Population in 2019 trebled that of 1990, reaching 116,886 inhabitants, while the number of tourists grew by almost 500%, reaching 1.8 million tourists per year (ISTAC, 2021b, 2022).[5] This massive growth was based on the increase of the accommodation capacity (60.236 bed spaces by 2022, 58% of them four- and five-star hotels) through the development of up-market seaside resorts, in most cases in places with no prior human use (at most, in some cases, the land was used for extensive livestock farming). As the island is closest to the African coast, Fuerteventura meets some of the main criteria for the development of sea and sun tourism in which the Canary Islands had already been specialized by the 1990s: long, white sandy beaches and sunny and dry weather throughout the year. However,

[4] The existing sources do not provide data on tourism development for the islands of El Hierro, La Gomera and La Graciosa. Data of tourism Offer in these islands suggest, nevertheless, that the scale of tourism development is even lower (ISTAC, 2022).

[5] In 2016 and 2017, more than 2 million tourists visited Fuerteventura.

Table 12.1 Activities in the Canary Islands

		Island of stay											
		Lanzarote		Fuerteventura		Gran Canaria		Tenerife		La Palma		Total	
		%	Tourists	%	Tourists	%	Tourists	%	Tourists	%	Tourists	%	Tourists
Activities in the Canary Islands (multiple choice question)	Sports	14,4%	362,681	17,5%	291,111	12,4%	457,143	12,4%	622,829	14,6%	34,410	13,4%	1,768,173
	Nature	8,6%	216,977	9,0%	149,233	8,6%	319,870	11,6%	583,218	35,1%	82,761	10,3%	1,352,059
	Sea	9,9%	248,762	12,3%	204,431	9,4%	349,317	9,4%	472,466	15,0%	35,277	10,0%	1,310,252
	No activities practiced	75,6%	1,906,729	72,3%	1,199,777	77,0%	2,849,070	75,2%	3,791,039	56,0%	131,810	75,1%	9,878,425
	Practice activities	24,4%	614,939	27,7%	459,339	23,0%	849,058	24,8%	1,249,343	44,0%	103,749	24,9%	3,276,427
	Total	100,0%	2,521,668	100,0%	1,659,115	100,0%	3,698,127	100,0%	5,040,382	100,0%	235,559	100,0%	13,154,852

this geographical situation also means strong winds and maritime currents, making many of the beaches not as attractive for the traditional sea and sun tourism.

On the contrary, the very little rainfall means that water supply for both tourists and local populations requires costly (both economically and environmentally) water desalination plants (Baños González et al., 2019). Because of all of these factors, during the first two decades of tourism development in the Canary Islands, Fuerteventura could not attract investments like those that started tourism development in Tenerife and Gran Canaria. However, during the 1990s, a series of improvements in the infrastructure of Fuerteventura (water desalination plants, airport capacity, roads, etc.) and the increasing barriers to entry in the main tourist areas of Gran Canaria and Tenerife directed significant investments in the tourism sector to the island; these investments have helped to expand the current accommodation capacity of Fuerteventura.

It is within this context that the orientation of Fuerteventura toward sport tourism (water sports tourism, mainly) can be seen as a case of making a virtue out of a necessity. Strong winds and sea currents make many of Fuerteventura's best beaches less suited for a more traditional family sea and sand tourism, but ideal for windsurfing, surfing, and other water sports, and therefore, for types of tourism based on these sports. As these sports gained popularity in European countries, tourists who traveled with the main goal of practicing them arrived in many parts of the island of Fuerteventura and a wide variety of firms who catered to them, such as wind, kite and surf schools, shops, and rental businesses or thematized bar and restaurants developed in places along the islands. This organic development of water sports tourism was also helped by the support given by local authorities, with a significant effort to hold events and championships. Since 1986, Fuerteventura hosts one of the best windsurfing competitions on the Professional Windsurfing Assocation's (PWA's) World Tour (Egli, 2022; *SurferToday.com*, 2022).

According to the exploitation of the data provided by the Encuesta del Gasto Turístico 2019, 27.0% of tourists visiting Fuerteventura, a total of 459,339 tourists, engage in sports activities. Therefore, it could be said that water sports tourism has not even remotely come to constitute the main tourist product of the island. This is in large part due to the fact the insertion of the island in the international tourist markets continues to be closely linked to the large intermediaries and their standardized formulas: Most tourists arrive in Fuerteventura through intermediaries (tour operators, travel agencies or the like provide 70.0% of flight reservations and 77.5% of accommodation reservations) and more than half (57.1%) of them purchase all-inclusive hotel packages. Sport tourism and activities at sea appear here as one more element in a process of "*complexing the sun and beach through the incorporation of a diverse set of natural and cultural activities and products*" (Marrero Rodriguez & Santana Talavera, 2018, p. 291). These authors have studied the sector of maritime-recreational activities on the island, highlighting among its features the informality that accompanies the sector due in part to the personal trajectories of the entrepreneurs, who go from being amateur practitioners to tourism entrepreneurs, and in part to the ambiguity of the regulatory framework of the activities and the scarce activity of inspection and sanction. They also point out that most of these companies are owned by foreigners, frequently from the same countries that send tourists to the island, who came to Fuerteventura as tourists to practice their hobby and settled there taking advantage of their language skills and their contacts with the circuits of their countries of origin.

In the case of La Palma, the changes during this period have not been so spectacular as those of Fuerteventura. The population of the island, 82,671 inhabitants in 2019 (ISTAC, 2021a), has remained practically the same as it was in the 1990s, even though currently it shows symptoms of aging and other problems associated with the decline of rural areas. The growth in the tourist's arrival, even if it has been important (currently, 230,648 tourists), lags well behind from the significant figures of the tourist islands. The island's accommodation offer stood at 7539 beds in 2019, 48.8% of them four- or five-star hotel beds, showing marked stability (or stagnation) during the period in which the hotel boom occurred in Fuerteventura, but in the island of La Palma, the accommodation offer in rural houses has a certain role, which was used in 2019 by 9.2% of the tourists who chose La Palma compared with 5.9% in the Canary Islands.

As in the case of Fuerteventura, in La Palma, the geomorphological and climatic specificities of the island have influenced tourism development. Being the westernmost and youngest among the Canaries, it has an abrupt orography and the biggest rainfall in the archipelago. The greater availability of water has favored export agriculture, mainly bananas and other tropical crops such as avocado, and therefore agricultural employment on the island maintains a substantially higher weight (9.0%) than on all the islands (2.7%). The dynamism of the island's agriculture, sustained and promoted via direct and indirect public support, has hindered the development of tourism in La Palma. Unlike what happens in other Canary Islands, in La Palma, the most appropriate coastal soils for tourism development are also the most productive and profitable for agricultural production. This fact has hindered, even from the insular regulations on land use, the development of accommodation and tourist areas. Likewise, various public initiatives for rural development have favored the appearance of rural accommodation, in rural areas in the interior of the island, on or between farms, with highly fragmented ownership.

By 2021, the rural accommodation houses in La Palma were 1974, with a significant degree of dispersion throughout the island territory (INE, 2022b). The orientation of the tourism development of the island toward rural and nature tourism has been based on the agricultural and sparsely urbanized landscape of the island, as well as on the uniqueness of its natural spaces; La Palma has a National Park (Caldera de Taburiente) and the island as a whole has been declared a UNESCO Biosphere Reserve. This orientation has entailed a considerable investment in the recovery, conditioning, signaling, and maintenance of a network of trails based on the island's traditional communications network. La Palma's network of trails crosses the different natural spaces and currently comprises 948 km in 86 marked trails of variable length and difficulty, and there are also seven mountain bike routes totaling another 120.9 km (Cabildo de La Palma, 2022). The island has been positioned on the international extreme sports circuits with an international ultramarathon event, the Transvulcania Trail Race, which has had several thousand participants in its editions prior to 2020, with an important presence of professional runners and international participants (Medina Hernández, 2018; Transvulcania by UTMB®, 2022). All this considered, it can be said that the island's tourism authorities' effort to market La Palma as the

"Hikers Island" (Cabildo de La Palma, 2022) has been quite successful. According to data provided by the Tourists Expenditure Survey, it can be estimated that currently, 44.0% of tourists visiting the islands engage in outdoor and nature activities. It should be considered, however, that even though the rate of active tourists in La Palma (44.0%) is certainly the highest among the Canaries since the island barely reaches 300,000 tourists per year, tourists from La Palma account for only 6.9% of the 1.4 million tourists who carry out nature activities in the archipelago.

Conclusions

This chapter aims to provide an overview of sport tourism development in the Canary Islands. In order to truly understand this phenomenon, it must be borne in mind that the archipelago is the biggest tourism destination in Europe. Therefore, even if figures such as 13.0% of the total number of tourists who engage in sport activities might not seem all that impressive, the total volume of sport tourism in the islands is quite significant. Sport tourism development in the islands started in the beginning of the twenty-first century, as one in a series of strategies of rejuvenation in mature destinations such as the Canary Islands. Prior to the 2000s, sport was seen by tourism authorities in the islands as just a marketing strategy, a way to advertise the islands. But following the strategy to rejuvenate and qualify the islands as novel tourism destinations, an idea that impregnated the new regulations of the early 2000s, such as the moratorium (Hernández-Martín et al., 2015; Inchausti-Sintes & Voltes-Dorta 2020; Simancas Cruz, 2015), the development of certain types of sport tourism, such as golf tourism, gained attention (Bianchi, 2010).

Unlike other destinations in Europe, such as the Balearic Islands, where cycle tourism plays a key role (Rejón-Guardia et al., 2018), or Algarve and Andalucia, which have become specialized in golf tourism, sport tourism in the Canary Islands is just another strategy, not the only one, and certainly not the most salient, to enrich the tourist experience in a destination that was originally developed under a traditional sea and sand model. Theme parks, gastronomy, luxury, and the renewal of urban areas are other strategies that complement this type of strategies. Some particularities of the Canaries also play a key role in limiting the development of sport tourism, such as the scarcity of some resources, like vast extensions of lands required to develop a great number of golf courses, the situation of the Canaries being closest to that of islands states such as Malta (Markwick, 2000).

The low seasonality of tourism makes it less attractive in the Canary Islands to cater to sport tourism when the demand is lower, as it happens in many destinations in the Mediterranean. Therefore, sport tourism must be seen as a means for the qualification of the destination. Sport tourism provides a complement, not the main theme to the general idea for the tourism industry in the Canary Islands being that "*We want to sell tourists 'sport activities' to get more revenues from them.*" Therefore, the development of sport tourism in the Canary Islands must be interpreted as an adaptation and diversification of the sun and sand tourism that permits it to persist, in line with previous experiences in other destinations such as the Balearic Islands (Aguiló et al., 2005).

Regarding Fuerteventura and La Palma, in the first case, the demographic and economic growth of the island due to tourism development has been based on immigration. Jobs created by tourism development, and particularly those jobs linked to sport tourism, have been occupied by the foreign population. Currently, the population born outside the island largely surpasses the population born in Fuerteventura. In La Palma, a much smaller tourism development has not been able to overcome the demographic stagnation typical of rural areas but has prevented it from becoming much more serious due to the fact that it has received immigrants, many of them lifestyle migrants, who have directly or indirectly contributed to tourism development. In the case of Fuerteventura, the orientation toward windsurfing, surfing, and other water sport tourism has made possible the development of a significant network of small locally based companies (although mostly owned by foreigners), a fact that has been favored by the equipment requirements of these sports. The need for professional guidance and monitoring these sports require has implied the creation of a significant number of qualified jobs in this subsector (Marrero Rodriguez & Santana Talavera, 2018). On the contrary, in La Palma, the types of activities that are more developed, mostly hiking and mountain races, require little equipment and rarely require professional assistance, such as guides. Therefore, the development of a specialized offer that could provide jobs for locals is much smaller. It should be considered. However, the maintenance (cleaning, conditioning, repairs, fire surveillance, and emergency assistance) of the trail network is carried out with public resources through various local employment programs that contribute to avoiding the emigration of the population living in many areas. All this considered, we think that the study of sport tourism in the Canary Islands provides an insightful illustration of the development of sport tourism in mature destinations, and it shows that the role of sport tourism within the broader tourism system has to do with diversification. Further research will be certainly useful to evaluate how successful the strategy followed in the Canary Islands has been beneficial to its sustainable development in the future.

References

Agarwal, S. (2002). Restructuring seaside tourism: The resort lifecyle. *Annals of Tourism Research, 29*(1), 25–55. https://doi.org/10.1016/S0160-7383(01)00002-0

Aguiló, E., Alegre, J., & Sard, M. (2005). The persistence of the sun and sand tourism model. *Tourism Management, 26*(2), 219–231. https://doi.org/10.1016/j.tourman.2003.11.004

Baños González, I., Martínez Fernández, J., Ángel Esteve, M., & Pérez Cutillas, P. (2019). Evaluación de la sostenibilidad en el sector socioturístico de Fuerteventura (Islas Canarias). *Cuadernos de Turismo, 43*, 69–96. https://doi.org/10.6018/turismo.43.03

Bargeman, B., & van der Poel, H. (2006). The role of routines in the vacation decision-making process of Dutch vacationers. *Tourism Management, 27*(4), 707–720. https://doi.org/10.1016/j.tourman.2005.04.002

Bianchi, R. (2010). Tourism restructuring and the politics of sustainability: A critical view from the European periphery (The Canary Islands). *Journal of Sustainable Tourism, 12*(6), 495–529. https://doi.org/10.1080/09669580408667251

Britton, S. (1982). The political economy of tourism in the third world. *Annals of Tourism Research, 9*(3), 331–358. https://doi.org/10.1016/0160-7383(82)90018-4

Butler, R. (2006). *The tourism area life cycle*. Channel View Publications.
Cabildo de La Palma. (2022). *Red de Senderos de La Palma*. Retrieved 19 Oct 2022 https://www.senderosdelapalma.es/
Domínguez-Mujica, J., González-Pérez, J., & Parreño-Castellano, J. (2011). Tourism and human mobility in Spanish archipelagos. *Annals of Tourism Research, 38*(2), 586–606. https://doi.org/10.1016/j.annals.2010.11.016
Egli, R. (2022). René Egli History. *René Egli Center Fuerteventura*. Retrieved 19 Oct 2022 https://www.rene-egli.com/team-history-information/rene-egli-history
European Commission. (2018). *Special Eurobarometer 472: Sport and Physical Activity - Data Europa EU*. Retrieved 19 Oct 2022 https://data.europa.eu/data/datasets/s2164_88_4_472_eng?locale=en
European Commission (2021), https://ec.europa.eu/regional_policy/policy/themes/outermost-regions_en
Gobierno de Canarias. (2022a). Golf durante todo el año. *Hola Islas Canarias*. Retrieved 19 Oct 2022. https://www.holaislascanarias.com/golf-durante-todo-el-ano/
Gobierno de Canarias. (2022b). "Navegar en mar abierto." *Hola Islas Canarias*. Retrieved 19 Oct 2022. https://www.holaislascanarias.com/navegar-en-mar-abierto/.
Gobierno de Canarias. (2022c). *Turismo Activo*. Retrieved 19 Oct 2022. http://www.gobiernodecanarias.org/turismo/dir_gral_ordenacion_promocion/TurismoActivo/
González Ramallal, M., Marrero Rodríguez, J., & Santana Turégano, M. (2010). Sport and tourism: A potentially conflictual relationship. The case of marinas in Tenerife. *PASOS. Revista de Turismo y Patrimonio Cultural, 8*(2), 265–276.
Hernández Luis, J., González Morales, A., Hernández Torres, S., & Ramón Ojeda, A. (2017). El impacto del turismo de masas en las Islas Canarias en el contexto de las reservas mundiales de la biosfera. *Cuadernos de Turismo, 40*, 363–387. https://doi.org/10.6018/turismo.40.309751
Hernández Martín, R. (2011). *La Cuenta Satélite Del Turismo de Canarias. Metodología. Metodología, Resultados y Reflexiones, En*. Fundación FYDE-CajaCanarias.
Hernández-Martín, R., Álvarez-Albelo, C., & Padrón-Fumero, N. (2015). The economics and implications of Moratoria on tourism accommodation development as a rejuvenation tool in mature tourism destinations. *Journal of Sustainable Tourism, 23*(6), 881–899. https://doi.org/10.1080/09669582.2015.1027212
Higham, J. (2005). *Sport tourism destinations: Issues, opportunities and analysis*. Elsevier Butterworth-Heinemann.
Hudson, S., & Hudson, L. (2014). *Golf tourism*. Goodfellow Publishers Ltd.
Huete, R., Mantecón, A., & Jesús Estévez, J. (2013). Challenges in lifestyle migration research: Reflections and findings about the Spanish crisis. *Mobilities, 8*(3), 331–348. https://doi.org/10.1080/17450101.2013.814236
Inchausti-Sintes, F., & Voltes-Dorta, A. (2020). The economic impact of the tourism Moratoria in the Canary Islands 2003–2017. *Journal of Sustainable Tourism, 28*(3), 394–413. https://doi.org/10.1080/09669582.2019.1677677
INE. (2022a). *Contabilidad Regional de España—Revisión Estadística 2019. PIB per cápita*. Retrieved 19 Oct 2022 https://www.ine.es/daco/daco42/cre00/b2015/pr_cre.xlsx
INE. (2022b). *Viviendas Turísticas En España*. Retrieved 19 Oct 19 2022 https://www.ine.es/dynt3/inebase/index.htm?padre=7690&capsel=7690
Ioannides, D., & Debbage, K. (1997). Post-Fordism and flexibility: The travel industry Polyglot. *Tourism Management, 18*(4), 229–241. https://doi.org/10.1016/S0261-5177(97)00019-8
ISTAC. (2021a). *Cifras Oficiales de Población. (Mayo de 2021)*. Retrieved 19 Oct 2022 http://www.gobiernodecanarias.org/istac/estadisticas/demografia/poblacion/cifraspadronales/E30245A.html
ISTAC. (2021b). *Encuestas de Alojamiento Turístico / Series Mensuales. Canarias Por Islas y Municipios Turísticos. 2009–2020-Pernoctaciones, Viajeros Entrados y Viajeros Alojados Según Islas de Residencia de Canarias Por Islas de Alojamiento de Canarias y Periodos*. Retrieved October 19, 2022: http://www.gobiernodecanarias.org/istac/estadisticas/sectorservicios/hosteleriayturismo/demanda/C00028A.html

ISTAC. (2022). *Encuesta Sobre Gasto Turístico*. Retrieved 19 Oct 2022 http://www.gobiernodecanarias.org/istac/estadisticas/sectorservicios/hosteleriayturismo/demanda/C00028A.html

Kotzeva, M., Brandmüller, T., & Önnerfors, A. (2019). *Eurostat regional yearbook: 2019 edition*. Publications Office.

Macías-Hernández, A. M. (2011). Canarias, 1800-2000: La singularidad de la historia económica isleña. *Historia contemporánea, 42*, 225–260.

Markwick, M. C. (2000). Golf tourism development, stakeholders, differing discourses and alternative agendas: The case of Malta. *Tourism Management, 21*(5), 515–524. https://doi.org/10.1016/S0261-5177(99)00107-7

Marrero Rodriguez, J., & Santana Talavera, A. (2018). La estructura de oportunidades empresariales en los destinos turísticos de sol y playa. El caso de Fuerteventura (Canarias). *Cuadernos de Turismo, 42*, 285–304. https://doi.org/10.6018/turismo.42.12

Martín Martín, V. O. (2000). *El turismo en el sur de Tenerife: de la renta agraria a la renta del ocio*. Cabildo Insular de Tenerife, Cabildo Insular de Gran Canaria.

Medina Hernández, A. (2018). *Turismo Deportivo En Canarias La Transvulcania Como Caso de Estudio*. Retrieved 19 Oct 2022 https://riull.ull.es/xmlui/bitstream/handle/915/12002/Turismo%20deportivo%20en%20Canarias%20la%20Transvulcania%20como%20caso%20de%20estudio.pdf?sequence=1

PROMOTUR. (2018). *Perfil Del Turista Según Aspectos Que Influyen En La Elección Del Destino (2017): Análisis Comparativo Turismo Deportivo*. Retrieved 19 Oct 2022 https://turismodeislascanarias.com/sites/default/files/promotur_aspectos_de_eleccion_analisis_comparativo_con_total_deportivo_2017.pdf

PROMOTUR. (2022). *Evolución anual del perfil del turista. Canarias. 2018–2021*. Retrieved 19 Oct 2022 https://turismodeislascanarias.com/sites/default/files/promotur_evolucion_canarias_2021.pdf

Rejón-Guardia, F., García-Sastre, M. A., & Alemany-Hormaeche, M. (2018). Motivation-based behaviour and latent class segmentation of cycling tourists: A study of the Balearic Islands. *Tourism Economics, 24*(2), 204–217. https://doi.org/10.1177/1354816617749349

Rodríguez González, P., & Santana Turégano, M. (2012). Los agentes sociales y la política urbanístico-turística: percepción y performatividad. El caso de las Directrices de Ordenación del Territorio y del Turismo de Canarias. *Investigaciones Turísticas, 3*, 56–82. https://doi.org/10.14198/INTURI2012.3.04

Santana Turégano (2003) *Formas de desarollo turístico, redes y situación de empleo. El caso de Maspalomas (Gran Canaria)*. Tesis doctoral).

Santana Turégano, M. (2006). Dependency and development patterns in tourism: A case study in the Canary Islands. *Tourism and Hospitality Planning & Development, 3*(2), 117–130.

Sheller, M., & Urry, J. (2016). Mobilizing the new Mobilities paradigm. *Applied Mobilities, 1*(1), 10–25. https://doi.org/10.1080/23800127.2016.1151216

Simancas Cruz, M. (2011). El fracaso de la renovación de áreas turísticas consolidadas -de litoral a través de la sustitución de la oferta de alojamiento obsoleta: la experiencia de las Islas Canarias. *Cuadernos de Turismo, 27*, 875–906. https://doi.org/10.6018/turismo

Simancas Cruz, M. (2015). *La moratoria turística de Canarias: la reconversión de un destino turístico maduro desde la ordenación del territorio*. Servicio de Publicaciones, Universidad de La Laguna.

Simancas Cruz, M., & Mañoso Valderrama, J. (2016). *La residencialización de las áreas turísticas de Canarias*. Promotur Turismo de Canarias. ISBN: 978-84-608-9300-4.

Sosvilla Rivero, S., Cova Alonso, D., & González Pérez, J. M. (2010). Canarias y la Política Regional Europea. *Papeles de Economía Española, 123*, 136–156.

SurferToday.com, Editor at. (2022). *A short history of the PWA World Tour. Surfertoday*. Retrieved 19 Oct 2022 https://www.surfertoday.com/windsurfing/a-short-history-of-the-pwa-world-tour

Transvulcania by UTMB®. (2022). *Transvulcania by UTMB®. Transvulcania La Palma Island*. Retrieved 19 Oct 2022 https://transvulcania.utmb.world/

Urry, J. (2012). *Sociology beyond societies: Mobilities for the twenty-first century*. Routledge.

Urry, J. (2016). *Mobilities: New perspectives on transport and society*. Routledge.

Chapter 13
Madeira Archipelago: Paths to a Sustainable Nature Sport Tourism Destination

Jorge Soares and Ricardo Melo

Introduction

Madeira archipelago is a Portuguese autonomous region situated in the Atlantic Ocean, integrated into the African plate, between 30° and 33° north latitude (almost the same latitude as Casablanca) and located around 1000 km from the European continent, southwest of Lisbon, just 500 km from the west coast of the African continent and approximately 450 km north of the Canary Islands.

Tourism today is more than just an economic activity; it is (or can be) one of the main promoters of a vision of sustainability for the planet. In this regard, over the years, the Autonomous Region of Madeira has implemented several programs and actions that aim to contribute with solutions so that the future of the region may come to be more green, inclusive, and equitable (Madeira Sustainable for all, 2023). The economic activity of Madeira is strongly based on the services sector, with tourist activity standing out as the engine of the region's economy, where sport tourism, particularly active and nature sport tourism activities (e.g., levadas walking, MTB, and canyoning) and sea and nautical sport tourism activities (e.g., diving,

J. Soares
University of Madeira, Madeira Island, Portugal

CITUR – Centre for Research, Development and Innovation in Tourism, Faro, Portugal
e-mail: jorges@staff.uma.pt

R. Melo (✉)
CITUR – Centre for Research, Development and Innovation in Tourism, Faro, Portugal

Coimbra Education School, Polytechnic Institute of Coimbra, Coimbra, Portugal

IRNIST – International Research Network in Sport Tourism, Lille, France

D. Van Rheenen et al. (eds.), *Sport Tourism, Island Territories and Sustainable Development*, Sports Economics, Management and Policy 24,
https://doi.org/10.1007/978-3-031-51705-1_13

surfing, and kayaking), represents key strategic products of the archipelago's sustainable tourism policy (Madeira Sustainable for all, 2023).

Despite all this, few studies have analyzed the sustainability of sport tourism in the territory, although some studies researched the demand, the offer, and the impacts of sport tourism products or services. Within the IRNIST research framework produced specifically for this project, outlining the contribution of sport tourism for local sustainable development in several small island territories, the main goal of this chapter is to present an overview of sport tourism in the Madeira archipelago. The chapter presents results essentially from the analysis of secondary data, resulting from scientific publications, as well as technical documents prepared with the support of governmental bodies.

The chapter is structured in six sections, including this introduction. The next section presents a characterization of the Madeira destination, including a geographical and socio-demographic analysis of the archipelago. The third section presents an analysis of the tourism industry, followed by a section where the focus is given to sport tourism in Madeira, in particular the offer dimension. The analysis of the sustainability of tourism in Madeira is presented in the fifth section, while, in the final section, the fundamental issues presented in the chapter are discussed and some suggestions are presented for the stakeholders involved in the sport tourism sector at Madeira archipelago.

Geographic and Sociodemographic Characteristics of the Madeira Archipelago

Madeira is an archipelago composed of the inhabited islands of *Madeira* and *Porto Santo* and the uninhabited islands of *Desertas* (which includes *Deserta Grande, Ilhéu Chão,* and *Bugio*) and *Selvagens* (which includes *Selvagem Grande* and *Selvagem Pequena*) (Fig. 13.1). Both sets of uninhabited islands are protected natural areas and are only inhabited by the nature guards of the *Instituto de Conservação da Natureza da Madeira* (Madeira Nature Conservation Institute).[1]

In 2021, the Autonomous Region of Madeira had 250,744 inhabitants, which means a loss of 6.4% of the population over the last 10 years (mainly influenced by the decrease in the birth rate), with an aging rate to 157 (157 seniors for every 100 young people), which demonstrates a tendency toward the aging of the population. In 2019, the gross domestic product (GDP) per capita was 5126.4 million Euros, with a GINI index of 0.36 and an unemployment rate of 7.1%, 2% lower than the previous year (Santos, 2019) (Table 13.1).

[1] In this chapter, we will analyze in more detail the islands of Madeira and Porto Santo, for reasons related to tourism and sports services. However, this does not mean that the uninhabited islands of the Madeira archipelago have no potential for tourist visitation; on the contrary, they bring together very interesting natural characteristics for ecological tourism and scientific research associated with marine life and birds.

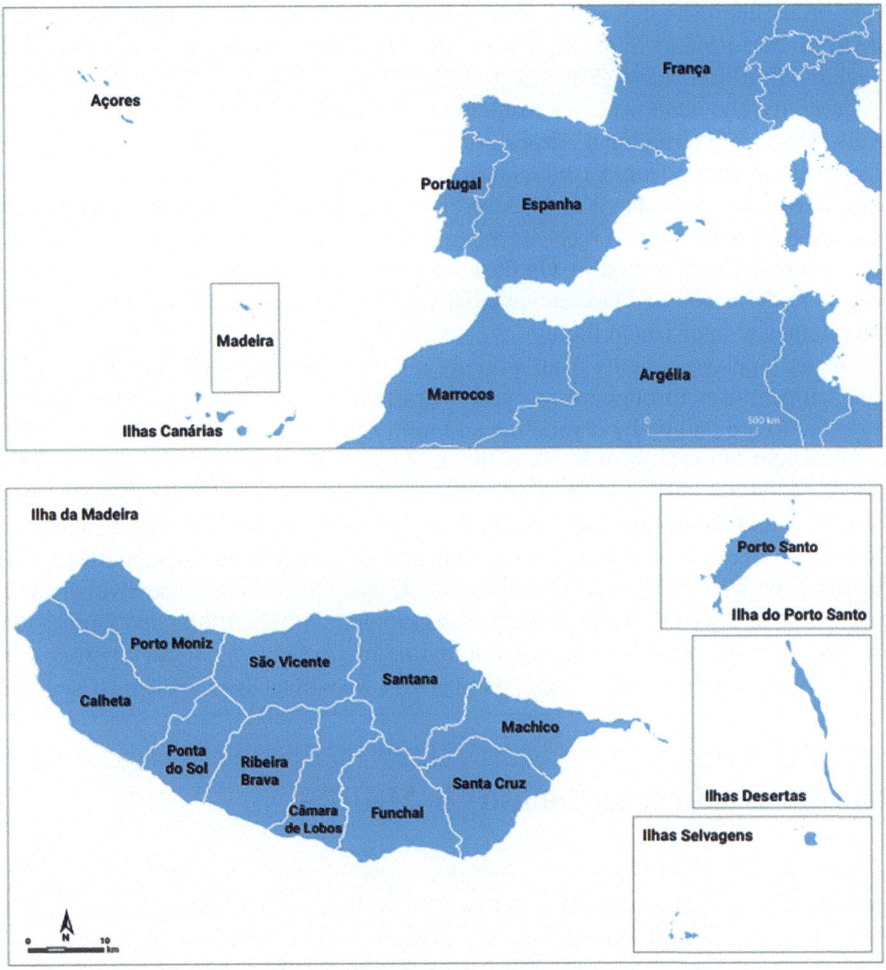

Fig. 13.1 The Madeira archipelago. (DREM, 2021a, b)

Table 13.1 Area, population (2021), and GDP (2019) in Madeira

	Madeira	Porto Santo	Total
Area (sq./km^2)	741.0	42.5	783.5
Population	245,594	5149	250,744
GDP per capita	–	–	20.2
GDP (Million Euros)	–	–	5126.4

The island of Madeira is the main island of the autonomous region and has an area of 741 km^2, 57 km long and 22 km wide. Funchal is its capital (also the capital of the autonomous region). In turn, the island of Porto Santo is situated 60 km northeast of the island of Madeira and has an area of 42.5 km^2, 11 km long and 6 km wide. Vila Baleira is its capital. Both islands are of volcanic origin, wherein the island of Porto Santo is geologically older (18 million years) than the island of Madeira (7 million years) (Silveira et al., 2012).

Currently, the main accesses from and to Madeira and Porto Santo are, respectively, Madeira International Airport, in Santa Cruz, and Porto Santo Airport. The plane trip, departing from Lisbon, takes approximately 90 min. However, it is also possible to access the islands of Madeira and Porto Santo by sea. The port of Funchal has a modern platform that receives many cruise ships (about half a million passengers disembark here annually). On the contrary, the island of Porto Santo has a connection to the island of Madeira via a daily ferry boat (passenger and cargo) and a daily domestic inter-island flight.

The archipelago benefits from a highly inviting subtropical climate, registering mild temperatures throughout the year. Furthermore, the islands of Madeira and Porto Santo are bathed by crystalline and mild waters. Due to the influence of the warm current of the Gulf of Mexico, the sea has temperatures that vary between 18 and 22 °C. Associated with this subtropical climate exists a diversity of winding mountains and valleys, which make up a dense and endemic forest, particularly in the interior areas of the mountains and in the North, called Laurissilva and which is classified, by UNESCO, as a World Natural Heritage Site (1999). The diversity and uniqueness of the valleys and mountains, associated with their flora and fauna, surrounded by a crystalline sea and rich in marine life, provide a natural environment that is highly sought after for tourist purposes, as we will analyze in the next section.

Tourism as a Strategic Industry in Madeira

In 2019, the travel and tourism industry accounted for 17.1% of the Portuguese total gross domestic product of USD 42 billion (37.5 billion euros). This indicator placed Portugal as one of the countries where tourism is one of the sectors that most contributes to the country's economy (Espírito Santo et al., 2021). According to Espírito Santo et al. (2021), Madeira is, just after the Algarve, the region of Portugal where tourism has the greatest impact on the economy. Adding to the direct and indirect influence, tourism represents about 37% of the region's Gross Domestic Product and more than a third of employment. The Madeira archipelago has been recovering from the losses of 2020, as the issuing markets open and is also preparing to be one of the summer destinations of choice for those who live on the continent. In 2017, RAM had 1194 tourist accommodations, which corresponded to 20.5% of the country's total. According to DREM (7/7/2022), tourists from Mainland Portugal and the Azores spent 979.0 thousand nights (+118.6% than in 2020 and +19.5% than in 2019), representing 25.1% of the hotel industry, which allowed the Portuguese market to stand out as the most important in this segment, followed by the British and German.

The island of Madeira has one of the lowest seasonality indices in Portugal, with a higher concentration of visits in August. According to DREM data for the year

2019, Madeira showed some stability in the number of guests and overnight stays over the 12 months. The month of January (month with the lowest indicators) accounted for 76,144 incoming guests and December 87,781. The highest month was August with 145,020 incoming guests. In the ACIF study (2015) on the strategic plan for tourism, the seasonality rate in 2014 was around 34% but should drop to 30% in 2020. On the contrary, the seasonality of Porto Santo is high and takes place between June and September of each year due to tourists from Madeira Island. The lowest month that received tourists was January with 958 and the highest month was August, which received 16,282 (DRE, 2020) (Table 13.2).

Contact with nature is the main reason for tourists to travel and visit Madeira (ECAM, 2010; ACIF, 2015). Madeira levadas walking is one of the main tourist products most sought after by tourists and constitutes an intersection of nature sport tourism with active and wellness tourism (Almeida et al., 2013; Quintal, 2010; Soares & Nunes, 2020). Hundreds of tourists visit Madeira every month to discover the landscapes and experience a hike through the mountains of Madeira (Laurissilva). Madeira's levadas are water channels built since the island's settlement (1425) that served to conduct and transport water for domestic consumption, agricultural irrigation, and mill water, and later, from the twentieth century, supplying hydroelectric power stations (Santos & Correia, 2017; Quintal, 2013). The levadas are accompanied by land routes that allow and encourage pedestrians (levadas walking). In the 2010 study (ECAM), the levadas walking represented the third most popular

Table 13.2 Madeira tourism statistics

Items	Madeira island	Porto Santo island	Madeira Archipelago
Tourists (number)[a]	1,487,123	103,759	1,590,882
Number of cruise passengers (number)[a]	594,974	Data not available	594,974
The average length of stay (number of days)[a]	5.14	4.58	5.11
Tourism intensity[a, b]	30.70	91.30	31.94
Tourism density (km^2/day)[c]	28.28	22.83	27.89
Tourist saturation index[d]	5.97	19.94	6.25
Tourists per area (km^2/day)	5.49	4.98	5.46
Total revenues (US Dollars, in millions)[a]	436,396,852	29,102,220	465,499,071
Tourism revenue as a direct contribution (% of total GDP)[e]	Data not available	Data not available	26.6% (2017) and 37% (2019)
Domestic travel agencies (number)[f]	149	5	154
Tourist beds (number)	35,754	32,949	2805

[a]Data retrieved from DREM
[b]Calculated by the number of overnight stays per population
[c]Calculated by the number of overnight stays per area (km^2)
[d]Calculated by the number of tourists per inhabitants
[e]Data retrieved from Vieira (2015)
[f]Data retrieved from Visit Madeira

activity among tourists in Madeira, after experiencing gastronomy and car touring around Madeira. In the study by ACIF and KPMG (2015), around 40% of the tourists who visit Madeira are interested in nature, with emphasis on walking the levadas.

According to the predictable growth of new market niches for tourists to Madeira, younger and more curious, in challenging experiences, by taking advantage of the strengths of the footpaths, levadas, canyons, landscapes, forest, and other elements of nature, they constitute competitive advantages in relation to another tourist offer (Vieira, 2016). The organizations of physical and sports activities in the mountains and at sea, such as levadas walking and diving, are unique experiences with breathtaking landscapes. This demonstrates how sport tourism, particularly nature sport tourism, is an important sector for the development of the Madeira archipelago. The following sections analyze the supply and demand of this growing sector in this region.

Sport Tourism in the Madeira Archipelago

Sport tourism is a relevant sector for the tourism industry of Madeira archipelago. Sport tourism activities are included in two (Sea and Nautical Tourism, and Active and Nature Sport Tourism) of the six strategic pillars of the sustainable touristic products identified (Madeira Sustainable for all, 2023). Despite data available about sport tourism supply and demand being very scarce, it is noted that since 2018, the number of jobs and companies was growing until the pandemic situation. In 2022, the growth has been exponential, with emphasis on small companies (less than 10 employees). In this regard, data collected from the website of the Regional Directorate of Tourism, which included the identification of tourism animation companies operating in RAM, identified 404 companies, 390 in the island of Madeira and 14 in the island of Porto Santo.

The tourism sector contributes directly 26–30% of the GDP of the region, and nature sport tourism and levadas contribute 25–40% of the revenues. By way of illustration, a company that provides a levadas walking guide service (2019) handled 13,800 tourists and generated revenue in the order of half a million euros. Although access to levadas/mountains is free, it is estimated that the tourist product levadas walking (accompanied by a guide and with minibus transfer) can generate revenues of over 5 million dollars a year. There are 268 companies offering levada guiding activities (Abreu et al., 2022). For example, for the Levada das 25 Fontes, in Rabaçal, Calheta, the price per person is 45.00€, including a tourist guide and minibus transfer. Moreover, as we analyzed, 30% (477,264) of tourists in Madeira make a levadas walking, making this the main sport tourism product of Madeira Island (ACIF, 2015). In fact, there are more than 40 levadas (the levadas are horizontal but have different degrees of difficulty and distance, from 2 to 20 km walking). Nature festival is a 1-week festival that includes an exhibition of the main activities that residents and tourists practice on the mountain and at sea, for example, walks along the paths and levadas inside *Laurissilva* forest (Quintal, 2010).

If, on the one hand, Madeira is very wooded and intertwined with valleys and streams, which carry water to the sea, especially in the higher areas facing north, the island of Porto Santo is devoid of natural spring water. On this island, a process of reuse of sea water through desalination is used. However, despite not having the vegetation or the levadas of Madeira, it has a sandy beach as its ex-libris (9 km of fine golden area that allows safe walks and swimming in crystal clear waters). In addition, the sand on this beach has chemical properties with very promising therapeutic effects (Batista, 2016). Furthermore, in April 2022, Porto Santo Beach was voted the best beach in Europe by the European Best Destinations website.

In addition to the diversified offer of active and nature sport tourism services, because of the work of tourist entertainment companies, Madeira Island has come to stand out as an island of sport tourism events. In the year of 2022, 60 national and international sporting events were planned and partially held, with the support of almost half a million euros from the regional public administration (DRD, 2022), many of them held in the context of sea and mountains, where the following stand out: for example, mountain running (Madeira Ultratrail Running) and others, in sports complexes such as swimming and water polo at the Funchal Olympic Swimming Pools (Silva, 2018). At an international level, sport tourism events, along with nature tourism, and particularly running tourism, have been gaining interest from local stakeholders and have contributed to a sharp growth in sport tourism (Melo et al., 2021). In this regard, Madeira offers some international sporting events that fall within the scope of small-scale/community sports events (Sean, 2011), planned for the islands of Madeira and Porto Santo (2021 and 2022), but which have relevant economic and tourism impacts.[2]

Although the island of Madeira is relatively small, its mountainous features consisting of valleys, peaks, streams, canyons, waterfalls, and hundreds of trails and washes, accompanied by a dense and diverse forest, represent a very favorable nature space for the practice of exercise, adventure sports, and physical and psychological well-being. In summary, we can quantify the following spaces for the practice of physical exercise and sport, which are increasingly associated with active and sports tourism (Quintal, 2010; DRJD, 2019): levadas walking (more than 40 in conditions to be covered); Paths and royal paths (more than 100, although most royal paths are inactive and need to be rehabilitated); canyons in the mountains (7); south coast of Madeira sea; tourist and nature diving spots (6); mountain bike trails (more than 20); paragliding spots (3); surfing and bodyboard spot (10); and Stand up Puddle spots (2).

Active and nature tourism is the most important tourist cluster on the island of Madeira. Levadas walking represents the tourist product most preferred by tourists to enjoy unique experiences in a dense and diverse forest and can generate dozens

[2]Additional information about sport tourism events in Madeira archipelago is available at: https://www.madeira-web.com/en/whats-on/madeira-events.html

of millions of euros per year. A small company that provides a guide service for levadas walking (2019) handled 13,800 tourists and generated revenue of half a million euros (Abreu et al., 2022).

From the point of view of analyzing the offer of artificial sports infrastructure, the islands of Madeira and Porto Santo have a diverse sports park that is used for the practice of sports by the 250,000 people from Madeira and Porto Santo. Although the regular physical activity index of the Madeiran and Portuguese populations is quite low (it is estimated to be below 30%, DGS, 2021), the two islands have a very extensive and diverse offer of artificial sports facilities (DRJD, 2019) that can be characterized by:

- Sports facility complex for various sports: 2 (Água de Pena- Santa Cruz; Ribeira Brava);
- Football stadiums that host games of the Portuguese championship league: 2, plus 3 that serve for training;
- Athletic track: 3;
- Olympic swimming pool: 1, supported by 25-meter swimming pool with 10 lines;
- Swimming pool 25 meters: 14;
- Golf courses: 3 (1 is in Porto Santo);
- Sports pavilion: 30;
- Tennis court: 18;
- Paddle court: 10-;
- Football field: 11, synthetic flooring; 9, natural grass; 3, clay field;
- Motocross track careers: 1;
- Kart track: 2;
- Skating rink: 2;
- Auto Model Track: 1;
- Big-game fishing marina: 5.

Some of these facilities and sports complexes have been used to host various sporting events of an international nature in the competitive federated sports sector, in addition to representing a sporting development factor for the generalization of sports practice by various groups and clubs of the Madeiran population.

In Madeira, the companies that offer sport tourism are named as tourist entertainment companies (which includes also cultural and environmental touristic activities). These companies organize and offer various physical and sports activities in nature for different segments of tourists. For example, there are companies that provide canyoning services on the stream of the mountains of Madeira that include experiences in canyons nature that require the ability to walk on the rocks and water, jump, climb, swim, slide, etc. There is also a local sport club that provides sport activities for local population. In a study carried out in May 2022 (Abreu et al., 2022), of 18 tourist entertainment companies on the island of Madeira, 12 responded that they included physical activity or physical exercise in tourist services. The main activities of the services pointed to levadas walking and hiking on the paths, mountain biking, and diving.

In 2019, each company served between 100 and 15,000 tourists, with the levadas and footpaths being the ones that handled the most tourists. This fact is due to the possibility of attending this activity every day of the year and to the different alternatives of levadas and paths, some facing the north and mountainous coast, others located further south. For example, the levadas walking service of a company that handled 15,000 tourists in 2019 was able to generate a total revenue of 500,000 euros. When asked at what time of the year the services were provided, most answers indicated the whole year, which confirms the non-seasonality of services provided in nature, although the months of July and August are the most sought after.

In Fig. 13.2, we present the main characteristics and strengths of the islands of Madeira and Porto Santo, associated with nature, and the identification of various physical and sports activity products that have integrated the service offer of tourist entertainment companies and responsible sports organizations by small-scale tourist sporting events.

For several years now, tourists' prospects of revisiting Madeira have been associated with nature, with emphasis on the levadas walking and whale watching, in addition to various activities carried out in a mountain and sea context, such as trail running, canyoning, mountain bike, paragliding, diving, and sailing. At the time, the same study warned to preserve the connection between the characteristics of the products or services marketed with the positioning of the tourist destination. Although golf and surfing are not among the main preferences of the surveyed tourists, they are niche markets that should not be undervalued. In 2021, the completion of 45,626 rounds on the three golf courses in Madeira archipelago represented a growth of 21.9%, but −30.2% than in 2019. However, in a year of recovery (2021), it was able to generate around 1.9 million euros in revenue (DREM, 2022a, b).

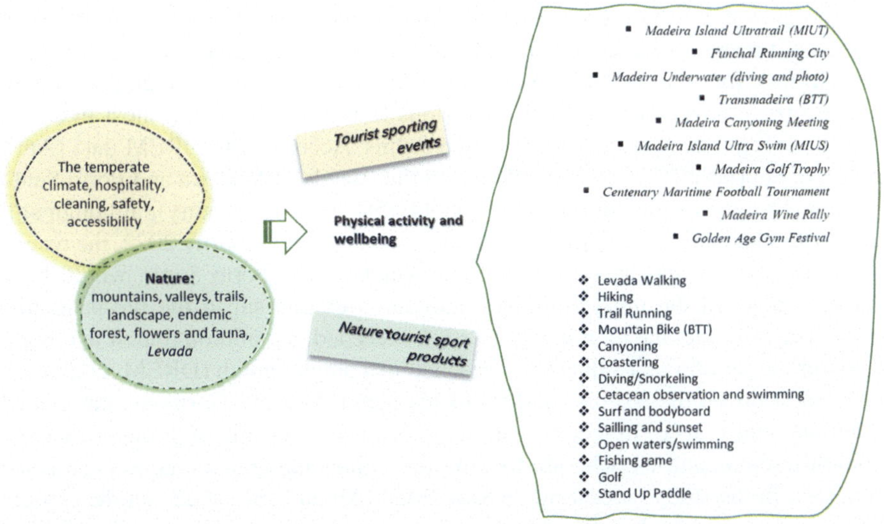

Fig. 13.2 Nature and sport tourism products in Madeira Island

The tourist destination trend on the island of Madeira clearly points to different market niches for active and nature tourism. The segmentation of the tourist profile can be done based on the active-passive or/and autonomous-in-company criteria, whether the reason may be associated with environmental observation and interpretation, or visiting and relaxing in nature, or even experiencing an adventure or challenge (Silva & Carvalhinho, 2017). The nature tourist on the island of Madeira must present a profile of visiting and relaxing experience, another of challenging and observing the natural landscape, in a sea and mountain context and for different age groups. Health and well-being, associated with physical exercise and relaxation, can also be important aspects to be valued in the offer of tourist services. Performing physical exercise in a natural context favors the relaxation process, meditation, and contributes to changing the person's mental and spiritual state (Wolsko et al., 2019). In the study on the impacts of the Golden Age Gym 8th edition held in Funchal between October 2 and 7, 2022, for a target audience over 55 years old, provisional data from the questionnaire to the participants and interviews carried out to the organizers indicated Madeira and the city of Funchal as excellent places to host events and active tourists of advanced ages, taking into account the favorable climate, the proximity and accessibility of the places, and nature. Well-being, associated with exercise, will be a perfect combination associated with nature and well-being tourism, and Madeira appears to be an excellent tourist destination. According to the ACIF study (2015), wellness tourism shows a trend in the growth of tourism and must be based on three pillars of sustainability: economic pillar, social pillar, and environmental pillar. Wellness tourism represents ≈6% of all domestic and international travel, but it is expected to grow at a rate of over 9% per year. This type of tourist adds greater local economic power and helps to leverage the preservation of traditional culture and natural assets, enhancing the sustainability of destinations in the long term.

Considering the sharp growth and the diversified offer of flights from different origins in Europe to Funchal-Madeira and vice versa, especially in terms of low-cost airlines, since April 2022, the island of Madeira has shown a high potential to increase the number of tourists from various European cities and, most likely, to start receiving much younger tourists due to this. According to DREM data (April 2022), in February 2022, tourist activity on the islands of Madeira and Porto Santo continued to grow quite significantly (about 85% of total capacity), although still below pre-pandemic levels. But more recent data, referring to July 2022, the results of tourist arrivals and accommodation in Madeira and Porto Santo would have already exceeded the most optimistic indicators of tourism in the pre-pandemic phase. For example, the month of July 2022 exceeded one million tourist overnight stays for the first time, a value never reached in a single month (DREM, 2022a, b), with emphasis on the issuing markets of mainland Portugal, Germany, the United Kingdom, and France. In this growth, there has been a profile of younger tourists, from different countries, who are looking for a challenging and unique experience in nature. In the study carried out in May 2022 (Abreu et al., 2022) on the characterization of sports and physical exercise services offered by tourist entertainment companies based in Madeira, it was found that levadas and footpaths are the most sought-after places for tourists to enjoy excellent landscapes and live unique experiences in a context of nature. Every day, there are a few hundred tourists looking for one of the most popular levada walking: Levada das 25 Fontes, which includes a waterfall located in the heart of the Laurissilva, a protected forest classified as a natural heritage of humanity.

Certify and Care for Nature: Key Factors for the Sustainable Development of Madeira as a Sport Tourism Destination

The islands of Madeira and Porto Santo are at a crossroads between the sharp growth of the tourist offer associated with nature and the need to preserve and regulate the most precise good that the tourist destination has—nature associated with the forest and the characteristics of the sea as a space of physical activity, sport, and tourism. This process includes the certification of the islands of Madeira and Porto Santo as sustainable tourist destinations through an external assessment carried out by the certifying entity Earth Check, one of the world's leading certification, consulting, and advisory groups for sustainable destinations and tourism organizations. The tourist destination audit (scheduled for December 2022) assesses the islands' degree of excellence through four pillars of sustainable development: environmental, cultural, economic, and social. These pillars include an independent assessment based on the following dimensions: energy use, water use, carbon emissions, chemicals, community and employee impact, paper use, and waste.

The DMO (Destination Management Organization), created by the Resolution of the Regional Government Council no. 51/2022, of January 31, 2022, under the dependence of the Regional Secretariat of Tourism and Culture, is the structure responsible for planning, managing, and supervising the certification process. Aware of the importance of developing integrated, participative, and continuous action, aligned with the Sustainable Development Goals of the 2030 Agenda, the various DMO Working Groups were consulted and, together, defined five priority SDGs and five other relevant SDGs (Fig. 13.3).

This process is in line with the guidelines of the European Union to evaluate the sustainable destination through a system of evaluation of tourist indicators, such as the ETIS (2016). The ETIS is a management, information, and monitoring tool specifically intended for tourism destinations. The objective of the ETIS is to contribute to improving the sustainable management of destinations and help stakeholders within to measure their sustainability management processes, enabling them to monitor their performance and progress over time. According to the local political leader responsible for tourism, the objective of certifying Madeira and Porto Santo as sustainable destinations is to achieve a better quality of life for the local community and minimize risks to safety and the environment with the use of tourist activities. The balance between the preservation of the properties of the environment and natural spaces, and the provision of tourist services, including

Fig. 13.3 Priority and other relevant sustainable development goals for tourism in Madeira. (Madeira Sustainable for all, 2023)

small- and medium-scale sporting events, associated with the forest and the sea, is a need that local authorities seem to be committed to.

Scientific studies in several tourist regions show that sport tourism events, even on a small scale, should deserve measures to safeguard nature and avoid negative impacts on the life of local species (Melo et al., 2021). In the organization of sporting events of international scope, several local impacts are considered that go beyond economic and touristic effects, particularly in small-scale sporting events (Taks et al., 2015; Melo et al., 2021). For the sustainable development of a (sport) tourist destination, which depends on the preservation of nature and its strongest points such as the typical forest and the mountains, points that differentiate it from other mass tourism destinations (such as the Canary Islands or the Algarve), it is strategically crucial to consider and assess the social, sporting, and environmental impacts of sporting events (Hover et al., 2016, Soares, 2019).

For example, a greater use of sports tourists' interaction with the quality of life of residents and for the development of sport can and should be strategically planned (Melo et al., 2020; Soares, 2022; Chalip, 2005). Unfortunately, regarding environmental concerns, Bučar et al. (2019) have demonstrated that there is no standardized set of criteria for green certification relative to environmental sustainability practice. Better results for the local quality of life and for sustainable development can also be achieved if local stakeholders and the public administration that oversees the tourist destination, integrate in their strategy an empowerment of the impacts of sporting events. In Fig. 13.4, a diagram is presented highlighting the various differentiating factors that characterize the strengths of the tourist destination of the

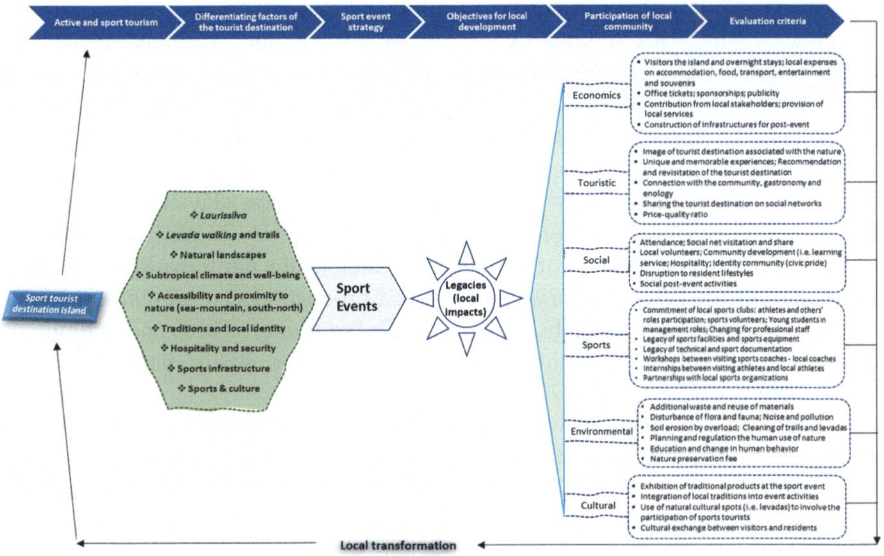

Fig. 13.4 Differentiating factors of Madeira tourist destination and local development strategy through the optimization of the impacts (legacies) of sport tourism events

islands of Madeira and Porto Santo and the various impacts that can be exploited from a strategy of organizing national and international sporting events.

It is possible and desirable to have an orderly planning of the use of natural spaces for tourism purposes, for the benefit of local communities, and in a sustainable way (Gavinho & Silva, 2017). In this context, we also highlight several community fund programs that the Madeira archipelago has used for environmental sustainability and nature conservation. One such program is LIFE (a financial instrument for environment preservation), which aims to contribute to the preservation and enhancement of the environmental area through the co-financing of projects of recognized value. The projects that have been supported are associated with the protection of the environment, the reforestation of burned areas, the enhancement of biodiversity, and the preservation of fauna and flora, for example, the species of the Sea Lion of the Desertas, who plays an important role in the preservation of Madeira's nature and life.

Another strong point of the sustainable development of Madeira's natural forest is the candidacy of "Levadas da Madeira" to World Heritage of Humanity, a process started in 2017. Under the responsibility of the Government's Regional Secretariat for Environment, Natural Resources, and Climate Change (SRAAC) Regional da Madeira, the first version of the Application of "Levadas da Madeira" to World Cultural Heritage that was delivered in February 2002 to the National Commission of UNESCO. This candidacy includes several activities aimed at involving citizens in valuing the natural life of Madeira's levadas walking, including environmental education actions with children and young people, various pastimes such as the plastic expression activity "Levadas in miniature," the "Levadas with soul," and "Taken with soul." This last activity is a record of memories, experiences, and curiosities of Madeirans, particularly of local seniors about the history and use of Madeira's levadas. Information about the traditions and memories of the levadas walking, including poetry and legends, can be found in the work of Santos e Correia (2017). In addition, there are several exhibitions and seminars that seek to publicize the natural activity and history of the levadas walking, nowadays one of the main tourist attractions and promotion of healthy physical activity. In planning a sustainable (sport) tourist destination, it is essential to include the participation of the local community in the offer of tourist services and products that represent a factor of identity and distinction (Gavinho & Silva, 2017; Perna et al., 2019) and to highlight the most important strengths of the Madeira brand: hospitality, natural heritage, climate, and ocean (Silva, 2017).

At a sea level, two projects were created that could have an impact on the number of diving tourists to observe different marine species, namely, the creation of the Artificial Reef spots in Porto Santo and the artificial reef in Cabo Girão through the sinking of two Portuguese small navy warships or corvettes that were deactivated and prepared specifically to host marine activities. The objectives of these projects are to provide the area with a bioecological mechanism capable of increasing the exploitable biomass in situ, leading to an increase in the number of catches in adjacent fishing grounds and, consequently, an increase in fishing income. Additionally, it gives companies and diving clubs the opportunity to explore the sites for tourism and sports purposes. For example, in October 2022, for the second time, at the Porto Santo spot, an activity of the World Championship of Underwater Photography and Video was held.

Finally, one of the political measures adopted by the Regional Government of Madeira to contribute to environmental sustainability relates to the application of fees for the use of nature, particularly in protected and classified areas such as nature reserves and natural heritage of humanity, which are excellent tourist spots. Thus, in 2017, Ordinance No. 30 of February 8 came to establish the rules and values of user fees by companies and for commercial purposes. However, this ordinance raised some objections on the part of businessmen and the local community, with the Regional Government changing the same diploma through Ordinance No. 124/2018 of April 4, creating a working group to analyze and apply the development of this measure. Of note from this amendment is the temporary suspension of various fees for tourist activities associated with footpaths, levadas walking and canyons, and adjustments to the prices of other services.

For example, the visit to the marine reef spot in Cabo Girão went from €5 per dive/person to €2. Considering the overuse of tourist sites (Majdak et al., 2022), explained in part by the exponential increase in tourist arrivals in the activity recovery phase (DREM, 2022a, b) and the consequent overload of the ecological footprint and erosion of some levadas and paths, such as Ponta de São Lourenço and Levada das 25 Fontes (Rabaçal), each of them can receive more than 1000 tourists in a single day, in a single path system (back-and-forth), it is expected that the introduction of regulatory fees for visiting certain sites will be generalized to many nature reserve sites. This procedure seems to be necessary considering the need to preserve nature, safeguard people's safety as a quality and sustainability factor (Silva & Carvalhinho, 2017) and also, an important recipe for the conservation of trails and enhancement of biodiversity. From the point of view of some local entrepreneurs who provide tourist services in the levadas walking and footpaths, there seems to be this sensitivity if it adds order to their use and contributes to greater environmental sustainability (Abreu et al., 2022).

Conclusions

The Autonomous Region of Madeira presents itself as a territory of natural and cultural diversity, awarded and valued by many. In 1995, the Desertas Islands received the status of Nature Reserve; in 1999, the Laurissilva Forest was recognized by UNESCO as a World Natural Heritage Site; the Municipality of Santana (2011) and, finally, the Island of Porto Santo (2020) received the distinction of UNESCO Biosphere Reserve (Madeira Sustainable for all, 2023).

The rugged relief and its slopes are the geological result of the archipelago, which has contributed to the formation of different types of landscape, vegetation, microclimates, and cultures. Besides its natural wealth, the archipelago also has a vast and rich, artistic (painting, sculpture, gilded woodcarvings, and tiles), and intangible cultural heritage, expressed in different popular manifestations such as handicrafts, gastronomy, wine-making, and the Region's activities and festivities.

The projects and initiatives encouraged in recent years demonstrate an innovative approach to the principles of sustainable development, which proves an increased concern in protecting and preserving the territory, while working on its promotion, namely: Strategy of the Turismo Sector at Autonomous Region of Madeira for the period 2022–2027; Circular Madeira Platform; Waste Strategy of the Autonomous Region of Madeira; Sustainable Porto Santo—Smart Fossil Free Island; Certification of Good Practices in the Management of Biological Risks and several projects included in the Life Programs (Madeira Sustainable for all, 2023).

The affirmation of sustainability in the archipelago of Madeira is seen as a long-term goal that involves several entities, to face several challenges, among them, the preservation of natural and cultural resources and decarbonization (Madeira Sustainable for all, 2023). Valuing and protecting resources will translate into the creation of authentic products, experiences, and activities that, together, will result in an increasingly unique and attractive offer and create a sustainable nature sport tourism destination for the future.

References

Abreu, J., Freitas, E., Mendonça, A., & Soares, J. (2022). *Tendências e desafios para o desenvolvimento de serviços turísticos associados à atividade física em contexto de natureza na Ilha da Madeira.* Resumo de comunicação oral: Congresso Nacional de Gestão do Desporto, 29 e 30 de Novembro 2022. Leiria, Portugal, pp. 10–13. https://www.apogesd.pt/Congressos/Congresso2022

ACIF. (2015). *Documento Estratégico para o Turismo na RAM (2015–2020).* Associação de Comércio e Indústria da madeira, em colaboração com a KPMG. Disponível em https://www.acif-ccim.pt/wp-content/uploads/2020/estudos/Doc_Estrategico_Turismo_RAM.pdf

Almeida, A., Soares, J., & Alves, A. (2013). As Levadas da Madeira no Contexto da Afirmação e da Confluência do Turismo de Natureza com o Turismo Ativo. *Revista Portuguesa de Estudos Regionais, 33*, 28–42. ISSN: 1645-586X.

Batista, J. (2016, May 28). *Porto Santo. A ilha onde as areias e as águas curam.* In Alexandra Coelho. Jornal Público. https://acervo.publico.pt/fugas/noticia/ilhas-psto-pri-a-ilha-onde-as-areias-e-as-aguas-curam-1733142. Acedido em 20 de abril 2022

Bučar, K., Van Rheenen, D., & Hendija, Z. (2019). Ecolabelling in tourism: The disconnect between theory and practice. *Turizam: međunarodni znanstveno-stručni časopis, 67*(4), 365–374.

Chalip, L. (2005). Beyond impact: A general model for sport event leverage. In J. Higham (Ed.), *Sport tourism destinations: Issues, opportunities and analysis* (Vol. XIV, pp. 226–262). Elsevier Butterworth Heinemann.

DGS. (2021). *Programa Nacional para a Promoção da Atividade Física—Portugal 2021.* Direção-Geral da Saúde, Lisboa. Disponível em https://www.dgs.pt/programa-nacional-para-a-promocao-da-atvidade-fisica/ficheiros-externos-pnpaf/relatorio_pnpaf_20211.aspx

DLR N.° 35/2008. *Decreto Legislativo Regional N° 35/2008/M, de 14 de agosto.* Estabelece o regime de protecção dos recursos naturais e florestais DA REGIÃO Autonóma da Madeira. Disponível em: https://data.dre.pt/eli/declegreg/35/2008/08/14/m/dre/pt/html

DRD. (2022). *Os 60 eventos desportivos apoiados pela administração pública regional 2022.* Disponível em https://www.rtp.pt/madeira/politica/mais-de-60-eventos-desportivos-apoiados-pelo-governo-video_96833. Acedido em 2 julho 2022.

DREM. (2021a). *CONTAS REGIONAIS—SEC 2010—BASE 2016.* https://estatistica.madeira.gov.pt/download-now/economica/contaseconomicas-pt/contaseconomicas-cr-pt/contaseconomicas-cr-emfoco-pt/send/55-contas-regionais-emfoco/14194-em-foco-contas-regionais-base-2016-2016-2020po.html. Acedido em fevereiro 2022.

DREM. (2021b). *Madeira em Números 2020.* Direção Regional de Estatística da Madeira. Funchal, Portugal. ISBN: 978-989-8755-76-6.

DREM. (2022a). *Estatísticas do Turismo da Região Autónoma da Madeira.* Resultados Provisórios, Funchal, 2002. https://estatistica.madeira.gov.pt/download-now/economica/turismo-pt/turismo-publicacoes-pt/send/40-turismo-publicacoes/15154-turismo-julho-pe-e-junho-po-de-2022pdf.html. Acedido a 4 outubro 2022.

DREM. (2022b). *Estatísticas do Turismo e Estatísticas dos Transportes Janeiro a Fevereiro 2022.* https://estatistica.madeira.gov.pt/download-now/economica/turismo-pt/dashboards-pt/category/634-dashboards.html. Acedido em 24 de abril 2022.

DRJD. (2019). *Carta de Instalações Desportivas Artificiais da Região Autónoma da Madeira 2018.* Direcção regional de Juventude e Desporto, Governo Regional da Madeira. https://www.madeira.gov.pt/drd/Estrutura/DRD/ctl/Read/mid/10885/InformacaoId/41684/UnidadeOrganicaId/30/CatalogoId/0. Acedido em 4 de abril 2021.

ECAM. (2010). *Estudo do Gasto Turístico—Madeira.* Empresa de Consultoria e Assessoria Empresarial da Madeira, S.A. Disponível em https://www.visitmadeira.pt

Espírito Santo, H., Caballero, J., Constantin, M., Köpke, S., & Binggeli, U. (2021). *A recuperação do turismo começou, mas levará anos até Portugal recuperar totalmente. O que poderão fazer os principais intervenientes no setor?* McKinsey Company. https://www.mckinsey.com/featured-insights/europe/travel-recovery-in-portugal/pt-PT. Acedido em 20 Sept 2022.

ETIS. (2016). *European tourism indicator system—Toolkit*. European Union, Luxembourg. Available at https://ec.europa.eu/docsroom/documents/21749

Gavinho, E., & Silva, M. (2017). Espaços Naturais: O Desafio de Planear um Destino Turístico. In *Planeamento e Desenvolvimento Turístico* (pp. 451–464). Coord. Francisco Silva e Jorge Umbelino. Editora Lidel.

Hover, P., Dijk, B., Breedveld, K., Enkeren, F., & Slender, H. (2016). *Creating social impact with sport events*. Mulier Institute, Utrecht University, Netherlands.

Madeira Sustainable for all. (2023). *Certificação*. Available at https://sustainableforall.visitmadeira.com/certificacao/

Majdak, P., Mosz, J., & Almeida, A. (2022). Overtourism: The impact of tourism on Madeira's social and economic environment. *Studia Ecologiae Bioethicae, 20, 2*, 1–14. https://doi.org/10.21697/seb.2022.09

Melo, R., Van Rheenen, D., & Gammon, S. (2020). Part I: Nature sports: A unifying concept. *Annals of Leisure Research, 23*(1), 1–18.

Melo, R., Sobry, C., & Van Rheenen, D. (2021). *Small scale sport tourism events and local sustainable development. A cross-national comparative perspective* (Sports economics, management and policy, D. Coates, Series Editor). Springer.

Perna, F., Custódio, M., & Oliveira, V. (2019). Local communities and sport activities expenditures and image: Residents' role in sustainable tourism and recreation. *EJTHR, 9*(1), 49–59. https://doi.org/10.2478/ejthr-2019-0006

Quintal, R. (2010). *Levadas and Footpaths of Madeira* (4ª edição). Edição Francisco Ribeiro.

Quintal, R. (2013, May). *Levadas da Madeira, monumentos criados por heróis anónimos*. Impactum, Imprensa da Universidade de Coimbra. Available at http://hdl.handle.net/10316.2/26345

Santos, F. (2019). *Retrato da Madeira*. PORDATA, Edição 2019. Fundação Francisco Manuel dos Santos, Lisboa.

Santos, P., & Correia, F. (2017). *Levadas da Madeiras. Uma Antologia Literária*. Imprensa Académica, Associação Académica da Universidade da Madeira, Funchal, Portugal.

Sean, G. (2011). Sports events: Typologies, people and place. In *The Routledge handbook of events* (pp. 104–118). Routledge.

Silva, F. (2017). Planeamento Turístico nos Espaços Insulares. In *Planeamento e Desenvolvimento Turístico* (pp. 465–480). Coord. Francisco Silva e Jorge Umbelino. Editora Lidel.

Silva, A. (2018). Políticas públicas e eventos desportivos: O caso dos eventos da Federação Portuguesa de Natação organizados na Madeira. *Motricidade, 15*(S2), 29–32. https://doi.org/10.6063/motricidade

Silva, F., & Carvalhinho, L. (2017). Turismo na Natureza e de Aventura. In *Planeamento e Desenvolvimento Turístico* (pp. 259–274). Coord. Francisco Silva e Jorge Umbelino. Editora Lidel.

Silveira, A., Prada, S., Ramalho, R., Madeira, J., Fonseca, P., Canha, E., & Brilha, J. (2012). *Inventariação do Património Geológico da Ilha da Madeira* (414 p.). Secretaria Regional do Ambiente e Recursos Naturais—Relatório Final. https://geodiversidade.madeira.gov.pt

Soares, J. (2019). Evaluation indicators of the socio-sport impact of the national swimming championship. *Motricidade, 15*(S2), 19–22. https://doi.org/10.6063/motricidade.18360

Soares, J. (2022). *Contributo para um modelo de avaliação dos impactos sociais e desportivos dos eventos desportivos de cariz turístico*. Resumo de comunicação oral. Seminário Desporto e Ciência, Universidade da Madeira, 31 de março e 1 de abril 2022.

Soares, J., & Nunes, N. (2020). Levada walks and canyoning as mountain sport products in nature tourism. *European Journal of Tourism, Hospitality and Recreation, 10*(1), 41–55. https://doi.org/10.2478/ejthr-2020-0004

Taks, M., Chalip, L., & Green, C. (2015). Impacts and strategic outcomes from non-mega sport events for local communities. *European Sport Management Quarterly, 15*(1), 1–6. https://doi.org/10.1080/16184742.2014.995116

UNESCO. (1999). *Laurissilva of Madeira*. World Natural Heritage of Humanity. Available in https://whc.unesco.org/en/list/934/

Vieira, A. (2015). *As Levadas. Os Caminhos da água na Madeira* (pp. 1–59). Cadernos de Divulgação do Centro de Estudos de História do Atlântico, Funchal. Disponível em https://issuu.com/milsumav/docs/cadernos_09_levadas. Acedido em 01 Dec 2022.

Vieira, E. (2016). *Potencial de Crescimento e Desenvolvimento do Turismo Jovem na Região Autónoma da Madeira*. Universidade de Aveiro, Departamento de Economia, Gestão, Engenharia Industrial e Turismo.

Wolsko, C., Lindberg, K., & Reese, R. (2019). Nature-based physical recreation leads to psychological well-being: Evidence from five studies. *Ecopsychology, 11*(4), 222–235. https://doi.org/10.1089/eco.2018.0076

Chapter 14
Sport Tourism in Small Island Developing States: A Case Study of the Barbados Sport Tourism Industry

Rudolph Alleyne, Antonio Alleyne, Neela Cezair, and Wayne Charles-Soverall

Introduction

Tourism is one of the major industries in small island developing states. Sport tourism has been widely regarded as one of the fastest-growing niche markets in recent years (Sporting Barbados, 2022). It is defined as "any form of activity or passive involvement in sporting activities, carried out randomly or organized for non-commercial or commercial reasons, which require leaving the home or place of work" (Standeven & De Knop, 1999).

Sport-related tourism has become a popular niche of the tourism sector over the past two decades (Kiani et al., 2019). It is quickly becoming a viable subdivision of the tourism industry in developed and developing markets and a component of the global entertainment industry. "According to the World Tourism Organization (WTO), sport tourism accounts for approximately 12–15 million international arrivals each year, with a growth forecast of around 6 per cent per year in the short term" (Sporting Barbados, 2022). In addition, it represents 10% of the world tourism industry, with a turnover of approximately USD$600 billion. Moreover, one of the world's leading research agencies has forecasted that between 2018 and 2023, the sport tourism industry would grow incrementally by 32.3% annually, thus translating to an additional USD$6.12 billion in absolute terms (Sporting Barbados, 2022).

This product continues to grow in small island developing states like those in the Caribbean, with the hosting of mega events such as the International Association of Athletics Federations (IAAF) World Championships hosted by Jamaica in 2002, the 2006 World Cup Golf Tournament hosted by Barbados, the 2007 Cricket World Cup

R. Alleyne · A. Alleyne · N. Cezair (✉) · W. Charles-Soverall
University of the West Indies, Cave Hill, Barbados
e-mail: rudolph.alleyne@cavehill.uwi.edu; neela.cezair@cavehill.uwi.edu; antonio.alleyne@cavehill.uwi.edu; wayne.charles-soverall@cavehill.uwi.edu

D. Van Rheenen et al. (eds.), *Sport Tourism, Island Territories and Sustainable Development*, Sports Economics, Management and Policy 24,
https://doi.org/10.1007/978-3-031-51705-1_14

and the 2009 Twenty/Twenty World Cup hosted by Barbados, and the FIFA Men and Women Under 17 World Cup hosted by Trinidad and Tobago in 2001 and 2010, respectively (Joachm, 2010). Sport contributes approximately 2% of gross domestic product (GDP), while tourism contributes 4–6% of GDP (Sinclair, 2005) of economic activity. This may be considered by some as insignificant, but the contribution of sport tourism to small island developing states represents an essential contributor to economic activity for these territories. These contributions may include jobs, income opportunities, increased visitor spending, increased exposure to local culture and interaction with locals, enhanced community pride, and the promotion of gender-neutral opportunities. The purpose of this chapter is to explore how sport tourism can contribute to sustainable economic development in Small Island Developing States (SIDS). The authors apply an empirical approach for the case study analysis as the methodology for exploring the phenomenon of sport tourism. This approach allows for multiple facets of the phenomenon to be revealed and understood based on the case as the unit of analysis.

Literature Review

Tourism is considered a dynamic industry (Sengel et al., 2022) and one of the major industries in small island developing states. Island nations or small island developing states (SIDS) may differ economically and socially from developed or developing countries and often suffer problems of insularity and vulnerability (Lim & Patterson, 2008). However, they also experience benefits related to their geographical locations, unique cultures, and international investment, which often aid in their overall competitiveness. Many of these small island states seek to capitalize on their natural comparative advantages—inviting climate, water tourism, compelling architectural heritage, an assortment of culinary offerings, and a friendly and welcoming society—to explicitly and deliberately target tourism as a primary economic development option (Lewis-Cameron & Roberts, 2010). Sinclair (2005) highlights the increased desire for small island states to diversify their traditional product (sun, sand, and sea), hence focusing on exploring different tourism experiences. Thus, hosting successful internationally recognized sporting events could represent a sustainable development support system for small islands (Lim & Patterson, 2008). This is also echoed by Sinclair (2005), presenting sport tourism as one medium to showcase the small island experience worldwide.

Generally, sport tourism includes travel to participate in a passive sports holiday (such as a specific sports event or a sports museum) or an active sports holiday (such as sailing, diving, cycling, or golf). It may also involve mixed instances where either sport or tourism is the dominant activity or reason to travel (Ritchie & Adair, 2004). Higham and Hinch (2018) contend that sport tourism comprises a wide variety of niche market segments, each of which has distinct features that require further empirical research. One related topic is destination competitiveness which is under-researched in small island developing states. Sport tourism destinations must comprehensively understand their competitiveness (Maravic et al., 2015) because increasing competition in sport tourism causes a more in-depth understanding of the distinct perceptions of the niche market (Higham, 2005). In addition, destination competitiveness is affected by various factors (Crouch, 2011), including that there is no one-size-fits-all approach (Khin et al., 2014), and each destination emphasizes different factors (Mulec & Wise, 2013). Moreover, a comprehensive approach is needed to address small islands' multidimensional issues (Azzopardi & Nash, 2016). Furthermore, sport tourism helps the overall development of sport and tourism and occupies a distinct economic niche in administration and marketing (Tichaawa et al., 2015). Standeven and De Knop (1999) described persons traveling for business or non-commercial reasons that necessitate travel, away from home and work locality, participating casually or in an organized way as sport tourism.

Sport tourism, more specifically, has been seen as a viable option for the smaller islands, as they have sought to develop a niche market tourism strategy to mitigate the negative impacts of mass tourism-oriented development on the islands. Butler (2008) writes that islands in the Mediterranean, Caribbean, and South Pacific are significant destinations. For example, Bonaire in the Dutch Caribbean is one of the most successful cases of sports market development (Pieters & Gevers, 1995). Additionally, Malta exemplifies a changing market position that has moved away from "mass" tourism to a "new trend" toward special interest travel (Bull & Weed, 1999; Higham, 2005). This country has developed a range of sporting activities to diversify its tourism product offering. Sugden (2007) notes that Cuba is known for its residential coaching camps in many elite competitive sports.[1] Cyprus' aspiration to be a highly regarded golfing destination could be seen in its heavy investment in the sport (Boukas & Ziakas, 2013). Likewise, Sinclair (2005) underscored high-profile events from the Caribbean that currently impact the international tourism market: Antigua's Sailing Week, the Annual November Motor Racing Championship in Guyana, The Cockspur Gold Cup Horse Racing (now known as the Sandy Lane Gold Cup) event in Barbados, The Tour de la Martinique, the Central American and Caribbean Bodybuilding and Fitness Championship, and last, but certainly not least, test cricket matches in the Caribbean. Barbados has shown its cricketing prowess by producing the most renowned international players per capita and is therefore considered the cricketing capital of the world (Martin, 2022).

Various research studies indicate that international tourism is an essential source of revenue and employment for the host nation, resulting in increased foreign exchange earnings. Sporting tours draw visitors to their favorite sporting event, facility, or destination worldwide (Gozalova et al., 2014). However, one of the drawbacks of international tourism is that it presents itself as a divisive topic among the locals. Previous literature has highlighted a significant challenge in the globalized sports market, where local and national brands suffer as a result of international franchises (Zhang et al., 2018). Deccio and Baloglu (2002) reason that the increased

[1] Athletics, boxing, basketball, volleyball, and gymnastics, and sport events for the general population such as golf, marathons, and cycling.

economic activity occurs due to constructing and/or improving sports facilities that are required to host these events. The literature shows that this drawback has perceived benefits as locals can utilize the new and improved sports facilities for recreational and community activities (Kim et al., 2006).

It is crucial to note that Zhang et al. (2018) linked the importance of sporting infrastructure to the growth of the sports industry. The authors further expressed that "the most viable means to secure the needed sporting infrastructure, is to host a mega event" (Zhang et al., 2018). The seminal work of Higham (1999) acknowledged the need for host communities to attract sporting events that complement their scale, infrastructure, and resource capabilities. He notes that sporting events usually operate within existing infrastructures, require minimal investments of public funds, and can curtail the adverse seasonal effects. Therefore, sporting events may significantly impact a host nation's economic growth (Idrees et al., 2021) through inward investment, increased tourist arrivals, and the development of the service sector (Weed & Bull, 2004).

The economic impact of sporting events is the net change in an economy, resulting from a sporting event, caused by activities involving the acquisition, operation, development, and use of sports facilities and services (Lieber & Alton, 1983). These may increase visitor spending, public spending, employment, and generate revenue. Specifically, the economic impacts are direct, indirect, and induced/rippled effects. Direct effects are those purchases needed to meet the increased demand of visitors for goods and services. Indirect effects are the multiplier effects caused by recirculating the initial spectators' dollars. Induced effects are the increase in employment and household income that result from the economic activity fueled by the direct and indirect effects, i.e., employees of impacted businesses spending some of their salaries and wages with other companies in the host community (Howard & Crompton, 1995).

As an emerging topic, the literature has reported that hosting mega events presents opportunities for societal, economic, and cultural development. Horne and Manzenreiter (2004) detailed that mega events can bring an expansion of sport facilities, renovation of urban infrastructure, and the introduction of new sports. Kim et al. (2006) add to the list of Horne and Manzenreiter (2004) by detailing the possibility of positive economic impacts such as increased tax revenues, new employment opportunities, and additional sources of income. Opposing these views are numerous scholars that challenge the actual economic growth in developing countries that host mega events (Deccio & Baloglu, 2002; Kim & Morrison, 2005; Matheson Baade, 2006). However, several studies on the economic effects of sporting events or activities have found significant positive impacts. Baim (1994) found positive employment effects from MLB and NFL teams in 15 cities in the USA. Sport Tu (2005) found a significant positive impact from the FedEx Field (Washington) on real estate prices in the surrounding neighborhood, as did Ahlfeldt and Maennig (2009, 2010a, b) for three arenas in Berlin, Germany.

Jones (2001) identified the nature of the social as well as the economic impact of hosting the mega sporting 1999 Rugby World Cup in Wales. A review of the literature by Jones (2001) revealed that mega sports events have significant social impacts, such as stress on property markets and local housing on the one hand. But these events have important positive social implications as they contribute to developing a sense of worth in the community. The author concluded that hosting mega sports events was a success for the host nations, but the economic benefits of hosting the event are still uncertain. Findings by Giampiccoli et al. (2015), which focused on the impact of hosting FIFA World Cup 2010 on the economic growth of South Africa, indicated that this mega sporting event had been a success because it contributed to boosting the tourism industry in the short-term. It was found to be sustainable, less costly, and positively impacted the host region's economy.

In 2006, Matheson assessed the economic impact of sporting events by reviewing the related literature. The author observed the effect of mega sporting events on local, national, and regional economies. He recommended different ways to maximize the net benefits of hosting such events in host regions. First, limiting the new infrastructure's expenditures to compensate for an event's cost will increase the benefits. Second, the local officials of the host regions should understand that hosting many small events may generate more net benefits than the strategy that leads to larger significant events because crowding is less likely to occur in smaller sporting events. Therefore, instead of focusing on hosting mega events, smaller island nations could host numerous small events in the region that will increase the net benefits of significant events (Coates & Depken, 2011; Maennig & Zimbalist, 2012). This is supported by a similar study conducted by Kavetsos and Szymanski (2010) using data that covered three important events including the Olympic Games, the FIFA World Cup, and the UEFA European Championship from 12 European countries on the impact of hosting mega sports events on the national well-being.

In 2010, Lorde, Greenidge, and Devonish used residents' perceptions to measure the impacts of the 2007 ICC World Cup on Barbados. The authors acknowledged that the residents' perception alone did not present a complete image of the event's impact. However, with the employment of two theories, the social exchange theory and the prospect theory, they were able to analyze Barbadians' perceptions of the 2007 ICC World Cup. The social exchange theory explains that "locals are likely to engage if they believe that the benefits of hosting the mega event, in this case, ICC World Cup, outweighed the cost and will help the community in the future" (Allen et al., 1993). The prospect theory conveys "if there is a difference in pre- and post-perceptions, residents modify their perceptions, hence changing their attitude and engagement to mega events in the future" (Kahneman & Tversky, 1979). The authors concluded that Barbadians felt that the benefits did, in fact, outweigh the cost of hosting the 2007 ICC World Cup (Lorde et al., 2010). This mega event, hosted by multiple small island developing states, was deemed a success in the minds of Barbadians.

The potential of hosting mega sporting events to foster sustainable development in small island states requires further investigation. Given the importance of the tourism industry and inherent resource scarcity in small island developing states, any sporting tourism agenda must optimize the use of available resources to achieve the range of benefits. In addition, small island developing states may suffer from a lack of economies of scale, high transportation and communication costs, and costly public administration and infrastructure establishment (Lim & Patterson, 2008). Therefore, the overreliance of the small island economies on traditional tourism questions the competitiveness of the local economy without harming the individual cultural dynamic.

During a 2010 CARICOM conference held in Barbados that focused on Sport Tourism, it was reported that some small island states seem unable to tap into the viability of the sport tourism market (Joachm, 2010). One example highlighted was the presence of numerous world-class cricket facilities around the region that should translate into hosting international competitions and teams regularly. Joachm argues that the lack of proper marketing and management of the facilities would "suggest to the international cricketing fraternity the lack of readiness to offer consistent high-quality facilities for top level competitions" (Joachm, 2010). In other words, the sport tourism offerings of small island states would not have a major impact if "the facilities are at one level one year, and of a worse or declining level the following year/s" (Joachm, 2010).

Tourism functions as a significant industry in small island states, with sport tourism as an emerging and viable sub-sector. With tourism being one of the major economic earners for many small island states, sport tourism should be promoted as a meaningful contributor to sustainable economic development. In the words of Joachm (2010), "the Caribbean can and must engage in a systematic strategy aimed at facilitating the exploitation of the vast sport tourism potential that exist." This approach can be adopted by small island developing states as they seek to develop their sport tourism product.

Methodology

This section discusses the empirical approach taken to investigate the Barbados case. The data that has been used is time series data that covers the years from 1998 to 2020. Model data for dependent and independent variables were acquired from the World Bank databank. Information regarding sporting events at the regional and international level was collected from Sporting Barbados magazine—an annual publication.

Acknowledging the data limitations, an ordinary least square (OLS) estimation technique was employed to examine the impact of these sporting events, as an indicator of sport tourism, on the economic growth of the host island Barbados. Following the economic theory, this study aligns with the Cob-Douglas production function that describes the relationship between inputs and output, i.e., growth is a function of (K) capital and (L) labor—$(\_K,\_L)\_ = \_\ \_bKaLC$. For this model, the control variable is gross fixed capital formation per capita (gross fixed capital formation/population), and the focus variable is the number of international/regional events held annually. This research also singles out the number of non-domestic cricketing events held each year. Due to its dominance as a mega sport in the country and the island being labeled a cricketing capital with many famous international players (Martin, 2022), this secondary investigation is to observe whether the sport offers significant positive impacts on the economic growth of Barbados over the years under review.

Worrell (2012) highlighted the importance of generating foreign exchange through trade for small open economies. According to the World Tourism Organization (2010), due to the increasing number and rising investment in developing tourist-related industries, travelers have become an essential source of economic progress because they contribute to the creation of employment opportunities, the development of infrastructure, and earning of revenues through the export of local goods and services. Thus, as a robustness check of the impact of sporting events, this study replaces capital within the model with trade openness (as % of GDP) to verify results.

The following regression model dictates the assessment formed within this study:

$$DPpcg\_ = \_\_\beta 1\_\_ + \_\_\beta 2\_GFCpct - 2\_ + \_\_\beta 3\_SEt + \_\_ut\_$$

where the $GDPpc$ _represents the gross domestic product per capita growth rate; $GFCFpct-2$_ _represents gross fixed capital formation with a 2-year lag; the inclusion of lags is to minimize concerns about simultaneity bias (Hyland et al., 2021); SEt _indicates the number of tourists impacting events that Barbados hosted for the years, and ut _is the error term. To reemphasize the position of the WTO (2010), tourists-related contributes to revenue generation and job creation, which lead to the economic growth of developing nations. Overall, the primary purpose is to investigate the current strength of high-level sporting events to boost Barbados' earning potential, as events encourage infrastructural (foreign) investment in support of the local communities and bring foreign exchange earnings to the local economy.

As a result of the limited data series (1998–2020), the hypothesis test statistic (t-statistic) is used.

Theoretical Implications

A significant econometric outcome is not the sole criterion by which to measure the impact of sport tourism or major sporting events on the small island developing state of Barbados and others like it. Several factors influence the impact of sporting events on the economy. Given the limitations expressed in economic theory and the applied economic model, caution is necessary when interpreting the findings. In this context, attention should also be placed equally on the social, cultural, political, environmental, and management factors that maximize the net benefits of sport tourism and hosting sporting events. Thus, this study has significant implications for practitioners and policymakers.

Estimation and Results

This section discusses the estimated methods and results of the regression model that assesses the impact of non-domestic sports events on the growth of the Barbados economy.

Stationary Test

Using a time series data structure requires the validation of stationarity in the data. This study employs the popular Augmented Dickey-Fuller and the Phillips-Perron tests for unit-root presence. The results for each test taken at levels and first difference are presented in Table 14.1.

The results above indicate the presence of a unit-root/non-stationary in both control variables at levels with the ADF and PP tests. According to the results of the tests at levels, the *p*-values of the dependent variable and the focus variable are below 5%, which rejects the null hypothesis that these two variables possess unit roots. At first difference, the ADF and PP tests statistics for the measure of per capita capital and trade openness are greater than the critical values, thus indicating stationarity of all variables.

Regression Results

The OLS estimation was performed through Stata 14 to observe the impact of international and regionally hosted sporting events on the economic growth of Barbados (see Table 14.2).

The results in Table 14.2 indicate that gross fixed capital formation per capita significantly positively affects the growth of GDP in Barbados across both estimations. However, neither the summary of all internationally/regionally hosted

Table 14.1 Summaries of the results of each test carried out

	Augmented Dickey-Fuller (ADF)		Phillips-Perron (PP)	
Variables	Levels	First difference	Levels	Second difference
GDP per capita growth	**0.0008**		**0.0008**	
Gross fixed capital Formation per capita	0.7551	**0.0022**	0.7254	**0.0031**
Trade openness	0.8804	**0.0004**	0.8062	**0.0003**
Sporting events	**0.0207**		**0.0259**	

Table 14.2 Results of Ordinary Least Squarenational (OLS)

Variables	Coefficient	*t* statistics	VIF
Gross fixed capital formation per capita	14.457*	1.95	1.14
Lags (2)	*(7.424)*		
International and regional sporting events	0.011	0.13	1.14
	(0.086)		
Constant	−0.917	−0.49	
	(1.887)		
Observations	**23**		
R-squared	**0.165**		
Gross fixed capital formation per capita	15.082*	2.03	1.11
Lags (2)	*(7.432)*		
International and regional cricket events	0.227	0.45	1.11
	(0.505)		
Constant	−1.536	−0.79	
	(1.949)		
Observations	**23**		
R-squared	**0.172**		

sports nor non-domestic cricket events has indicated any significant influence on the growth of the Barbados economy. Based on the above literature, there is evidence that sport tourism and hosting non-domestic sporting events could be used to enhance economic growth in small developing nations like Barbados. However, realizing such potential requires effective planning and management of the niche areas. It may be prudent to target key sporting events to serve under that sport tourism banner (where the burden of infrastructural spending is minimal), and significant private investment is provided.

The authors have also checked the existence of multicollinearity through the VIF test. VIF for all variables is less than 10, meaning that our model has no multicollinearity.

Robustness Checking

The initial regression was repeated as a robustness check, replacing capital with trade openness as the control variable. The results in Table 14.3 indicate all expected results, with positively significant effects from trade openness as a key driver of growth. Again, the current trend of hosting high-level sporting events on the island offers no significant measurable impact.

Table 14.3 Results of robustness check

Variables	Coefficient	*t* statistics	VIF
Trade openness	0.3330*	1.76	1.00
	(0.189)		
International and regional sporting events	−0.051	−0.60	1.00
	(0.086)		
Constant	1.214	0.68	
	(1.797)		
Observations	**23**		
R-squared	**0.165**		

Discussion

We have examined the relevant factors identified in the literature that influence the relationship between sport tourism and mega sporting events and their impact on the economic growth of small island developing states such as Barbados. Our analysis found no measurable evidence that sport tourism and hosting non-domestic sporting events have enhanced economic growth in small developing nations like Barbados. However, realizing positive gains may require adequate planning and management in niche areas of these sporting events. It may be prudent to target key sporting events that can attract tourists and significant private investment.

Based on our research, there are several gaps in the literature on factors that influence the impact of sport tourism on small island developing states. First, there is an over-emphasis on the economic impact, but it is not the only important aspect that requires consideration. Second, there is little recognition that sport tourism comprises a wide variety of niche market segments, each with distinct features that require further research. Such niche market segments have not been duly considered in sport tourism in the case of small island developing states. Third, the effects of destination competitiveness in small islands appear to be under-researched. According to Crouch (2011), destination competitiveness is affected by various factors, including the approach of no one-size-fits-all (Khin et al., 2014), and each destination emphasizes different factors (Mulec & Wise, 2013). A comprehensive approach is needed to address small islands' multidimensional issues (Azzopardi & Nash, 2016). Finally, sport tourism helps the overall development of sports and tourism and occupies a distinct economic niche in administration and marketing (Tichaawa et al., 2015). Therefore, for the Barbados case, before discarding the possibility of a viable sport tourism sector, the values and benefits to the economic, social, and cultural fabric of the economy must all be considered.

The effects of hosting any mega tourism sporting event will have an impact on the social and cultural landscape, bringing value to the socioeconomic development of a country. An increase in tourist arrivals, increased visitor spending, employment, improved infrastructure, and new sporting facilities, though marginal, still adds value to the economic landscape. Hosting these events may also encourage the local sporting fraternity to adopt a more practical and internationally accepted approach to hosting sporting events and the overall management of the product. This will eventually aid in developing both the local sporting industry and the sport tourism product of small island developing states. As with the example of the 2007 Cricket World Cup in the Caribbean, the research revealed that the locals felt that the benefits of hosting this event outweighed the cost of hosting it.

Conclusions

This research paper focused on examining the impact of sport tourism and hosting sporting events on the economic growth of a host small island developing nation, namely, Barbados. To this end, we used time series data from 1998 to 2020 and observed the impact of sporting events on the economy of Barbados. We used an OLS technique to examine the impact of international and regionally hosted sporting events on the economic growth of Barbados. Various research studies indicated that international tourism is an essential source of revenue and employment in the host nation through increased foreign exchange earnings. Therefore, it can be said that sporting events may have a significant impact on the economic growth of a host nation's economy.

However, in cases such as Barbados, governments and practitioners should give due consideration to the requirements of the sports infrastructure, thus avoiding unnecessary investment to reduce the cost of sports events, especially the mega ones. A transparent and precise assessment of the significance of mega sports events is needed if effective, efficient, and sustainable planning can be achieved to avoid huge costs associated with the events. The results suggest that the economic impact of hosting sporting events is not always found to be positive as the predicted economic benefits. Hosting sporting events can help improve human capital including knowledge and skills that can be improved in the service provided through the training of hospitality personnel in the tourism sector.

Policy Implications

Practitioners and policymakers should consider sport tourism destinations such as small islands as unique niches rather than simply focusing their attention on tourism in general and sport tourism as a small component of that larger picture. In the final analysis, there is little or no real focus on the unique niches provided by sport tourism destinations like Barbados. A great deal of time and financial resources are allocated by policymakers to infrastructural development in sport tourism, especially in relation to its economic, environmental, and technical impacts. However, the evidence suggests that little effort is devoted to effective planning and management of the niche areas to achieve the desired returns on investment. In this context, it may be useful to identify destination competitiveness factors that can create a coherent policy framework to guide destination development. This overarching goal requires the coordination of policies and strategies that are currently implemented in a silo fashion by the relevant stakeholders. We suggest that the policymakers consider providing several incentives to the tourism industry to strengthen the brand for an interesting and strong destination. In this context, emphasis should be placed on institutional development to attract more investment, promotion of the destination's unique attributes, and the development of a communications network to promote the story of the brand through destination management and branding.

Practical Implications

We suggest that there are different ways in which to maximize the net benefits of hosting sporting events such as limiting the expenditures on new infrastructure in order to compensate for the cost of an event that will in turn increase the benefits of the sporting events. In addition, sport tourism destinations must comprehensively understand their competitiveness because increasing competition in sport tourism causes a more in-depth understanding of the distinct perceptions of the niche market. Moreover, sport tourism destination managers of small islands need to organize priorities better (Corneal, 2020) in order to enhance their destinations' competitiveness and provide tourists with a more accurate depiction of the destination and its unique appeal. To this end, destination management organizations should consider how best to craft their strategies based on features and competitive advantages that may differ from destination to destination. Hence, greater attention paid to such details will be critical to the success of the destinations in question.

Limitations and Future Research

Although this research study highlights the economic impact of sport tourism, from a Barbados perspective, there were several limitations. The limited data set was a factor in choosing a suitable method for this paper, in accordance with traditional economic theory. The acquisition of an appropriately larger data set can facilitate alternate qualitative and/or quantitative approaches with multiple developing nations. In this context, future research may focus on sport tourism development and destination competitiveness on small islands as other areas which should be the focus of government. Perhaps there should be a focus on a destination competitiveness model or a comprehensive model to address small islands' multidimensional issues. In addition, future research should focus on measuring the socio-cultural impact as well as the environmental impact of sport tourism and hosting mega sports events in Barbados since economic impacts are not isolated from the wider environment that shapes the eventual incomes.

References

Ahlfeldt, G., & Maennig, W. (2009). Arenas, arena architecture and the impact on location desirability: The case of 'Olympic Arenas' in Prenzlauer Berg, Berlin. *Urban Studies, 46*(7), 1343–1362.

Ahlfeldt, G., & Maennig, W. (2010a). Impact of sports arenas on land values: Evidence from Berlin. *The Annals of Regional Science, 44*(2), 205–227.

Ahlfeldt, G., & Maennig, W. (2010b). Stadium architecture and urban development from the perspective of urban economics. *International Journal of Urban and Regional Research, 34*(3), 629–646.

Allen, L. R., Hafer, H. R., Long, R., & Perdue, R. R. (1993). Rural residents' attitudes towards recreation and tourism development. *Journal of Travel Research, 21*, 27–33.

Azzopardi, E., & Nash, R. (2016). A framework for Island destination competitiveness—Perspectives from the Island of Malta. *Current Issues in Tourism, 19*, 253–281.

Baim, D. V. (1994). *The sports stadium as a municipal investment*. Greenwood Press.

Boukas, N., & Ziakas, V. (2013). Exploring perceptions for Cyprus as a sustainable golf destination: Motivational and attitudinal orientations of golf tourists. *International Journal of Sport Management and Marketing, 14*(1/4), 39–70.

Bull, C., & Weed, M. E. (1999). Niche markets and small Island tourism: The development of sports tourism in Malta. *Managing Leisure, 4*(3), 142–155.

Butler, R. (2008). Islands. In M. Lück (Ed.), *Encyclopedia of tourism and recreation in marine environments* (p. 254). CABI.

Coates, D., & Depken, C. (2011). Mega-events: Is the Texas Baylor Game to Waco what the Super Bowl is to Houston? *Journal of Sports Economics, 12*(6), 599–620.

Corneal, A. (2020, January 28). *Sports tourism in the Caribbean*. Caribbean Insight.

Crouch, G. I. (2011). Destination competitiveness: An analysis of determinant attributes. *Journal of Travel Research, 49*(3), 10.1177/0047287510362776.

Deccio, C., & Baloglu, S. (2002). Nonhost community resident reactions to the 2002 Winter Olympic: The spillover impacts. *Journal of Travel Research, 41*(1), 46–56.

Giampiccoli, A., Lee, S. S., & Nauright, J. (2015). Destination South Africa: Comparing global sports mega-events and recurring localized sports events in South Africa for tourism and economic development. *Current Issues in Tourism, 18*(3), 229–248.

Gozalova, M., Shchikanov, A., Vernigor, A., & Bagdasarian, V. (2014). Sports tourism. *The Polish Journal of Sport and Tourism, 21*, 92–96.

Higham, J. (1999). Commentary—Sport as an avenue of tourism development: An analysis of the positive and negative impacts of sport tourism. *Current Issues in Tourism, 2*(1), 82–90.

Higham, J. E. S. (2005). *Sport tourism destinations: Issues, opportunities and analysis*. Elsevier.

Higham, J., & Hinch, T. (2018). *Sport tourism development* (3rd ed.). Channel View Publications.

Horne, J. D., & Manzenreiter, W. (2004). Accounting for mega events: Forecast and actual impacts of the 2002 Football world cup finals on the host countries Japan/Korea. *International Review for the Sociology of Sport, 39*(2), 187–203.

Howard, D. R., & Crompton, J. L. (1995). *Financing sport*. Fitness Information Technology, Inc.

Hyland, M., Djankov, S., & Goldberg, P. K. (2021). *Do gendered laws matter for women's economic empowerment?* Women's Economic Empowerment Research Initiative. Retrieved from https://www.piie.com/sites/default/files/documents/wp21-5.pdf

Idrees, M. A., Bibi, M., Naveed, J., & Khan, A. (2021). Sports tourism and its impact on economic growth of a developing nation. *Sustainable Business and Society in Emerging Economies, 3*(4), 617–628.

Joachm, S. (2010). *Caribbean sport tourism can be viable*. The Offical Website of St. Vincent and The Grenadines Olympic Committee.

Jones, C. (2001). Mega-events and host-region impacts: Determining the true worth of the 1999 Rugby World Cup. *International Journal of Tourism Research, 3*(3), 241–251.

Kahneman, D., & Tversky, A. (1979). Prospect theory: An analysis of decision under risk. *Econometrica, 47*(2), 263–291.

Kavetsos, G., & Szymanski, S. (2010). National wellbeing and international sports events. *Journal of Economic Psychology, 31*(2), 158–171.

Khin, E., Daengbuppha, J., & Nonsiri, P. (2014). Destination competitiveness: A structural model for measuring attributes competitiveness of Bagan, Myanmar. *International Journal of Business Economics and Law, 4*, 51–59.

Kiani, M. S., Nazari, L., & Shahbazpour, L. (2019). Sport tourism and sustainable locak development for host cities for sporting events. *American Journal of Sport Science, 7*(1), 7–10.

Kim, S. S., & Morrison, A. M. (2005). Change of images of South Korea among foreign tourist after the 2002 FIFA World Cup. *Tourism Management, 26*, 233–247.

Kim, H. J., Gursoy, D., & Lee, S. B. (2006). The impact of the 2002 World Cup on South Korea: Comparisons of pre- and post-games. *Tourism Management, 27*(1), 86–96.

Leiber, S. R., & Alton, D. J. (1983). Visitor expenditures and the economic impact of public recreation facilities in Illinois. In S. R. Leiber & D. R. Fesenmeier (Eds.), *Recreation planning and management*. Venture Publishing.

Lewis-Cameron, A., & Roberts, S. (2010). Small Island developing states: Issues and prospects. In A. Lewis-Cameron & S. Roberts (Eds.), *Marketing Island destinations: Concepts and cases* (pp. 1–8). Elsevier.

Lim, C. C., & Patterson, I. (2008). Sport tourism on the islands: The impact of an international mega golf event. *Journal of Sport & Tourism, 13*(2), 15–133.

Lorde, T., Greenidge, D., & Devonish, D. (2010). Local residents' perceptions of the impacts of the ICC Cricket World Cup 2007 on Barbados: Comparisons of pre- and post-games. *Tourism Management, 32*(2), 349–356.

Maennig, W., & Zimbalist, A. (2012). What is a mega sporting event? In W. Maennig & A. Zimbalist (Eds.), *International handbook on the economics of mega sporting events* (pp. 9–14). Edward Elgar Publishing.

Maravić, M. U., Bednarik, J., & Lesjak, M. (2015). Slovenian sport tourism destinations and their competitiveness. *Annales Kinesiologiae, 6*(1), 59–67.

Martin, A. (2022, March 19). *'It's in the blood': How Barbados became cricket's ultimate hotspot.* Retrieved from The Guardian https://www.theguardian.com/sport/blog/2022/mar/19/its-in-the-blood-how-barbados-became-crickets-ultimate-hotspot

Matheson, V. (2006). *Mega-events: The effect of the world's biggest sporting events on local, regional, and national economies* (Economics Department Working Papers. Paper 68).

Matheson, V. A., & Baade, R. A. (2006). Padding required: Assessing the economic impact of the super bowl. *European Sport Management Quarterly, 6*, 353–374.

Mulec, I., & Wise, N. (2013). Indicating the competitiveness of Serbia's Vojvodina Region as an emerging tourism destination. *Tourism Management Perspectives, 8*, 68–79.

Pieters, R., & Gevers, D. (1995). A framework for tourism development on fragile Island destinations: The case of Bonaire. In T. Baum & M. V. Conlin (Eds.), *Island tourism: Management principles and practice* (pp. 123–132). Wiley.

Ritchie, B. W., & Adair, D. (2004). *Sport tourism: Interrelationships, impacts and issues.* Channel View Publications.

Sengel, U., Çevrimkaya, M., Genç, G., Iskın, M., Zengin, B., & Sarıısık, M. (2022). An assessment on the news about the tourism industry during the COVID-19 pandemic. *Journal of Hospitality and Tourism Insights*. https://doi.org/10.1108/JHTI-05-2020-0072/full/html

Sinclair, D. (2005). Sports education—A priority for Caribbean sports tourism. *International Journal of Contemporary Hospitality Management, 17*(6), 536–548.

Sporting Barbados. (2022). *Sporting Barbados magazine*. Hiltop Publications Ltd.

Standeven, J., & De Knop, P. (1999). *Sport tourism*. Human Kinetics.

Sugden, J. (2007). Running Havana: Observations on the political economy of sport tourism in Cuba. *Leisure Studies, 26*(2), 235–251.

Tichaawa, T. M., Moyo, S., Swart, K., & Mhlanga, O. (2015). The perceived impacts of sport tourism events: Residents' perceptions in East London, South Africa. *African Journal for Physical Health Education, Recreation and Dance, 21*(4:1), 1218–1232.

Tu, C. C. (2005). *How does a new sports stadium affect housing values? The case of FedEx field.* University of Wisconsin Press Journals Division.

Weed, M., & Bull, C. (2004). *Sports tourism: Participants, policy and providers.* Elsevier.

United Nations World Tourism Organization (2010). *UNWTO Annual Report 2010: A Year of Recovery.* Accessed at https://www.unwto.org/archive/global/publication/unwto-annual-report-2010

Worrell, D. (2012). *Policies for stabilization and growth in small very open economies* (Occasional Paper, No. 85). Group of Thirty, Washington, DC.

Zhang, J. J., Kim, E., Marstromartino, B., & Qian, T. Y. (2018). The sport industry in growing economies: Critical issues and challenges. *International Journal of Sports Marketing and Sponsorship, 19*(2). https://doi.org/10.1108/IJSMS-03-2018-0023

Chapter 15
Sport Tourism in French Polynesia and Its Dependence on Air Transport

Philippe Bachimon and Anthony Tchekemian

Introduction

In the tourism sector, the cultural resources of a territory are a key driver of competitiveness. In French Polynesia, this sector represents a significant weight in the local economy[1]; in 2018, before the COVID-19 pandemic, it contributed to 8% of the gross domestic product (GDP). It should be remembered that French Polynesia is

[1] In 2018, in the context of the economic crisis linked to the COVID-19 pandemic, tourists spent 65 billion Pacific francs (Xpf) or about 541 million (excluding international air transport) during their stay, which averaged 14.7 days. This expenditure represents five times the amount of exports of local products. Each tourist spends an average of Xpf 300,000 per stay (i.e., about Xpf 2500). Compared with other overseas collectivities, this analysis confirms the important share of tourism in the GDP and its added value of French Polynesia. See website of the Presidency of French Polynesia, "Compte rendu du Conseil des ministres", Wednesday 29 July 2020, Bureau de la communication, 6 p., online 29 July 2020, [URL: https://www.tntv.pf/tntv/wp-content/uploads/2020/07/Conseil-des-ministres-29-juillet-3.docx], accessed 24 August 2022.

P. Bachimon (✉)
Joint Research Unit (UMR) 228 Space for Development, University of Avignon (French), Avignon, France
e-mail: Philippe.bachimon@univ-avignon.fr

A. Tchekemian
Department of Letters, Languages and Humanities, University of French Polynesia, Puna'auia, French Polynesia

UMR 241 Oceanic Island Ecosystems & UMR 228 Space for Development, Tahiti, French Polynesia

International Judo Federation, Eligible for the 2023 Pacific Games, Budapest, Hungary
e-mail: anthony.tchekemian@upf.pf

D. Van Rheenen et al. (eds.), *Sport Tourism, Island Territories and Sustainable Development*, Sports Economics, Management and Policy 24,
https://doi.org/10.1007/978-3-031-51705-1_15

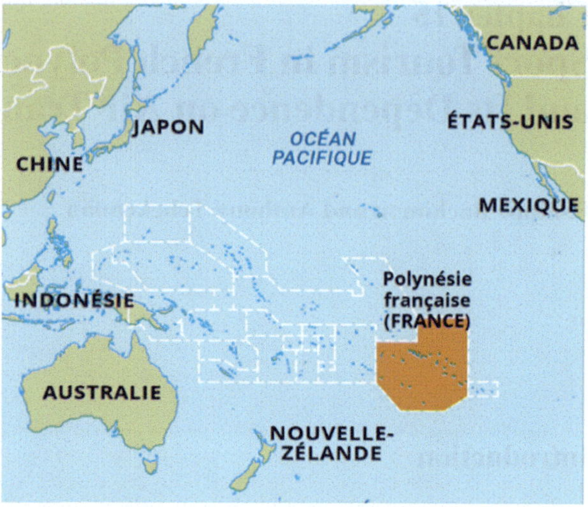

Fig. 15.1 French Polynesia in Oceania, in the South Pacific. (Source: Encyclopædia Universalis France)

located in the southern hemisphere, in the centre of the Pacific Ocean, i.e., on the opposite side of the world to metropolitan France. Depending on the season, there is a time difference of between 11 and 12 h between Paris and Papeete, which is approximately 15,700 km away. The closest continents are Australia, 6000 km away, and North America, about 6500 km. Asia is 10,000 km away and Europe is 18,000 km (Fig. 15.1).

French Polynesia is an Overseas Collectivity (COM)[2] with a "*very advanced, quasi-political administrative autonomy*" (Moyrand, 2009: 3), within the French Republic.[3] It is made up of 121 islands, 75 of which are inhabited, spread over 2.5 million km^2 of ocean and only 3793 km^2 of land, according to the latest global census of the Institute of Statistics of French Polynesia (ISFP) in 2017.[4] These islands are grouped into five archipelagos which include the "high" islands of

[2] In the 1930s, the use of the adverb "overseas" became widespread: in 1934, the Ecole Coloniale was renamed the Ecole Nationale de la France d'Outre-Mer (ENFOM); in 1935, the Musée des Colonie became the Musée de la France d'Outre-Mer. Thus, in 1930, "overseas" became a euphemistic way of referring to the colonies: "France d'outre-mer" was strictly a synonym for French colony, until 1958. This is confirmed "in the 1958 Constitution, which distinguishes the French people from the overseas peoples, to whom the right to free determination is recognized" Gay (2018: 2).

[3] The official geographical code of French Polynesia, assigned by the National Institute of Statistics and Economic Studies (INSEE), is 987.

[4] Cf. ISFP website, "Databases", "Population Censuses", "2017", "Presentation", "Geography" [URL: https://www.ispf.pf/bases/Recensements/2017/Presentation/g%C3%A9ographie], accessed 19 June 2021.

volcanic origin, such as those of the Society,[5] the Marquesas,[6] the Australs,[7] and some of the Gambiers[8] and then the so-called "low" islands, mainly in the Tuamotus[9] and the Gambiers. The latter, without relief, are atolls of madreporic origin, crowned with corals that can form a lagoon (Fig. 15.2).

The Polynesian population, numbering 275,918 in 2017, is mainly concentrated in the Windward Islands (207,333 in 2017). Two-thirds reside in Tahiti, the largest island (1042 km^2), particularly in the urban centre of Papeete, comprising the communes of Punaauia, Faa'a, Papeete, Pirae, Arue, and Mahina. In Polynesia, the climate is punctuated by two seasons: one hot and rainy, which extends from November to April, and the other, relatively cool and dry, from May to October. However, the climate varies according to latitude, relief, and exposure to the prevailing winds. Due to their location, geomorphological configurations, climatic particularities, and representations as "paradise destinations" (Gay, 2018), the Polynesian islands and atolls offer the possibility of practising a diversity of sport tourism activities throughout the year. Thus, hiking, surfing, diving, sailing, deep-sea fishing, and mountain biking (MTB) testify to the potential of these small island territories (Fig. 15.3).

With accommodation structures present on some 40 islands, 34 of which have an airport, 55 are served by air or sea (Figs. 15.1 and 15.3). Thus, tourism enables more than 3800 businesses to operate while employing 5500 people in the accommodation structures alone. However, only 40 islands are impacted by a sometimes episodic tourist activity. According to the Ministry of Tourism and Labour (2020: 13), the disparities in flows, services, reception facilities, and points of interest are linked to both the nearly 100 years of development history of French Polynesia[10] and to the

[5] The Society Archipelago comprises two subdivisions: the Windward Islands, with a surface area of approximately 1195 km^2 (consisting of four high islands, Tahiti, Moorea, Maiao, Mehetia, as well as the atoll of Tetiaroa), and with 207,333 inhabitants, in 2017; and the Leeward Islands, of 408 km^2 (with five high islands, Huahine, Raiatea, Taha'a, Maupiti, and Bora-Bora, and four atolls, Manuae, Maupihaa, Motu One, and Tupai), with 35,393 inhabitants. With a surface area of approximately 1600 km^2—as much as a department such as Guadeloupe or Essonne—this archipelago is the main settlement area (88% of inhabitants) with 242,726 inhabitants (ISFP, 2017).

[6] In the Marquesas, an archipelago with a surface area of 1056 km^2, only 6 of the 12 islands are inhabited: Nuku Hiva (2951 inhabitants in 2017, spread over 387 km^2), Ua Pou (2213 inhabitants on 105 km^2), and Ua Huka (674 inhabitants on 83.4 km^2), in the north and Hiva Oa (2243 inhabitants on 316 km^2), Tahuata (653 inhabitants on 61 km^2) and in the south Fatu Hiva (612 inhabitants) on 83.4 km^2), in the north and Hiva Oa (2243 inhab. on 316 km^2), Tahuata (653 inhab. on 61 km^2), and in the south Fatu Hiva (612 inhab. on 80 km^2), i.e., a total population of 9346 inhabitants (ISFP in 2017).

[7] In 2017, the population of the Austral Islands, spread over 145 km^2, was 6965. Since 2012, it has increased by 1.8%, compared with 8.4% between 2007 and 2012 (ISFP).

[8] The Gambiers archipelago covers an area of 31 km^2 and has a population of 1592 (ISFP, 2017).

[9] The 850 km^2 Tuamotu archipelago is home to 15,289 people (ISFP, 2017).

[10] For example, the construction of the Faa'a airport coincided with the end of the political career of Pouvana'a A Oopa (1895–1977), whose great struggle had been the refusal of nuclear tests, and with the establishment of the Pacific Experimental Centre (Regnault, 2016). Thus, as early as 1957, the French government chose two territories to host its nuclear tests: the Algerian Sahara and

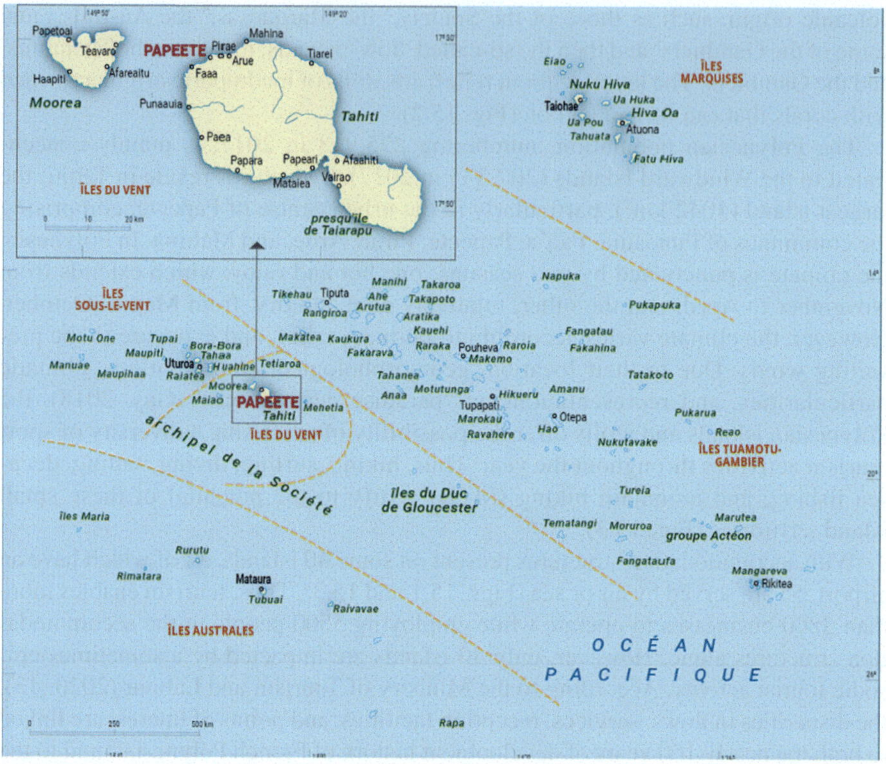

Fig. 15.2 French Polynesia is divided into groups of islands: Austral, Leeward, and Windward, forming the Society archipelago, Tuamotu-Gambiers, and Marquesas. (Source: Encyclopædia Universalis France)

evolution of trends, demands, and expectations of tourists from the outbound markets. The Polynesian tourist territories have developed according to multiple criteria that create a specific attractiveness: the types and capacities of accommodation (which define the desired clientele), the international commercial relations and networks, the services and activities on offer, the promotion of events and points of

---

the Tuamotu islands in the French Pacific. However, French Polynesia did not yet have an airport, unlike the Algerian colony, which was therefore chosen as a provisional site until French thermonuclear tests were conducted (Regnault, 1993). However, France's official decision was taken in 1962, after the completion of the Faa'a airport, the construction of which had been decided in May 1957, i.e., 3 months after the conclusions of General Ailleret's report, which recommended the Tuamotu archipelago for high-power tests. Moreover, in 1958, the first French nuclear test had not yet taken place in the Sahara. The Faa'a airport and the port infrastructures were therefore built not for economic or even tourist purposes, but for military logistical support linked to general policy issues.

Fig. 15.3 Surfing scene as seen through our eighteenth-century Western eyes. (Source: Bishop Museum)

interest for tourists, the reputation and the information available (particularly digital), and the ease of access and service.[11]

No singular island in French Polynesia is a destination in itself. During their stays, international tourists visit several islands, with different modes and ranges of accommodation. These visited territories, in their variety, diversity, and complementarity, define French Polynesia as a tourist destination. Some of these islands are more frequented than others, with a possible feeling of tensions and tourist pressures on the local inhabitants' way of life. Others, on the contrary, do not receive sufficient numbers of visitors to allow the development of infrastructures that would benefit the indigenous population (energy, drinking water, sanitation, services, and mobility, etc.) (Figs. 15.4 and 15.5).

In French Polynesia, the economic and social context is fragile[12]: in 2018, the unemployment rate was 14.5% of the population; one inhabitant out of two has a job and 16% of these jobs are precarious contracts or training courses.[13] However, since

[11] Although other types of services exist: by private jets, helicopters, or nautical means.

[12] Poverty, unemployment, discrimination on the basis of origin, and social and territorial inequalities are all problems in French Polynesia. On average, these families live on an income equivalent to the minimum wage, i.e., Xpf 159,092/month, since 1 May 2022. Two-thirds of this budget is spent on food, housing, and transport. These expenses correspond to 40% of the budget among the richest (Presidency of French Polynesia, 2021: 18).

[13] In French Polynesia, less than 52% of Polynesians of working age are employed, according to a study by the Institute of Statistics in French Polynesia (ISFP) covering the year 2018. This employment rate is lower than in mainland France, where it stood at 65.9% in the same year. However, in French Polynesia, the reality of unemployment is similar to that of the French Overseas Departments and Regions (DROM), where unemployment rates exist (French Guiana: 19.3%; Guadeloupe: 23.4%; and Reunion: 24.5%), in these Outermost Regions (OR), which are therefore an integral part of the European Union. It should be remembered that even though it is an overseas collectivity

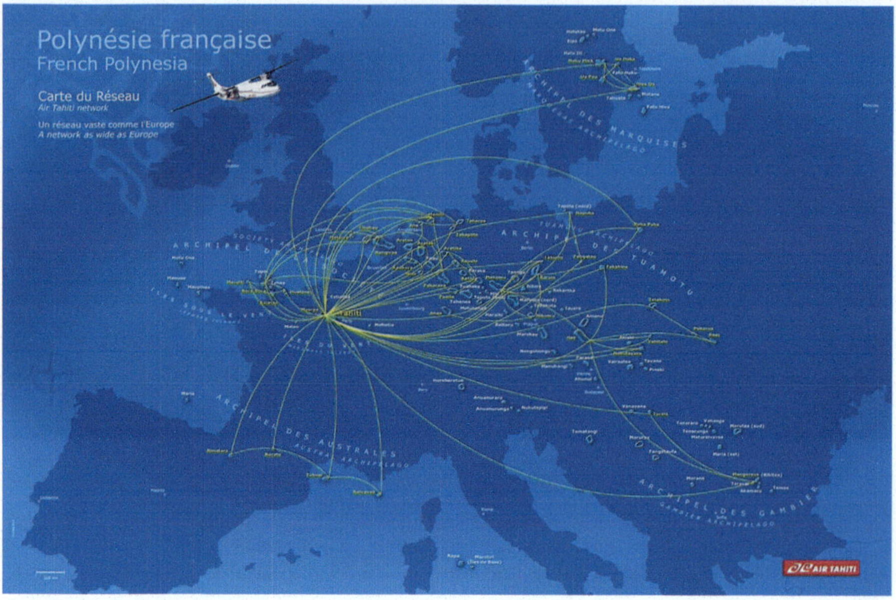

Fig. 15.4 The air network in French Polynesia, as extensive as Europe

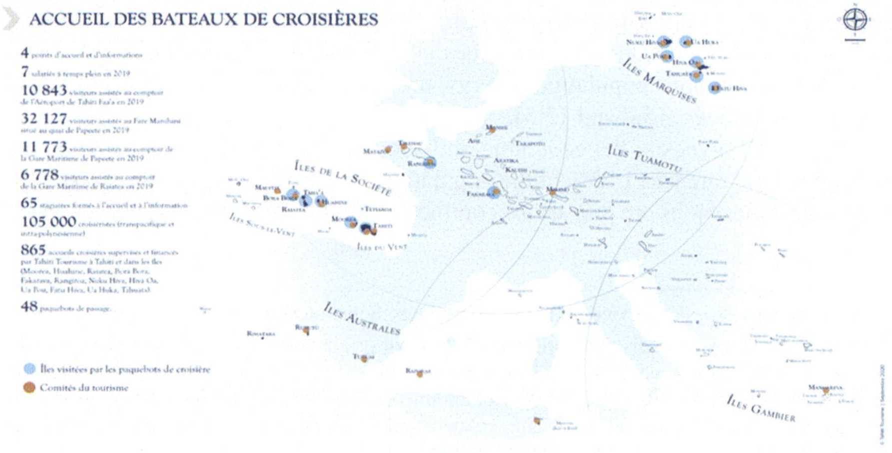

Fig. 15.5 The reception of cruise ships in French Polynesia, for the year 2020

the COVID-19 crisis, tourism is one of the few sectors which, in 2022, is growing significantly. In fact, in the first quarter, French Polynesia welcomed more than 30,000 tourists[14] or 60% of the 2019 total.[15] Bookings to this destination have increased by more than 60% in March 2022[16] and by more than 75% in June 2022[17]; to such an extent that luxury hotels are recruiting staff (receptionists, barmen, cleaning staff, etc.),[18] and airlines are looking for flight personnel[19] while reinforcing their fleet.[20] Confident, the tourism professionals emphasise the remoteness of the destination in order to attract more metropolitan people, in search of adventure and exoticism in a healthy and social environment that has stabilised. The functioning of the tourist activity is therefore synonymous with good health for the islands. That is, tourism generates jobs for the whole population, at all levels, according to Melinda Bodin, President of the General Federation of Polynesian Entrepreneurs (Fig. 15.6).[21]

---

within the French Republic, there is no unemployment benefit in French Polynesia. Cf. ISFP website, "En 2018, un Polynésien sur deux est en emploi", Points études et bilans de la Polynésie française, Ed. ISFP, n°1193, 4 p., put online on 30 December 2019, [URL: https://www.ispf.pf/content//uploads/1193_Emploi_2018_719dada89a.pdf], accessed on 24 August 2022.

[14] Customers from North America (more than one in two tourists this quarter, with 60% of the total number of tourists) and from Metropolitan France (the second largest market in French Polynesia with 30% of the total number of tourists) represent almost 90% of the number of tourists. Land-based commercial accommodation accounts for 60% of the clientele, compared with a quarter for the floating sector (7300 tourists). Notably, 535,000 overnight stays were consumed this quarter, which is 80% of the 2019 result. Excursionists are making a comeback after an absence of 2 years. See ISFP, "Tableau de bord Tourisme 2022-T1", *Points Conjoncture*, "Tourisme", first quarter 2022, 22 p., put online on 21 August 2022, [URL: https://www.ispf.pf/content//uploads/TBD_2022_T1_f287280079.pdf], consulted on 22 August 2022.

[15] In 2019, before the health crisis, tourist numbers were around 237,000.

[16] Cf. article by Mazerolle J.-M., "Tourisme: les réservations reprennent pour la Polynésie", *Polynésie La Première*, published on 18 March 2022, [URL: https://la1ere.francetvinfo.fr/polynesie/tahiti/polynesie-francaise/tourisme-les-reservations-reprennent-pour-la-polynesie-1259699.html], accessed on 25 August 2022.

[17] Cf. *Polynésie La Première* website, "Journal", broadcast Thursday, 25 August 2022, [URL: https://la1ere.francetvinfo.fr/polynesie/programme-video/la1ere_polynesie_journal-polynesie/diffusion/3788038-edition-du-mercredi-24-aout-2022.html], accessed 25 August 2022.

[18] For example, the *Saint Régis* and the *Four seasons* in Bora-Bora, the *Intercontinental* in Faa'a, the T*e moana Tahiti resort* in Punaauia, and *Le Brando* in Tetiaroa. Cf. *Tahiti Infos* website, "Job dating at the hotel school", article by Marau Biret, posted on 28 March 2022, [URL: https://www.tahiti-infos.com/Job-dating-au-lycee-hotelier_a208159.html], accessed on 5 September 2022.

[19] Cf. *Polynésie La Première* website, "Air Tahiti Nui has recruited 4 new pilots", online 28 February 2022, [URL: https://la1ere.francetvinfo.fr/polynesie/tahiti/polynesie-francaise/air-tahiti-nui-a-recrute-4-nouveaux-pilotes-1243120.html], accessed 5 September 2022.

[20] Cf. *Polynésie La Première* website, "Redécollage du tourisme, Air Tahiti renforce sa flotte", online 14 June 2022 [URL: https://la1ere.francetvinfo.fr/polynesie/tahiti/polynesie-francaise/redecollage-du-tourisme-air-tahiti-renforce-sa-flotte-1294252.html], accessed 5 September 2022.

[21] Cf. *Polynésie La Première* website, "L'invitée café: Mélinda Bodin", interviewed live by Ibrahim Ahmed Hazi, broadcast on 24 August 2022, online 24 August 2022, [URL: https://la1ere.francetvinfo.fr/polynesie/tahiti/polynesie-francaise/l-invitee-cafe-belinda-bodin-24-08-2022-1315204.html], accessed 24 August 2022.

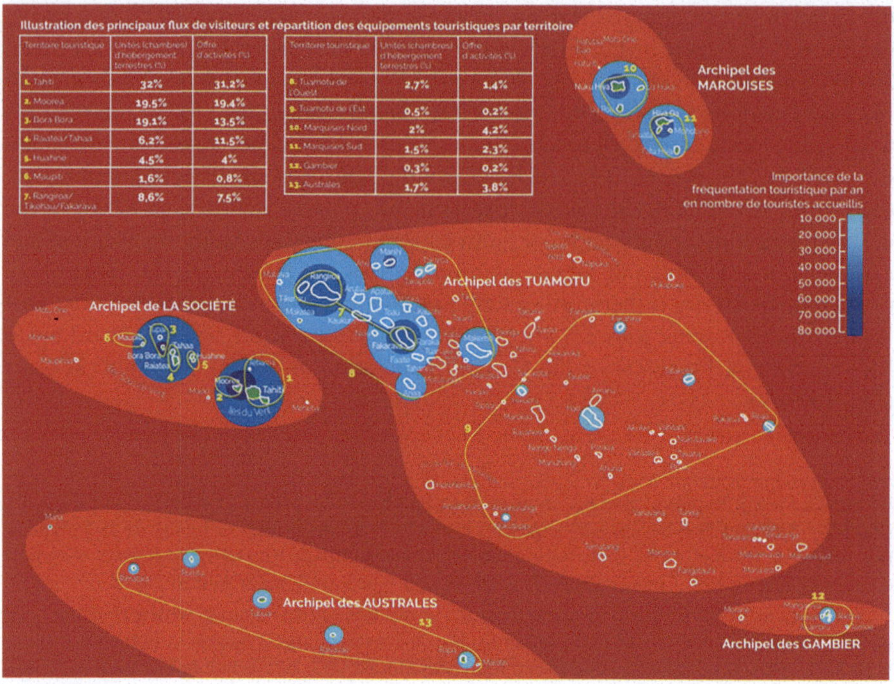

Fig. 15.6 Main visitor flows and distribution of tourist facilities in 2020 in the five archipelagos of French Polynesia. (Source: Ministry of Tourism and Labour, 2020: 13)

Tourism development in French Polynesia is not homogeneous and uniform. In the Society Archipelago, the Windward Islands have two tourist territories: Tahiti (n°1 on the map), the island of international arrivals at Faa'a airport, and, to the west-north-west, Moorea (n°2), known as the sister island, 17 km apart, but separated by a deep channel that in some places is more than 1500 m deep. On Tahiti, several internal divisions can be envisaged: the urban area incorporating part of the east and west coasts, the interior of the island, the peninsula of Taiarapu in the south-east of the island (called *Tahiti Iti*, the "little Tahiti"),[22] and then the atoll of Tetiaroa, famous for having been the property of Marlon Brando in 1966, and for having hosted a luxury hotel, the Brando Resort, with a private airstrip since 2014.

The Leeward Islands are made up of four tourist areas: Bora-Bora (3), which has a concentration of top-of-the-range hotels, plus the atoll of Tupai, the islands of Raiatea and Taha'a (4), which are 4.5 km apart and share the same lagoon, and Huahine (5) and Maupiti (6). In the Tuamotus, three groups are considered tourist areas: the first one includes the atolls of Rangiroa, Fakarava, and Tikehau (7); the second one, called "Western Tuamotus" (8), includes 14 atolls and tourist islands,

[22] The peninsula of Taiarapu is attached to the whole island (*Tahiti nui*, "big") by the isthmus of Taravao. It is shared between the communes of East Taiarapu and West Taiarapu (which today includes the associated communes of Teahupoo).

from Manihi in the north to Anaa in the south and from Mataiva in the west to Makemo in the east; and the third one, called "Eastern Tuamotus", includes six atolls which are very dispersed. The Marquesas have two groups of tourist territories: the "northern" one, built around Nuku Hiva, the main town (10), and the "southern" one, around Hiva Oa, where small museums dedicated to Paul Gauguin and Jacques Brel, as well as their tombstones, are located. Then, despite their remoteness, the islands of the Gambiers (12) and the Australs (13) are tourist territories in their own right.

Different types of tourism contribute to territorial development, whether through infrastructures or social representations (Gay, 2018), to the promotion of the destination, and to its anchoring in the collective memory (Bachimon, 1990). We therefore observe a diversification of markets both geographically[23] and demographically.[24] By consulting the Tahiti Tourism website,[25] we notice that in French Polynesia, outdoor activities certainly occupy a place in the offer but are secondary with regard to the contemplative well-being, the seaside, and the honeymoon theme. By the expression "outdoor", we mean a mix of outdoor activities (surfing, hiking, diving, horse riding, paragliding, parachuting, etc.) and motorised activities, in an exotic environment, on land or at sea (quad, 4 × 4, Jet ski, motorised foil-surfing, electric bike, microlight aircraft,[26] etc.). The sporting experience in the true sense of the word is present but subordinated to others as a secondary theme. It is largely entrusted to subcontractors as a supplement to packages. This type of service, which is very rarely offered in the all-inclusive hotel package, may nevertheless appear as a niche product developed by a few private operators.

Indeed, a sectoral approach could be taken between "outdoor land-based tourism", "water-based and underwater tourism", "aerial activities", "sport tourism", "cultural or well-being tourism", or even "memorial or religious tourism". However, all of these sectors are not sufficiently structured and the various actors concerned do not have a clear and unanimous vision of the desirable evolution of the sector. Moreover, the dynamics of renewal and evolution of the tourist offer as a whole are weak. These are individual initiatives that often come up against a lack of trust, both from marketing networks and intermediaries and from administrations, which are embarrassed by proposals deemed to be far-fetched and unorthodox, or from financing networks, which are reluctant to support and encourage entrepreneurial risk, and which are poorly trained to evaluate it.

---

[23] Many infrastructures have been built for the American market.

[24] For example, the luxury market and the green tourism market, but also cultural tourism, especially for the Japanese, or family tourism (for Americans) and honeymoon tourism with Bora-Bora as the main destination.

[25] *Tahiti Tourisme* is the tourism promotion body or Destination Marketing Organisation (DMO) of French Polynesia, better known as *Tahiti Et Ses Îles*. It manages marketing, advertising, public relations, promotions, events, and all marketing programs for the destination. Its mission is to increase the number of visitor arrivals to *Tahiti Et Ses Îles*.

[26] In French, we use the acronym ULM, for Ultra-Light-Motorized, especially paramotor and pendular (class 2).

Nevertheless, despite this context, sometimes thanks to support programmes set up by the Polynesian government, some offers stand out, such as initiation to sailing pirogues (at the traditional Polynesian culture learning centre), accrobranche, seaplanes, or climbing sites. But for the most part, these initiatives remain small in relation to the tourism industry. Apart from the creation of botanical parks, the Fare Natura of Moorea, water trails, and a few rare marked and maintained hiking trails, there is still no framework and concerted objective for the evolution of the tourist activity offer. Moreover, many service providers have had new regulations imposed on them, new titles necessary for the operation of their structures, the legitimacy of which is sometimes badly perceived. In fact, this regulatory framework sometimes constitutes an obstacle or even a hindrance to accessing the tourism professions for young Polynesians, and few bridges exist with sports associations or clubs.

Active, original leisure holidays are developing in all markets and remain an important growth factor: this is the case for lovers of sports, Polynesian dance, tattoos, and outdoor activities, etc. These tourists are also repeaters,[27] those who will come back several times and form lasting links with the destination. Since 2015, the offer of tourist activities has expanded, but in a disparate way according to the islands considered, and with recurrent typologies, little originality, and creativity, despite some innovative initiatives. Our study therefore focuses on the sports activities offered to national and international tourists coming to French Polynesia, as well as to locals. Among these activities, we were particularly interested in surfing, a sport that attracts a specific number of tourists, in diving, seen as a "product" subcontracted in particular by hotels, as a specific offer, and then in more occasional activities such as paddling,[28] feeding,[29] hiking, or trekking.[30] The survey on which this study is based was conducted throughout the first half of 2022. It consisted of consulting statistics from the Government of French Polynesia,[31] international and

[27] A repeater is a tourist who visits the same destination at least twice.

[28] Paddleboarding, also known as stand up paddling or SUP, is a water sport in which the paddler stands on a board and propels himself with a paddle.

[29] In diving, feeding consists of feeding the fish.

[30] A trek is a hike lasting several days in high or medium mountains, in areas that are not easily accessible.

[31] The law of the Country n°2012–17 SAGE specifies that the SAGE Sustainable Development Plan (PADD) of the General Development Scheme (SAGE) of French Polynesia sets out, at the scale of French Polynesia, the way in which the Country wishes to see its territory evolve in accordance with the principles of sustainable development. It organises the articulation of the archipelagos between them. The PADD includes in particular the layout of the major infrastructure facilities of territorial interest on the scale of the country in terms of education, culture, health, information and communication, transport of goods and people, energy, and sports. Moreover, according to the SAGE (2019: 37) book II *"projet d'aménagement et de développement durable du fenua & schéma d'archipel des Tuamotu-Gambier"*, edited by the Ministry of Housing and Spatial Planning in charge of inter-island transport, specifies that in French Polynesia, culture, heritage and water sports are essential differentiation vectors for its image as a tourist destination, in addition to the beauty of its terrestrial and underwater landscapes. However, the richness of Polynesian heritage is mainly intangible (Mā'ohi myths and cosmogony, music and dance, crafts, *patutiki*, *va'a* practice, etc.), rather than tangible, as the *marae* remain today essentially as archaeological remains, when

local airlines,[32] notably Air Tahiti Nui (ATN),[33] Air Tahiti (AT),[34] the Tourism Service (TS),[35] as well as Tahiti Tourism,[36] the Directorate of Youth and Sports, the Union of French and Francophone Associated Airports (UAF & FA),[37] the Institute of Statistics of French Polynesia (ISFP),[38] the Tahitian Federation of Underwater Sports (TFUS),[39] the Polynesian Union of Diving Centres, and local tourism operators. From a qualitative point of view, the analysis is essentially based on semi-directive interviews conducted with actors in the sectors concerned, such as the French Polynesian Olympic Committee (FPOC), the Tahitian Surfing Federation (TSF), the Polynesian Union of Diving Centres (PUDC), the Union of Hiking Guides in French Polynesia (SGRPF),[40] 12 service providers, and 13 tourists. It is

---

they have not been destroyed or swallowed up by vegetation. However, many marae have been restored or rebuilt. The Government supports the communes and associations in order to enhance the heritage.

[32] Because of their collection of information on passengers travelling with specific equipment (surfboards, fishing rods, diving equipment, etc.).

[33] For example, the information extracted from their database indicates the percentages of tourists travelling with sports equipment (as long as a specific note is included in the traveller's file). However, their definition of a "tourist" is a passenger who has issued a ticket outside of French Polynesia.

[34] The difficulty for Air Tahiti is to know, among all the excess baggage, what is sports equipment or not, because the company does not identify it. However, very few tourists travel with sports equipment, given the 23 kg baggage allowance. Indeed, divers do not come to French Polynesia with their diving equipment. As for surfers, the size of the hold in Regional Transport Aircraft (RTA) does not allow for the transport of large boards. The same is true for bicycles.

[35] The TS is a service of the administration, under the authority of the Minister in charge of tourism. Its missions are framed by Order No. 585 of 3 May 2012 on the creation and organisation of the service. However, the TS only collects data on tourist accommodation, not on passenger flights and their activities.

[36] *Tahiti Tourisme* is the tourism promotion body or Destination Marketing Organisation (DMO) of French Polynesia, better known as *Tahiti Et Ses Îles*. This institution manages communication, advertising, public relations, promotions, events, and all marketing programs for the destination. Its mission is to increase the number of visitors to French Polynesia. To do this, *Tahiti Tourisme* has a team of about 40 people at its headquarters in Papeete, and nine representatives, including a subsidiary in Los Angeles, working in 20 international markets. In this respect, the TS and *Tahiti Tourisme* can collect information from hotels and guesthouses offering tourist activities to their clients.

[37] Including, Results d'activités des aéroports français, 2021, *Statistiques de trafic*, Ed. UAF&FA, 47 p.; Union des Aéroports Français, 2022, *Note trafic*, trimestre 1, Ed. UAF&FA, 2 p.; Union des Aéroports Français, 2022, *Flash trafic*, trimestre 1, Ed. UAF&FA, 1 p. However, the only statistics collected are annual and concern passenger traffic, commercial and non-commercial movements, freight, and post.

[38] The ISFP collects passenger declaration forms on arrival, possibly specifying the nature of their stay in French Polynesia. Globally, for the share of tourist passengers, the ratio is about 70% on international flights and 30% on domestic flights. Then, some studies carried out by the ISFP inform us, for example, on the number of tourists practising golf, diving or other sports activities.

[39] For their sports-related data (number of members, sports events, etc.).

[40] The SGRPF is one of the tourism stakeholders, which in May 2022 was among the actors of the framing committee for the new strategies for Inclusive and Sustainable Tourism (called *Fa'ari'ira'a Manihini*), driven by the local Government.

also based on participatory observation carried out in situ with guides and tourists (13 interviews).

The interview methodology developed consisted of exploring three hypotheses that underpin our approach, namely: first, that the mythical Teahupo'o wave—the site of which will host the surfing events during the Paris Olympic Games in the summer of 2024—would represent a flagship product, which would largely "escape" the tourist activity, but which would contribute to the sporting reputation of Tahiti and more widely of French Polynesia; that thematic holidays would be organised by tour operators, such as diving and deep-sea fishing; while other activities, practised only in Polynesia, would be proposed more as animations and less as sports activities. Thus, the main line of research consisted in understanding the articulation of local practices, through popular sports, such as the Polynesian outrigger canoe (called *va'a* in Tahitian)·[41] sport fishing, and others, with the tourist offer (Fig. 15.7).

Indeed, if, for example, the *va'a* mobilises crowds of local spectators during famous races, such *as Hawaiki nui va'a,*[42] *Vodafone Channel Race,*[43] and then *Faa'ati Moorea,*[44] it still does not attract tourism. This leads us to wonder what effect the surfing event, which will be held in Teahupo'o during the 2024 Paris Olympics (more than 15,700 km away), will have on tourism in French Polynesia, as *va'a* will be present there for the first time as a competitive sport (Figs. 15.8 and 15.9).

On the contrary, the most popular sports take place in a more specific competitive environment:

- or that of the South Oceanic hemisphere (traditional *mā'ohi* sports and games), such as Polynesian stilts, javelin throwing, stone lifting, fruit-carrying race, tra-

[41] In Tahitian, *va'a* means "boat" or "canoe", but also in Hawaiian. There are several types of Polynesian canoes: V1 (*va'a hoe*), V3 (*va'a toru*), V6 (*va'a ono*), and V12 (*va'a tauati* or double canoe). In French Polynesia, the *va'a* is practiced as a sport and important competitions are organized.

[42] The *Hawaiki nui va'a* is an annual international Polynesian canoe competition held in October or November, depending on school holidays. Its name means "the great Hawaiki", in reo mā'ohi, and refers to a mythical island, the cradle of Polynesian civilisation from which the entire Polynesian triangle was colonised. The 3-day race links the islands of Huahine, Raiatea, Taha'a, and Bora-Bora, in the Leeward Islands archipelago, between the open sea and the lagoon. It was created in 1992 for men's teams, then in 1995 for women's teams (*va'a hine*), and in 1998 for juniors (*taure'a*). During the 28th edition (2019), the race brought together more than 140 crews, including 5 foreign teams, i.e., 2500 rowers, 30 officials, 30 rescue workers, and 1200 volunteers. In addition to television viewers, the race is followed each day by around 300 recreational boats.

[43] The *Vodafone Channel Race* is a traditional *va'a* race held in August. It is reputed to be the toughest in the world. In 2022, for the third edition, about 30 crews (composed of six rowers) from the five Polynesian archipelagos were to sail from Tahiti to Moorea (85 km).

[44] The V6 *va'a* race, *Faa'ati Moorea*, takes place every year in July, as part of the Heiva races. In 2020, it will bring together some forty crews, including Hawaiians and New Zealanders, on a 93-km circuit, which starts in Tahiti and goes around Moorea.

Fig. 15.7 Poster advertising a hiking guide in Huahine, Leeward Islands, Society Archipelago, French Polynesia. (Photo credit: Bachimon, 2022)

Fig. 15.8 The Hawaiki nui *va'a* race, in French Polynesia, gathers on average more than 3000 people in Huahine, Raiatea, Taha'a, and Bora-Bora (Leeward Islands, Society, French Polynesia). (Photographic credit: Buquet, 2019)

Fig. 15.9 Many people, on land and at sea, welcome the rowers of the Hawaiki nui *va'a* race, upon their arrival in Bora-Bora (Leeward Islands, Society, French Polynesia). (Photo credit: Byers, 2021)

ditional wrestling, *va'a* (at the famous *Hawaiki Nui* competition), or folk dances (called *heiva*),[45] coconut husking, up to the art of tattooing (*tātau*, in Tahitian);
- or an environment influenced by the tutelary powers, ancient or not, which explains the presence of sports such as rugby, boxing, athletics, football, swimming, weightlifting, volleyball, basketball, cycling, gymnastics, karate, judo, golf, diving, sailing, tennis, bi- and motocross, skateboarding, dance, taekwondo, bodybuilding, and so on. Up to those introduced more recently in French Polynesia, from the United States, such as beach soccer, body-building, stretching, pole dancing, and mixed martial arts (MMA) (Figs. 15.10 and 15.11).

Now many of these sports make up the Pacific Games, the first of which were held in 1963 in Fiji. They followed a decision in 1959 at the South Pacific Conference in Papua New Guinea. They were called the "South Pacific Games" and were initiated by the Trustee Powers, which included Australia, the United States, France, New Zealand, and the United Kingdom. From then on, some "tourists" were de facto potential athletes, who travelled to the islands to compete.

Thus, there are few connections between the two areas described. Rarely, with the exception of surfing and the big wave at Teahupo'o, do major sporting

[45] In Tahitian, *heiva* defines a pastime, an entertainment or a physical exercise, but also a dance or an assembly gathered to dance. Stopped after the arrival of the first Protestant missionaries from the London Missionary Society in the eighteenth century, along with other traditions and customs, because they were considered indecent and unchristian, the first *heiva i Tahiti* was officially organised in 1881, named *Tiurai* at the time. Since then, it has become an annual event, which takes place over several days in each archipelago, bringing together children, women and men of all ages, ethnic and social backgrounds and all levels, amateurs and enthusiasts alike. Several dance groups and schools compete and represent the different islands. The scenography stages a fable, a legend, a historical fact... During the performances, the dancers are dressed in traditional costumes, entirely hand-made, and often made from plants, decorated with flowers, plants, and shells. The traditional firewalk (*umu ti*, in Tahitian) opens this festival, which in addition to dance performances, called *'Ori Tahiti*, is punctuated by traditional singing competitions (*himene*), as well as traditional sports competitions.

Fig. 15.10 Traditional mā'ohi wrestling to be featured at the opening of the 2024 Paris Olympics. (Photo credit: Tu'aro Maohi, 2022)

Fig. 15.11 Five Polynesians competed at Epic Fighting 31 (MMA) in 2016 in the United States (from left to right, top: Haunui Otcenasek and Teva Paulet; bottom: Henri Burns, Raihere Dudes aka *The Islander,* and Tamahau Mc Comb). (Photo credit: SportsTahiti/Tefau, 2016)

competitions attract a large international audience. In fact, the distance between the islands and the rest of the world raises the problem of absolute dependence on air transport, which in terms of sustainable development is hardly viable and sustainable. First, we will look at two key activities (surfing and diving) to examine the issues raised above. Second, we will look at the marginality of hiking, and third, we will provide some explanations for this state of affairs.

Surfing and Diving as Key Activities

Sport as a primary motivation for a stay in French Polynesia hardly concerns surfers, especially during sports events such as international competitions. This is a niche clientele, that of the tours. Several sporting events take place in Tahiti: *The Tahitian Pro* (in August) brings together about 250 people from France, the United States, Brazil, Australia, South Africa, and to a lesser extent Costa Rica and Barbados (in the Caribbean Sea). Two other competitions, the Qualifying Series (the *Challenger Series*) and the *Champion Tour* (where 18 girls and 36 men compete), attract competitors and their teams. According to one of the members of the FPOC, *"the organisation of the surfing events during the Paris 2024 Olympic Games should bring to Tahiti a thousand people (one third of whom will be the competitors and their supervisors). The site will then be fitted out to accommodate everyone. And the first world record will be broken: the one for the greatest distance between the location of an Olympic event and its host city, Paris"*. Twenty-four countries will be competing, compared with 18 at the Tokyo Olympics in 2020. The hotel centre at the *Puunui* site on the peninsula, which has become a tourist wasteland (Bachimon, 2012), is expected to reopen on this occasion to accommodate the Olympic village, while the adjoining marina will be redeveloped for the competitions to be held offshore on the Teahupo'o wave, one of the world's top spots, and to host one of the three fan zones (Fig. 15.12).[46]

Far from being a competitive sport, surfing is a very different experience for the amateur surfer, usually a beginner, who makes his or her first surfing experience, whether on a paddle or a bodyboard.[47] In addition to the famous site of Teahupo'o, reserved for experienced surfers, the beaches of Taharu'u[48] in the south of Tahiti and Papeno'o[49] in the north of the island, with their regular waves, are accessible to

[46] Fan zone allows a large number of spectators to be at the heart of the competition, to enjoy a space for leisure and entertainment during the festivities.

[47] Bodyboarding is a water sport that is practiced with a short and flexible board on which the practitioner slides lying down or kneeling.

[48] In Tahiti, in the commune of Papara, the Taharu'u site is known for its surf spot. Since 2014, it has been home to a leisure park created for a family audience that regularly visits the site. During the year, the site is used for sporting and cultural events (*matarii ni'a*, or "the rising of the Pleiades" in Tahitian, marking the beginning of the season of plenty, but also dance performances, etc.).

[49] Papeno'o is one of the communes associated with Hitiaa O Te Ra, located in the northeast of Tahiti Nui (the main part of the island). Papeno'o is known for its black sand beaches, surf spots, and wild valleys. From one of them flows one of the most important rivers. It is called *Vaitūòru* (Tahitian for "waters that rise and swell" or *Vaitūàru* ("water that rises like a rough wave") or *Ha'apai'ano'o* ("gathering of all waters", as in the nineties, this name referred to the river and the district of Papeno'o). Nowadays, in everyday language, Tahitians refer to the river of the Papeno'o valley—Maroto. Shaped like a vast cirque upstream, the valley gradually narrows as it approaches the ocean, with two bottlenecks dividing the valley into three parts. Fed by more than thirty tributaries, the river has the largest catchment area of 89 km^2 (representing about 10% of the total area

Fig. 15.12 The famous wave stele at Hava'e Pass, Teahupo'o, a former commune associated with West Taiarapu, on the peninsula, French Polynesia. (Photo credit: Tchékémian, 2016)

beginners. Surfing is one of the few sports in French Polynesia where it is common to see travellers unloading their boards in hand luggage (Fig. 15.13).

Half of the clientele is American and the other half European (mainly French and to a lesser extent Italian). The American clientele is generally young (20–35 years old on average) and well-to-do, and they often come on honeymoons to large Polynesian hotels (the Intercontinental in Tahiti, the Saint Régis in Bora-Bora, etc.). European tourists are older (35–50 years old) and stay in guesthouses or small hotels for more varied reasons: to contemplate the exotic landscapes, the fine sandy beaches, and turquoise lagoons, but also to explore archaeological sites, to discover crafts, dance, music, gastronomy, and Polynesian culture by staying with local people. Nowadays, for beginners, there are "surf packages" offered in guesthouses located near the spots, such as those located on the Tahitian peninsula, *Tera Nui*, *Bonjouir*, *Reva*, and so on near Teahupo'o. On the contrary, the *va'a* does not yet

of Tahiti) and the highest average rainfall of 6600 mm/year. Following a north-south flow, its length is 23 km (according to the IGN map of Tahiti). Thus, the valley records the highest average flow, 11 m^3/s, and up to 2200 m^3/s during cyclone Veena, from 7 to 13 April 1983. It should be noted that between December 1982 and April 1983, the El Nino phenomenon caused six cyclones in French Polynesia during the warm season: Lisa (11–13 December 1982), Nano (20–27 January 1983), Orama (22–27 February 1983), Reva (6–14 March 1983), Veena (7–13 April 1983), and William (15–21 April 1983). Nowadays, the river of the Papenoo—Maroto valley is one of the most dangerous in French Polynesia: the water level can suddenly rise by several meters and trap hikers and swimmers. See article entitled "Papenoo, la dangereuse", published in *Polynésie La Première*, on 5 January 2022, [URL: https://la1ere.francetvinfo.fr/polynesie/tahiti/polynesie-francaise/papenoo-la-dangereuse-1197067.html], accessed on 27 August 2022.

Fig. 15.13 Water sports are becoming increasingly popular in French Polynesia, particularly at the Taharu'u site in the commune of Papara (PK 39), Tahiti. (Photo credit: Tahiti Infos/SB)

attract tourists. It is true that not one hotel on the coast offers its guests free access to this type of boat, but it is rarely used because of the complexity of sailing with an outrigger for the uninitiated. The hotel has often replaced it with sea kayaks, which are more manoeuvrable and unsinkable and are widely used for short lagoon tours.

In Tahiti and Moorea,[50] diving is the most visible sport, and to a lesser extent in the Tuamotus, where the waters of two atolls, Rangiroa and Fakarava, are in the "top ten" sites worldwide. Tahiti is a flagship destination for lagoon diving and enjoys a worldwide reputation—it has only the Maldives as a direct competitor—even though other destinations are emerging in this niche. There is a clear divide between snorkelling (underwater trekking with fins, mask, and snorkel) which can be practiced by the average tourist and scuba diving which requires a previous level of experience (knowing that some clubs offer introductory courses). If snorkelling is in the chain of construction of the offer made by the big hotel groups for the international and local clientele, in particular by the Van group, owner of many intermediate class hotels, scuba diving remains more confidential. Indeed, for this activity, a diploma is required for divers and accompanying instructors. The classic tourist

[50]The island of Moorea, known for its tourist appeal, saw an increase in visitors in 2018 with 111,126 visitors for 216,268 in French Polynesia, putting it in third place, after Tahiti (155,331 tourists in 2018) and Bora-Bora (112,367 in 2018), according to Bernardo et al. (2018).

operator for scuba diving is the diving "club". It is more often linked to guesthouses than to large hotels.

For snorkelling, which has a wider clientele, private operators (such as fishermen with motor boats) offer a day on site (like the best known of these sites, the *Lagon Bleu* of Rangiroa) all inclusive (boat, meals, dives, walks, weaving, etc.). If we stick to scuba diving, two types of clubs stand out:

- First of all, the "five leaders",[51] who, in order to satisfy their clients, have a dozen or so supervisory divers, substantial equipment, and several boats, and whose offer is diversified in terms of services and sites;
- then between "15 and 20 intermediaries", to which can be added the more modest ones (a single diver and a boat), which have a hedonistic approach and remain more confined to a site (generally a portion of an island) and a host.

Since 2020, the sector has begun to organise itself into a diving union. The spin-offs of this activity remain modest, except when considering their indirect aspect and except for a few destinations such as Fakarava where they are estimated to represent 80% of the added value. However, we cannot speak of a "diving" product. For example, it is not offered to cruise passengers (it is only offered by two cruise ships). Only 10% of the clientele (70% of which is international, i.e., non-local) goes through tour operators (compared with 80% in the American market). The Clubs offer inter-island packages. This niche market targets three customer segments: the rare "professional" divers, the enlightened amateurs (with a sports product combining, for example, trekking, diving, and kitesurfing[52]), and the "honey mooners", according to the President of the Polynesian Union of Diving Centres (PUDC).

Hiking on the Margins of the Sports Activities Offered to Tourists

In Tahiti, hiking, widely praised as an approach to the authenticity of the landscape and culture, is also neglected, apart from a few spots such as Mount Aorai, which rises to 2066 m, the third highest peak in Tahiti after Mount Orohena (2241 m) and

[51] Currently, the five largest clubs in Tahiti are: *Ecole de plongée de Tahiti* (located in Arue), *Eleuthera plongée Tahiti* (in Punaauia), *Fluid plongée Tahiti* (Punaauia), *Topdive* Polynésie (which, in addition to their presence at the *InterContinental* Resort in Tahiti, has dive centres in Moorea, Bora Bora, Rangiroa, Tetiaroa, North and South Fakarava, as well as partners in Tikehau, Raiatea and Huahine).

[52] Kitesurfing or kiteboarding, which can be translated as aerotow surfing or flying board, is a water sport that consists of moving on the surface of a body of water, with a board while being pulled by a kite, called a wing or sail. This sport took off in the early 2000s.

Mount Pito Iti (2110 m), the Vaipahi water garden in Taravao and the Maroto relay,[53] the starting point for hiking, which allows you to contemplate a remarkable waterfall and its basin. In the islands, only in Moorea (including Mount Rotui and the tour of the caldera) and Bora-Bora (with the World War II cannon sites) is there any form of organisation of tourist accompaniment (guides, 4 × 4 s, trails, etc.), not forgetting a few possible hikes on Tahiti, such as the Faautaua waterfall (in Papeete) (Fig. 15.14).

In French Polynesia, there are 23 guides, up to date with their contributions, to which are added eight guides in training with the Directorate of Youth and Sports,[54] according to the President of the syndicates of hiking guides in French Polynesia, currently 32 guides exercise this professional activity, without being able to "live decently". They hold a transport licence (in 4 × 4) and bring their clients to the site, like the Safari agencies which only transport their clients in all-terrain vehicles. These guides occasionally work with local, national, or international agencies, but it is more generally through word of mouth and their reputation on the web (Facebook page, website) that they build up a clientele, "*with little help from the Polynesian government and its tourism department*", according to the President of the hiking guides' union in French Polynesia.

A guide with an activity that can be considered economically viable will carry out in the "high" season, from May to October (i.e., during 6 months) about four outings per week (including some on the lagoon) with an average of four clients per outing, i.e., a total of 300 clients per year. For a price of about 3500 Pacific francs (XPF) per person (i.e., 29.63 euros), this represents an income of about 120,000

Fig. 15.14 Panoramic view of Mount Rotui, Moorea, surrounded by Opunohu (left) and Cook (right) bays, from the lookout tourist site, above the Opunohu agricultural high school. (Photo credit: Tchékémian, 2014)

[53] The refuge is located in the centre of the island, 18 km from Papeno'o. A 4 × 4 is necessary to travel through the valley and to cross the fords of the rivers; unfortunately, accidents, sometimes fatal, are regularly reported due to high floods. If necessary, service providers offer 4 × 4 excursions in this valley.

[54] A Polynesian Professional Certificate of Outdoor Physical Activity Guide, mention hiking, and a Polynesian Professional Certificate of Outdoor Physical Activity Leader offered by the Institute of Youth and Sports of French Polynesia.

XPF/month (i.e., 1015.78 €). The use of agencies (generally *Tahiti Nui travel*), but also hotels (such as the *Intercontinental* and the *Hilton*), results in a commission of around 20% on the price charged. Whereas the shipowners, including the *Paul Gauguin* and the *Wind-Star* schooner, explicitly mentioned by our interlocutor, demand up to 40% commission, which of course limits the subcontracting activity with the cruise operators (even if the rates charged are higher). The hiking clientele is composed of:

- 80% of American couples on honeymoon who in the trek are looking for an immersion in tropical nature, a discovery of the interior of the island, of the fauna and flora, in particular the observation of luxuriant plants,[55] of birds.[56] This can consist of simple walks from stages carried out in 4 × 4, but also in sports activities (hiking, horse riding, mountain biking, quad biking, canyoning[57] etc., for example, for the discovery of the lavatubes of Hitiaa, if it is a question of discovering the wilder sites (off the beaten track);
- The remaining 20% are older Europeans who practice hiking.

However, the major routes are poorly marked and almost inaccessible to the uninitiated. For example, the popular tour of the Tahitian peninsula requires a boat ride or even a night on a beach, with the risk of being attacked by aggressive hermit crabs (*Coenobita* sp.). On the island of Huahine, which is part of the Leeward Islands in the Society Archipelago, the marking of hikes has been limited to a few trails, such as the one leading to the Ana Iti Territorial Domain, which is built on a hotel wasteland. In Bora-Bora, there are a few walks that are almost exclusively by 4 × 4 or quad. On the Tuamotus, there are a few atoll "tours" (often in fact the motu of the main village). In the Austral Islands, Rurutu is a bit of an exception with its multiple routes. In the Marquesas, leaving the track and the road is a real challenge except on some small islands (like Ua Pu).

---

[55] Such as the tree fern (order *Cyatheales*), the frangipani tree, called "tipania" in French Polynesia (*Plumeria* sp.), the flamboyant (*Delonix regia*), the hibiscus (*Rosa sinensis*), the porcelain rose (*Etlingera elatior*), the Tahitian tiare (*Gardenia tahitensis*), the coconut palm (*Cocos nucifera*), the pandanus (*P. tectorius*), tamanu (*Calophyllum inophyllum*), Tahitian jasmine or "pitate" in Tahitian (*Oleaceae*), ylang-ylang (*Cananga odorata*), breadfruit tree (*Artocarpus altilis*), whose fruit is called "uru" in Tahitian, mango tree, papaya, guava, banana, noni, and so on. On this subject, in the Society's archipelago, flower production concerns particular types of flowers, such as *Musaceae*, *Marantaceae*, and *Zingiberaceae*, but also orchids, pandanus, and anthurium, which are the main plants exploited in French Polynesia.

[56] Various birds, called "manu" in Tahitian, are commonly observed in Tahiti, such as the Tahitian monarch (Pomarea nigra), Gould's harrier (*Circus approximans*), Society salangane (*Aerodramus leucophaeus*), Kuhl's loris (*Vini kuhlii*), nonette (*V. peruviana*), and ultramarine (*V. ultramarina*), but also marine birds, such as frigatebirds, boobies, petrels, terns or "kaveka" (ultramarina), but also seabirds, such as frigatebirds, gannets, petrels, terns or "kaveka". See the website of the *Société d'Ornithologie de Polynésie*, "Manu", which works to protect the wild birds of Tahiti, its islands and their habitats [URL: https://www.manu.pf/], consulted on 6 July 2022.

[57] On the island of Tahiti, canyoning is practiced in the largest valley of Tahiti, the Papeeno (Purau, Vai Taporo, Putoa, Mamao), then on the east coast of the island (Faananu, Hitia and the Lavatubes).

To sum up, walking in Polynesia is an obstacle course in every sense of the word (technical and legal difficulties, risks of dog bites, falls that can sometimes lead to death[58]) even though we are on islands without any particular danger (no snakes or other dangerous animals, or even toxic plants).[59] And this while on the paper of the flyers the offer (photographic at least) seems large and varied (lava-tubes, marae hidden by the vegetation, plateau of orange trees and Orofero, Valley of the hunters of Papara, the Diadem, etc.) (Fig. 15.15). But these are often almost (or have become) inaccessible sites. The hiker, or even the simple walker, returns frustrated from islands where there is practically no signposting or marking.

Fig. 15.15 Remains of a "marae", a Polynesian sacred place, present on the "archaeological discovery and ethnobotanical interpretation path", in the Opunohu valley in Moorea, Windward island, Society archipelago, in French Polynesia. (Credit: Tahiti Tourisme, 2019)

[58] As in March 2022, when a 34-year-old American tourist fell to his death while hiking on Mount Rotui, in Moorea. Cf. article (see Lienard et al., 2022).

[59] It should be noted that in French Polynesia, apart from the *Scolopendra*, a carnivorous and venomous arthropod of the Myriapoda family, also called "hundred-footed", which can measure 20 cm in length and inoculate a painful venom, there is no deadly or even dangerous endemic animal. This is not the case in the regions where venomous marine animals are found, such as Stonefish (*Synanceia verrucosa*), scorpionfish (Scorpaenidae), conefish (Conidae), octopus or octopus (molluscs), stingray (Aetobatis narinari), sea anemone, jellyfish, spiny starfish (*Acanthaster planci*), moray eel, scorpionfish (*Scorpaenopsis gibbosa*), sea urchins (Echinoderms), fire corals (Cnidaria—Coelenterata—*Millepora platyphylla*), corals (Madrepores), sea cucumbers (Echinoderms—*Holothuriidae—Bohadschia*), and leeches (*Synapta maculata*), not to mention the deep-sea sharks that penetrate them.

A Destination Where the Sports Offer Remains Marginal

All the operators stress that the added value of sport in the tourist offer is low and that the profit margins that are hoped for are almost non-existent. Also, the prices charged seem to them to be insufficient to ensure good profitability. For example, a diving trip is marketed at 70 € (8059.50 XPF), a price equivalent to that charged in mainland France, even though there are hidden costs (distance from the diving sites, travel between the islands, etc.). Paradoxically, for a Californian, or even a metropolitan, a diving trip in French Polynesia remains a quite competitive product, even taking into account the extra cost of air transport. So, it could represent an attraction (a product of appeal). For the moment, there is little official communication in the open air (cf. in July 2022, *Tahiti tourisme*, the official body for the promotion of the territory, is just starting to mention it). *Air Tahiti*, a public company, which has a quasi-monopoly on inter-island air transport, and, in fact, a role as a promoter of remote destinations, does not have a well-established policy on the subject.[60] In October 2021, the Polynesian Council of Ministers approved a draft decree granting an air carrier licence to a low-cost company, *Motu Link Airline*. This company will serve the archipelagos: first, the most touristic and frequented islands[61] and then the others.

By boat, inter-island transport is also possible. A ferry service serves the nearby islands, such as Tahiti—Moorea or Tahiti—Maupiti. More picturesque, cargo ships link the Leeward Islands, the Tuamotu, Gambier, Marquesas, and Austral Islands. However, in the same archipelago, fast boats or rowboats provide round trips. The link between Tahiti and Moorea is one of the busiest. It transports Moorea residents who work in Tahiti, offering a range of timetables and a comfortable service at a competitive price, with 10 rotations per day. Two companies offer round trips from the port of Papeete, the Aremiti (including the Aremiti 5, a fast ship that makes the crossing in 30 min, and the Aremiti Ferry 2, which makes the crossing in 40 min). The crossing offers beautiful scenery of the lagoon and the coasts of Tahiti and Moorea, where during the season, whales can be seen. For example, in the Tuamotus, particularly Fakarava and diving, service providers regret the lack of flights. Only Rangiroa, as the archipelago's capital, has more frequent air links. These have led to the development of seven diving clubs, which, in conjunction with guesthouses, have enabled operators to offer packages (stay, air transport, and diving) to local

[60] Air Tahiti offers regular flights from Tahiti to a network of 47 islands and atolls, including a 7-min flight from Tahiti to Moorea. Travellers are allowed a baggage allowance of 23 or 46 kg. Divers have an additional 5 kg allowance on presentation of their card or proof of identity, according to the *Tahiti Tourisme* website, "D'une île à l'autre" [URL: https://tahititourisme.fr/fr-fr/preparer-son-voyage/se-rendre-a-tahiti/deplacement-inter-iles/], accessed on 27 August 2022.

[61] Initially, *Motu Link Airline* plans to serve the following six airfields: Huahine, Raiatea, Bora-Bora, Rangiroa, Tikehau, and Fakarava. See article entitled "Motu Link Airline: a new airline to serve the islands", *Polynésie La Première*, posted on 13 October 2021 [URL: https://la1ere.francetvinfo.fr/polynesie/motu-link-airline-une-nouvelle-compagnie-aerienne-dans-le-ciel-polynesien-1127758.html], accessed on 27 August 2022.

tourists (resident rate 10% lower than the international rate). This is also a way of building customer loyalty which appeals to the clientele.

Tourist sports activities are often subject to local reticence, which sometimes limits their practice drastically. In French Polynesia, the best example is hiking. Its practice comes up against private property, whether it is individual plots, subdivisions, or undivided land,[62] which is exercised in a very restrictive sense in terms of right of way. Owners are generally opposed to the marking of routes; they are increasingly enclosing their land and letting unsupervised dogs roam on it, which are increasingly of the mongrel variety. Therefore, the first risk for the hiker is the dog bite and in general big scares. On Tahiti itself, it is very difficult to find a starting point that is open and accessible to hikers.

The guides and associations supervising the hike are forced to verbally negotiate rights of way which are rarely respected by the owners in the long term. If it is willingly granted (and often without financial compensation), obtaining it can also require long weeks of negotiation, generally facilitated by the town halls, all the more so when communal associations manage the peripheral areas coveted by the guides. This is notably the case in Papara de la Taharu'u, where the areas are managed by a group of hunters, as well as in Punaauia, on the Orange Tree Plateau—one of the most popular hikes in Tahiti—managed by the Orange Pickers Association.[63] These restrictions on accessibility appear to be very protectionist in the eyes of the service providers, who are essential intercessors in maintaining hiking insofar as there is no signposting or public policy for maintaining (and even less for creating) paths.

The risks incurred by hikers (dog bites first of all, but also bad falls) translate into high insurance premiums for guides when crossing private land. This leads them more and more to restrict the offer to the public domain, which is itself very limited. This creates a sense of mystery in terms of the tourist experience, given the difficulty of reaching the interior of the main island. While for access to the beaches, the communes and the Polynesian government have massively acquired coastal concessions in order to re-establish access to the sea, which had a tendency to become privatised and rarefied, nothing has been done in the mountains where the hiker is

[62] The indivision of family land is the result of social habits; it contributes to the unavailability of land by making it difficult or impossible to develop. Moreover, in French Polynesia, land tenure is a sensitive issue, due to the numerous cases of undivided land within Polynesian families. To remedy this, the Government has created a land court to hear all cases of undivided ownership to relieve the current court in Papeete. Because of these land problems, it is rare to find public land, offering access to all. Cf. Decree n°2017–1474 of 16 October 2017 on the organisation and functioning of the Land Court of French Polynesia, Lexpol, French Polynesia.

[63] In the latter two cases, there is a symbolic toll charged by the associations. For the Papara hunters, it is 5000 Xpf (43.51 €). Every year, the association of orange carriers of Punaauia accompanies visitors from the Punaruu valley to the Tamanu plateau, known as the orange tree plateau. Hikes are organised by the orange carriers' association for a fee of Xpf 1000 (€8.70) per hiker (over 13 years old). The trek can be done in one day (between 8 and 10 h walk there and back) or over 2 or more days. The toll is justified by a contribution to the maintenance of the paths, the pruning and the cleaning of waste.

confronted most of the time with the aggressiveness of attack dogs (category 1) and guard and defence dogs (category 2).[64] For the actors, these attacks, as well as the uncertainty as to the authorisation to cross private property, are not without consequences for the "profession" of guide and the development of the "hiking" activity. To take a telling example, Lake Vaihiria, the largest lake in Tahiti, was forbidden access for many years by a barbed wire fence, reinforced by a pack of dogs, put up by a landowner located downstream, who opened it in exchange for a large toll. In the end, the only central crossing of the island was rendered impassable.

Then, in French Polynesia, there is practically no accommodation located in the mountains. Two refuges[65] exist on the Aorai, renovated in 2020; but the first one was damaged in January 2021.[66] These are basic accommodations, dating from the 1990s. The Papara hunters' and orange pickers' associations of the Punaruu and Orofero plateaus can offer accommodation, but this remains uncertain and very precarious. Even if, conversely, amateurism adds to the tourist experience the stamp of a great and unique conviviality.

As for diving, it is not subject to the same rejection as hiking, even if the conflicts of use in the lagoon are far from over, especially since the application of the anti-feeding legislation[67] in 2017, accompanied by the reinforced control of marine protected areas. Indeed, the feeding of wild species has been prohibited since the publication of the Law of the Country n°2017–25 of 5 October 2017 relating to the environmental code of French Polynesia.[68] This law prohibits the feeding of sharks

[64] According to the classification of article L.211–12 of the metropolitan rural code, completed by the decree n°HC 1927/DRCL of 22 December 2009, establishing the list of types of dogs likely to be dangerous [URL: http://www.polynesie-francaise.pref.gouv.fr/Demarches-administratives/Loisirs/Regles-pour-chiens-dangereux/Les-principales-dispositions-du-code-rural/1.-La-liste-des-chiens-susceptibles-d-etre-dangereux], consulted on 16 July 2022. Let us add that in French Polynesia, there are regularly stray dog attachments, whether on certain beaches or in urban areas, as recently in Tahiti, Maupiti, and Bora-Bora, where the municipal council is putting in place measures to fight against their spread (sterilisation campaign, construction of a shelter and census of owners in order to raise their awareness). Cf. article in *Polynésie La Première*, 24 July 2017, [URL: https://la1ere.francetvinfo.fr/polynesie/bora-bora-tente-reguler-sa-population-canine-495933.html], accessed 16 July 2022. In 2020, the attack on an octogenarian woman who was taking her daily walk around a stadium in Pirae (Tahiti) died following an attack by a pack of five dogs (pitbulls and bull terriers). See article in *Polynésie La Première*, 12 May 2020, [URL: https://la1ere.francetvinfo.fr/polynesie/tahiti/pirae/femme-tuee-molosses-pirae-831866.html], accessed 16 July 2022.

[65] The first refuge, called fare Mato, is located at an altitude of 1400 m, on Mount Aorai, culminating at 2066 m, and the second, *fare Ata* is at 1800 m.

[66] Cf. article published in *La Dépêche de Tahiti*, 19 January 2021 [URL: https://actu.fr/faits-divers/le-premier-refuge-de-l-aorai-saccage_38836939.html], accessed 17 July 2022.

[67] In 2019, the President of the Moorea Tourism Committee counted about 30 boat service providers who went daily to the spots to feed the stingrays and sharks, such as the one in Tiahura (Moorea), the most coveted. In addition to these, there were eight jet-ski providers who included this feeding activity in their programme of exploration of the island.

[68] Since 2017, following accidents involving sharks in Polynesian waters, particularly in the Moorea lagoon, feeding fish (rays, sharks, and others) is strictly prohibited. Only observation is allowed: bathers must not try to attract the animal, they must let it come to them. Thus, the law

and rays, practised by operators on scuba diving sites (particularly at the exit of Opunohu Bay in Moorea and at the Rangiroa "aquarium"). In fact, it is still practiced to attract a maximum number of fish as tourists pass by. Indeed, the experience of contact with small, harmless sharks (usually black tips, *Carcharhinus melanopterus*, and white tips, *Carcharhinus albimarginatus*) and stingrays (*Dasyatis pastinaca*), which are also harmless, remains a highlight of diving in the lagoon. However, this poses a recurring problem in terms of eco-development, particularly in the off-season (November to March), when the service providers no longer feed the fish, in the absence of tourists.

Since May 2018, the environmental code regulations applied in French Polynesia have been tightened. It is now forbidden to practice whale-watching (a successful tourist attraction) in lagoons, bays, and passes. Outside of these areas, a distance of 100 m for whales and 30 m for dolphins must be respected. Under these conditions, launching seems to be compromised. Therefore, in order to observe marine mammals, service providers must have a permit, granted by the Directorate of the Environment. The protection of species desired by tourists for the (quite involuntary) spectacle they offer is similar to that of marine parks, such as *Marinelands*, which in Europe are in economic difficulties (or even closed permanently) after controversies linked to ethical issues (including animal mistreatment). However, in French Polynesia, trafficking in protected species continues, particularly in turtle meat.[69] The same applies to the capture of sharks for their fins, which continues to have an impact on the resource, even if the vigilance of the *mūto'i* (municipal police officer, in Tahitian) on this last point is high.

Conclusion

Addressing the question of the relationship between "sport and tourism" remains ambiguous in French Polynesia, as we have been able to grasp throughout the preceding approach. Quantitative data on this "intersection"—which does not represent a "hybridization", as it appears on Reunion Island (Folio, 2009)—remain incomplete, both on the part of the local tourism administration (*Tahiti Tourisme*) and on the part of operators. We are therefore left with approximate estimates. Moreover, it would be appropriate to consider the operational delimitation that could be proposed on what distinguishes a well-being holiday (which characterises the honeymoon product, for example) from a more sporting (more active) holiday which may be foreign to it, but also complementary.

prohibits *"attracting wild species to oneself in any way whatsoever, in particular by gestures, noise or promises of food, when this practice is likely to constitute a nuisance for other users of the area or to attract predators"*.

[69] Cf. article published in *TNTV News*, 21 April 2022 [URL: https://www.tntv.pf/tntvnews/polynesie/justice/9-personnes-jugees-pour-braconnage-de-tortue/], accessed 18 July 2022.

French Polynesia is still perceived and marketed as a sensory experience, an exotic atmosphere and environment. The actors multiply the references to what concerns the senses, and not what is similar to the extreme. The wording is eloquent, and there is no shortage of sensory qualifiers for the destination: *"paradise"*, *"zen"*, *"honeymoon"*, *"sun"*, *"sandy beaches"*, *"lagoon"*, *"turquoise waters"*, *"sweet scents"*, *"vanilla"*, etc. Approaches to the landscape on land or at sea are rather "soft", without too much physical effort on the part of tourists (4 × 4 safari, snorkelling, boat trip, panoramic flight in a helicopter or plane over Tahiti, etc.), which marginalises the experience linked to effort and risk. The extreme is rarely part of the experience and is hardly evoked in the memory. The majority of tourists do not come to practice sports activities, but rather to experience sensations of well-being and relaxation. The result is a niche of elitist tourism, which is constantly moving upmarket. An overnight stay in a stilt *fare*[70] in the lagoon in Bora-Bora can cost around €3000 (342,566 XPF) and on average costs €1800 (205,540 XPF). It is the quality of the service and the privacy (strict privacy, strict confidentiality) that are at stake, more than the access (unlikely, by the way) to possible intense sports activities. Finally, apart from the practice of golf—of which French Polynesia has two, one in Tahiti, the other in Moorea—which benefits from targeted themed stays, the entry through sport remains quite marginal.

Achieving the objectives of sustainable tourism is therefore not automatic; it requires relevant choices and a set of voluntary and targeted measures on the part of stakeholders. These measures must include voluntary actions within the framework of specific (tourism sector) or general (multisectoral or macroeconomic) policies that promote the three pillars of sustainable development (social, economic, and environmental). However, the ageing of the clientele, which includes more and more retired people, the splitting of holidays, the reduction of working time, and the diversification of the leisure offer have an impact on the relationship between tourism and sport as far as Polynesia is concerned. An increase in the range of accommodation for shorter stays is proving to favour discovery excursions, more than diving courses or hiking, to take just two examples. Tourists are certainly more aware of environmental issues, but their practices remain below this concern if we evoke the carbon balance induced by a stay in French Polynesia.

References

Bachimon, P. (1990). *Tahiti entre mythes et réalités. Essai d'histoire géographique* (390 p.). Editions du Comité des Travaux Historiques et Scientifiques.

Bachimon, P. (2012). Les friches touristiques en Polynésie française—Révélateur d'une crise de la destination et forme de résistance au tourisme international. *Via*, 1–2012, mis en ligne le 16 mars 2012. https://doi.org/10.4000/viatourism.1314. consulté le 15 septembre 2022.

[70] In the Tahitian language, fare refers to a house, a hut, a domicile, a dwelling, or even a building. With a qualifier, it refers to all kinds of buildings, according to the online dictionary of the Tahitian Academy. The "fare sur pilotis" are very comfortable bungalows, built in natural materials (coconut wood, bamboo, and pandanus), generally including a semi-piloted terrace.

Bernardino, V., Dupont-Teaha, N., Hoang-Sacault, D., Likong Chi, M., Mathieu, R., Natier, S., Periou, C., & Rakkah, M. (2018, juin). Polynésie française. *Rapport annuel*, Ed. IEOM, 94.

Folio, F. (2009). Réalités et singularités du tourisme réunionnais: entre utopie et motifs d'espoir. *Les Cahiers d'Outre-Mer*, n°245, 7–33.

Gay. J.-C. (2018, mai-juin). Les outre-mers européens. *La Documentation photographique*, n°8123, 63.

Lienard, K., Lee, T., Garot, P.-E. (2022, May 26). Mort d'un randonneur à Moorea. *Polynésie La Première*. Posted on 26 May 2022. https://la1ere.francetvinfo.fr/polynesie/tahiti/polynesie-francaise/mort-d-un-randonneur-a-moorea-1287940.html. Accessed on 24 Aug 2022.

Ministère du Tourisme et du Travail, en charge des relations avec les institutions. (2020). *Fāri'ira'a Manihini 2025. Le tourisme en Polynésie française en 2020: chiffres clés, contexte détaillé, tourisme inclusif* (36 p.). Papeete, en ligne. https://www.fm25.pf/sites/default/files/2020-09/fm25_contexte_sept2020_v5.pdf. consulté le 29 août 2022.

Moyrand, A. (2009, novembre). Les systèmes de gouvernement des collectivités françaises du Pacifique (Nouvelle-Calédonie et Polynésie française). *Revue générale des collectivités territoriales*, Ed. du Papyrus, n°46, 73–83.

Présidence de la Polynésie française. (2021). *Les objectifs de développement durable en Polynésie française. Rapport de suivi des ODD en Polynésie française* (50 p.). mis en ligne le 26 juillet 2021. https://www.ispf.pf/content//uploads/rapport_odd_2021_59bd4a3aa7.pdf. consulté le 4 septembre 2022.

Regnault, J.-M. (1993). *La bombe française dans le Pacifique. L'implantation, 1957–1964* (186 p.). Ed. Scoop édition.

Regnault, J.-M. (2016). *Pouvana'a et de Gaulle. La candeur et la grandeur* (416 p.). Ed. Api Tahiti.

SAGE (2019) Projet d'aménagement et de développement durable du fenua & schéma d'archipel des Tuamotu-Gambier", edited by the Ministry of Housing and Spatial Planning in charge of inter-island transport, French Polynesia, p. 37.

Chapter 16
Sport and Tourism: Tahiti (Polynesia) as Case Study

Thibaut Aussant

Tourism is important for many small island societies. Despite the strong tourism image associated with them, the relationships that local societies have with this phenomenon cannot be considered uniform, for beyond their common insularity, these communities are obviously different, given their own historical background and contemporary situation. They are not all concerned by tourism, and for those that are, this is in different ways and to varying extents. Overlooking that fact can lead to a hurried generalisation which, as Guibert and Taunay (2018: 15) write when discussing common sense assertions on "tourism", "constitutes an impasse and potentially leads to errors of interpretation".

For all that, and without denying the diversity of the situations mentioned, the existence and development of tourism practices and services are a characteristic shared by many island areas. Likewise, the questions that arise are often common to all these territories, where they resonate in a particular way. Owing to the colonial past, fragile environmental context, and sometimes precarious economic or social situations that can characterise a certain number of them, the issues of tourism—whose "neo-colonial" and "environmentally destructive" dimension is often denounced—carry particular significance. For island territories, therefore, tourism certainly seems to be a relevant lens through which to analyse societal dynamics and recompositions.

That said, how can the study of tourism related to the field of sport contribute to an understanding of the societal issues facing island territories? Amid the (long-term) massification of tourism—the first witnesses of which are islands—its practices and services have become more diverse. The concern for differentiation and the mindset of distinction to which it responds have already been highlighted (Réau,

T. Aussant (✉)
Laboratory ESO (UMR CNRS 6590), University of Angers, Angers, France
e-mail: thibaut.aussant@etud.univ-angers.fr

D. Van Rheenen et al. (eds.), *Sport Tourism, Island Territories and Sustainable Development*, Sports Economics, Management and Policy 24,
https://doi.org/10.1007/978-3-031-51705-1_16

2011; Cousin & Réau, 2016: 31–54). Indeed, the rise of sport tourism is a phenomenon that the academic community has sought to characterise for several decades (Sobry, 2004). There is just as much merit in analysing tourism associated with sport as there is in other "forms" of tourism. Far from ironing out the tensions and euphemising the relations of domination that tourism engenders (through the supposedly intrinsic virtues of the sports world), an analysis of the issues at stake in sport tourism provides an equally pertinent opportunity to "increase and enrich the knowledge of the facts of tourism" (Pigeassou, 2004: 33–34).

Within the context of the broad inter-island comparison that this book aims to achieve, our contribution therefore focuses on the case of a Polynesian island, Tahiti, and more generally on the archipelago to which it belongs, namely, the Society Islands. How significant are sports-related tourism practices and services there? How did they develop? Who were the initiators? What role did, and do, public policies play? And finally, what are the implications of its development (in economic, socio-cultural, and environmental terms)? To answer these questions, this chapter will be divided into several parts. An introductory section will begin by giving a reminder of the natural, historical, and contemporary context of Tahiti and the adjacent islands. Then, the development of tourism practices and services associated with sports will be presented. In a third part, the issues raised by this form of tourism will be examined—rather than its "impacts", as this approach may induce a "ballistic vision" (Picard, 2001) that often omits the role and participation of local society in the tourism system (Blondy, 2010). The conclusion will be an opportunity to briefly mention the surfing event of the 2024 Olympics which will take place in Teahupo'o in Tahiti, as well as, of course, the tourism issues around it.

This work is based on current research, conducted as part of a sociology thesis on the issues surrounding the 2024 Olympic surf contest scheduled in Tahiti. This chapter draws on a variety of material. Tourism documentation and press articles related to this topic have been collected and analysed. During initial fieldwork in November 2021, various observations were made and interviews were held with relevant stakeholders (surfers, environmental associations, politicians, inhabitants, tourism institutions, the surfing federation, etc.).

General Context: Tahiti and the Society Islands

Understanding the conditions and scope for the development of sport tourism in Tahiti, as well as its specific features (its particularities, its limits, etc.), can only be done by looking at it within a given temporal and spatial context. This is true for any research object in fact and perhaps especially those that are too often essentialised (Guibert, 2021), which only the conscientious work of historicisation can make intelligible. Thus, regarding our subject of interest, it is obvious that the development of sports practices (and their touristification) is essentially understood in light of the political, social, cultural, and economic contexts in which they are embedded. Moreover, whilst the characteristics of the environmental milieu may not seem relevant in theory to understand the establishment or wider expansion of some sports practices, they do appear to have an obvious influence on others. Without ultimately determining the conditions and scope for the development of such sports,[1] they nevertheless structure them in part. We therefore suggest beginning with a review of the physical and geographical aspects of the island in question. As we will see later, these features do have an impact: they may allow or restrict the development of sports practices and services and consequently their use in a tourism context.

Geographical and Natural Indicators

The Society Islands are located in the South Pacific Ocean, in a region commonly called Polynesia, a few thousand kilometres from the nearest mainland. They are made up of two groups of islands: the Windward Islands (including Tahiti, the largest, and Moorea) and the Leeward Islands (including Bora-Bora, Raiatea, and Huahine). These are mostly high (volcanic) islands, although there are also a few atolls, which are much less populated (and much less concerned with tourism) (Fig. 16.1).

Tahiti is the largest of these islands and home to most of the population (nearly 200,000 of French Polynesia's 280,000 inhabitants live there today). And yet, it is barely 60 km long. Tahiti nui (the main part) is generally distinguished from its peninsula, Tahiti iti. The current urban area, around the capital, stretches for

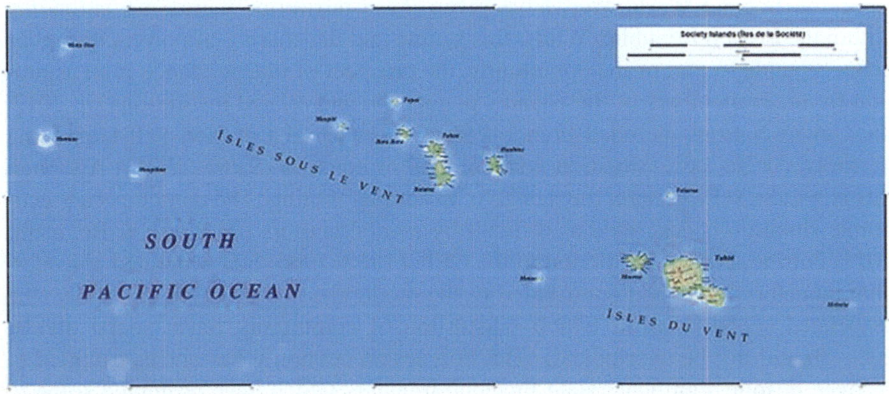

Fig. 16.1 The Society Islands, with the Leeward ("sous le vent") Islands at the northwest and the Windward ("du vent") Islands at the southeast, with Tahiti as the main island

[1] This would amount to accepting a geographical determinism which ignores the fact that human activity takes advantage of, adapts to, and bypasses the "opportunities" or "constraints" presented by its environment. More specifically, as far as sports are concerned, it would be to overlook that skiing is nowadays practised in the mountains as well as on artificial snow and surfing in the ocean or in wave pools.

kilometres to the north and south and concentrates most of the population. The rest of the island is more rural.

Human activities are mainly gathered along a rather narrow coastline—a mere few hundred metres long—surrounded by mountains and the lagoon. There is a road that goes around the island (except along some parts of the peninsula). As on most other islands, the mountains are very steep and quite high (the highest elevation is 2241 m). This makes some places difficult—if not impossible—to access. It is heavily forested. A lagoon, a few hundred metres wide, encircles most of the islands. It is occasionally interrupted by channels that communicate with the ocean, and disappear in rare places, allowing the waves to break on the shore. The water and air temperatures are roughly equivalent, between 25 and 30 degrees. The climate is tropical and humid; there are two seasons, one dry and one rainy.

Political and Economic Situations

Tahiti is the political, economic, and cultural centre of French Polynesia, whose name reflects its colonial past. Probably populated around the eleventh century, the first contact these island societies had with the Western world occurred at the end of the eighteenth century. Evangelised by English Protestant missionaries, these islands then came under France's control during the nineteenth century—as a protectorate (1842) and then through annexation (1880) in the case of Tahiti. Its current situation as a French regional government is the result of statutory changes which, since the Second World War, have given these territories increasing autonomy within the French Republic. Whilst this statute and the sharing of powers it implies are acceptable to part of the community, the prospect of independence is an aspiration for other members of the island's population and related institutions (in 2013, the United Nations included French Polynesia on the list of Non-Self-Governing Territories[2]). So, is its situation post-colonial or colonial (Saura, 2021)? Although this is not the subject here, the debate is worth mentioning. The contemporary economic situation is described as difficult by several authors (Al Wardi & Regnault, 2011; Poirine, 2011). The period of the CEP[3] (1962–1996) had generated considerable monetary flows and contributed to the modernisation of Tahitian society (the associated social costs of which—distortion of community links, social divide, etc.—should not be overlooked). The Polynesian economy has since struggled to recover, despite significant state financial transfers which have prompted some to speak of an "artificial economy". Tourism is the main sector of activity, accounting for 8% of GDP in 2018 (amounting to 620bn CFP francs[4]).

[2] https://www.un.org/dppa/decolonization/en/nsgt/french-polynesia

[3] (Centre d'Expérimentation du Pacifique). This is a nuclear experimentation centre.

[4] Source: ISPF (Institut de la Statistique de la Polynésie française—French Polynesian Institute of Statistics).

Tourism in Tahiti

The significance and implications of so-called "sport" tourism practices and services can only be understood by considering what tourism in Tahiti represents more broadly. We have mentioned its relative importance to the economy, but this should not be our sole consideration. The stakes involved are necessarily wider—spanning social, cultural, and environmental aspects alike.

Tourism grew in Tahiti in the 1960s, driven by several favourable factors. Apart from the international context—with tourism becoming increasingly accessible to and popular among the Western public and the development of long-haul commercial flights—it was above all the opening of an airport on the island in 1961 that sparked a surge in tourist numbers. The 1500 or so tourists recorded in 1959[5] increased exponentially over the following decade. Tahiti was an expensive tourist destination but associated with an alluring image whose roots can be traced back to the myths of the "noble savage" and the lustful *vahine* (which means "woman" in Tahitian). Tourist numbers rose at a steady pace at first. Large hotel complexes subsequently developed across the island, a sign of the authorities' ambition to turn it into a luxury tourism destination (Blanchet, 1987). Public policies pinned their hopes on this sector providing the stimulus needed for the post-CEP Polynesian economy to recover. And it did quickly become the leading economic sector, but its growth has continued to be fraught with challenges from the end of the century through to today. Most recently, the upswing in the situation since the beginning of the 2010s has not withstood the COVID crisis.

Sport and Tourism in Tahiti and Its Islands: An Overview

Quantifying the Importance of Sport (in) Tourism

Against this backdrop, sport represents a worthwhile communication medium for promoting tourism, in that it can showcase a natural Polynesian environment idealised by tourists. But how important really is it? What weight does it have in tourism practices and services?

Some market studies have attempted more or less precise estimations, by analysing the perceptions and motivations of tourists visiting French Polynesia, including with regard to sports practices. Thus, according to a study by Hawai'i Pacific University,[6] the image of French Polynesia is largely correlated to "sports activities" among tourists (with variations according to the different emitting markets). Another

[5] Source: "40 ans d'économie polynésienne", ISPF publication 1261, 2021.

[6] Its findings are outlined in a document from the Ministry of Tourism, "Stratégie de développement touristique de la Polynésie française (2015–2020)", p. 12.

statistical source[7] cites a number of 8–9% for whom sport (often water sports) would be one of the main expectations of their stay. However, these indicators should be interpreted with caution. Although quantitative methods give the reassuring impression of precision and veracity, it is nevertheless advisable to question the methods of data production to ensure these figures are reliable and to avoid "false precision" (Guibert & Taunay, 2018: 16). Accordingly, to estimate the weight of tourism related to the field of sport, the object must first be defined, which raises the question: what is meant by tourism and sport?

The polysemy surrounding these terms explains the difficulty in defining certain social realities through classifications. Definitions of sport vary (is it necessarily a physical activity? competitive?) and are the subject of scientific or institutional disagreement (Defrance, 2011: 97–108). The same applies to tourism—does this depend on spatial (distance), temporal (length of stay), or situational (the reason) criteria? These questions are not only of theoretical interest but also raise very concrete questions in the present case study such as when does pleasure sailing become sport? In addition to the predominant international tourism, should we be interested in a form of domestic tourism (and if so, under what conditions)? These are all questions that need to be addressed in order to grasp the full breadth of the subject, the myriad possible approaches, and the potential renewal of its issues and thus to avoid being locked into pre-established interpretative frameworks. The review by Van Rheenen et al. (2017) of the different approaches and definitions of sport tourism in the field of academic production clearly demonstrates their relative diversity. Indeed, beyond the polysemy of each of these words, their association or "their combination multiplies the possible meanings of this term [sport tourism]" (Sobry, 2004: 13). The realities to which this form of tourism seems to correspond in Tahiti thus depend on the variables considered. As recommended by several authors (including Bouchet & Bouhaouala, 2009), we will therefore consider sports-related tourism from the perspective not only of tourists who *practise* sport but also of *spectators* (of a sports event) or *visitors* (to a sports site).

A Recent and Rather Limited Development: Sector and Stakeholders

Sports-related tourism practices and services have only recently developed in Tahiti and the Society Islands and, as we shall see, in a rather limited way. By the 1960s, a few tourists, with sufficient funds and experience in their particular sport, had probably travelled as far as Tahiti, which they saw as a virgin playground ideal for practising their discipline. In 1967, for example, a few Californian surfers visited to

[7] "Points forts de la Polynésie française, 2018", *ISPF.*

test the Tahitian waves.[8] However, this was obviously only a tiny proportion of tourists, the majority of whom were not yet familiar with this type of sport. Although some sports began to develop and gain official status in Tahiti (the 1970s saw the creation of the first surfing clubs, and the hiking scene became more organised around associations, etc.), they were still not aimed at tourists at this time. As such, the sports offering for tourists was almost non-existent until the end of the century in Tahiti, which had only positioned itself as a sports destination for an elite few.

Diving was certainly the earliest activity to be structured: the first centres appeared in the 1980s, on the initiative of expatriates from mainland France. The number of centres has grown steadily. Across French Polynesia, there were 20 in 1995, 30 in 1999,[9] and 54 in 2020,[10] thus rapidly forming the most important and lucrative sector of tourism related to sport. 15 to 20% of international tourists apparently take up this activity, according to a report by the Economic, Social and Cultural Council of French Polynesia,[11] for an estimated expenditure of 300 m CFP francs.[12]

Without reaching the same proportions, other sports have become available to tourists since the end of the 1990s. At the instigation of entrepreneurs—again often expatriates from mainland France with the requisite diplomas, otherwise Tahitians trained in France owing to a lack of training programmes in Tahiti[13]—a tourism offering relating to sports practices is gradually being structured. With schools, rental service providers, and specialist shops developing on varying scales, it is gradually becoming possible for the average tourist to access these sports, through the provision of equipment and supervision. This is the case of surfing (for which Tahiti is now clearly known following the media coverage of the Teahupo'o wave in the 1990s), but also, more marginally, windsurfing, kite surfing more recently, and SUP and kayaking as far as water sports are concerned. The mountains are also a popular setting (albeit very relative) for sports. A few tourists sometimes take up hiking and trail running, as well as canyoning, paragliding, and mountain biking. The development of these different practices is very uneven, however, as is their touristification.

[8] Fourteen surfers from the WindanSea Surf Club arrived in 1967 to surf in Tahiti. This episode is mentioned by the first Tahitian surfers and recounted in a film and a book.

[9] Two thirds of which are located in the Society Islands. "Le tourisme en Polynésie française: monographie, années 1995–1999", ISPF.

[10] "Les plongeurs professionnels ont leur syndicat", *TNTV*, 21 July 2020.

[11] "L'avenir de la Polynésie française face à une gouvernance durable de son patrimoine marin", CESC report, January 2015.

[12] "L'économie bleue en Polynésie française", CEROM (Comptes économiques rapides pour l'outre-mer) document, August 2016.

[13] This is the case, for example, of the first paragliding instructor in French Polynesia.

Public Policies for Sport and Tourism: Structuring and Promoting

It has been highlighted through numerous studies that "from a perspective of territorial enhancement or identification, territorial or localised development, the strategies of local politicians participate in the construction and production of the sports and tourism offer established in their territory" (Guibert, 2016: 1). What role do public policies play in the development of sport tourism in Tahiti?

The government of French Polynesia and the departments concerned (such as the IJSPF) have been supporting this organisation by opening certifying courses, aimed at professionalising and "oceanising" this sector. For example, since 1997, there has been a course available for training hiking guides. More recently, in 2020, a Polynesian professional certificate for outdoor physical activity guides was created, with a clearly stated tourism objective.[14] Despite the increasing choice of such training courses, French people from the mainland are still over-represented in this professional sector. The administrative bodies in charge of structuring tourism services are aware that "the oceanisation of executives and operators must be a shared objective for the sector", as it could otherwise "lead to the brutal exclusion of populations in difficulty with the conventional educational systems, to the benefit of exogenous professionals who would have the required qualifications or diplomas".[15] This policy of oceanising tourism employment also concerns sport, as demonstrated by the creation in 2017 of a Polynesian diving certificate.

In conjunction with these structuring actions, a policy for the promotion of sport tourism in Tahiti is tentatively being developed. Aware of the tourism opportunities that a territorial identification with sport can represent,[16] the bodies in charge of promoting the destination abroad are showing an interest in the subject. This can be seen in a quick analysis of the contents and communication campaigns of the EIG

[14] "This desire on the part of the country is the result of the upturn in the tourism sector, which has been observed for several months, and which has been accompanied by the development of services for supervising outdoor physical activities, and in particular jet skiing, quad biking, snorkelling, rowing (kayaking, stand-up paddling) and e-biking. The development, quality and safety of these outdoor physical activities require a real professionalisation of the management", report of the Council of Ministers meeting on Wednesday, 26 February 2020, *Communication Office of the Presidency of French Polynesia.*

[15] "Fāri'ira'a Manihini 2025: Le tourisme en Polynésie française en 2020", Ministry of Tourism and Labour document, in charge of relations with institutions, p. 33, 2020.

[16] The document "Stratégie de développement touristique de la Polynésie française 2015–2020" of the Ministry of Tourism returns several times to the opportunities of positioning the archipelago as a sports destination. Here are some examples that show the willingness to obtain such a territorial identification: "the data collected during this study allowed to identify key factors of success for the Polynesian tourism development [among which] the diversity of sports activities" (p. 33) "What ambitions for the French Polynesian destination? … A paradise, yes, but a sports one" (p. 89).

Tahiti Tourisme.[17] The documentation displayed on its website as well as the short videos it posts over the various social networks showcase the wide range of sport activities on offer. For example, the cover of the "Tahiti & Moorea travel guide 2022" features a surfer on a crystal blue wave (Fig. 16.2).

Other, more proactive means are also used to promote sport to tourists. Tahiti Tourisme finances "press trips" for the foreign media, which are intended to inspire the creation of editorial content in the foreign press. For 2019 and the diving sector alone, Tahiti Tourisme's Director of International Operations counted "30 press trips dedicated to or including diving which generated 113 articles mentioning diving in *Tahiti and her Islands*".[18] Sport tourism can also be promoted through other tourism stakeholders—such as an airline—and other channels, such as the sponsorship of sport-related events abroad.[19]

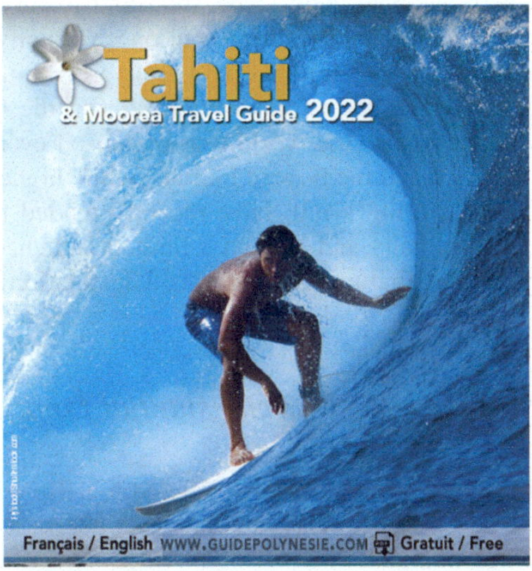

Fig. 16.2 The "Tahiti and Moorea travel guide 2022." (Source: Tahiti tourism)

[17] Created in 1992, this Economic Interest Grouping, financed in large part by the Polynesian government, is in charge of promoting the destination of "Tahiti and its islands". "The Group acts in close coordination with the ministry in charge of tourism and its supervisory department", *Statuts du GIE Tahiti Tourisme*, p. 3, 2017.

[18] As well as 4 emailing campaigns, 132 posts dedicated to diving on *Tahiti Tourisme*'s social networks, participation in five shows dedicated to diving, … see the collaborative project "Making Polynesia a 'diving destination'" on the Internet, Fāri'ira'a Manihini 2025.

[19] Recently, for example, Air Tahiti nui and the EIG Tahiti Tourisme have joined forces to organise surfing events in France—cultural and sports events such as the Brest Surf Film Festival in 2019 or the Maider Arosteguy surfing competition in Biarritz in 2022—to promote Tahiti as a leading surfing destination.

Sport Events and Their Impact on Tourism

The development of sports events and competitions is also considered to be a way of boosting tourism, by attracting participants and/or spectators. Although these are mainly locals (sometimes from neighbouring islands) in Tahiti, the competitions also appeal to a small number of international tourists whom the organisers, in collaboration with the public authorities, try to attract through communication campaigns and advantageous airfares. The statistics kept by the organisers of the main sport competitions in the Society Islands give an indication of the (relatively small but growing) number of foreign participants, as follows (Fig. 16.3).

In addition, some of these competitions, with their international reputation and established popularity, also attract spectators. This is notably the case for the Hawaiki Nui Va'a (an outrigger canoe competition) and the Tahiti pro (a professional surf competition). However, the visitors who come are almost exclusively domestic, either residents or tourists already on-site. For the Hawaiki Nui Va'a, which has been held since 1992 and is very popular in Tahitian society, several dozen additional flights are put on, especially to the Leeward Islands, which rally around this exciting race in a festive community spirit. The village of Teahupo'o also experiences a tourist boom during the few days of the Tahiti Pro, a surfing competition on the world tour organised every year since 1997. The hype around this wave and the flow of tourists, while highest on the days of the competition and during the big swell, have recently tended to continue year-round. Now known internationally and beyond a limited audience of surfers, the wave has acquired a reputation that makes it a tourism destination in itself.

Structural and Circumstantial Limits to Development

The development of sports-related tourism practices and services in Tahiti, which we have so far endeavoured to describe, must nevertheless contend with barriers of different kinds.

	Existence since …	*Numbers of foreign participants in 2019*
XTerra (triathlon)	2015	unknown
Moorea marathon (running)	1988	around 40 (among 400 participants)
La Ronde tahitienne (cycling)	From 2011 to 2019	227 (among 629 participants)
Tahiti Pearl Regatta (sailing)	2004	15 foreign crews (among 50 crews)
Hawaiiki nui va'a (outrigger canoe)	1992	14 foreign crews (among 141 crews)

Fig. 16.3 The main sport events in the Society Islands and the part of tourists among their participants

The natural conditions of the environment are an initial hindrance. Beyond the postcard image with the mountains and the lagoon, which may suggest a vast arena for all sorts of sports, many of the sites for practising sports can in fact be hard to reach and even dangerous, especially for inexperienced tourists. Most of the surfing spots are reef breaks, for example. Far from the coastline, difficult to surf, and dangerous for beginners, the latter can only give this sport a go at the two beach breaks that quickly become overcrowded. Also, the scarcity of strong winds used to be detrimental to sailing sports until the more recent innovations of kitesurfing and foiling—which can be practised in light winds. Several parameters also limit the wider uptake of mountain sports. The Polynesian mountains are very steep and relatively difficult to access due to the lack of trails. There is both a legal and cultural obstacle here: the Polynesian land tenure system makes the land essentially a private property rather than a public domain. This drastically limits its access, and perhaps especially for less well-informed tourists, for whom mountain sports such as hiking or trail running are therefore not very accessible.

There are also political factors. The competent institutions and agents of the economic sector deem the means for structuring and promoting the sport tourism offering set up by the authorities to be insufficient. In 2015, the CESC emphasised French Polynesia's weak and uncertain positioning in the diving market, apparently owing to the "absence of promotion".[20] As regards surfing, various association leaders point out the mismatch and inconsistency between tourism promotion policies proposing territorial identification through this sport and the reality of local services: surf schools are mostly itinerant and without permits or licenses, surf beaches are unequipped or not laid out with surfers in mind (for example, with embankments built, etc.), and so on. The lack of service providers and the inadequate structuring (if at all) of certain sports from a tourism perspective (reported in 2014 for example)[21] is still true about *va'a* today.[22] Finally, the general cost of the destination and the high cost of local services (+20% for diving, for example, according to the CESC) obviously hamper the growth of this tourism sector in a competitive market that is disadvantageous to Tahiti.

---

[20] "L'avenir de la Polynésie …", *op.cit*, pp. 18–19, 2015.

[21] "Développer le sport et le tourisme sportif en Polynésie par la formation: le point sur le partenariat Etat-Pays-Air Tahiti Nui", Tahiti-infos, 30 October 2014.

[22] At the request of the Polynesian government, a study on the subject was produced in 2016 by a consultancy firm to identify the barriers and drivers for the tourism development of *va'a* ("Assistance technique au développement touristique des activités nautiques autour du va'a en Polynésie française"). However, 4 years later, the finding is unequivocal: "The case of the development of tourist discovery services for va'a, for example (as a differentiating and original marker of a Polynesian experience), which could have been provided by existing sports clubs, federal supervisors, or even experienced practitioners, is significant [of the failure]. The additional income that could have been generated, for the benefit of local populations or sectors, will not be generated because the required supervision, when it comes to tourists, becomes insurmountable for the existing structures", *Fāri'ira'a Manihini 2025: Le tourisme en Polynésie française en 2020*, document from the Ministry of Tourism and Labour, in charge of relations with institutions, p. 33, 2020.

The development of sport tourism practices and services is therefore rather limited in Tahiti and the Society Islands. Most sports are only available to tourists with sufficient funds and experience, for whom the multiple constraints mentioned above can be overcome. They are typically able to afford the high cost of the destination, and to take on challenging terrains, with the help of potential local acquaintance networks. These tourists who come to practise their discipline (sometimes for a competition) are thus in a minority. The majority of tourists, on the contrary, might perhaps include sport in their stay (either by practising it or by attending a sport event), but it is not a main motivation for them.

What Kind of Sustainability? Economic, Social, and Environmental Issues Around Sport Tourism

The focus on "sustainability" is not new to the academic field, "irrigated by policy prescriptions and contexts" (Zaccai, 2007). Although it does not constitute a scientific paradigm, the sustainability approach is nonetheless a frame of reference that is used increasingly—an interesting point about the autonomy/heteronomy of the academic field—if more or less explicitly. This includes in studies on tourism. Considering the sustainability of sports-related tourism practices and services means taking an interest in the environmental and socio-cultural issues surrounding them, to understand the multiple facets of a social reality. We will see that, far from being "sustainable" by nature because it is sport-related, this form of tourism represents a relevant lens through which to grasp the different issues raised by tourism in general in Tahiti.

Economic Outcomes and Their Distribution

The economic contribution of sport tourism practices is not easily identifiable. Except for diving (see footnote 12), we do not have any statistics that allow us to gauge this more precisely. Without being able to provide an accurate figure, it is nevertheless possible to make an estimation. These incomes vary and correspond to supervision and rental services, equipment purchases, registration fees for competitions, and so on. Indirect expenditure (accommodation, transport, catering) should be taken into greater or lesser account depending on the situation; for example, for tourists for whom sport was the reason for the trip (whether to practise it or for a competition)—i.e., "sport tourists" (Bessy & Naria, 2005: 5).

However, these economic benefits are unequally distributed socially, and often ethnically too. We mentioned earlier the over-representation of nationals from mainland France in many sports-related tourism facilities—such as surfing schools, diving centres, and so on.—because of the lack of training courses in Tahiti. This situation has improved unequally depending on the sport. For example, in 2017, at the start of the very controversial reform concerning the qualifications required to work as a diving instructor in French Polynesia, there were only three Polynesians working in this profession on the territory,[23] whereas the sector generated dozens of jobs locally. Socio-cultural factors may explain this in some cases—scuba diving is practised more by tourists and mainland French residents than by the Tahitian population for example. This argument does not hold for surfing, however, which is very popular among young Tahitians, and yet few locals have become qualified professional instructors to date (due to the lack of training courses).

How is sport tourism (whether in terms of its practitioners or providers) perceived by the local population? The following example, without claiming to be representative, nevertheless outlines the social issues related to the development of these tourism practices and services.

The Example of the "Sapinus Boys": A Complicated Relationship with Surfing and Scuba Diving Tourism Development

Participant observation[24] at the Pointe des Pêcheurs[25] in Punaauia, along with semi-structured interviews with young surfers living in the neighbourhood, yielded several insights into the complicated relationship that a certain number of them have with the development of sports-related tourism practices and services. Indeed, there is a surfing spot (reef break) as well as diving sites of interest (for whale watching among other attractions). The Pointe des Pêcheurs is also a popular hangout among neighbourhood locals who, in the evening and on weekends, meet there to have a drink, listen to music, have "little barbecues, *'uru* …'. According to one of the youngsters who goes there every day, "on the whole island, here in Tahiti, this is perhaps […] the last little place where the Polynesian community can still gather together to do this kind of thing".

However, this cohabitation between the local community and the world of tourism (whether practitioners or service providers) has become increasingly difficult in

[23] I tend to say to them [to the detractors of the reform] "what local job? […] Today, clearly, among the number of instructors working in Polynesia, there are 3 Polynesians, the others work internationally [where the diploma that can be obtained in Polynesia is recognised]", interview with Nicole Bouteau, Minister of Tourism. "La réforme de la plongée votée par l'Assemblée", *Radio 1 Pf*, 9 November 2017.

[24] These "participant observations" took place over a short period of 1 month (in November 2021), but nevertheless at a regular frequency (up to several times a week).

[25] Also known as Nu'uroa or Sapinus, this district is located on the west coast of Tahiti in the municipality of Punaauia.

recent years. Amid a growing number of foreign surfers and diving boats,[26] a small number of young local surfers have sought to control and restrict—or even prevent—access to the sports sites in question for surfers and tourism providers, through various and more or less aggressive intimidation measures. M. and T., the two protagonists behind these restrictions, who go to great lengths to ensure they are respected, are about 25 and 35 years old, respectively. They have lived in the city since 'childhood' and have been familiar with the place and its people for a long time—an autochthony that is used as a capital (Retière, 2003) in their social relations with the various users. One is a shop assistant, and the other was a former tattoo artist and is now unemployed. Both strongly claim their Ma'ohi identity and do not shy away from voicing their political opinion ("I'm not going to lie to you, I'm a separatist"). Their commitment is not expressed through overt support for the corresponding party, however ("you know now, politicians, they're thieves, they're liars!", "we can't really say we're experts when it comes to politics, we're not really into such things"), but rather through "direct" actions ("we've quit waiting for something to happen and now we act directly", "there's too much talk—it would be better to act, act directly").

In their eyes, this tourism development is in no way positive for the local community. They, along with their close family and friends, do not gain from it and are well aware that the benefits do not reach them or Polynesians at large.

> M.: "[…] you see, it's like the excursions that take place, the whale watching excursions, the diving excursions—they're not organised by a local, but by someone from the mainland".
> T.: "Polynesians don't have doing business in mind. When you look at the community, well, they're always at the bottom!".

On the contrary, they are quick to point out the problems that this development causes to the local natural and socio-cultural environment. For example, they believe that diving tourism providers are damaging the local environment, in various ways: corals are destroyed when the diving providers' boats drop anchor, and whale-watching services disturb the latter for example. But for them, it is above all the socio-cultural balance that is being undermined by such tourism development. The attitudes of relative indifference, disdain, and even contempt shown by the ever-increasing number of foreign and mainland resident surfers, in their view, are stirring up feelings of territorial dispossession likely to revive memories of a colonial situation.[27] Controlling access to the surfing zone is thus considered a means of reappropriation (Walker, 2011), so "that it remains as it has been for a long time—

[26] "Before, there were just one or two diving-boats there … now, if you come on days when it's calm and sunny at the weekend, man, there are like—six, seven, ten!". T., interview 17 November 2021.

[27] "I wouldn't say it's only for the identity of the spot. It's a lot of things. I mean in relation to the French State, to the mainland population, to France, right? … Bear in mind that 70 years ago, our grandparents, our *matahiapo*, our elders, were forbidden to speak their language, the native language, the Tahitian language, otherwise they would be punished, sent to the back [of the classroom], be beaten up, those kinds of things. You see, you can't admit that". M., interview 17 November 2021.

i.e. the Polynesians' [property]". These actions obviously serve personal interests—a limited number of surfers in the line-up makes it easier to enjoy the waves. They are also a way for locals to engage in concrete actions that the political class (including the pro-independence party) does not manage to carry out according to them.

A More Sustainable Form of Tourism?

The issues raised by sport tourism, far from being specific to this sector, are in many ways indicative of the questions that are more generally encountered regarding tourism in Tahiti.

The unequal distribution of the economic profits generated by sport tourism thus echoes what is happening across the tourism sector at large on the island. In the early decades of tourism's growth in Tahiti, it was already noted that "only a part of the wages bill generated by the tourism industry goes to Polynesians" (Robineau, 1974: 8). While things may have changed since then, the previous observations made about the distribution of economic benefits linked to sport tourism temper the hopes of the author quoted above who, in Marxist terms, advocated a better redistribution of "the ownership of the means of production".

Furthermore, while tourism practices have often been depicted as detrimental to local communities (in a discourse that is sometimes Manichaean and simplistic), other reified discourses have, on the contrary, ascribed to *sport* tourism practices all the attributes of *sustainable* tourism. The practice of sport, naturally inclusive and conducive to social cohesion, would therefore by its very nature favour harmonious cohabitation between tourists and locals. This is a widely shared idea but is by no means a given and does not necessarily reflect the social reality observed, for it turns out that tourists and the Tahitian population (this is different for expatriates from mainland France who have kept their sports habitus) do not necessarily practise the same sports. And when they do, their encounters may or may not be amicable. Often, they reproduce the same avoidance strategies that exist more generally in other tourism activities—as is the case for beach practices (Blondy, 2013).

Finally, the environmental issues associated with sport tourism practices and services also require a more detailed analysis than a simplistic dichotomy—that tourism would be a vector of environmental degradation or, on the contrary, sport contributes to environmental awareness. There is a wide range of tourism practices and services relating to sport, and their relationship with their natural environment is just as diverse. Nevertheless, it should be noted that, except for diving, the anthropic pressure generated by these activities remains low, since they require little

T. "We talk to them, we explain to them, they don't want to understand, they are above us ... And when you know what these people have done to our people, you go crazy".

M. "That's what I'm saying, what they did to our people and on top of that they don't want to respect ..." interview 17 November 2021.

or no construction and only generate a relatively small number of visitors in Tahiti's case.

The 2024 Olympics and Its Implications for Tourism

These various issues surrounding sport tourism are, incidentally, more topical than ever in Tahiti. On the eve of the 2024 Olympic Games, for which Teahupo'o (a village on the Tahitian peninsula) is gearing up to host the Olympic surfing contest, the relationship between the tourism and sports fields is even more in the spotlight. The tourism dimension of so-called *mega-sport events*, of which the Olympic Games are the highest profile, has been the subject of numerous case studies that highlight the interest they spark in the various host countries and local authorities. Although tourism incomes from the Olympic Games are variable and far from homogenous (Chappelet, 2018), they are nonetheless a central part of the political discourse as well as a major argument to justify their organisation. The French Polynesian government's agenda is no exception: the organisation of the Olympic event is indeed perceived and presented as a driver for tourism in the short and long terms, on the scale of the island as well as for the village of Teahupo'o. However, whilst the latter's inhabitants do not reject tourism (it provides a not insignificant complementary source of income for many of the people whose livelihoods mainly revolve around fishing and farming), the ways in which this expected and hoped-for tourism should be managed sometimes give rise to divergent opinions, as each of the different social stakeholders concerned—village residents, tourism representatives, environmental associations, local councillors, etc.—try to assert the interests they are defending. The ongoing tensions go far beyond the definition of solely the Olympics event and broadly bring to light the discord over larger topics, which obviously include tourism.

Conclusion

The relationship between tourism and sport in Tahiti is a relevant subject of interest. Not only because of their relative development over the last decades and years but also in light of the questions that these two activities raise when considered together. In this chapter, we have discussed the development (and limits thereof) as well as the importance of sports-related tourism practices and services, by studying how public policies use them for the purposes of territorial identification or tourism in connection with sports events. We have also underlined how the social, economic, and environmental issues raised by this so-called sport tourism could shed light on the issues of tourism in a more general way. Although this study focuses on Tahiti, there is no doubt that many of the questions raised are also common to other small island territories. The broad analysis this book undertakes through the prism of sport tourism provides insight into the similarities, differences, and specific features between these territories.

References

Al Wardi, S., & Regnault, J.-M. (2011). *Tahiti en crise durable: un lourd héritage*. Éd. de Tahiti.

Bessy, O., & Naria, O. (2005). Les enjeux des loisirs et du tourisme sportif de nature dans le développement durable de l'île de la Réunion. In *Management et marketing du sport: du local au global* (pp. 307–339). Septentrion.

Blanchet, G. (1987). Le tourisme en Polynésie française ou le rêve à l'épreuve des faits. In *11th symposium*. SEFANRIT, Bordeaux.

Blondy, C. (2010). *Les territoires touristiques polynésiens : une lecture géographique de la participation de la société locale au système touristique*. PhD thesis, Bordeaux 3.

Blondy, C. (2013). Pratiques de la plage en Polynésie française. *Mappemonde, 3*(111), 1–17.

Bouchet, P., & Bouhaouala, M. (2009). Tourisme sportif: un essai de définition socio-économique. *Téoros. Revue de recherche en tourisme, 28*(2), 3–8.

Chappelet, J.-L. (2018). Les retombées touristiques des Jeux Olympiques d'aujourd'hui. *Entreprises et Histoire, n° 93*(4), 90–104.

Cousin, S., & Réau, B. (2016). *Sociologie du tourisme* (New ed.). la Découverte.

Defrance, J. (2011). *Sociologie du sport* (6th ed.). la Découverte.

Guibert, C. (2016). Introduction au dossier "Sport, tourisme et production de territoires". *Mondes du Tourisme, 11*, 10.4000/tourisme.1002.

Guibert, C. (2021). *Les mondes du surf. Transformations historiques, trajectoires sociales, bifurcations technologiques*. MSH Aquitaine.

Guibert, C., & Taunay, B. (2018). *Tourisme et sciences sociales: postures de recherches, ancrages disciplinaires et épistémologiques*. L'Harmattan.

Picard, M. (2001). Bali: vingt ans de recherches. *Anthropologie et Societes, 25*(2), 109–127.

Pigeassou, C. (2004). Le tourisme sportif: une réalité sociale aux contours incertains. In *Le tourisme sportif*. Presses universitaires du Septentrion.

Poirine, B. (2011). *Tahiti: une économie sous serre*. Harmattan.

Réau, B. (2011). *Les Français et les vacances: sociologie des pratiques et offres de loisirs*. CNRS.

Retière, J.-N. (2003). Autour de l'autochtonie. Réflexions sur la notion de capital social populaire. *Politix. Revue des sciences sociales du politique, 16*(63), 121–143.

Robineau, C. (1974). *L'économie tahitienne et le tourisme*. Presented at the Seminar on Tourism at the East-West Center, Honolulu.

Saura, B. (2021). *Des Tahitiens, des Français. Essai sur l'assimilation culturelle en situation coloniale consentie*. Au vent des îles.

Sobry C. (2004) Le tourisme sportif. Presses universitaires du Septentrion, https://doi.org/10.4000/books.septentrion.117333

Van Rheenen, D., Cernaianu, S., & Sobry, C. (2017). Defining sport tourism: A content analysis of an evolving epistemology. *Journal of Sport & Tourism, 21*(2), 75–93.

Walker, I. H. (2011). *Waves of resistance: Surfing and history in twentieth-century Hawai'i*. University of Hawai'i Press.

Zaccai, E. (2007). Développement durable et disciplines scientifiques. *Natures Sciences Societes, 15*(4), 379–388.

Chapter 17
Sport Tourism and Local Sustainable Development in Small Island Territories: The Case of Kiribati

Kiali Molu, Stephen Pratt, and Sandra Tarte

Introduction

Kiribati is one of the world's Least Developed Countries (LDCs), a status given by the United Nations to low-income countries that are highly vulnerable to economic and environmental or natural disaster shocks. Geographically, Kiribati consists of the Gilbert Islands, Phoenix Islands, Line Islands, and one isolated volcanic island, Banaba. The island archipelago spreads over an area of 5 million km^2 and has a total land area of 810.5 km^2, with a coastline of 1140 km. Its highest point is only 2 m above sea level. Only Banaba Island, rich in phosphate, is 80 m above sea level (Chambers & Chambers, 1975; Kiribati National Statistics Office, 2020b). Christmas Island (Kiritimati) is known as the largest coral atoll in the world (Fig. 17.1). The capital of Kiribati is Tarawa. The nationality of the people of Kiribati is called I-Kiribati.

In 2020, the Kiribati population stood at 119,438 (Kiribati National Statistics Office, 2020b). The country's population density per square kilometer is 165, making its capital one of the world's most densely populated areas. The average household size is 5.6 persons, with an average of 6.5 persons in urban areas and 4.7 persons per household in rural areas. The 2020 Population and Housing Census reported there were more females than males in the total population (Males = 58,904

K. Molu · S. Tarte
Discipline of Government, Development and International Affairs, School of Law and Social Sciences, University of the South Pacific, Suva, Fiji
e-mail: s02001730@student.usp.ac.fj; sandra.tarte@usp.ac.fj

S. Pratt (✉)
Department of Tourism, Events & Attractions, Rosen College of Hospitality Management, University of Central Florida, Orlando, FL, USA
e-mail: Stephen.pratt@ucf.edu

D. Van Rheenen et al. (eds.), *Sport Tourism, Island Territories and Sustainable Development*, Sports Economics, Management and Policy 24,
https://doi.org/10.1007/978-3-031-51705-1_17

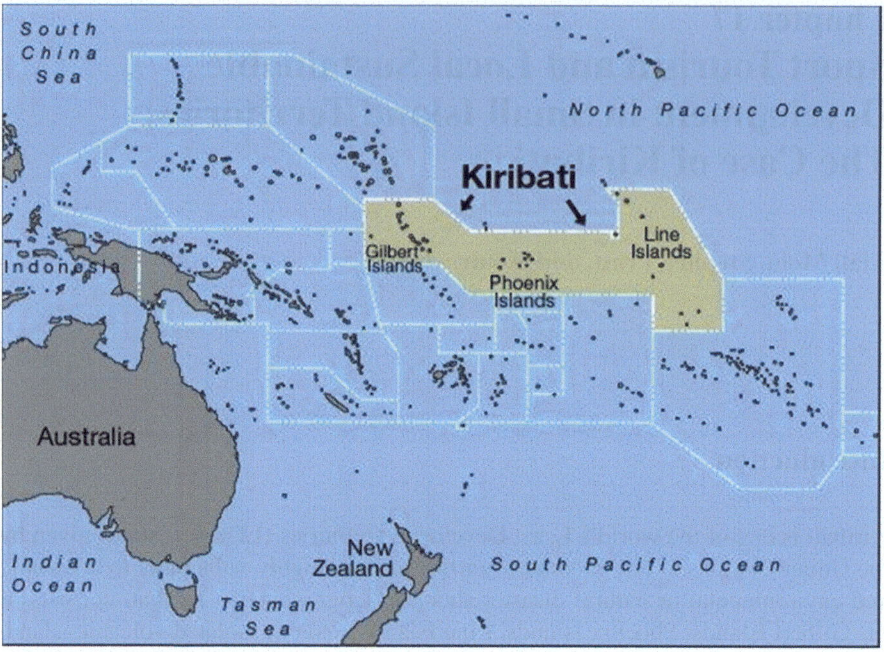

Fig. 17.1 Map of Kiribati

(49.3%) and females 60,534 (50.7%)). A total of 70,441 (59.0%) i-Kiribati live in urban areas and 48,997 (41.0%) reside in rural areas. Kiribati is a predominantly Christian country with 58.9% of the population identifying as Roman Catholic and another 21.2% and 8.4% identifying with the Kiribati Uniting Church and Kiribati Protestant Church, respectively. Other Christian denominations and religions include The Church of Jesus Christ of Latter-Day Saints, Seventh-Day Adventist, Assemblies of God, Baha'i, Jehovah's Witness, Muslim, and All Nations. Kiribati is a relatively young country. The median age is 22.9 years. Over a third (35.9%) of the country is 15 years or under, and only 3.8% of the population is aged 65 years or older.

In 2020, Kiribati's GDP stood at AUD 262,88 million and its gross national income (GNI) was AUD 490,49 million (Kiribati National Statistics Office, 2020a). GDP per capita and GNI per capita are AUD 2,201 and AUD 4,107, respectively. Kiribati's major revenues are sourced from its foreign reserve fund called the Revenue Equalization Reserve Fund (RERF), which in 2000 held more than AUD 658 million (Pretes & Gibson, 2008). Like other small island developing states, Kiribati has a large public sector so, in terms of GDP by industry, public administration and defense comprise 19.8% of GDP, followed by agriculture and forestry (17.1%) and fishing (9.1%).

Fisheries through the Vessel Day Scheme (VDS) and observer program are also a major contributor to the nation's economy. Fishing fleets from South Korea, Japan,

China, Taiwan, and the United States pay licensing fees to operate in Kiribati territorial waters, and these produced revenues were worth USD 207.1 million (91% of GDP) in 2015 (Webb, 2020). Challenges faced by the country include its highly dispersed populations, remoteness to major markets, lack of arable land, a narrow economic base, and the dual problems of sparse outer island communities and heavy overcrowding in the capital (Webb, 2019).

Politically, Kiribati is a democratic republic with a unicameral legislature, where the President is the Head of State, and the Head of the Government is elected by a democratic nation (Commonwealth Secretariat, 2022). The legislature has 44 members elected for 4 years by universal adult suffrage and one ex-officio member who is the Attorney-General. The Cabinet consists of the President, Vice-President, the Attorney-General, and 15 Ministers who are all appointed members by the President from his/her political party.

Kiribati was previously known as the Gilbert Islands and, together with the Ellice Islands, now known as Tuvalu, became a British protectorate in 1892, in the hope of eradicating slave raids and incessant tribal warfare (Macdonald, 1982). In the early 1900s, at one of its islands called Ocean Island, phosphate was discovered, which sparked a surge of British interest in the area, and more islands in this Pacific archipelago were placed under the British protectorate (Chambers & Chambers, 1975). For many years, phosphate was the prime source of income for Kiribati. In 1979, these phosphate deposits were exhausted. Kiribati gained independence from the British administration in 1980, 2 years after Tuvalu officially gained independence from the colony (Macdonald, 1982).

Geographic and Natural Indicators

Kiribati has 33 islands in total, 20 of which are inhabited. These inhabited islands exist within three island groups of the Gilbert Islands, where the population is concentrated, the Line Islands, and the islands of the Phoenix group. Kiribati is 3870 km wide from east to west and 2050 km in length from north to south. It is the only country that straddles the International Date Line and spans the four hemispheres (Brown, 2017). While the land area is small, Kiribati has a large Exclusive Economic Zone (EEZ) of 3,500,000 km^2 (Foster & Macdonald, 2021).

The country's climate is maritime equatorial or tropical (Bedford, Macdonald, & Munro, 1980). Temperatures range from 27 to 32 °C. The average precipitation in the Gilbert group ranges from 3000 mm in the north to 1000 mm in the south, although all of the islands experience periodic droughts (Foster & Macdonald, 2021). The most rain falls in the season of westerly winds, from November through March. From April to October, northeast trade winds prevail. The low-lying coral atolls are most vulnerable to the effects of climate change and especially sea-level rises. Several uninhabited islands have already been inundated and the Government of Kiribati is looking into several strategies to ensure their citizens' safety, with the possibility of relocation of some or all of the islands of Kiribati becoming uninhabitable (Foster & Macdonald, 2021). Coconut palms are abundant. Breadfruit and pandanus also are grown. Apart from fish, protein in the diet comes from pigs and chickens. Tarawa hosts the International Airport for the Gilbert Islands, and Kiritimati has a major international airport too. Interisland shipping is provided by the government, and most islands are linked by domestic air service.

Tourism and Sport Tourism on the Island

Tourism is not a major part of the Kiribati economy. Tourism earnings as a percentage of GDP were 5.1% in 2019. On the supply side, 52 accommodation places are hosting 525 rooms and the total number of employees working in the tourism sector was 449 in 2019, of which 36.5% were females (SPTO, 2020).

The type of tourism found in the Pacific islands is contingent on numerous factors, including the attractions on offer, distance, ease, and cost of access from tourism-generating markets, their economic buoyancy, and general perceptions of health, safety, and development in the destination area (Harrison, 2004). For instance, Fiji improved both air and sea transportation to ease access by tourists to the country; by contrast, Kiribati has fewer international visitors due to its difficulty to reach, limited quality accommodation, fewer quality restaurants, and irregular air service.

In 2019, the year before the COVID-19 pandemic, Kiribati received 7917 international visitor arrivals. Notably, 70.3% of these visitors visited Tarawa, while the other 29.7% visited Kiritimati (SPTO, 2020). Since 2011, there has been an upward trajectory, growing at an average of 22.2% between 2011 and 2019, albeit growing from a small base (Fig. 17.2). International tourism receipts in 2019 were USD 9.2 million; tourism receipts per arrival were USD 1,164.

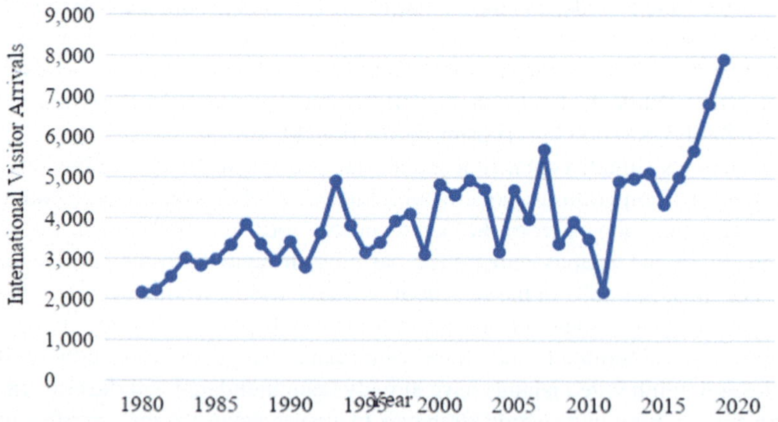

Fig. 17.2 International visitor arrivals to Kiribati in 1980–2019

The two main destinations in Kiribati offer very different locations. The capital city of Tarawa in the Gilbert Islands is primarily a business destination. In 2019, 56% of visitors to Tarawa visited for commercial reasons, with a further 32% visiting for holidays. The three top source markets for Tarawa are Australia, New Zealand, and the USA (Tourism Authority Kiribati, 2020). Kiritimati in the Line Islands is predominantly a holiday destination (71%) rather than a business destination (16%), with the top three source markets being the USA, Australia, and New Zealand (Tourism Authority Kiribati, 2020). Kiritimati is known as a world-class fishing destination and is heavily promoted in the USA and Australia. The USA tops the list of visitors to Kiritimati due to the atoll's location in proximity to Hawaii, a 3-hour flight. Additionally, Kiritimati recorded Chinese arrivals in the last quarter of 2019, not long after Kiribati switched its diplomatic ties to the People's Republic of China from the Republic of China (Taiwan). In Kiritimati Island and Fanning Island, apart from game fishing and other water activities, bird watching is also very popular as these islands are home to many migratory bird species. With the COVID-19 pandemic travel restrictions, as in many other places in the world, there was a major disruption in arrivals to Kiribati. With tourism in general being at its early stages of development, sport tourism has significant potential, but there is very little, if any, sport tourism currently.

Future Trajectories for Sport Tourism and Sustainable Development on Kiribati

While sport was not directly included under any of the Sustainable Development Goals, the opening declaration of the 2030 Agenda for Sustainable Development Goals contains a statement that: "Sport is also an important enabler of sustainable development goals. We recognize the growing contribution of sports to the realization of development and peace in its promotion of tolerance and respect and the contribution it makes to the empowerment of women and young people, individuals, and communities as well as to health, education, and social inclusion objectives" (United Nations, 2015, p. 3).

Sport for development in the developed world has contributed to achieving sustainable development goals in the areas of health, tourism, nation-building, peace and conflicts, gender equality, poverty alleviation, and economic growth (Commonwealth Secretariat, 2017). According to Andreff (2001), investing in sports not only contributes to a country's economic success but also contributes to a happy and healthy population. Likewise, the building of an infrastructure to support sporting events will only enhance economic growth by providing jobs and contributing to the alleviation of poverty.

Sport has long had a positive impact in many areas of life, and both sport and physical activity can make a substantial contribution to the well-being of people in developing countries (Sherry et al., 2017). SPC (2018) reported that exercise, physical activity, and sport have long been used in the treatment and rehabilitation of communicable diseases and non-communicable diseases (NCDs). Physical activity for individuals can build resilience to avoid diseases, and for nations, it is a cost-effective method to improve public health across the population (Andreff, 2006).

Despite the acknowledgment by academics and international agencies like the United Nations that sport is important for development, for low-lying small island developing states in the Pacific such as Kiribati, sport is less prioritized. Other issues of national concern take precedence. Education is perceived as a driver for economic success and is prioritized in the Kiribati National Development Plan (Government of Kiribati, 2019). Further, as Kiribati is vulnerable to the impacts of anthropological global warming, climate change has become one of the important priorities for the Government. Because of climate change and rising sea levels, development efforts have focused more on climate change mitigation and adaptation.

In regard to global health trends, 80% of cardiovascular disease deaths occur in low- and middle-income countries (World Health Organization, 2017). The World Health Organization suggests a significant change in lifestyles, such as an increase in physical activity and a decrease in dependency on processed food will help decrease the incidence of NCDs, resulting in 41 million deaths each year since 2005. This can be achieved through healthy living and active lifestyles through physical activity and sports participation. Due to limited food choices, which contribute to poor health outcomes and poor governance of food security programs, the incidence of NCDs has continued to rise in Kiribati. In 2015, 49.7% and 46.9% of all presentations to health facilities were of people with confirmed or suspected diabetes and hypertension, respectively (Cauchi et al., 2021). The effect of climate change has exacerbated the problems of food insecurity and adverse health experiences of communities in Kiribati. Kiribati is particularly at risk of food insecurity due to its limited land availability, population pressures, and low-lying topography making Kiribati vulnerable to sea-level rise (Cauchi et al., 2019).

Given the above statistics, the Kiribati Government has prioritized the issues of climate change, human resource development, and education as reflected in their national long-term plans (Government of Kiribati, 2019). In Kiribati, sport development comes under the Ministry for Women, Youth, Sports, and Social Affairs. A sport policy was approved by the Government in 2019; however, the annual budget dedicated to sport development is limited. Over 11 years from 2009 to 2019, the sports sector received AUD 11,580,868. Despite the delay in formulating the sport policy, the completion of the policy is significant because it demonstrates a recent commitment to sports development.

As part of Goal 3.2, Overcoming the Challenges in Improving Health, in the Kiribati Development Plan 2020–2023, the Government seeks to promote engagement in sport (Government of Kiribati, 2019). However, the plan acknowledges numerous barriers to participating in sporting activities. These barriers include the following:

(i) The perception that sport is an optional recreational activity rather than being essential to physical and mental well-being.

(ii) Sports in Kiribati are typically male-dominated.
(iii) Poor quality sport facilities.
(iv) Lack of sporting equipment.
(v) Lack of experts/professionals in different sports.
(vi) Lack of capacity building for sport associations.
(vii) Lack of investment in sport by Government and development partners.

The Role of the Indigenous Population Within the Organization of the Sport Tourism Sector on Kiribati

Sport means different things to different indigenous people and groups. For many indigenous peoples, the primary purpose of sport tourism is to build self-esteem among indigenous people while also maintaining their culture and traditions (Nelson, 2009). In the 20202 Census, almost all (95.7%) of the population identified as i-Kiribati, while 3.8% identified as mixed Kiribati ethnicity (Kiribati National Statistics Office, 2020a). Therefore, any discussion about the indigenous population applies to the country as a whole. Culturally, sport in Kiribati has always played a major role in community bonding and strengthening of culture through social interaction during national sporting events such as the "Te Runga" (Government of Kiribati, 2019).

Sport Tourism Sites and Spaces of the Island

While Kiribati is blessed with many natural tourism resources, especially water-based resources of oceans and lagoons, the calm and turquoise waters are suitable for scuba diving, fishing, and other water sports, in which sport tourism can be developed. However, Kiribati is lacking in tourism facilities and tourism services (Xu, 2019). Economic development is generally low. For example, Kiribati's Human Development Index for 2019 was 0.630, ranking it 134 out of 189 countries and territories.

Investment in the tourism sector, either foreign direct investment or domestic investment, remains limited. Transportation infrastructure is limited. Roads are narrow and mostly sandy. Sea transportation is the dominant mode of travel, but travel is slow and infrequent. Air access is limited with flights into Kiribati departing from Fiji, Tuvalu, and Hawaii (Xu, 2019). However, the Tourism Authority of Kiribati noted on their website that the tourism sector has improved tremendously in the last decade or two, with an increasing number of accommodation providers for tourists, certified scuba diving personnel, registered hired boats for sport fishing, and handicraft markets. There is also a community willingness to host tourists and showcase their indigenous culture and traditions through dances, as well as provide home accommodation for tourists to experience how locals live their daily lives.

Kiribati has several sports facilities, but their conditions are in poor shape and do not fulfill international standards, thus making it difficult to host regional games. Officially named the Reuben K. Uatioa Stadium, the Bairiki National Stadium is located in South Tarawa, Kiribati (Fig. 17.3). It is the home of Kiribati's national football team and can seat around 2500 fans. It has never hosted an international football match because the field surface is sand rather than grass. The sand surface is one of the reasons Kiribati is not, at present, a member of FIFA. The stadium is currently under redevelopment to make it a FIFA standard stadium.

The Betio Sports Complex is also located at the Kiribati Institute of Technology, Tarawa (Fig. 17.4). The complex has facilities for indoor sports and outdoor games. The building of the complex was funded and completed by the Republic of China (Taiwan) which handed over the sports complex to the Kiribati Government in 2007.

The Kiribati Government has made commitments through its long-term development plan for the upgrading of two stadia in Tarawa. In the Outer Islands, the standard of sporting facilities is very poor, whereas some islands do not have proper sporting facilities at all. Therefore, school premises are common locations for sport, as they seem to be the only places with playable grounds.

Fig. 17.3 Bairiki National Stadium. (Source: https://oceaniafootballcenter.org/2020/01/01/kiribatis-new-stadium-is-being-built/)

Fig. 17.4 Betio Sports Complex. (Source: Playmakerstats.com)

Organized sports are common and soccer is the most popular in the country, according to the Kiribati National Olympic Committee (KNOC). Kiribati has sports federations in volleyball, basketball, boxing, archery, badminton, canoeing, tennis, taekwondo, surfing, and touch rugby. Schools and island communities compete in athletics every year during a national sporting event called "te runga."

In 1998, Kiribati became a member of the Commonwealth Games; in 2004, it became a member of the Olympic Games. The interest of I-Kiribati, particularly among the younger generation, in weightlifting was spurred after the Glasgow Commonwealth Games in 2014, where a Kiribati athlete won a Gold medal in the sport for the first time. During Independence Day celebrations, the Department of Sport and the KNOC co-organize various sport activities and events for all communities in the capital in which to participate. These tournaments are separate from parallel sporting events run by single federations around the same period.

Conclusions

Sport tourism has the potential to contribute to sustainable local development in small island states and territories globally. However, some of the characteristics of small island states and territories, including lack of land, small populations, limited investment, and distance from major source markets can be barriers to the development of sport tourism. Sport tourism in Kiribati is at a formative and ideation stage. The development of sport tourism can include all the potential benefits of tourism, including employment creation, income generation, foreign exchange, revival of culture, and protection of the environment (Pratt, 2022), coupled with the benefits of sport which include increases in physical activity, improved health, and lowering of NCDs (Andreff, 2006).

At present, tourism is in its infancy and sport is yet to be developed fully within this island archipelago. Nevertheless, progress is underway. According to the Tourism Authority of Kiribati, since 2010, there has been an increase in the number of local people venturing into tourism after noting its socio-economic benefits individually and to its local communities. There are now more accommodations that cater to tourists and more locals providing services on sport tourism such as game fishing, scuba diving, and sailing. Tourism has begun to focus on local arts and crafts while also providing tours to communities to experience communal living.

Kiribati has recently developed a sport policy, whereby funds have been allocated from the national budget to grow sport, including the improvement of sport infrastructure. This can improve the welfare of the local community through sport opportunities, showcase local culture and traditions, as well as attract potential tourists. There is significant potential in sport tourism in Kiribati but much to do for this plan to become a reality.

References

Andreff, W. (2001). The correlation between economic underdevelopment and sport. *European Sport Management Quarterly, 1*(4), 251–279.

Andreff, W. (2006). Sport in developing countries. In W. Andreff & S. Szymanski (Eds.), *Handbook on the economics of sport*. Edward Elgar.

Bedford, R., Macdonald, B., & Munro, D. (1980). Population estimates for Kiribati and Tuvalu, 1850-1900: Review and speculation. *The Journal of the Polynesian Society, 89*(2), 199–246.

Brown, C. (2017). The coastal challenges of adaptation for sea level rise in Kiribati. In *Coasts, marine structures and breakwaters* (pp. 235–244).

Cauchi, J. P., Correa-Velez, I., & Bambrick, H. (2019). Climate change, food security and health in Kiribati: A narrative review of the literature. *Global Health Action, 12*(1), 1603683. https://doi.org/10.1080/16549716.2019.1603683

Cauchi, J. P., Bambrick, H., Correa-Velez, I., & Moncada, S. (2021). 'White flour, white sugar, white rice, white salt': Barriers to achieving food and nutrition security in Kiribati. *Food Policy, 101*, 102075. https://doi.org/10.1016/j.foodpol.2021.102075

Chambers, K., & Chambers, A. (1975). A note on the Ellice referendum. *Pacific Viewpoint, 16*(2), 221–222.

Commonwealth Secretariat. (2017). *Sport for development and peace*.

Commonwealth Secretariat. (2022). *Kiribati*, Date Retrieved 4 April 2022. Retrieved from https://thecommonwealth.org/our-member-countries/kiribati

Foster, S., & Macdonald, B. (2021). Kiribati. *Encyclopedia Britannica*, Date Retrieved 4 April 2022. Retrieved from https://www.britannica.com/place/Kiribati

Government of Kiribati. (2019). *Kiribati development plan 2020–2023*, Date Retrieved 6 April 2022. Retrieved from https://policy.asiapacificenergy.org/sites/default/files/Kiribati%20Development%20Plan%202020-2023.pdf

Harrison, D. (2004). Tourism in Pacific Islands. *The Journal of Pacific Studies, 26*(1 & 2), 1–28.

Kiribati National Statistics Office. (2020a). *Gross domestic product,* Date Retrieved 4 April 2022. Retrieved from https://nso.gov.ki/statistics/economy/gdp/

Kiribati National Statistics Office. (2020b). *Kiribati 2020–2021 population and housing census data,* Date Retrieved 4 April 2022. Retrieved from https://nso.gov.ki/census/kiribati-2020-2021-population-and-housing-census-data/

Macdonald, B. (1982). *Cinderellas of the empire: Towards a history of Kiribati and Tuvalu.* ANU Press.

Nelson, A. (2009). Sport, physical activity and urban indigenous young people. *Australian Aboriginal Studies, 2009*(2), 101–111.

Pratt, S. (2022). Can tourism contribute to the sustainable development goals? Evidence from Pacific Island countries & territories. *International Marketing Journal of Culture and Tourism, 1*(1), 45–61.

Pretes, M., & Gibson, K. (2008). Openings in the body of 'Capitalism': Capital flows and diverse economic possibilities in Kiribati. *Asia Pacific Viewpoint, 49*(3), 381–391.

Sherry, E., Schulenkorf, N., Seal, E., Nicholson, M., & Hoye, R. (2017). Sport-for-development in the South Pacific region: Macro-, Meso-, and micro-perspectives. *Sociology of Sport Journal, 34*(4), 303–316.

SPC. (2018). *Health through Sports in the Pacific: (Re)Discovering the joys of a healthy body,* Date Retrieved 6 April 2022. Retrieved from https://www.spc.int/updates/blog/2018/08/health-through-sports-in-the-pacific-rediscovering-the-joys-of-a-healthy-body

SPTO. (2020). *2019 annual review of visitor arrivals report.* Retrieved from Suva: https://pic.or.jp/ja/wp-content/uploads/2019/07/2018-Annual-Visitor-Arrivals-berReportF.pdf

Tourism Authority Kiribati. (2020). *Kiribati for travellers: A review of the tourism sector,* Date Retrieved 6 April 2022. Retrieved from https://southpacificislands.travel/kiribati-tourism-industry-review-2020-report/

United Nations. (2015). *Unga reports on social development 2015,* Date Retrieved 6 April 2022. Retrieved from https://www.un.org/development/desa/dspd/publications/united-nations-reports-on-social-development/un-ga-reports-on-social-development-2015.html

Webb, J. (2019). *Kiribati's unique economic structure,* Date Retrieved 4 April 2022. Retrieved from https://apo.org.au/node/249526

Webb, J. (2020). Kiribati economic survey: Oceans of opportunity. *Asia & the Pacific Policy Studies, 7*(1), 5–26. https://doi.org/10.1002/app5.297

World Health Organization. (2017). *Cardiovascular disease,* Date Retrieved 6 April 2022. Retrieved from https://www.who.int/news-room/fact-sheets/detail/cardiovascular-diseases-(cvds).

Xu, J. (2019). Thinking on exploitation of tourism resources and development of tourism industry in the Republic of Kiribati. *Asian Agricultural Research, 11*(12), 42–49. https://doi.org/10.19601/j.cnki.issn1943-9903.2019.12.010

Chapter 18
Japan: The Sustainable Development of a Small Peninsula: The Case of Irago Triathlon

Misuzu Toba

Introduction

Atsumi, with Irago at its heart, is famous in poetry as a scenic spot in Japan, even mentioned in the *Man'yoshu* and *Nihonshoki*, collections of waka poetry from the ancient and Nara periods (Atsumi Peninsula Local Research Group, 1997). Even today, the Atsumi Peninsula is rich in nature and natural beauty and has great potential to attract tourists and migrants. However, not even all Japanese can accurately point to it on a map. Sporting events are expected to contribute significantly to raising the profile of such a remote area and promote regional development.

Specifically, this chapter focuses on the Irago Triathlon, which makes use of the geographical location and landscape of the region. In the relatively short history of triathlons, the Irago triathlon has a long history spanning 35 years. It is also one of the largest in Japan, with nearly 1000 participants. The triathlon was canceled in 2020 and 2021 due to the COVID-19 pandemic; this chapter refers to the 2021 event plan and an overview of its implementation from when it was first hosted in 1987–2019. In all, 925 people participated in the 2019 triathlon, and the capacity was expanded to 1200 people for the scheduled 2021 plan.

This chapter was written primarily based on interviews the author conducted with the stakeholders, including the Tahara City Hall, the Irago triathlon governing body, the Aichi Triathlon Association, the fish market, and local residents. It also simultaneously referred to about 35 years of regional and local newspapers such as *Chunichi Shimbun* and *Higashiaichi Shimbun* as well as statistics from Aichi Prefecture and Tahara City along with the history of the event.

M. Toba (✉)
Kwansei Gakuin University, Nishinomiya, Japan

D. Van Rheenen et al. (eds.), *Sport Tourism, Island Territories and Sustainable Development*, Sports Economics, Management and Policy 24,
https://doi.org/10.1007/978-3-031-51705-1_18

Characteristics of the Atsumi Peninsula

Atsumi Peninsula is located on the southern tip of the Aichi Prefecture in central Japan, surrounded on three sides by the sea: Mikawa Bay to the north, the Pacific Ocean to the south, and the Ise Bay to the west. It is a long, narrow peninsula stretching about 30 km east to west with a coastline of about 100 km. It has an administrative area of 191.11 km^2, almost all of which is under the jurisdiction of Tahara City. Only on its eastern side does it have a land border with Toyohashi City (Tahara City Statistics, 2021). In 2003, Tahara Town was merged with Akabane Town to form Tahara City, and 2 years later in 2005, Atsumi Town was merged with it to form the new Tahara City.

It is difficult to reach the Atsumi Peninsula using public transportation. The Bullet train takes about 1 h from Osaka to Nagoya, the capital of Aichi Prefecture, in central Japan. Tokyo to Nagoya is a journey of about 1 h and 40 min, so Tokyo to Irago takes about 3 h, and involves changing trains in Toyohashi. Osaka to Irago via Nagoya takes about 3.5 h. The railroad station closest to Cape Irago, which is located at the tip of the peninsula, is the Mikawa-Tahara Station about 30 km away. The Mikawa-Tahara Station is on the eastern border of the Atsumi Peninsula, so traveling to Cape Irago, which is on the western border, requires taking the bus, which is an hourly service.

Due to the unique topography of the peninsula, which juts out into the sea, the Atsumi Peninsula is very windy throughout the year. It is also known as the Windy Cape, and it is a region of concern for salt damage or sand damage to crops due to the strong winds. Therefore, many mangrove forests have been planted to protect the area from tidal damage and flying sand. However, the region is mild all year round due to the influence of the Kuroshio Current in the Pacific Ocean, and it is close to major shopping centers, namely, the Nagoya metropolitan area, the Tokyo metropolitan area, and the Kansai region. Above all, the abundant and stable supply of water from the Toyokawa irrigation system from 1968 has ensured a thriving production of vegetables and flowers that visitors to the Atsumi Peninsula can see in the farms across the region, the most representative example being cabbage.

Thus, the Atsumi Peninsula is a thriving region for vegetables, flowers, and livestock farming. However, in considering sustainable development, the fact of the serious shortage of successors in agriculture due to the aging population and fewer children cannot be ignored. In a previous study (Toba, 2017), the author conducted interviews with mini-tomato farmers and pig farmers in the region and clarified the current situation and institutional problems of having to rely on technical intern trainees from China and Vietnam, who have cross-cultural backgrounds, and the existing concerns of communication with them despite Japanese language training.

Atsumi is surrounded by the sea on three sides and has long been a thriving fishing region. In 1966, the Atsumi Fish Market was established in Irago Harbor, on the border of the Pacific Ocean and Ise Bay. However, since the late 1970s, the fishing industry declined throughout Japan, including in this region. In this context, a shift was observed from capture fishing to aquaculture and cultivation fishing.

Nevertheless, with the lack of successors due to the declining birth rate and the large influx of cheap imported fish, difficulties have arisen (Atsumi Town History Compilation Committee, 2005). In an interview with the author, the representative director of the Atsumi Fish Market said, "We have tried everything, including octopus, mullet, and shrimp farming. We have tried so many things, and none of them has worked." This is because although these are all technically possible, they are not profitable. Thus, more operations focused on fishing species that matched customer needs, and as a result, the number of fishermen engaged in catching Japanese littleneck clams and geoducks increased. According to Tahara City, agricultural workers accounted for 59.5% of the total in 1960 but declined to 28.4% in 2015. For fisheries, it dropped from 5.5% to 1.2%. Secondary and tertiary industries flourished, replacing these primary industries. The expansion of the consumer economy, mainly in the service industry, led to significant numbers of young farmers and fishermen who disliked heavy work leaving those industries.

According to the census (October 1, 2015; see published data from Aichi Prefecture), the total population of Tahara City is 62,364 against Aichi Prefecture's 7,483,128. The neighboring Toyohashi City has a population of 374,765. The population of the Atsumi area of Tahara City, located in the western part of the Atsumi Peninsula, accounts for about 30% of the total population of Tahara City. The population density of Tahara City is 326.3 (people/km^2): a large difference if compared with Nagoya City, which has a total population of 2,295,638 with a population density of 7032.1 (people/km^2). The population aged 65 years or above in Tahara City is 25.7% (25.6% according to Tahara City statistics), but the aging of the population is progressing steadily, and the figure is expected to soon hit 30%.

In terms of schooling facilities, the declining birthrate—also connected to the Irago triathlon, which we will come back to later—has had a remarkable impact. The media coverage of the event focuses on the volunteer activities of the students of the local Fukue High School, but here we focus on the changes in the number of "all students." All students of Fukue High School started to participate as volunteers from the third iteration of the triathlon (1989). The number reported at that time was "all 830 students," which by the 13th iteration (1999) had dropped to "all 510 students" (*Chunichi Shimbun*, August 15, 1989; September 5, 1999). Including Fukue High School, Tahara City has three prefectural high schools; the total number of students in 1990 was 2786, which dropped to 2138 in 2000 and 1424 in 2020, nearly half of what it was 30 years ago (Tahara City Statistics, 2021).

Although there is a significant gap between the years of measurement, it is clear here that the number of students in Tahara City as a whole is on a downward trend, so even if all students volunteer, the number of students involved in the tournament is decreasing. Another example of a middle school that has seen a decline is Iragomisaki Junior High School. The taiko drumming club of this school performed every year on the eve of the Irago Triathlon. However, the school closed down at the end of March 2019 and was merged with Fukue Junior High School. The club remained as a local club, but another alma mater of local residents disappeared (*Chunichi Shimbun*, February 8, 2018). As can be seen from the above, it is not enough to simply glorify the involvement of local students. For sustainable development, it is necessary to go beyond the framework of the local community and also encourage a wider age group to get involved in the event. In this context, it is quite appropriate that City Hall is calling for volunteers to participate in the triathlon, not only for residents of the city, but also for those living outside the city, and for those over 15 years of age who cannot qualify for the entry requirements as athletes (at least 18 years old).

Background and History of the Irago Triathlon

The Triathlon was conceived in San Diego, California, USA, in 1974, but triathletes considered the first "Ironman Triathlon" (about 226 km), held on Oahu Island, Hawaii, in 1978, the prototype of the competition. The first triathlon competition in Japan was held on the Kaike Onsen beach in Tottori Prefecture in 1981. It had about 50 participants and was held over a distance of 109 km. In 1985, triathlons were held on Miyako Island in Okinawa Prefecture; Lake Biwa in Shiga Prefecture; and Amakusa in Kumamoto Prefecture; attracting much attention (Japan Triathlon Union, 2008; *Chunichi Shimbun*, September 16, 1994). The first Irago triathlon was then co-organized by Atsumi Town and the Chubu Triathlon Federation in 1987. The town hall seems to have followed the example of other regions in developing the attraction of the triathlon for regional development. However, according to the Tahara City Sports Division, there was no strong relationship with other regions, and the Chubu Triathlon Federation approached Atsumi Town Hall at the time to hold the tournament because "an environment suitable for holding the Games was in place." Moreover, according to the executive director of the NPO, the town office refused to host the triathlon. The Atsumi Town Chamber of Commerce and Industry was approached and persuaded to hold the event on the grounds that it would attract spectators in September and promote the local community, and the mayor at the time finally took the decision to host the event. Further, the Aichi Prefectural Police did not give permission to use a national road for the bike course, so the town had a hard time setting it up. Without a sufficient budget, they had to set up tents and other preparations almost entirely on their own.

The initial reluctance of the town hall and prefectural police was revealed through several interviews conducted by the author. The media at the time reported that "the entire town was actively involved" in the first race, which was a "large scale race hosted by the entire local community," with approximately 900 athletes participating, and that "almost all town employees were also involved, serving as backstage staff and putting in a great deal of effort into the event." It also highlighted the success of female athletes invited from abroad, such as twin sisters from the United States, and the event was reported extensively as an international event, as signified by the name "International Japan Cup Triathlon in Irago" (*Higashiaichi Shimbun*, September 7, 1987; *Tokai Nichinichi Shimbun*, September 7, 1987).

Since the second triathlon (September 4, 1988), former Atsumi Town was the sole host of the event, but the town's attitude toward it as "the biggest event of the year" and its positive organizational efforts can be confirmed only a year and a half later. In mid-March 1989, the former Atsumi Town established a secretariat in the town hall, and in April, an action committee was formed, beginning preparations over a 6-month period. The aim was to take advantage of the arrival of athletes from all over The world and to promote the "Flowers and Tides of Cape Irago" so as to revitalize the town. The total cost of the event was approximately 30 million yen. The financial resources can be broken down into approximately 7 million yen in town fees, participation fees, and corporate sponsorships.

The third triathlon held in 1989 also received a subsidy from the Japanese government (through the Regional Revitalization Event Support Project). Although the number of Japanese triathlon competitors was low, the 900 athletes for the third competition were selected through a lottery out of the nearly 3000 applicants. There were participants from all over Japan, from Miyagi in the north to Oita in the south, fulfilling the town hall's hopes. About 1400 locals supported the 1989 triathlon. The volunteers that year included over 900 individuals from Fukue High School (830 students, 40 teachers, and 45 PTA officials) in the former Atsumi Town, about 200 employees of the town office, and people from various organizations such as the Chamber of Commerce and Industry, agricultural and fisheries cooperatives, the Tourist Association, and the Japan Sport Association (*Chunichi Shimbun*, August 15, 1989).

The Aichi Triathlon Association was established in 1988, and in the following year, 1989, it sponsored the first "Gamagori Orange Triathlon," held in Gamagori City in Aichi Prefecture. The event's name was based on the orange cultivation that flourishes in the city. It had around 300 participants, less than half that of the Irago Triathlon held in the same prefecture.

In 1989, the International Triathlon Union (ITU) was established with a view to the future Olympic Games. In 1994, it was decided that the triathlon would be adopted as an official event at the 2000 Sydney Olympics. In the same year, the Japan Triathlon Union (JTU) was established, integrating organizations and associations from all over the country against the backdrop of the rapidly growing number of athletes. It integrated the Japan Triathlon Federation (JTF) and the Japan Triathlon Association into one organization. This integration further strengthened organization, operation, management, dissemination, and enhancement (*Tokyo Shimbun*, July 26, 1997). For example, the number of members of the Aichi Triathlon Association increased from 90 members in 1990 to 1269 in 1995 (Aichi Triathlon Association, 2020). According to an interview with the managing director of the Aichi Triathlon Association (who is acquainted with Les McDonald, founder of the ITU, or International Triathlon Union, and its President from 1989 to 2008), this was when "the triathlon, a sport that until then had been practiced as a form of self-discipline and a challenge to one's limits, was transformed into something to spectate and entertain." To maintain spectator interest, a reduction in competition time was also considered.

In 1996, the first Tokyo Bay International Triathlon was held, the first event of its kind in the Tokyo metropolitan area. It was a 100 million yen-scale event. However, due to problems related to water quality, the event was limited to designated athletes of the Japan Triathlon Union and competitors invited from overseas who were proficient in swimming; there was no open recruitment for general athletes (*Tokyo Shimbun*, July 8, 1997). The venue, Odaiba, is an island near Tokyo Bay created through land reclamation. It was decided conversely that the Lake Biwa triathlon hosted in the region would be held for the last time in 1997. As implied by the name "Ironman Triathlon Japan," the event was held for a much longer distance (226.3 km), and for a longer time than the official Olympic event (51.5 km). However, when the renewal of the contract for the use of the "Ironman" name came up at that year's tournament, Shiga Prefecture listened to the wishes of the cities and towns involved and agreed to pursue new regional promotion measures in lieu of the event. In response to the idea of continuing the event by increasing the participation fee (35,000 yen), many expressed the view that resident and volunteer interest in the event was dropping, that it was difficult to gather volunteers, that the effect of the regional promotion was nowhere near as impactful, and that road regulations were very difficult to enforce. Another reason for the decision to cancel the event was that the economic downturn had reduced corporate sponsorships, while the burden on the prefecture and municipalities had simultaneously grown significantly (*Chunichi Shimbun*, June 18, 1997; July 23, 1997).

After such a move outside the prefecture, Tahara City and Atsumi Town merged in Aichi Prefecture in October 2005. At the Legal Absorption Conference in 2004, the mayor of Tahara and the mayor of Atsumi Town mentioned that the number of participants in the triathlon was increasing due to the merits of triathlons and the natural beauty of Irago. Thus, it was unanimously decided that the triathlon was to be continued after the merger, but the 2005 event was the last to be held by Atsumi Town. At the time, Atsumi Town alone was in a dire financial situation that would have made it impossible to budget approximately 33 million yen for the triathlon in 2006. After the merger, the secretariat of the competition action committee established under the Atsumi Town Hall was moved to Tahara City. The 2006 event was organized by the action committee which consisted of the city, its Chamber of Commerce and Industry, and the Atsumi Chamber of Commerce and Industry (*Chunichi Shimbun*, September 28, 2004; September 6, 2006).

Since 2013, the Irago Triathlon Action Committee, the secretariat of which is under the Tahara City Hall, has entrusted the event's management to an NPO. They cooperate with the Aichi Triathlon Association to manage the Irago Triathlon, and Tahara City provides subsidies to support the event. The event also receives support from parties that were visible before the change in tournament management, including Aichi Prefecture, Tahara City Chamber of Commerce and Industry, Atsumi Chamber of Commerce and Industry, Atsumi Peninsula Fisheries Promotion Council, Aichi Minami Agricultural Cooperative, Atsumi Peninsula Tourism Bureau, and the Chunichi Shimbun. The following year, in 2014, the "Sports Division" was newly established in the Tahara City Education Board, which was responsible for the management of sports facilities and the general promotion of sports. This organizational change shifted the Irago Triathlon Action Committee Secretariat, previously under the Tahara City's "Commerce, Industry, and Tourism

Division," to the "Sports Division" in 2014. However, as in Atsumi Town, some Tahara City employees continue to participate in the event as volunteers.

According to the NPO, the cost of the 2021 triathlon came to 44 million yen, but the company continues to operate in the black. It is supported by the participation fees (¥28,000 for the A-type race and ¥24,000 for B-type) and handling charges (¥1000 for both A-type and B-type). Although the event does not offer refunds in case of cancellation due to inclement weather, for the COVID-19 pandemic, the management refunded appropriately 40% of the fees after deducting expenses for athletes, and supplies for athletes and volunteer participants who had already purchased them were mailed to them at great expense of time and effort. However, "since it is a one-day event, sponsors are hard to come by, and the event relies on local influential figures." For example, one of the main sponsors is an integrated agricultural support enterprise located in Toyohashi City, Aichi Prefecture. The company is over 110 years old, and is committed to "making agriculture a sustainable and hopeful industry." Many costs are involved in running the event, including the costs of setting up tents and chairs from rental companies, as well as security. Several security guards are stationed in the transition area to guide athletes and prevent the theft of bicycles, which are often expensive, costing more than one million yen. To keep costs down, the NPO prepares banners on their own printing facilities. In addition, their staff mow grass and otherwise maintain the courses for over a week before the tournament for volunteer-like remuneration. Before the management was commissioned, a temporary road was built on the beach (about 380 m) to serve as a transition area, but to reduce costs, the course was changed accordingly, including incorporating the usage of hotel parking lots. Thus, through such a thorough review of problems and improvement efforts, the Irago event has continued as the most popular event for triathletes after Sado, Yokohama, and Ishigaki.

During this period, from 2010 to 2018, the number of members of the Aichi Triathlon Association saw another increase (from about 1000 to 1891). The following circumstances can provide the background for this. First, the popularity of the triathlon, which had been on a downward trend after the 2000 Sydney Olympics, began to change after 2006. Successive events in the UK and other foreign countries saw a swell in the number of participants to several thousand, and specifically in Japan, the establishment of the Tokyo Marathon in 2007 triggered a running boom that has provided a tailwind for triathlons as well. The event held on Ishigaki Island, Okinawa Prefecture, in April 2009 attracted a flood of about 2000 participants, nearly twice as many as in the past. There were approximately 300,000 domestic triathlon enthusiasts in 2009, and the event attracted so much attention that the Japan Economics Newspaper ran a special feature on triathlons that reported that "in particular, the number of businessmen in their 30–50 s are taking up the challenge."

Due in part to the influence of TV programs that aired celebrities participating in triathlon challenges from 2008 to 2009, some events have been reaching their recruitment capacity like marathons. In 2009, about 250 triathlons were conducted in Japan with a capacity of about 300–500 participants. In 2010, the Japan Triathlon Union (JTU) established a new event for beginners, which led to the "fourth triathlon boom" according to the JTU Executive Director. The event was made easier so that beginners could comfortably participate in day trips by reducing the distance and introducing a relay format that divides the race between three people. The number of triathlons hosted domestically increased by nearly 30 in addition to the approximately 250 already conducted, about a quarter of which were held in the Tokyo metropolitan area. Even more effective was the sale of starter kits for beginners, priced at about 100,000 yen on the Internet. Participating in a triathlon requires a bike, a wetsuit, and shoes for both running and biking, and the total costs easily exceed 100,000 yen for expensive items. It is estimated that during this time, the potential enthusiasts numbered somewhere near 1.5 million (*Japan Economics Newspaper*, October 29, 2010). However, beginning in 2020, many events were canceled due to the prolonged COVID-19 pandemic, and the membership of the Aichi Triathlon Association has seen little growth.

Overview of the Irago Triathlon

The Irago Triathlon is hosted in the region around Cape Irago in Tahara City. There is no fixed date for the event, and it is held on the first or second Sunday of September. Since the competition includes swimming, it is scheduled to be closer to high tide. This period falls shortly after the summer holidays when tourism is at its peak and the heat is intense. Registration and vehicle inspections are held on Saturday, the day before the event, which is followed by the opening ceremony and the eve of the triathlon. The weekend schedule makes it relatively easy for school students and workers to participate, but athletes are required to stay in the area for at least one night or visit the host city for two consecutive days. They are also required to pay an entry fee and handling charges.

The first triathlon had three types of courses (112 km half-type, 56 km quarter-type, and 25.5 km eighth-type) prepared, but due to several falls and cases of sunstroke, the second and then subsequent events only had two types of courses. In other words, the longer distance A-type (middle) and shorter distance B-type (standard) are the two kinds of races. Athletes choose to compete in one, and swim, bike, and run the distances of that event within their time limits. Although athletes miss out on the appeal of enjoying different paths, circuit tracks were included in the course for cycling to enhance safety since the second triathlon. For reference, in 2021, the A-type race was planned for a distance of 90.25 km and the B-type race for a distance of 53.5 km. The time limit for the run, which is the third leg of the triathlon, was set to 6 h and 30 min after the start for A-type, and 4 h and 30 min for B-type. Participants for both types must be healthy individuals, 18 years of age or older on the day of the tournament, and competitors of the A-type race must be JTU members. Both types are open to both professionals and amateurs. Some athletes (including invited athletes) are seeded based on their performance.

Athletes compete for time on a course that takes them around the tip of the Atsumi Peninsula from the Ise Bay side to the Pacific Ocean side. To review whether the event takes advantage of the unique nature and scenery of the Atsumi Peninsula, it is necessary to examine its details. First, participants swim north of Irago Port near the Ise Bay exit, then ride bicycles along the flat coast-facing paths and cabbage fields during planting, and finally run on an undulating course with mountains and sandy beaches overlooking the Pacific Ocean. The Atsumi Bicycle Path, which is used as the running course, was developed by the prefecture with state aid and opened to traffic in 1986. The bicycle path connected tourist attractions such as the Cape Irago Lighthouse, Koijigahama Beach, the "Yashinomi" (Coconut) song monument (the song, which was written by Shimazaki Touson, is well known in Japan), and the Irago Flower Park in Atsumi Town.

Traffic regulations are in place for the event itself. For example, the fish market, which is normally closed on Saturdays, is asked to remain closed for the entire weekend for the triathlon. In addition, the course runs next to in-use fields, which inevitably means that the operation of agricultural equipment is prevented, in lieu of which the farmers are compensated. The course changes every year, roads under construction, and so on, also make it imperative to change the course. When the author visited the NPO in March 2022, the management group was considering changes to the bike course, including parts of the farm roads, for the following year to address the above problems.

Awards are given to the first three finishers by gender and age for both A- and B-type races. Specifically, between the "24 years and under" and "70 years and over," there are categories starting from "25 to 29 years" and so forth, in groups of 5-year increments. The 2021 triathlon roster had two men in their 80 s in the A-type race. As an example from the 2014 competition, the overall first-place winner received an original design championship trophy from the prefectural governor and a major regional newspaper, half a dozen melons, an invitation to the following year's competition, and an article with an interview in a major local newspaper. There are no monetary rewards.

Athlete Profiles and the Triathlon's Appeal

According to the "2019 Irago Triathlon Athlete Survey" (239 respondents, about 26% of the 925 participants), 7% of the athletes lived in Tahara City, 52% in Aichi Prefecture (outside Tahara City), and 41% outside Aichi Prefecture. Interviews with the NPO indicated that "70–80% of the athletes come from Aichi Prefecture," so it is important to note that more athletes from outside Aichi Prefecture responded to this question in the survey than from within Aichi Prefecture, including Tahara City. Nevertheless, there appear to be far fewer local participants than those who come from outside the city or prefecture, despite the inconvenience of transport accessibility, which is not surprising in light of the population ratio. Despite the dissatisfaction expressed by prospective participants not selected through the lottery process, the establishment of the "Tahara Citizen Quota" in 2016, along with the "Hometown Tax Quota," is significant.

Approximately 33% of the athletes are first-time participants, and the majority are repeat visitors. Among them, 21% participated for the sixth time or more in 2019, the 33rd year of the event. The use of lodging is a key economic benefit of the event, and 67% of the respondents reported that they stayed overnight in the city. However, excluding those who stayed at friends' or relatives' houses or in their cars, about 60% of the respondents used lodging facilities in the city, contributing to it economically. Including those who stayed overnight outside the city, what is characteristic is that 61% of the players stayed overnight with two or more people and 34% of them stayed with a group of five or more people. It is not clear whether the respondents were staying together with athletes, spectators, colleagues, or relatives. However, the triathlon has obviously become an opportunity to attract a large group of people, although only for a short period of time. The tourist attractions they visited included not just the sites close to the triathlon venue but extended as far as the Mt. Zao observatory, over 30 km away. Athletes are assumed to be using their own cars, which they also use to transport their bicycles, to stop at multiple tourist destinations in a much shorter time and more efficiently than by using local public transport. Promoting the switch to environmentally friendly cars, along with the use of bicycles in the future, might revitalize the town while maintaining the natural environment.

The following athlete profiles were found in the regional newspapers *Chunichi Shimbun* and *Higashiaichi Shimbu*n over about 30 years. For example, "four members of the local fire department participated" and "two people from the local police department in charge of road security participated." The four firefighters were cheered on by uniformed colleagues who had dispatched an ambulance near the finish line in case of an emergency. Staff from a prominent sponsor also participated. There was also a case where a student from a school for the deaf in Toyohashi City participated, saying that he was confident because he had swum at the beach since elementary school. Thus, the event is significant in encouraging social participation by people with disabilities and foreigners, a small number of whose participation was verified.

What is particularly appreciated about this event is the "homey" atmosphere, which may seem contrary to the grueling nature of the triathlon. This is closely related to the fact that various family members and work colleagues come together every year to participate in the event. In addition, the event permits mothers who dropped out halfway through the race to hold hands with their daughters to reach the finish line. There are also many innovations in management. For example, a few moments after the run starts, messages cheering the participants on are read out. Messages are collected at the registration desk the previous day, left by the athletes' family members, colleagues, and friends. In addition, many local residents volunteer to help manage the event, creating a family-like atmosphere. They handwrite individual messages on each of the paper cups handed out to athletes after the finish line.

In a survey conducted by Tahara City of the athletes in 2017, when asked what they liked about the event, the most common response was "roadside support" from volunteers (*Chunichi Shimbun*, May 15, 2018). This was still true in the survey conducted after the 2019 triathlon mentioned above. After "cheering," the responses mentioned the provision of ice water and fruit available at the multiple aid stations set up along the course. Although theoretically the beautiful landscape of the Atsumi Peninsula in all its glory should be the best selling point of the Irago triathlon, athletes who are tired from the long hours of the competition do not necessarily have time to enjoy it.

Almost 13,000 spectators (based on the number of visitors to tourist attractions, Tahara City Statistics) thronged to see the 2019 triathlon and cheer on the athletes. There, what the athletes from all over the country appreciated most was the cheering of the local students. They do not cheer for athlete numbers but rather look up the full name of the athlete and use that for cheering.

The organizers actively encouraged the local community to support the event, which has led to the creation of this "cheering" attraction that is consistently assessed as better than the previous night's celebrations, fireworks display, and each of the swimming, biking, and running courses. Specifically, the former Atsumi Town, which hosted the tournament, asked the students of the only high school in the town (Fukue High School) to volunteer in the tournament. This is why the school has been involved in volunteering since the first triathlon in 1987. Only the track and field club was involved in the first iteration, but the following year, about 50 people from all sports departments volunteered. Their role was to cheer the participants on and hand out drinks and water along the course on the day of the event. Although some students were somewhat uncomfortable in the beginning, the students' response was better than expected. Therefore, the high school decided to make it a school event for the third iteration of the event, one in which all school students would participate. The relationship between the school and the triathlon continued over the next 30 years.

The students' participation in the triathlon as volunteers positively contributes to their education. It also increases satisfaction in athletes and spectators and contributes to regional PR. Nor is that all. They are the future leaders of the region, which is crucial for sustainable development in remote areas. Since the minimum age for participation in the triathlon is 18 years, students in the third year of high school or first year of college and above can participate, according to the Japanese educational system. That is to say, the system works so that there are students who first get involved as volunteers and can then go on to participate in the event. In fact, there is the case of a woman from former Tahara Town who participated for the first time while she was a student, and has since participated consecutively afterward (*Chunichi Shimbun*, September 7, 1998). There is also a 15-year-old first-year high school student who wants to participate in the triathlon after she starts working because she participated for the first time as a volunteer (*Chunichi Shimbun*, September 12, 2017). These examples make it clear that even if they are to leave the region in the future, they can return with friends and colleagues and contribute to its revitalization. However, it is necessary to implement further measures to solve problems that even the former Atsumi Town faced during the period of high growth when it was frequented by tourists, namely, the transformation from a "transit-type tourist destination" to a "stay-type tourist destination," and implement measures for population increase by retaining young people (Atsumi Town History Compilation Committee, 2005).

Impact of Sporting Events

Aichi Prefecture is working to attract more tourists by enhancing its attractiveness through sporting events, as well as strengthening local PR by holding product and tourism fairs at the venues. At the Irago Triathlon specifically, free clam soup is provided to participants and spectators at the finish line, finishers are given free Atsumi Peninsula melons, fresh local vegetables are offered on the eve of the triathlon, and a fireworks display is held to facilitate interaction with locals. The private sector, tourism organizations, Aichi Prefecture, and municipalities take the lead in these efforts (Aichi Prefecture Higashi-Mikawa Vision Council, 2015).

According to the NPO, the clams are purchased from the Atsumi Fisheries Association, and the Women's Club of the Chamber of Commerce and Industry prepares the clam soup. They prepare about 240 melons as well. During the 2001 triathlon, the Fisheries Association provided it free of charge, hoping to promote the deliciousness of the clams from Atsumi. About 30 people prepared clam soup for about 2000 people from about 250 kg of clams freshly caught in Fukue Bay the day before the event. After the tournament, direct orders were also received from distant athletes (*Chunichi Shimbun*, September 11, 2001). In 2008, a direct sales event was held at the Atsumi Fish Market in conjunction with the Irago Triathlon and was well received (*Tohnichi Shimbun*, December 21, 2008). On the eve of the triathlon, paper lanterns decorated with supportive messages were created with the cooperation of fourth to sixth graders from 17 elementary schools in Tahara City (*Tohnichi Shimbun*, September 3, 2015).

Are the various initiatives to attract tourists contributing to the sustainable development of the region? The course around the coastal area of Cape Irago is also a tourist route and is sometimes littered with empty cans and post-eating garbage. There is also much floating litter on the Irago Beach, the point where the swim starts, and on the bike path along the sandy beach. When Atsumi Town hosted the event, town officials, along with local tourism staff, conducted a 20-km cleanup before the event (*Chunichi Shimbun*, September 5, 1997). Such activities are still necessary in recent years, and members of the Cape Irago Tourism Council, which consists of lodging facilities, restaurants, and other facilities in the area, have been conducting regular monthly cleanup activities. Before the Irago triathlon is held, they clean up the site of the competition and the places where the prior celebrations take place (*Tohnichi Shimbun*, September 4, 2015). In other words, promoting the beauty of the landscape and attracting tourists is negatively impacting the nature of the region with artificial objects. At the same time, government officials, local residents, and people working in facilities related to tourism are making concerted efforts for preservation to attract tourists and make it beautiful. Here, we should note that it is not always man-made objects that damage nature. Landscapes change or are damaged even when left in their natural state.

For example, the natural sandy beaches on the Irago Port coast were lost year after year due to beach erosion caused by waves, and coastline retreat became a major issue. To restore the sandy beaches, Aichi Prefecture constructed breakwaters and jetties (submerged breakwaters) as measures against coastal erosion. According to the prefectural government, "Many people visit the Irago Port beach daily, and marine recreational activities such as swimming and fishing are popular throughout the year. Further, approximately 110,000 people use it annually as a stopover for various events, such as the triathlon, guests of nearby hotels, and tourists to Irago Port." Thus, in explaining the necessity of implementing the coastal project (coastal environment improvement project), the prefecture also makes reference to the triathlon and counts its participants among the users of the beach. To put it another way, this is one of the reasons used to convince residents of the use of taxpayer funds and to motivate the government to preserve nature.

It is assumed that hosting sports events will economically benefit the lodging facilities and roadside stations in the vicinity of the event, but the NPO managing the triathlon has never carried out expert surveys, so there are no concrete figures. Interviews with lodging facilities whose parking lots serve as sites for athlete registration and vehicle inspection confirmed that September is a busy season and that early reservations are necessary, but there is no way to verify whether the guests are there for the triathlon. There are also skeptical voices among local store owners about the reported number of visitors because sales do not seem to increase with the increasing number of visitors on the days of the triathlon.

Conclusion

While attracting visitors for regional promotion, how should the relative lack of accessibility be addressed? This is a challenge faced by islands and peninsulas all over the world. It takes a long time to travel to Atsumi Peninsula from major cities such as Tokyo and Osaka. In addition, even though the peninsula is small, extending only about 30 km in the east-west direction, it is difficult to efficiently travel around the tourist destinations scattered there in a short time by relying solely on public transportation. However, as an environmentally adaptive sport, triathlon competitions that take advantage of the location and terrain are precisely suitable for such areas in terms of the course setting. Even the fact that the first triathlon in the Tokyo Metropolitan Area (the first Tokyo Bay International Triathlon) was held in 1996, around 10 years after the first Irago Triathlon (in 1987), points to this. Other areas that hosted triathlons in the 1980s have geographical similarities with Irago. Many of these annual events have been discontinued.

Nonetheless, it is also true that the competition itself is a great burden for athletes who have to transport their bicycles. What the peninsula can offer these athletes is beautiful natural scenery. Event organizers involved in the Irago Triathlon have begun to use drones to film and publicize the event for regional promotion. Triathlons are also very suitable for this technology, which provides extensive aerial views of the vast ocean and mountain landscapes. Although the local communities that host the tourists suffer some natural destruction, increased consciousness of this issue can lead to the preservation of the local environment.

The waters in the aforementioned Tokyo venue stand in stark contrast to Irago, which has preserved its pristine natural beauty. There used to be several beaches in Tokyo Bay, the water quality of which was an issue during the pre-2021 Tokyo Olympics qualification competition (August 2019). The beaches disappeared around 1950 due to landfills and deterioration of water quality. In 2010, the water quality of Tokyo Bay was confirmed to have improved, thanks in part to the efforts of a local NPO to revive sea bathing, and a triathlon was held for the first time with public participation. At the same time, however, there were several beaches that allowed bathing but prohibited swimming and immersing faces in the water, and swimming was limited to competitions and other occasions (*Tokyo Shimbun*, August 5, 2021; September 17, 2010).

The case of Atsumi Peninsula also confirms that there was a value added through the creative ingenuity of the people involved using limited funds and human resources. For example, local school students volunteer to participate in the event and cheer the participants on. The initiative will contribute to regional development, and the City's (previously the Town's) policy of involving local youth, who will be local leaders in the future, can be considered a success. On the contrary, the financial support from the prefecture, which is the primary sponsor of the triathlon, is insufficient. According to the Tahara City Sports Division, first-place finishers in previous years were usually given trophies from the prefectural governor and a major regional newspaper, but here as well, the prefectural government only lends its name, and the city arranges for the purchase of the items.

This chapter also noted that the year after the first triathlon was held in Tokyo, the regional Lake Biwa triathlon was discontinued. Again, in interviews of the prefectural governments, the involved municipalities mentioned the burden of gathering volunteers and road regulations, and above all, the high economic burden and the limited effect on regional development. What can be done to prevent Irago, which is also located outside the Tokyo metropolitan area but has a much lower degree of name recognition than Lake Biwa, from following the same path? The answer is as follows: we must closely examine its economic benefits so that the local residents and those involved in the event can grasp the advantages of regional development. Like the Lake Biwa triathlon, which we have reviewed, the Miyakojima triathlon was first held in 1985. It had to be canceled for three consecutive years during the COVID-19 pandemic and even faced a fatal accident despite good weather during the swimming event of its 18th iteration in 2002. It is now scheduled to continue from 2023 onward. The inhabitants of Miyakojima, which has a population of approximately 10,000 less than the total population of Tahara City, have been working together for a year to prepare for the event, and many have volunteered to participate on the day of the event. Before the merger of the five municipalities in 2005, a city official of the time asserted that the triathlon played a major role in revitalizing the region, saying, "The community came together to ensure the success of the event, which helped to develop the confidence to work together in solidarity to face challenges" (*Chunichi Shimbun*, April 26, 1995; *Tokyo Shimbun*, April 22, 2002).

The number of triathletes and spectators participating in the Irago Triathlon may be small compared with the total population of the Aichi Prefecture, which includes the large city of Nagoya. However, in a region with a declining birthrate and aging population, the number is by no means negligible. The annual event is confirmed to include many participating families and work colleagues, and many triathletes stay overnight with a group. From this, the triathlon event was identified as an opportunity to attract a large group of people for a very limited period of time. This suggests that increasing the number of participant slots could attract a much larger number of people in addition to the increased number of triathletes. Nonetheless, the lottery selection system has failed to draw a number of people greater than the capacity in previous years. And there is still scope to increase the capacity. The question of expanding the participant quota in the future while simultaneously preserving the natural environment and securing human and financial resources is an important key to regional development.

Acknowledgments In writing this chapter, I would like to express my deep gratitude to Mr. Noriyoshi Sakakibara, Managing Director of NPO With, Mr. Kazuaki Watarai, Chairman of the Board, Mr. Takao Yoshida, Managing Director of the Aichi Triathlon Association, Mr. Yukihiro Kiyota, Representative Director of Atsumi Fish Market, and the personnel of the Sports Division of Tahara City Hall for taking the time to participate in the interviews and research.

References

Aichi Prefecture Higashi-Mikawa Vision Council. (2015). *Higashi-Mikawa promotion vision: Regional promotion using sports events*. Higashi-Mikawa Vision Council. https://www.pref.aichi.jp/uploaded/attachment/35245.pdf (Cover, p.10); https://www.pref.aichi.jp/uploaded/attachment/35246.pdf (pp. 11–16).

Aichi Triathlon Association. (2020). *Documents of the annual general meeting of members*. Aichi Triathlon Association. http://www.tri-aichi.com/ata/2020.html

Atsumi Peninsula Local Research Group (Ed.). (1997). *Atsumi Peninsula and literature*. Atsumi Peninsula Local Research Group.

Atsumi Town History Compilation Committee (Ed.). (2005). *Atsumi Town history, contemporary edition*. Atsumi Town.

Japan Triathlon Union. (2008). *JTU Web Magazine for triathletes*. Japan Triathlon Union. https://archive.jtu.or.jp/

Tahara City. (2021). *Tahara City statistics*. Tahara City. http://www.city.tahara.aichi.jp/_res/projects/default_project/_page_/001/009/085/toukei3.pdf

Toba, M. (2017). Chinese interns at farms in Japan. *Ewha Womans University Journal of Social Sciences, 30*(1), 43–68.

Chapter 19
Sport Tourism, Sport Recreation, and Sustainable Development in Singapore

Guy Fontaine

Singapore, a Small Island But a Global Leader

Singapore, Identity Card

Singapore holds a strategic position between West and East Asia. At the Southern tip of the Malaysian Peninsula, 137 km north of the Equator, the waters of the China Sea and Indian Ocean meet. The total surface area of Singapore was 581.5 km^2 in 1959 and is today 719 km^2 thanks to the "land reclamation" policy. Singapore is 604 km^2 in size, 42 km from East to West, and 23 km from North to South. The island has 150 km of coastline.

Almost all of Singapore is within 15 m of sea level, and the highest point is Bukit Timah, 164 m in altitude and built of magmatic rock and granite. The northwest is mostly made of sedimentary rock, forming hills and valleys, while the east is sandy and flat. Singapore has no natural lakes or rivers, but reservoirs and water towers have been built to store water. Its climate is tropical (rainforest), with high humidity and an average rainfall of 2370 mm. There are no marked seasons, contributing to its tourist appeal.

The population of Singapore was 1.8 million in 1965 (independence), 4.2 million in 2001, and 5.4 million in 2021, of which 75% are residents (87% citizens and 13% foreign nationals but permanent) and 25% non-residents/expats. This latter group is comprised of 1.6 million (work or study in Singapore—not including tourists) inhabitants, of which 19% are Australian, 9% American, 6% Indian, and 1% French, with a median age of 41.8 years. Notably, 16% of the population is over 65 years old. The ethnic composition of residents is 75% Chinese, 13% Malay, 9%

G. Fontaine (✉)
University of Reunion Island (Retired), Saint-Denis, Reunion, France
e-mail: guyhfontaine@orange.fr

D. Van Rheenen et al. (eds.), *Sport Tourism, Island Territories and Sustainable Development*, Sports Economics, Management and Policy 24,
https://doi.org/10.1007/978-3-031-51705-1_19

Indian and 3% other. One-third of the population (33%) aged 25 years and over has earned a university degree. In terms of the density of the population, there are 7500 inhabitants per square kilometer; more than 50% of the island residents live in 100 m^2 flats of the Housing and Development Board (HDB). Since 1960, the HDB has provided well-connected, unique, and sustainable housing estates, where ample commercial, community services, and educational institutions are easily reachable by the transport system, expressway, and Mass Rapid Transit (MRT). These planned estates are self-contained spaces, concentrated in nine planning areas, notably Bedok (BD), Tampines (TM), and Jurong West (JW), each with more than 250,000 residents.

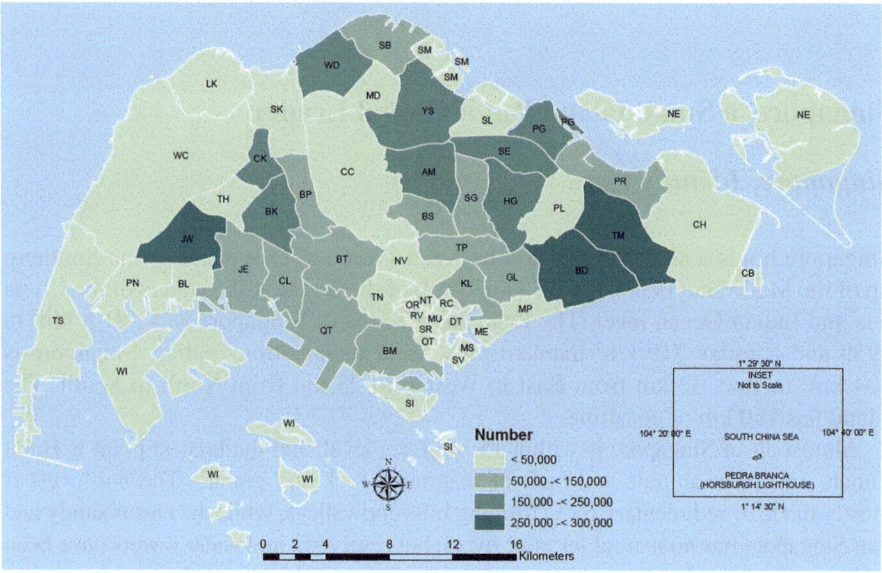

Source: Population Trends 2021, Singapore Department of Statistics https://www.singstat.gov.sg

The government of Singapore is a parliamentary republic, based on the principles of the separation of powers. This political system guarantees the secularity of the state and recognizes a set of democratic rights and freedoms. Since 1990, the President of the Republic is elected by universal suffrage. And yet, political power resides in the hands of the party in power. Since 1959, the *People's Action Party (PAP)* has held power. The party was founded by the charismatic leader, *Lee Kuan Yew*, who served as Prime Minister from 1965 to 1990, as Senior Minister from 1990 to 2004, and as Minister Mentor from 2004 to 2011. This longtime leader died in 2015. Since 12 August 2004, the Prime Minister is *Lee Hsien Loong*, son of Lee Kuan Yew. He is the third incumbent, following his predecessor *Goh Chok Tong, who* stepped down as prime minister. On the economic front, the PAP advocates a minimal social policy for the poorest while maintaining a particularly low tax rate for its citizens.

Singapore, City-State, and "Asian Dragon"

Singapore was a British trading post beginning in 1819. Singapore separated from Malaysia to become autonomous in 1959 and then independent in 1965. The geo-strategically positioned city-state of Singapore looks like a small island compared with large countries such as China, India, Russia, the United States, and France. And yet, since its independence until today (i.e. in only 57 years), this country has gone from being a "Third World" city-state to an "Asian dragon" and a major global economic player.

Singapore has the seventh largest gross domestic product (GDP) per capita in the world at US$ 72,795, just ahead of the United States at $70,249. It is now the fourth most competitive financial center in the world (behind New York, London, and Hong Kong, with Shanghai in fifth place) and the fifth most competitive economy in the world (behind Switzerland, Sweden, Denmark, and the Netherlands). It is the world's busiest container transshipment port, connected to over 600 ports in 120 countries, moving a total of 37.5 million TEUs in 2021. At any given time, there are approximately 1000 ships in the Port of Singapore; a ship arrives or departs Singapore every 2–3 minutes.

Politically secure, this city-state is ruled by a republican regime of "soft" rule or "illiberal" democracy (Singh, 2011), or a "communal democracy" (Chua, 1995), based on "Asian values." With an attractive regulatory and fiscal framework, Singapore is a preferred location for the region's headquarters: it has attracted over 40,000 international companies, including 7000 multinationals, which now have their regional or global headquarters there, a movement that is amplified by the current problems in Hong Kong.

Tourism, a Key Economic and Social Driver

Singapore's economic success is based on three sectors: financial, insurance, and business services (29% of GDP in 2019), trade and logistics (24% of GDP), and manufacturing (21% of GDP), specializing in electronics, biomedical, precision engineering, and chemicals—including petrochemicals. However, since 1990, boosted by the Singapore Tourism Board (STB), tourism has played a key role in enhancing Singapore's status as a dynamic global city that attracts capital, business, and talent. The tourism sector contributed about 4% to Singapore's GDP in 2020. "It also enhances the quality and diversity of leisure options for residents and help to create a living environment that Singaporeans are proud to call home" (STB, 2020).

Singapore may not have the "charm" of other countries such as Thailand, Vietnam, Indonesia or even Australia, but it still manages to regularly attract over 15 million international visitors each year (pre-COVID pandemic). A total of 19.12 million international travelers visited "Uniquely Singapore" in 2019, from all over the world. In comparison, Australia recorded 9.4 million international visitors in the same year. In 2019, total tourism receipts amounted to S$27.69 billion (US$20 billion). Visitors from mainland China (3.627 million), Indonesia (3.109 million), India (1.48 million), Malaysia (1.221 million) and Australia (1.143 million) were Singapore's top five international visitor-generating markets in 2019, accounting for 55% of all international tourists.

Tourists from Japan (0.884 million), the Philippines (0.829 million), South Korea (0.646 million), Vietnam (0.592 million), Thaïland (0.528 million), Hong Kong (0.489 million) and Taïwan (0.426 million) complete the significant presence of the Asian markets. However, the US (0.729 million) and UK (0.607 million) markets are also strong contributors to Singaporean tourism. Tourists from Europe come primarily from Germany (0.400 million) and France (0.200 million). Singapore's success as a tourist destination is strongly linked to its ability to be connected to the world, especially by air. Two key elements for this are Changi Airport, often nominated as the best airport in the world, and Singapore Airlines, the flagship carrier of Singapore, consistently ranked among the world's best airlines to 60 destinations and 30 countries. Singapore Airlines is a member of Star Alliance, which has codeshare partnerships with more than 36 other airlines. Changi Airport welcomes an armada of 84 world airlines and 69 million passengers, with 302,000 flight movements in 2019.

COVID-19 has heavily impacted Singapore's tourism. Global tourism to the island decreased from 19.2 million international arrivals and S$27.6 billion in 2019 revenue to 2.7 million arrivals and S$4.7 billion in 2020, when dealing with crises. "Singapore's tourism sector has had to fight for survival in 2020. Our tourism businesses have displayed immense resilience and adaptability throughout this difficult period, reinventing their business models and leveraging technology to find solutions in a COVID-19 world," noted STB chief executive *Keith Tan*. Tourism businesses played a key role in Singapore's battle against COVID-19, Hotels offered their properties for government quarantine, swab isolation and stay-home notice dedicated facilities, serving more than 80,000 people on stay-home notice with the support of more than 2300 frontline hospitality workers as of the end of last year. 45 tourist attractions, 270 hotels and 1686 tour itineraries have received approval for reopening in safety. More than 300,000 Singaporeans used "SingapoRediscoversvouchers," while 7000 tourism businesses received support to retain local employees.

While the pandemic has hit tourism in the city-state hard, it has also allowed officials to look ahead to the "day after," when Singapore rebounds to a full recovery. According to *Mr Chaly Mah*. Chairman, Singapore Tourism Board (in Tourism Board Annual Report 2020–2021),

> This pandemic has given us a once-in-a-generation opportunity to redefine ourselves as a safe, innovative, and trusted destination—and we must seize it. We should not aim for "business as usual" and go back to the way things were before the pandemic. The key challenges and opportunities facing Singapore's tourism sector are several, but four stand out. First, we have to reinforce Singapore's position as one of the world's most trusted, valued, innovative and safe destinations for business and corporate travelers. Secondly, we need to integrate wellness offerings and experiences into a modern city like Singapore so that it can be transformed into a leading urban wellness haven. Thirdly, Singapore has to become one of the world's top sustainable urban destinations. Lastly, we have to make Singapore a thriving market for great technology-solutions providers and tech-savvy tourism companies.

What Does the International Tourist Do in Singapore?

With an average tourist stay of 4 days, it is clear that Singapore remains more than ever "the interface between holiday and business" (Fontaine, 1999). In 2019, international tourists came to Singapore for holidays (58%), visiting friends (15%), business (10%), (5%) for MICE, and others (12%) (medical and sport events notably). It should be noted that 51% of visitors had already visited the city-state previously. Over half of the tourists (53%) choose to stay in one of the island's 412 hotels, including 77 Five-Star hotels. Notably, 16% of visitors stayed with friends.

The main items of expenditure for international tourists are shopping (15%), accommodation (20%), restaurants (9%), guided tours, entertainment and games (20%), and 35% air, sea, and inland transport, for business and medical purposes.

What About Sport Tourism?

Singapore is well-known for the quality and sharpness of its long-term planning and for a sustainable economic and social development growth. It was understood very early that in a globally interconnected world, and with the never-ending economic competition it faces, it must find innovative yet effective ways to continue growing and continue keeping its "top economy" ranking. In other words, it must continuously reinvent itself. As such, sport tourism is now an important component of the sustainable growth strategy of Singapore.

Jean Ng, director of sports for STB, stated in May 2018, "while the sports tourism industry in Singapore is still in a fairly nascent state, we believe there is enormous potential to tap into interest-based travel to power growth. Attracting and anchoring concepts with strong premium content, such as world-class sports events, is a key part of our strategy to develop Singapore as a leading leisure destination."

The Singapore Grand Prix, launched in 2008, was the first world-class event to open Singapore to sports enthusiasts. Generating some S$150 million in revenue per year, it has attracted 250,000 international visitors in 7 years and the city-state has just signed a new lease until 2028. "Since the first Formula 1 night race,

Singapore has other successful bids that enhance its image and put the country on par with other cities competing in the sports tourism arena" (*C Tan*, Research analyst at Euromonitor).

"Singapore develops a series of achievable plans each year to build its image as an international sports hub" (*C Tan*): WTA Finals (women's tennis, the world's top eight players qualify), which attracted nearly 130,000 spectators in 2017, Barclays Asia Trophy (football), Southeast Asian Games in 2018 (11 countries, 49 sports), FINA Small Pool Swimming World Cup 2015, 2021 and Sevens World Series Rugby Championship. All these world-class events boost Singapore's tourism revenues and, through the international media, convey the image of a dynamic city-state.

The 2015 WTA Finals at the Singapore Indoor Stadium from 26 October to 1 November 2015 was the 45th edition of the event since 1972 and the first in Singapore. The tournament was run by the Women's Tennis Association (WTA) as part of the 2015 WTA Tour. Singapore was the ninth city to host the WTA Finals and will host the event for at least 5 years. For the 2015 event, the competitors were divided into two groups, representing the colors of the flag of Singapore: the red group comprised *Simona Halep*, *Maria Sharapova*, *Agnieszka Radwabska,* and *Flavia Pennetta*, while the white group included *Garbine Muguruza*, *Petra Kvitova*, *Angelique Kerber*, and *Lucie Safarova*.

STB has partnered with event creator Lagardère Sports to leverage any event throughout the year by creating, co-creating, or curating content that attracts diverse audiences and tourism stakeholders. Sporting events are programmed both on court and also off court, opening the WTA Finals not only to tennis fans but also to occasional visitors, who are prolonged by off-court activities (culinary, paintings, etc.). For example, people followed Simona Halep in her yoga sessions in the Sky Park of Marina Bay Sands (MBS), Carolina Wozniacki playing traditional Malay drums with local artists NADI Singapura. The elite players learn about different parts of Singapore and then act as ambassadors to promote their experience.

This is also what happens with the Formula 1 Grand Prix, where STB also organizes a multi-faceted and dynamic living experience around the race. In the week leading up to the Grand Prix, the city "roars" with exciting experiences in the main tourist areas, high-energy entertainment in Marina Bay, shopping extravaganzas in Orchard Road, and a vibrant nightlife along the Singapore River.

E-Sports and Tourism

Increasing tourism receipts and potential visitors does not necessarily mean getting tourists to physically fly into Singapore. It can be done using digital technology, to showcase events and generate eyeballs (and indirectly bring Singapore, to the world). One example is Singapore-based Group ONE Holdings (ONE), Asia's largest global sports media platform with a broadcast to over 150 countries around the world. According to Nielsen, ONE ranks among the world's top 10 largest sports media properties for viewership and engagement, thanks to its marquee sports properties ONE Championship and ONE Esports. ONE Championship averaged 29 million global viewers per event in 2019 for its mixed martial arts (MMA), grappling, kickboxing, and Muay Thai events—for which it sold physical tickets, but it also streamed online on platforms such as YouTube. In cumulative reach, ONE Championship ranked #3 behind the English Premier League (#1) and UEFA Champions League (#2) and ahead of UFC (#6) in 2020. ONE Championship also ranked #4 overall in total video views in 2020, with over 6.1 billion across Facebook, Instagram, and YouTube, just behind the UFC (#3) and the NFL (#2).

External/Population: A Couple

In Singapore, the double standard approach of the external event and the population appears to be real. The sporting event is conceived as a means of attracting attention to the country, of bringing in tourists from outside, but at the same time of transmitting an "energy" to Singaporeans. According to a 2015 survey by SportSG, Singapore has seen a 30% increase in regular tennis play. This figure is highlighted by the launch of the Active SG Tennis Academy in 2016 and the Singapore Tennis Festival in 2017, both of which have created more opportunities for young tennis players to play the sport.

Thus, through sport, STB has highlighted the potential opportunities that Singapore has to offer to both visitors and residents. More than mere competitions, sport provides a vehicle to connect "different worlds," professionals, fans, and the population by conveying "emotion" rather than "sporting results."

This coupling is also beneficial for the development of sport in general. Singaporeans practice many sports activities on a daily basis, taking advantage of the development of many indoor and outdoor sports infrastructures. Singaporean athletes, who have emerged from this mass practice, participate in international competitions every year. In 2016, swimmer *Joseph Schooling* won the gold medal at the Olympic Games ahead of Michael Phelps in the men's 100 m butterfly. Singapore also won a total of 17 medals (5 gold, 2 silver, and 10 bronze) at the 2002 Asian Games held in their backyard (Busan, Singapore).

Sport tourism is one of Singapore's key components of its continuous reinvention, and it plays a sustainable critical role at two levels:

(a) Economically: to continue to position Singapore as a global, vibrant city and hub and continuously attract international visitors and increase tourism receipts. One good example is the Singapore F1. Singapore's estimated annual cost for hosting Formula 1 is \$100 million. Host organizers Singapore GP Pte. Ltd. and the Singapore Tourism Board reportedly paid \$35 million per Grand Prix in the previous contract, which ran from 2017 to 2021. The race has reportedly generated more than \$1.1 billion in tourism receipts, according to event organizers.

(b) Socially: to entertain local residents and give them the tools and platforms to live a healthier, happier, longer life—in other words, live better through sports. Sport Singapore invested over $60 million in the past decade on the renovation and upgrading of facilities. Its world-class sport centers annually record over 11 million users, accessing 26 swimming complexes, 24 fitness gyms, 17 stadiums, 18 indoor sports halls, and 11 tennis centres to name a few. The impact: in 2015, 26.4% of the Singapore population exercised at least three times a week. It grew to 35.9% in 2017 and is estimated to be close to 40% in 2021 (compared with less than 30% for Americans).

Sport Tourism, Leisure, and Environment

"To achieve First world standards in a Third World region, we set out to transform Singapore into a tropical garden city, to involve in clean and green movement" (Yew, 1998). Lee Kuan Yew's concept helped guide a set of regulations for many years from creating a sense of public awareness about cleanliness to implementing ambitious projects that have helped to create a rich urban landscape. Singapore, also known as the "Green City," is one of the top cities in the Green City Index. This concept is the very basis of sustainable development where any project, any establishment, and any activity must be in line with this sustainable development plan. Such planning is utilized for all tourist activities and infrastructures. Thus, tourism and sports leisure, in relation to sustainable development in the small city of Singapore, must be understood above all in relation to the concept of the "green city" and its applications in the field.

Singapore offers 400 parks, four nature reserves, and an island network of corridors linking the main parks and reserves. As depicted in Figure: The Park Connector Network (PCN) Map p. 321, is a 300-km island-wide network of linear green corridors linking major parks and nature areas across Singapore. Built and regularly expanded by the Singapore National Parks Board (NPB), the PCN is a favorite with cyclists, hikers, casual joggers, and walkers, enhancing the green leisure of Singaporeans as well as external tourists who can enjoy the extraordinary spectacle of nature, including the 260 "heritage trees" and "nature" roads, therapeutic and botanical gardens, that complete Pulau Ubin. This island, located off Singapore's northeastern coast, is a tranquil island sanctuary for nature lovers and day trippers. It is also a window into Singapore's past and home to diverse habitats and a variety of wildlife. NPB has also developed the 36-km Coast-to-Coast Trail, taking visitors, both local and tourist, through a variety of parks, nature areas, and places of interest.

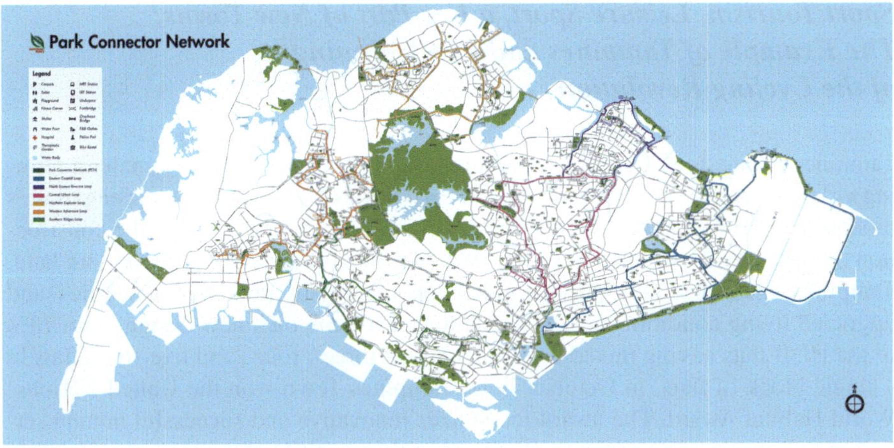

Source: https://www.nparks.gov.sg/gardens-parks-and-nature/park-connector-network (1 cm = 4.2 km)

The East Coast Park, a "Recreation for All," is one of Singapore's most sought-after inner-city gateways. The park offers an assortment of recreations with sight-seeing, dining, and outdoor activities. Spread over a sprawling area of 185 hectares with a scenic coastline, stretching across 15 km, the park attracts more than seven million tourists annually, and more than three million residents for leisure. East Coast Park is an excellent outdoor sporting zone for those who seek some action on their trip. The myriad of sports the park offers include beach volleyball, cable skiing, cycling, hiking, and a variety of water sports. Among the adventure activities available at the East Coast Park, the Xtreme SkatePark allures tourists who desire action leisure.

Since 1972, the Sentosa Development Corporation (SDC) has redeveloped the islands of Sentosa and Pulau Brani into a game-changing tourism and leisure destination, taking advantage of their unique island charm, geographical qualities, as well as their proximity to the city, 15 minutes from the CBD and shopping districts. Sentosa is Asia's premier leisure destination and Singapore's first island resort. The 500-hectare island resort is home to an array of themed attractions, an award-winning spa, lush rainforests, golden sand beaches (reclaimed sand), sports facilities, resort accommodation, world-class golf courses, a deep-water yachting marina, and luxury residences, as well as Asia's first Universal studio—making Sentosa, a vibrant island resort for business and leisure. In 2019, Sentosa attracted 19 million visitors, including 4.7 million local residents. It is in this kind of development where sport tourism and sports leisure activities are successfully planned and experienced by both domestic and international visitors.

Sport Tourism, Leisure Sport, a Key Part of New Towns: The Example of Tampines and the Beginning of the Cycling Revolution

Tampines Estate, located in the eastern region of Singapore, was originally a farming village. A massive development in recent decades has made it one of Singapore's modern new towns and also a popular retail and leisure destination. The construction of Tampines New Town began in 1979. As the government acquired more land, Tampines entered a new phase of development, with improved amenities and upgraded living conditions in 1984. The flats also took on a new design, with 75% of the HDB flats having mixed dwelling units of three-, four-, and five-room flats in a single block of flats. In October 1992, Tampines Town won the United Nations' World Habitat Award. The award recognizes innovative and successful human settlements. Tampines Town has been designated as a regional center since the early 1990s and is one of the most developed regional centers serving the eastern part of Singapore today.

Officially launched on 6 August 2017, Tampines Hub, envisioned as a "hub by the residents for the residents," is Singapore's first integrated community and lifestyle hub, containing a variety of sports facilities, a regional library, hawker center, public service center, art theatre, and other facilities under one roof. As part of the efforts to transform Tampines into Singapore's first "eco-town," an outdoor linear garden was launched at Tampines Hub in February 2021. The garden has a water feature, speakers that play recordings of birds chirping, and close to 40 types of flowers and herbs. With features that engage the five senses, the garden serves as a therapy space for elderly residents.

Another first for the area, Tampines was selected as the place to inaugurate the now familiar shared footpath scheme. The trail, developed between May 2007 and May 2008, offers 1.2 km of specially modified footpaths (widened to 2 m). The network utilizes almost 250 wardens to police the trail and help any confused pedestrians and cyclists. Deemed an urban success and in 2010, Tampines was named the first cycling town in Singapore and remains one of the few places on the island where cyclists can legally bike on the pavements.

Tampines walking and cycling town. (https://www.lta.gov.sg/content/ltagov/en/upcoming_projects)

Land Transport Authority
We Keep Your World Moving

Mobilization to a Sustainable Tourism

It is in this environment that sport tourism and sports leisure activities are expressed in Singapore in relation to the sustainable development policy for which the country has made it a prime target. The tourism sustainability program aims to support tourism businesses and their sustainability journeys. At all stages of their development, this sustainability project helps to build the workforce capabilities and develop innovative solutions to capture opportunities from sustainable travel.

The desire to set an example is reflected in numerous concrete measures. In the hotel industry, the MBS, the nation's flagship of the Singaporean hotel industry, has been leading the way for several years now. The famous five-star establishment has reduced its water consumption by more than 10% since 2015, by recovering condensation from some 3000 air-conditioning units, saving no less than 80,000 L every day. MBS is also committed to responsible sourcing of seafood products by supporting the fishing industry towards more sustainable practices.

Singapore stands out on the map of responsible destinations because of the sector's proven commitment to reducing their carbon footprint. These stakeholders are going further by making the environment and its preservation a new axis of tourism development, multiplying attractions aimed at sustainability. The issue of sustainable food is also at the heart of the sustainable development project. Several tour routes are intentional about their eco-friendly development, such as Wok & Stroll, Veg This City or Let's Go Tour Singapore. These efforts are also focused on the theme of water management. Business tourism is particularly targeted by the green dynamic, and various specific packages have appeared in the MICE segment. The "It's Easy Meeting Green" offer proposed by Marina Bay Sands includes a dedicated support service to better integrate and enhance the sustainable aspect of the event. The "Meets Lite, Feels Right" package launched by Pan Pacific Hotels Group promotes sustainable refreshments in addition to a "farm to table" working lunch while insisting on the elimination of single-use plastics and offering fixed "stations" for note-taking.

Singapore predicts that sport tourism will play an important role in this innovative plan by setting out three main axes. It must be developed taking into account the sustainability of natural resources, including biodiversity, a nature-based sport tourism which must defend the preservation of the natural environment, without which it cannot be sustained. Preservation of the landscape and the avoidance of visual pollution must also be taken into account and tourism fees must be used for the protection of natural resources. It is also about enhancing the interaction between Singaporeans and visitors, whether it be in the sports activity itself, in the dedicated spaces created such as the fan zones, or in the programs that aim to ensure that residents and visitors alike are able to participate together and enjoy the planned activities and events. Sport tourism should also aim to integrate people with disabilities, both residents and visitors, in the products offered, as well as in their accommodation and transport.

"Made in Singapore" has become the rule for all stakeholders, developers, event organizers, travelers, and practitioners, both external and local, to put sustainability at the heart of their actions to make Singapore the ultimate sustainable destination.

Conclusion

Singapore, a country devoid of natural resources and long considered a place of "business tourism" or "transit tourism," has been able to continuously grow by reinventing itself. This innovative development has occurred by widening its definition of tourism to include new growth-generating, awareness-building, and positioning-strengthening pillars such as sports tourism. Now more than ever the city-state is a global economic hub, a multicultural city, and a world sports and lifestyle destination. Singapore is therefore a model when it comes to applying a "pragmatic" vision of sustainable development, ensuring that any tourist and leisure approach must take into account the benefits and progress of the country and its people as a whole.

References

Chua, B. H. (1995). *Communitarian ideology and democracy in Singapore*. Routledge.
Fontaine, G. (1999). Singapore, tourism and world. *Asia Pacific Journal of Tourism Research, 4*(1), 30–38.
Singh, B. (2011). Singapore: Maintening the balance between prosperity, social growth and gradual democratization. *International Journal of Comparative Politics, 18*(1), 105–122. De Boeck.
STB. (2020). *Singapore Tourism Board annual report 2020–21*. STB.
Yew, L. K. (1998). *The Singapore story: Memoirs of Lee Kuan Yew*. Pearson Ed Asia.

Chapter 20
How Sport Tourism Helps to Improve the Sustainable Development of Outermost Island Destinations and Their Resilience: The Study Case of La Réunion Island

Philippe Jean-Pierre and Olivier Naria

Introduction

Sports activities and tourism activities are infinitely linked. The possibility of practicing a sport while on vacation has become an essential criterion for tourists when making their travel plans. This is especially the case for younger tourists.[1] This phenomenon is not new. The theoretical and empirical literature analyzing this link has been abundant for well over two decades now.

Following the seminal contributions of Kurtzman (1995) (see also Kurtzman & Zauhar, 1995; Gammon & Robinson, 2003), it has been widely recognized that sport tourism may play a crucial role in the development of tourism destinations (Fredline, 2005; Briedenhann, 2011). Sport tourism has become a key factor in the attractiveness of the region and local authorities are now investing more in this area. It is also known that tourism development is a complex process and that sport tourism development can also generate negative impacts on tourism destinations, including on their communities (Fredline, 2005; Orams, 2005).

In addition to the complexity of these linkages, there is growing interest in the study of the holistic sustainability of the sport tourism sector (Gibson et al., 2012;

---

[1] For instance, according to a study by Union Sport et Cycle, in France, two-thirds of French people (67%) have declared having practiced a sport activity during their vacations; this is especially true for those under 35 years old, which tends to think that "sport tourism is going to be a long-term trend."

P. Jean-Pierre
CEMOI, University of La Réunion, Saint-Denis, Réunion
e-mail: pjp@philippejeanpierre.fr

O. Naria (✉)
Department STAPS, Université de La Réunion, Le Tampon, ïle de la Réunion, Réunion, France
e-mail: olivier.naria@univ-reunion.fr

D. Van Rheenen et al. (eds.), *Sport Tourism, Island Territories and Sustainable Development*, Sports Economics, Management and Policy 24,
https://doi.org/10.1007/978-3-031-51705-1_20

Melo et al., 2021; Mazza & Sobry, 2022). Indeed, more recent studies recognize that sport tourism could be seen as an indispensable tool to reduce the negative impact of tourism on the environment. For example, by emphasizing physical activity and active mobility, slow tourism helps limit the pollution of transportation. Another similar observation stipulates that the development of outdoor activities allows for a better distribution of vacationers and therefore could minimize the environmental impact of the concentration of tourists in one place (e.g., overcrowding and carrying capacity).

Nonetheless, much of the research has focused on large events, large countries, or large mainland areas (Newland et al., 2021; Carneiro et al., 2016). Unfortunately, there has been limited research on sport tourism and small island destinations (SIDs). Yet, all the concerns we mentioned previously are particularly pressing for these SIDs. Indeed, for these destinations, many studies highlight the uncertainty surrounding the tangible effects of tourism in the long term and the decreasing economic returns of tourism impact (Jean-Pierre & Perrain, 2021). Much more than continental destinations, SIDs have to cope with numerous vulnerabilities (Klöck & Fink, 2019), such as the impact of tourist activities on the environment with coastal erosion, pollution of the sea, damages caused to reefs, and so on. The case of the SIDs is all the more relevant because islands have the opportunities to produce a wide range of sport activities. These vulnerable territories may be tempted to develop these activities to spur their regional or global attractivity without taking into account all the consequences of this strategy on a fragile destination. Unfortunately, with the exception of some recent contributions to the literature, such as Naria et al., 2022), there is limited research that refers to the sport tourism impact on small island destinations, particularly on isolated islands, even though these concerns are particularly prescient for small island economies (SIEs).

From this perspective, the aim of this chapter is to fill this gap in the literature and to delve more deeply into the analysis of the ambiguity of the relationship between sport tourism and SIDs. It is also intended to contribute to a better understanding of the conditions required to reinforce the sustainability of sport tourism in these ecoterritories. Indeed, are sport activities a sufficient vehicle to develop tourism destinations, particularly when they seek to move beyond their traditional positioning in the global market? In the specific case of SIDs, what are the key factors of success for developing such activities, as they are often isolated and characterized by low economies of scale due to their small size? Under which conditions does outdoor or nature sport tourism positively impact these destinations? What are the implications for the inhabitants and stakeholders of the destination regarding sustainability and sustainable development goals?

In order to answer these questions, we explore the case of La Réunion Island, a French outermost region in the Indian Ocean. Indeed, La Réunion is a small island destination seeking greater notoriety for a better branding strategy, based on a combination of different specializations by mixing traditional activities with novel and innovative approaches. During the last 25 years, several sport events have allowed this destination to become one of "THE" destinations in some specific activities, such as trail running ("Grand Raid," also known as "Diagonale des fous"), mountain biking ("Megavalanche"), and surfing championship tours. Thanks to its geographical specificities, La Réunion is characterized by many amenities, positioning this destination as a hot spot for sport and tourism. Despite the fact that famous activities have managed to develop during the past 25 years, these success stories are not guaranteed in the future. The impact of these different activities on the territory could compromise the attractivity and resilience of this destination both regionally and globally.

As a case study, La Réunion may be of interest to other like territories that combine similar specificities. Indeed, for remote islands, sport helps destinations improve their notoriety and boost their brand and their attractiveness. For instance, "Island Games," created in 1979 for the Indian Ocean Islands, has helped these destinations open themselves to each other in terms of cultural and economic aspects. Nonetheless, to help the destination improve its performance, investing in sport tourism requires the territory to pursue a smart specialization strategy (which is broader than the concept of smart destination) and to cope with ambitious goals regarding sustainable and social issues. Due to the small size of their economies, sport tourism likewise requires these territories to select the appropriate business model adapted to the particular destination.

To pursue this goal, this chapter will be organized as follows. The first section briefly presents the extant literature on the relevance of sport tourism in the development of small island destinations, focusing on their characteristics and the vulnerabilities they must face. The second section of the chapter will provide an empirical study on the case of La Réunion Island, targeting the sport tourism activities implemented in this territory. A specific focus will be made on the trail activities. This second step will highlight best practices and future implications which may help similar destinations to develop more sustainable sport tourism activities while improving the adaptive capacity of the destination.

Sport Tourism and Destinations: An Ambiguous Effect?

The fact that sport tourism may play a crucial role in the development of tourism destinations is largely recognized by the literature (Kurtzman, 1995; Briedenhann, 2011; Fredline, 2005). Sport tourism raises heightened expectations among governments and public decision-makers. However, this field of activities can also provide some disappointing results. This debate, relative to the pros and cons of this form of development, tends to realistically measure the true impacts of tourism on these economies.

Like Many Other Tourism Activities, Sport Tourism Can Generate Ambiguous Impacts on Destinations

If we consider sport tourism as but one of many fields of tourism activities, we can then analyze the consequence of the development of this specific area. This allows us to mobilize the framework used to study the multiple impacts of tourism on economies.

On one hand, sport tourism, like other tourism activities, is thought to contribute significantly to economic growth by improving efficiency in the allocation of production factors or by expanding the capital accumulation and the arrival of foreign direct investments. This assumption is supported by an expansive and diverse extent of research into the determinant of the correlations (cf. Dwyer & Chulwon, 2003; Holzner, 2011; Brida et al., 2009, 2016; Pablo-Romero & Molina, 2013). Nowak (2007) showed that expanding tourism exports can increase the volume of inputs, mainly through capital accumulation These exports enable the financing of imports of foreign capital goods, which leads in turn to an increase in capital formation and growth.

On the other hand, for another large group of researchers, the real impact of sport tourism and tourism activities appears to be more ambiguous. Indeed, several studies highlight the uncertainty surrounding the real effects of tourism in the long term and the decreasing economic impact of tourism.

The traditional vision of "beneficent tourism" can be thwarted by the detours of economic flows, operating to the detriment of the economy relying on this sector. This is the case for many small island economies in the Caribbean and the South East Pacific, which see a large part of the economic benefits of tourism leave their territory through significant economic leakage (Archer, 1984; Wilkinson, 2004). Destinations affected by this leakage may become less and less dynamic, leading to stagnation or even a decline in tourist arrivals.

The negative externalities of tourism raise questions about the true effects of tourism in the long term and about a possible form of slowdown of the impact of tourism on the economies concerned. Thus, the link between tourism and growth may be viewed as non-linear, as it depends on the comparison between the benefits associated with the development of this sector and the negative externalities that it induces.

Some studies highlight the uncertainty about the real effects of tourism in the long term and the declining economic impact of tourism (Adamou & Clerides, 2009; Po & Huang, 2008; Chang et al., 2010; Wang, 2011). In some cases, tourism has become progressively less dynamic, leading to stagnation or even decline. Indeed, this activity follows a "product life cycle," following the model of Butler (1980), who emphasizes the life cycle of the tourist area and the different stages involved in the development of a tourist destination. Finally, other works have focused more on the negative externalities generated by tourism, such as the possibility of deindustrialization, contraction of well-being, and *impoverishment*

(Copeland, 1991; Chen & Devereux, 1999; Hazari et al., 2003; Nowak et al., 2003; Brida et al., 2009; Chao, 2006; Schubert & Brida, 2008).

The degressive impact of tourism on economic growth highlights the possibility of different growth regimes depending on the stage of tourism development, implying a non-linear relationship between tourism and economic growth. This question of non-linearity, and more specifically the existence of regime change phenomena, is also the subject of much work in growth theory to study the factors of convergence between countries (Jean-Pierre, 1997).

Beyond these purely economic aspects, the impact of sport tourism must also be analyzed through the prism of sustainability, which includes several dimensions. For example, we can naturally think about the effects of sport tourism activities on the environment that implies that at a certain level, the causalities may call into question the benefits initially sought (Maupertuis & Gianonni, 2005; Cerina, 2006; Gomez et al., 2008). In this sense, the reactions of the populations in many cities or countries against mass tourism and the recent development of the concept of overtourism have also to be considered (Andersson et al., 2015; Fernández & Sánchez Rivero, 2009; Fredline, 2005; Gibson et al., 2012; Kožić & Mikulić, 2014). Additionally, we cannot ignore other dimensions such as technological and policy factors. Moreover, it is widely recognized that sport and tourism play a vital role in the global economy today, in terms of both economic and social contributions.

In that perspective, the literature provides several approaches and ways of defining the concept of sustainable tourism (Jean-Pierre & Perrain, 2021). Most of them make a transition from the concepts of sustainability or sustainable development. Despite several studies exploring this issue, they tend to focus on a limited number of aspects of sustainability, rather than looking at it from a multidimensional dimension.

Furthermore, the discussion on sustainable development in the field of sport tourism (Andersson et al., 2015; Fredline, 2005; Gibson et al., 2012) has largely relied on highly developed countries and large destinations. Few researchers refer to sport tourism's impact on small, isolated, and remote island destinations despite that these concerns are also particularly pressing for small island economies (SIEs[2]). Among the major contributions of sport tourism and SIEs, we highlight the following literature:

[2] Over the past 30 years, the United Nations system has supported the cause of small island developing States as countries meriting special attention because of the intrinsic, serious disadvantages they generally face. Island destinations are particularly vulnerable to natural disasters. For the case of the Pacific Island countries, for example, the World Bank (2006) reported 207 major natural disasters (157 of them windstorms) between 1950 and 2004. These affected about 3.5 million people in the region with an economic damage cost of over 6.5 billion of US dollars. In the Maldives, the combined direct and indirect costs to tourist resorts and loss of Government revenue from reductions in tourist arrivals as a result of the 2004 Indian Ocean tsunami has been estimated to be in excess of US 300 million of dollars. Overlooking the importance of tourism in island hazard and vulnerability assessments has far reaching implications, including biased calculations of human casualties.

- Economic effects: Kartakoulis and Karlis (2002) (Cyprus); Elcock (2005) (Barbade); Bull and Weed (2009) (Malta), Fadda (2020) (Sardinia); Bastien et al. (2022) (French Guyana)
- Empowerment impacts: Naria (2010) (La Réunion), Buckley et al. (2017) (Maldives); Towner (2016) (Carribean Is.)
- Innovation and ecosystemic schemes: Achilleos et al. (2021) (Cyprus)
- Management issues, landing spaces, and scarce resources: Bessy and Naria (2004) (La Réunion); Martel and Rey (2012) (Corsica); Naria and Sherwin (2013) (Seychelles); Guiltat and Taglioni (2015) (La Réunion); Alonso and Medina (2019) (Canarias Is.), Naria et al. (2022).
- Marketing strategies and attractiveness: Bessy and Naria (2004) (La Réunion); Seraphin and Gowreesunkar (2020) (Martinique).

While these contributions help to address some of the issues regarding the sustainable development of sport tourism in SIDs, they remain focused on traditional and more singular impacts and do not include more systemic approaches analyzing the way these economies have changed as a result of the development of the sport tourism sector. It seems therefore useful to delve more deeply to fill this gap in the literature, particularly for these outermost and insular economies.

Is Sport Tourism an Opportunity for Small and Outermost Island Destinations?

Answering this question is all the more important as small and remote island economies that choose sport tourism as a vehicle for development may be faced with two situations.

On one hand, investing in this tourism specialization can support growth. This is a possibility as proposed by the theory of tourism-led growth hypothesis studied by McElroy (2006); McElroy and Parry (2010) when they specifically focused on these kinds of economies and explained how many SIEs overcame their size disadvantage and remoteness, often seen as fatalities, and became the most sustained growth engine in the postwar global economy for the growth engine in the SIEs.

On the other hand, the temptation of SIEs to develop sport tourism due to their natural assets could spur concerns on their development process as they are affected by several constraints (small market size, cost of transport, impact of climate change, etc.) and could see their situation worsen due to the negative externalities of tourism on their economies.

This possibility is not surprising and is consistent with many studies that highlight the uncertainty surrounding the real effects of tourism in the long term and the decreasing economic impact of tourism for these small destinations (Maupertuis & Gianonni, 2005; Cerina, 2006; Po & Bwo-Nung, 2008; Gomez et al., 2008; Adamou & Clerides, 2009; Chang et al., 2010). Much more than continental destinations, SIEs have to cope with many vulnerabilities, such as the impact of tourist activities on the environment with coastal erosions, pollution of the sea, damages caused to reefs or hinterland landscapes, etc. In the end, these impacts may call into question the benefits initially sought and the relevance in the choice of sport tourism specialization as a lever for the development of their economy.

The necessity to go deeper into the exploration of the impact of sport tourism on SIEs can be solved by exploring the experience of a small island which chose to develop sport tourism activities as one of the pillars of its tourism attractiveness. In that case, focusing on La Réunion may be interesting for other territories that combine similar specificities.

A Study Case of Sport Tourism and an Outermost Island: La Réunion

Research and Geographical Context

The area under analysis is La Réunion, a French and European outermost region, located in the Indian Ocean (see Fig. 20.1).

Located in the South-West Indian Ocean, 700 km East of Madagascar, La Réunion Island, with its subtropical climate, forms the Mascareignes archipelago with the islands of Mauritius and Rodrigues. La Réunion and Mayotte are the only French departments in the southern hemisphere. La Réunion Island is 9180 km from Paris.

With a total surface area of 2512 km^2, the island is bordered by 210 km of coastline, which is mostly inhospitable, apart from 40 km of sandy beaches. La Réunion's climate is subtropical and is influenced by trade winds. It changes significantly depending on the location and the relief. The average temperature is 20 °C on the coast and 14 °C in the interior plains. The cyclone season lasts from December to April. The population of La Réunion Island was estimated to be 865,000 in 2021. The density was 312 inhabitants per km^2. As a French department, La Réunion is under the authority of a prefect appointed by the government.

The island, which became a French department under the law of March 19, 1946, has a Regional Council and a Departmental Council. All national laws are applicable there. Only certain adaptations have been provided by the law. As a French Overseas Department, La Réunion is part of the European Union, within which it is an outermost region; as such, it benefits from "specific measures" that adapt community law to take account of the particular characteristics and constraints of these regions. These adaptations fall within the framework of the application of Article 349-2 of the European Union Treaty.

The isolation of the island, with its size and location, is often put forward as constraints. However, they can also be seen as advantages. For example, in several sectors, such as energy, agri-food and industrial symbiosis, the usual configuration of industrial or economic models is that of economies of scale, and therefore, large

Fig. 20.1 La Réunion Island, located in the South-West Indian Ocean. (Source: Iedom, La Réunion, Cerom)

dimensions. However, in several of these fields, Réunion Island has already demonstrated its ability to adapt "to the small," in what is usually designed for "the large." This does not happen by chance but is the result of organizational innovations and agile processes that can be seen as levers of resilience.

Although the island's initial situation was so precarious that we can speak of a kind of "success story" in longitudinal terms over the last half-century, the growth and development process has not made it possible to close all of the gaps in relation to mainland France (in 2019, the GDP per capita of 22,300 euros was 64% of that of mainland France) and remains dependent on several key variables, particularly national solidarity. Moreover, challenge such as climate change and other major changes to come, such as those resulting from the digital transition and the tensions in international value chains, highlight the many vulnerabilities in the functioning of this island economy.

Thanks to its geographical, geomorphological, and climatic characteristics, La Réunion Island, despite its small size, offers a wide range of sport tourism activities (see Fig. 20.2). Thus, with the exception of sports associated with snow, practically all "indoor" or "outdoor" sport tourism activities seem possible in Réunion Island, regarding land-based activities as well as those which involve aquatic environments and the contemplation of Réunion Island from above.

Moreover, a pleasant climate year round allows sport tourism activities to be carried out 365 days, without any calendar or seasonal break. Beyond the nature, the island's sporting orientation can be seen in its population. Indeed, the inhabitants love and practice sports and often enjoy their island through a sporting experience. This comparative advantage also benefits from the existence of high-level infrastructures that meet European and world standards and are open to both competitors and the population. These factors of attractiveness are not neutral and have allowed, during the last 25 years, this destination to become one of "THE" destinations in some specific activities like the now world-famous ultra-trail races, such as the "Grand Raid" (also known as "Diagonale des fous") or the "Megavalanche" for mountain biking. This dynamic has led the destination to position itself as a training base in the Southern hemisphere, in juxtaposition to the winter seasons in the Northern, for several sports competitions, notably in extreme sports (mountain biking, canyoning, paragliding, base jumping, ultra-trail running, etc.).

If this strategy of investing in sport tourism has been successful, much remains to be done to consolidate this positioning. In spite of the efforts made, particularly to attract many international events in sport tourism, La Réunion Island is still struggling "to be on the map." And at the same time, this strategy can also be a source of vulnerability. Indeed, the sport events cited above have to face many challenges after two decades: are they consistent with the strategy of this peripheral destination, whose ambition is to be a leading green destination? Are the business models of the stakeholders which support these sport events sufficiently resilient to go through crises like the COVID-19 pandemic? Despite the fact that famous activities have managed to develop during the past 25 years, their success stories are not guaranteed. At the end, the impact of these different activities on the territory could compromise the global attractiveness and the resilience of the destination. These

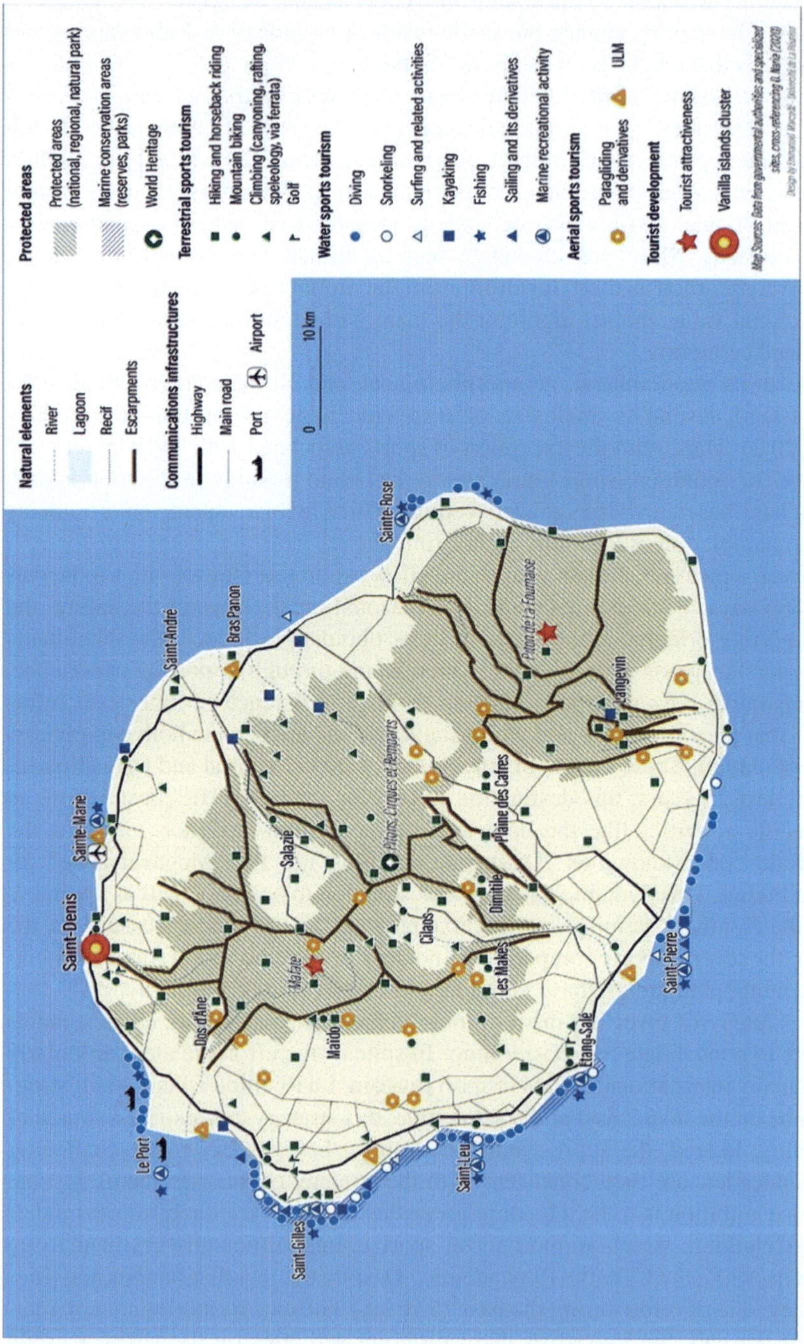

Fig. 20.2 Map of the territories of sport tourism

questions imply deepening the investigations on the impacts of sport tourism on a small isolated island economy, like La Réunion.

Sport Tourism in La Réunion: The Metamorphosis of an Economy

This overview of activities related to sport tourism sheds light on the dynamics of this island territory (see Table 20.1). Beyond the natural predispositions, important efforts in terms of investments on infrastructures, the increased competence of the actors, and the mobilization of partners have led to the performance achieved over the last 30 years. Three decades ago, La Réunion Island was not at all known as a sport tourism destination. But thanks to the organization of events that ensure a strong publicity for the island and important economic benefits, gradually this positioning has been built up to become an essential axis of the island's tourist attractiveness today. In this respect, it is not the fact of having the only or the most marvelous place for a sport practice that is the differentiating element (even if these remain exceptional), but rather the possibility of enjoying it almost simultaneously in the same period because of the short distances, the favorable climate, and the adequate infrastructures. In 1 week, intense emotions or strong experiences are within the reach of the tourist who has simply come to discover the island or who is looking to develop his practice more easily.

It is therefore a true success story that characterizes the development of sport tourism on the island. Of course, this evolution has benefited from the development of sport tourism throughout the world. But unlike other islands that have remained on the traditional dimension of seaside or mountain tourism, La Réunion has been

Table 20.1 Evolution of the main sport tourism facilities from 1989 to 2019

Activities		1989	2019
Paragliding (Number of spot of takeoff and landing)		10	74
Canyoning (spots)		10	210
Climbing (spot)		5	22
Horseback riding	Itineraries	10	48
	Kms of developed trails	40	110
Hiking	Kms of developed trails	700	800
	Lodges	20	232
Mountain biking (km of developed trails)		50	210
Canoë-kayak (spots)		1	10
Diving (Centers)		7	30
Surfing (Centers spots)		–	8
Sailing (Centers and storage facilities)		7	11

Sources: Bessy and Naria (2003), crossed data from Naria

able to mobilize its assets, despite the handicaps linked to its island position, to maximize the impact of this positioning around sport tourism.

The effects on the economy should also be highlighted. Indeed, sport tourism stimulates the economy directly through tourist spending but also leads to changes in the behavior of local tourists and in the practice of sports promoted by the events. In the case of La Réunion, trying to understand the direct and indirect economic effects of sport tourism is a difficult task, as surveys are rare in this area. We can cite the evaluation of the "SYPRAL" (the union of professionals of leisure activities) in 2014, which estimated the economic impact for the sport tourism and leisure industry in La Réunion to 63 million euros. In the same way, the impact of sport tourism in this island can be further refined by focusing on the effects of an emblematic event of sport tourism in La Réunion: "the Grand Raid" reputed to be one of the most difficult ultra-trail races in the world.

The Impact of Sport Tourism Through a Famous Ultra-Trail Event: The "Grand Raid"

More precisely, the "Observatoire du Tourisme de La Réunion" and the Grand Raid association conducted a study in 2015 to measure the economic impact of the Grand Raid in La Réunion. Even if this analysis dates back to 2015, it allows us to better seize the dimension of sport tourism in such a small island.

Thus, with a total of 5421 participants in 2014, including 1571 from outside, the study estimates that the total expenditure of competitors amounted to 9.4 million euros. This economic impact can be seen to be in direct relationship to the performance and tourist attraction of the island. We learn that word-of-mouth is the main trigger for participation in the event (82%) on the side of external runners. Word-of-mouth, free but not very controllable, is thus one of the main vectors of the visibility of the Grand Raid internationally. Nearly one participant out of two stays in a commercial accommodation, with a preference for bed and breakfasts, followed by hotels and seasonal rentals. In total, local participants and their support staff spend an average of €1154.10 between their preparation period, registration fees, and expenses. This budget rises to €3173.79 for external participants and their assistants, who stay on the island for an average of 17.9 days and must include the cost of the plane ticket and a car rental on site for more than three quarters (77%) of them. With 1571 outside competitors and 3850 local competitors in 2014, the estimated total expenditure amounts to €9.4 million, divided almost equally between the two categories.

These different economic data shed a first light on the importance of sport tourism for La Réunion Island. In order to better appreciate the stakes of this sector on the island, one could obviously put forward the priority of updating the studies and monetary evaluations. But the priority does not seem to us to be at this level. First, because we have already mentioned it: the evolutions in progress during the last years support the growth of the tourist sports activity in the enlargement of the proposed activities, the facilitation of the access to these activities and the reinforcement of the innovations and offers in terms of useful equipment for these activities. Therefore, there is no doubt that the direct and indirect economic effects have been strengthened compared to the studies of the late 2000s and 2015. Second, these quantitative evaluations do not shed light on a double issue of importance for sport tourism in an island territory: on one hand, there remains the issue of transformations in the capacities generated by the development of these activities; on the other hand, there is the issue of vulnerability through the consideration of the environmental dimension in the development of sport tourism in isolated island territories.

The Challenge of the Resilience of Sport Tourism in Insular Environments

In order to sustain sport tourism in an isolated and fragile island destination, two dimensions must be considered: the vulnerability of this sector and its capacity to adapt through its resilience.

Sport Tourism in Island Destinations: Are They a Source of Vulnerability?

The issue of the vulnerability of sport tourism in relation to sensitive island destinations has been studied from several perspectives.

The first level of vulnerability concerns the island territory itself. In this case, La Réunion Island is a perfect example of a territory exposed to multiple vulnerabilities: economic, energy, social, environmental, and climatic. These issues take into account the decision makers and stakeholders of the territory. Plans and schemes have been set up to consider the challenges of sustainable development and to reduce the entropy of economic development and population growth. We can cite, for instance, the waste management plan, the tourism development plan, the land use plan, and the national park diagram. However, despite these measures and ambitions, much remains to be done to support climate action, the various energy and social transitions, and the major changes linked to the circular economy in order to develop the territory in a more sustainable way.

In addition to this first level, there is a second level linked to the functioning of the sport tourism sector and its effects on the territory. The advent of this sector and its development is indeed not neutral. Certainly, as we have seen, it brings significant direct and indirect economic benefits. But in the opposite direction, it can be a source of vulnerabilities. Such is the case for the resources of the island territory. The multiplication of tourist sports activities puts the land, natural spaces, and places to the test, consumes water, and produces waste. This is not without effects on the environment of the sensitive island (see Martel & Rey, 2012; Alonso & Medina, 2019). This is one of the consequences associated with overtourism: if the flows are not managed, the host environment can be degraded. This risk is even greater when the island is small or when the spaces dedicated to sports events are limited. The role of local authorities and good governance is therefore crucial (see Bessy & Naria, 2004; Guiltat, 2011). But the most important threat at this level is the one linked to the potential change of behavior on the part of the clientele willing to reduce their carbon footprint. Remote islands are caught in the trap of the (carbon) means to access them. The "Flyskaam" phenomenon illustrates this threat. The same is true for the decisions of a number of competition stars who now choose to limit their travel by air. So what future for island sport tourism if moving to a sensational island would be associated with a "dirty" image?

Similarly, sport tourism has consequences for the population. Of course, they can benefit from the increased activity. But they can also be upset by the uncontrolled influx of sport tourists. This can be seen at several levels. First, there are possible nuisances and inconveniences related to the events. There are fewer spaces, conflicts of use, and reductions in moments of rest, whereas these island destinations often put forward this quality of quietude. Second, in terms of economic effects, because the demand from sport tourists leads to increases in the price of housing or consumer products, which impacts the purchasing power and well-being of the local population. Finally, at the level of cultural evolutions, sports events are often accompanied by the arrival of caravans of globalized industrial actors promoting their products, equipment, food, etc. This leads to changes in habits and customs. This also leads to changes in the habits and consumption behavior of local populations. This can result in health risks and lead to frustration. Here again, these risks are to be put in relation with the small dimension of the island territory. For if the island remains a territory of imports only, then the negative effects play out in full. If, on the contrary, the local market allows industries to set up and allow local production to replace imported goods, then the effects linked to the increase in income and the supply of work, including the mobilization of young people and women, can help mitigate the negative effects.

In addition to these effects, which are already known for other destinations, the development of sport tourism can also generate difficulties for the strategic positioning and attractiveness of the destination. Thus, the fact of banking on sport tourism will boost the image, attracting adventurers, thrill seekers or extreme sportspeople. On the contrary, this image of the destination can lead to a poor perception on the part of another part of the traditional clients who are attached to the choice of a restful, peaceful, and accessible destination without being a confirmed sportsman. There is indeed an image positioning issue here with tangible effects on customer commitment. Of course, they may be loyal to the destination. But they are not necessarily committed. They will not recommend it as they should. This is one of the difficulties encountered by those in charge of managing La Réunion's destination. It can be seen as a destination of high adventures, whereas it also wishes to position itself as a destination perfectly adapted for people in search of peace and serenity.

Finally, these risks of impacts of sports activities can increase the risk weighing on the economic model of the sports activity itself. As with the life cycle of destinations in general, products that are successful today may see their success dry up tomorrow if no efforts are made to renew them or keep them competitive. In Réunion, this is a risk that weighs on certain major events such as the ultra-trail competition, "the Grand Raid," whose benefits have already been described. However, these benefits may fade or stagnate. Indeed, after 30 years of existence, this one mobilizes the local population, but its potential is limited to the evolution of the population destined to evolve slowly in the next two decades. The external clientele is then the main lever of growth. But here again, this raises the concern of the impact on the controlled environments and natural spaces. Similarly, this competition remains confined to a niche and a milieu of mountain race specialists. Beyond that, the communication of the DMOs has difficulty in making the Grand Raid a symbol to identify Réunion Island on the map of world destinations, for all publics, following the example of the great annual sports competitions present in other countries. This clearly raises the question of the resonance of the activity and its ability to mobilize beyond the usual and initiated circles.

In spite of the numerous monetary economic impacts, the development of sport tourism activities in sensitive island territories highlights the dilemma faced by the territory and the stakeholders of territorial governance (the Government and those in charge of the Destination Management Organization). Should we promote sport tourism and develop its economic effects or should we put first the preservation of the environment in such a vulnerable territory? This dilemma often leads to contradictions between, on one hand, the will and promotion campaign desired by the governance, and, on the other hand, the territorial protection systems erected sometimes by the same local institutions and most often by the National Park or the Marine Reserve authorities.

Coping with these vulnerabilities and overcoming this dilemma are therefore crucial for the destination. This is the challenge of strengthening its resilience. In this context, the development of sport tourism has several levers that act positively on the resilience of the island destination.

Sport Tourism as a Vector of Resilience

Like sport, which is a lever for territorial development (see Bastien et al., 2022), sport tourism generates beneficial effects. But maximizing these benefits and reducing the risks implies working on the resilience of the territory and the tourism sector.

A first level of analysis is provided by the dynamics of transformation generated by the deployment of sport tourism within the destination. The example of La Réunion Island can be used again. Indeed, in the 1990s, at the dawn of the advent of sport tourism activities, it must be recognized that the means were rudimentary and the organizations not very professional. However, over the last 20 years, significant investments have been made by stakeholders to structure the activities and sectors, to equip them with international standards, to train the actors, etc. All in all, and following the example of cultural tourism, it is a real metamorphosis of an entire sector that has gone from an amateur organization to a professional operation in terms of both tools and actors' skills. In so doing, sport tourism has increased its capacity to adapt to the demands of tourist clients and local and international institutions, which can only make it more attractive and therefore more sustainable.

This capacity to adapt is also stimulated by the dimension associated with the ecological impact of sport tourism. The last 20 years have seen the emergence of numerous solutions and investments to reduce the impact of events on sensitive environments: reduction of plastic waste, mobilization of the principles of the circular economy, consideration of nuisances on sensitive sites and fauna, etc. Many challenges remain to be overcome, but it is clear that awareness and respect for standards is becoming the rule for development. This evolution is also to the credit of the public authorities and therefore of the governance attached to maintain the attractiveness of the destination of La Réunion on positions of high environmental requirements. The issue of good governance as a source of resilience is therefore crucial.

In that perspective, it is relevant to mention another benefit of sport tourism through its ability to help building bridges between remote destinations, to improve their notoriety, and to boost their brand, their attractivity, and their openness. For instance, "Island Games," created in 1979 for the Indian Ocean Islands, helped these destinations to open themselves to each other in terms of cultural and economic aspects.

A second level is linked to the fuel for strengthening this capacity to adapt. In addition to good governance, the role of which has just been highlighted, it is also the ability to mobilize the dynamics of innovation that becomes the cornerstone of change. Indeed, for the last 10 years, Réunion has been investing in all kinds of innovations, mobilizing a smart specialization strategy (S3 which is broader than the concept of smart destination) as desired by the European Union, one of the axes of which is smart tourism. This capacity for innovation strengthens the island's ability to meet the challenges of adapting to new demands, changes in customer requirements, the imposition of new standards, and the advent of new competitors. It is therefore an innovation at the service of the tourist supply and the experience of the tourist customer journey that supports this change.[3] The full potential of this dynamic has not yet been realized and invites us to consolidate the ecosystem while encouraging the players to give priority to these transformation levers through incentive mechanisms and financing that are commensurate with the challenges. Technical, managerial, and organizational innovations are therefore at the heart of the resilience of sport tourism on the islands in order to cope with ambitious goals regarding sustainable, economic and social issues and to select the best business model adapted to the destination as it is the case for the events industry.

[3] In the same vein, we can talk about the concept of the "vanilla islands" which can be seen as a cluster grouping together the islands of the Indian Ocean (Madagascar, Mauritius, La Réunion, and Seychelles). This initiative aims to both increase the visibility of these islands and also better publicize the complementarities of their offer on the World tourism market. Sports tourism, widely implemented in Réunion, is also one of the major segments put forward by this cluster.

A third level of analysis, and undoubtedly the most fruitful for strengthening the resilience of the destination, is that of the transformation of mentalities associated with the development of sport tourism in island territories that are often isolated and once cut off from the world. It is a question of taking stock of three decades of sporting events that have already shown that the destination is capable of organizing them like anywhere else in the world. It is therefore a form of pride. But this feeling of pride is also linked to the possibility of measuring ourselves against tourists and external champions. This exchange and these evaluations reduce this impression of the impossible, this vision of the difference and this feeling of frustration. On the contrary, comparing oneself stimulates the capacity to take up challenges in all human endeavors, not only social but also economic and technological compartments. In this respect, we can consider that sport tourism acts as a transfer of progress, not in terms of technology (like total factor productivity), as is the case with the exchange of goods and services, but in terms of mental attitudes, states of mind, pride, and courage, all of which are talents for creating innovative solutions and developing adaptation mechanisms.

All in all, we note that sport tourism promotes territorial development because beyond the economic and productive dimensions, it is also the behaviors and the games of actors that support the favorable impact of this sector on the resilience of the economy.

Conclusions

The aim of this chapter was to highlight the links between the development of sport tourism and the resilience of vulnerable island destinations. Our work has allowed us to recall the ambiguous role of tourism, which can be both a driving force for the development of a destination and also a stumbling block if the wrong choices are made. These risks were seen as being amplified in the context of vulnerable island territories due to their natural characteristics but also due to their isolation. An overview of the literature led to the observation that, in the face of the avalanche of work associated with the analysis of sport tourism in continental destinations, island destinations had received less attention, particularly island destinations that are largely isolated and remote, such as the French ultraperipheral regions.

Beyond the deepening of the vulnerable dimension of island territories, it is this geographical research gap that this chapter has sought to fill by mobilizing a case study: the Island of La Réunion, a French and European ultraperipheral region of the Indian Ocean. The expertise of the experience of this destination, positioned on sport tourism, showed that more than the visible productive and economic dimensions, it was also the dimensions of transformation that should be highlighted. Faced with the challenges of great vulnerabilities—ecological, climatic, economic, and social—related to these fragile and remote areas, this work has highlighted the less visible effects of sport tourism on the capacity of territorial adaptation through the dynamics of organizational and societal innovations and the transformation of attitudes conducive to strengthening the resilience of the sport tourism sector and its host area: the island territory. These elements also highlight the need to extend this work by updating and deepening quantitative evaluations of the effects of sport tourism on this type of sensitive territory, notably by mobilizing indicators of adaptive capacity (Jean-Pierre and Perrain, 2022) in order to improve the policy design dedicated to the resilience of the SIEs.

References

Achilleos, A., Makrominas, M., Markides, C., & Alexandrou, R. (2021). Promoting active sports tourism through technology and evaluating its economic impact: Experiences from Cyprus. *Journal of Sport & Tourism, 25*(3), 1–19. https://doi.org/10.1080/14775085.2021.1965009

Adamou, A., & Clerides, S. (2009). Tourism, development and growth: International evidence and lessons for Cyprus. *Cyprus Economic Policy Review, 3*(2), 3–22.

Alonso, V. L., & Medina, F. X. (2019). Impacts, counterparts and the promotion of sports tourism in the Canary Islands: The example of coastal sports in the Island of Fuerteventura. *Journal of Tourism and Heritage Research, 2*(2), 125–139. Retrieved from https://www.jthr.es/index.php/journal/article/view/45

Andersson, T. D., Armbrecht, J., & Lundberg, E. (2015). Triple impact assessments of the 2013 European athletics indoor championship in Gothenburg. *Scandinavian Journal of Hospitality and Tourism*. https://doi.org/10.1080/15022250.2015.1108863

Archer, E. D. (1984). Reassessing third world tourism: The case of Barbados. *The Tourist Review, 39*(2), 19–23. ISSN 0251-3102.

Bastien, J., Gouguet, J.-J., & Rougier, C. (2022). Sport et développement territorial en Guyane, une approche par la théorie de la base. *RERU, 4*, 613–632.

Bessy, O., & Naria, O. (2003). *Loisir et Tourisme de Nature et Sportifs à La Réunion: état des lieux, enjuex perspectives en matière de développement durable* (150p). Rapport de recherche pour le CCEE.

Bessy, O., & Naria, O. (2004). Sociologie des pratiquants de sport nature: l'exemple de La Réunion. *Cahier-Espace, en collaboration avec le Ministère de la Jeunesse, des Sports et de la Vie Associative, 81*, 52–78.

Brida, J. G., Schubert, R., & Wiston, S. (2009). The impacts of international tourism demand on economic growth of a tourism demand on economic growth of a small economy. *Tourism Management, 32*(2), 377–385.

Brida, J. G., Cortes-Jimenez, I., & Pulina, M. (2016). Has the tourism-led growth hypothesis been validated? A literature review. *Current Issues in Tourism, 19*(5), 394–430. ISSN 1368-3500. https://doi.org/10.1080/13683500.2013.868414

Briedenhann, J. (2011). Economic and tourism expectations of the 2010 FIFA World Cup: A resident perspective. *Journal of Sport & Tourism, 16*(1), 5–32.

Buckley, R. C., Guitart, D., & Shakeelah, A. (2017). Contested surf tourism resources in the Maldives. *Annals of Tourism Research, 64*(6), 185–199. https://doi.org/10.1016/j.annals.2017.03.005

Bull, C., & Weed, M. (2009). *Sports tourism as a diversification strategy in Malta, sports tourism* (1st ed.). Routledge. eBook ISBN 9780080942117.

Bulter, R. (1980). The concept of a tourist area cycle of evolution: Implications for management of ressources. *The Canadian Geographer, 24*(1), 5–12.

Carnairo, M. J., Breda, Z., & Cordeiro, C. (2016). Sports tourism development and destination sustainability: The case of the coastal area of the Aveiro region, Portugal. *Journal of Sport & Tourism, 1*, 1–27.

Cerina, F. (2006). *Tourism specialization and sustainability: A long-run policy analysis* (FEEM Working Paper 11.2006). https://doi.org/10.2139/ssrn.879705
Chang, C. L., Thanchanok, K., & McAleer, M. (2010). Estimation of a panel threshold model of tourism specialization and economic development. *Tourism Economics, 18*(1). https://doi.org/10.2139/ssrn.1583242
Chao, C. C. (2006). Tourism, Dutch disease and welfare in an open dynamic economy. *The Japanese Economic Review, 57*(4), 501–515.
Chen, L. L., & Devereux, J. (1999). Tourism and welfare in Sub-Saharan Africa: A theoretical analysis. *Journal of African Economies, 8*(2), 209–227.
Copeland, B. (1991). Tourism, welfare and de-industrialization in a small open economy. *Economica, 58*(232), 515–529.
Dwyer, L., & Chulwon, K. (2003). Destination competitiveness: Determinants and indicators. *Current Issues in Tourism, 6*(5), 369–414.
Elcock, Y. J. (2005). Sports tourism in Barbados: The development of sports facilities and special events. *Journal of Sport & Tourism, 10*(2), 129–134.
Fadda, N. (2020). Entrepreneurial behaviours and managerial approach of lifestyle entrepreneurs in surf tourism: an exploratory study. *Journal of Sport & Tourism, 24*(1), 53–77. https://doi.org/10.1080/14775085.2020.1726801
Fernández, J. I. P., & Sánchez Rivero, M. (2009). Measuring tourism sustainability: Proposal for a composite index. *Tourism Economics, 15*, 277–296.
Fredline, E. (2005). Host and guest relations and sport tourism. *Sport in Society, 8*(2), 263–279. https://doi.org/10.1080/17430430500087328
Gammon, S., & Robinson, T. (2003). Sport and tourism: A conceptual framework. *Journal of Sport & Tourism, 8*(1), 21–26.
Gibson, H. J., Kaplanidou, K., & Kang, S. J. (2012). Small-scale event sport tourism: A case study in sustainable tourism. *Sport Management Review, 15*(2), 160–170.
Gómez, C. M., Lozano, J., & Rey-Maquieira, J. (2008). Environmental policy and long-term welfare in a tourism economy. *Spanish Economic Review, 10*(1), 41–62. ISSN 1435-5477.
Guiltat S. (2011) Le rôle des sports côtiers dans le développement territorial de l'Ouest de La Réunion : entre nouvelle maritimité et jeux d'acteurs, Thèse de doctorat en STAPS.
Guiltat, S., & Taglioni, F. (2015). Le risque d'attaques de requins à La Réunion: Éléments d'analyse des attaques et contextualisation d'une gestion contestée. *EchoGéo*. https://doi.org/10.4000/echogeo.14205
Hazari, B., Nowak, J.-J., Sahli, M., & Zdravevski, D. (2003). Tourism and regional immiserization. *Pacific Economic Review, 8*(3), 269–278.
Holzner, M. (2011). Tourism and economic development: The beach disease? *Tourism Management, 32*(4), 922–933. ISSN 0261-5177. https://doi.org/10.1016/j.tourman.2010.08.007
Jean-Pierre, P. (1997). Sélection et tests de seuils de convergence. *Revue Economique*, 429–440. ISSN 0035-2764.
Jean-Pierre, P., & Perrain, D. (2021). *Tourism threshold on small islands economies* (Working Paper). Cemoi.
Jean-Pierre, P., & Perrain, D. (2022). *Adaptive capacity and resilience of destination, analysis and measurement with a tourism adaptive capacity index (TACI)* (Working Paper). Cemoi.
Kartakoulis, N., & Karlis, G. (2002). Developing Cyprus as a sport tourism destination: The results of a swot analysis. *Journal of Sport & Tourism, 7*(4), 3–17. https://doi.org/10.1080/10295390208718740
Klöck, C., & Fink, M. (Eds.). (2019). *Dealing with climate change on small islands: Towards effective and sustainable adaptation*. Springer.
Kožić, I., & Mikulić, J. (2014). Research note: Measuring tourism sustainability: An empirical comparison of different weighting procedures used in modelling composite indicators. *Tourism Economics, 20*(2), 429–437. https://doi.org/10.5367/te.2013.028
Kurtzman, J. (1995). Sports tourism categories revisited. *Journal of Sport & Tourism, 2*(3), 6–11.
Kurtzman, J., & Zauhar, J. (1995). Agency reports. Tourism sports international council. *Annals of Tourism Research, 22*(3), 707–708.

Martel, L., & Rey, T. (2012). *Les sports en Corse: Miroir d'une société* (Edition Albiana) (361p).

Maupertuis, M.-A., & Giannoni, S. (2005). *Environmental quality and long run tourism development a cyclical perspective for small island tourist economies*. Fondazione Eni Enrico Mattei (FEEM).

Mazza and Sobry (Eds) (2022). *Sport Tourism and Its Territorial Development and Opportunities*, Cambridge Scholars Publishing, 222 p.

McElroy, L. (2006). Small island tourist economies across the life cycle. *Asia Pacific Viewpoint, 47*(1), 61–77.

McElroy, J. L., & Parry, C. E. (2010). The Characteristics of Small Island Tourist Economies. *Tourism and Hospitality Research, 10*(4), 315–328. https://doi.org/10.1057/thr.2010.11

Melo, R., Sobry, C., & Van Rheenen, D. (2021). *Small scale sport tourism events and local sustainable development: A cross-national comparative perspective*. Springer.

Naria, O. (2010). Les activités de tourisme sportif et de nature comme outil de développement local: le cas de l'île de La Réunion. In *Comme un Parfum d'Iles* (pp. 251–259). PUPS.

Naria, O., & Scherwin, M.. (2013). *Tourisme, sports de nature et développement durable aux Seychelles*. https://doi.org/10.4000/etudescaribeennes.5642

Naria, O., Van Rheenen, D., Melo, R., & Sobry, C. (2022). Sport tourism, island territories and sustainable development: Approach to a methodological framework for research based on the insular context of the Indian Ocean. In B. Mazza & C. Sobry (Eds.), *Sport tourism and its territorial development and opportunities* (pp. 19–40). Cambridge Scholars Publishing.

Newland, B., Aicher, T. J., Davies, M., & Hungenberg, E. (2021). Sport event ecotourism: Sustainability of trail racing events in US National Parks. *Journal of Sport & Tourism, 1*, 1–31.

Nowak, J.-J., Sahli, M., & Cortés-Jiménez, I. (2007). Tourism, capital good imports and economic growth: Theory and evidence for Spain. *Tourism Economics, 13*(4), 515–536.

Nowak, J. J., Sahli, M., & Sgro, P. M. (2003). Tourism, trade and domestic welfare. *Pacific economic review, 8*(3), 245–258.

Observatoire du Tourisme de La Réunion. (2015). *Study with the Grand Raid association, 28 avril-28 mai 2015.*

Orams, M. (2005). Introduction to sport tourism impacts and environments. In J. Higham (Ed.), *Sport tourism destinations: Issues, opportunities and analysis* (pp. 248–259). Elsevier Butterworth-Heinemann.

Pablo-Romero, M. D. P., & Molina, J. A. (2013). Tourism and economic growth: A review of empirical literature. *Tourism Management Perspectives, 8*, 28–41. ISSN 2211-9736. https://doi.org/10.1016/j.tmp.2013.05.006

Po, W. C., & Bwo-Nung, H. (2008). Tourism development and economic growth, a nonlinear approach. *Physica A: Statistical Mechanics and its Applications, 387*(22), 5535–5542.

Schubert, S. F., & Brida, J. G.. (2008). *The dynamic effects of subsidizing the tourism sector* (MPRA paper, p. 16755).

Séraphin, H., & Gowreesunka, V. G. B. (2020). Unlocking changes for sport tourism products using the Blakeley-Visser model: An application to sports events in small islands: The case of Martinique. In *Post-disaster and post-conflict tourism* (1st ed.). Apple Academic Press. eBook ISBN 9780429276224.

SYPRAL. (2014). *Les loisirs de nature dans le secteur du tourisme, Propositions d'actions 2015–2020.*

Towner, N. (2016). How to manage the perfect wave: Surfing tourism management in the Mentawai Islands, Indonesia. *Ocean and Coastal Management, 119*, 217–222. https://doi.org/10.1016/j.ocecoaman.2015.10.016

Wang, Y. (2011). Research note: threshold effects on development of tourism and economic growth. *Tourism Economics, 18*, 1135–1141. https://doi.org/10.5367/te.2012.0160

Wilkinson, P. (2004). Caribbean tourism policy and planning. In D. Tual (Ed.), *Tourism in the Caribbean: Trends, development, prospects*. Routledge. ISBN 978-0-203-40269-6.

World Bank (2006). *The International Bank for Reconstruction and Development / The World Bank 1818 H Street*, NW Washington, DC 20433.

Chapter 21
Mauritius as a Sustainable Island Destination: Opportunities, Implications, and Challenges for Sport Tourism

T. Makoondlall-Chadee, P. Ramasamy-Coolen, Olivier Naria, and S. Bokhoree

Introduction

Tourism is a dynamic industry and has shown its economic, socio-cultural, and environmental potential over the past decades. Mauritius is a successful case example that evolved from a mono-crop economy to become a well-known island destination today, offering a panoply of tourism activities. Strategically well-situated in the Indian Ocean, Mauritius offers modern seaports and air connectivity (Ministry of Tourism, 2018).

Though tourism is considered a positive economic sector, it remains a sensitive and vulnerable sector and its damaging consequences must be addressed. Therefore, the idea of sustainable tourism arose globally to limit the negative impacts of tourism activities, which has become a desirable and politically suitable approach to tourism development. Accordingly, the tourism industry in Mauritius has witnessed an increasing shift toward adopting sustainable practices. Recently, through its Tourism Strategic Plan 2018–2021, Mauritius has worked toward becoming a "*leading and sustainable island destination.*"

Over time, the tourism products in Mauritius have innovatively extended to now include medical tourism, wellness and spa, MICE-weddings, shopping tourism,

T. Makoondlall-Chadee (✉) · P. Ramasamy-Coolen · S. Bokhoree
Department of Environment, Society and Social Sustainability, School of Sustainable Development and Tourism, University of Technology Mauritius (UTM), Port Louis, Mauritius
e-mail: tmchadee@utm.ac.mu; vrcoolen@utm.ac.mu; sbokhoree@utm.ac.mu

O. Naria
Department STAPS, Université de La Réunion, Le Tampon, île de la Réunion, Reunion, France
e-mail: olivier.naria@univ-reunion.fr

D. Van Rheenen et al. (eds.), *Sport Tourism, Island Territories and Sustainable Development*, Sports Economics, Management and Policy 24,
https://doi.org/10.1007/978-3-031-51705-1_21

heritage and cultural tourism, and sport tourism (e.g., kite-surf, golf, cycling, and trail) as major areas of development. Sport tourism is known to be part of island states, and Mauritius has seized this growth to further develop its sport tourism products and services to understand individuals, communities, and groups while positively supporting the local economic and environmental spheres. Among the land-based activities, Mauritius commonly offers nature trails, golfing, horse riding, and others. While maximizing the tropical climate, the island also offers a variety of water sports ranging from water skiing, stand-up paddles, and deep sea fishing, among many other activities.

Mauritius has hosted several international sports events and competitions such as the Ferney Trail, the Surf-ski Week, and the AfrAsia Bank Mauritius Open Golf tournament, among others. With the diversification of the local tourism product, Mauritius has been able to invest in and offer some well-sought-after air-based sports activities practiced both among tourists and locals, varying from activities such as paragliding, parachuting, aerobatics, and skydiving, among others.

This chapter consequently sets out to thoroughly understand the relevance and contributions of sport tourism for the local economy and further provides insights into the evolutionary journey of sport tourism in Mauritius. The chapter investigates the emerging economic, socio-cultural, and environmental opportunities and challenges arising from sport tourism in addition to the core sport tourism domain, ranging from hallmark events and outdoor recreation, among others. Furthermore, the work also references the socio-economic and cultural contributions that sport tourism can make toward meeting the Sustainable Development Goals (SDGs), namely, SDGs 1, 3, and 8, among others (World Tourism Organization, 2019).

Objectively a critical literature review was completed, whereby sport tourism as part of the Mauritian tourism product was the focal point. Accordingly, while adopting a secondary data analysis, local action plans, strategies, and research addressing sport tourism evolutions and contributions for Mauritius were studied. Through empirical observations, the chapter further reports and summarizes the specific characteristics of sport tourism in Mauritius relative to the socio-economic and environmental contributions.

The Republic of Mauritius

The Republic of Mauritius is a Small Island State located off the east coast of the African continent. The country consists of the main Island of Mauritius, the autonomous island of Rodrigues, the outer islands of Tromelin and Agalega, and the archipelagos of St. Brandon and Chagos, an Exclusive Economic Zone.

Geographical Context

Geographically, the main island is located on the Mascarene Plateau, an elevated underwater ridge that used to harbor volcanic activity, though today Mauritius only has dormant volcanoes (Dunn, 2009). It measures 61 km in length and 46 km in width and occupies a total land area of 1865 km^2. The island is ringed by coral reefs and features sandy beaches, as well as rich plains that rise dramatically from the sea (Saddul, 1995). The northern section is a plain that rises to a central plateau, with elevations ranging from 270 to 730 m above sea level (Nigel, 2010). The plateau is surrounded by minor mountains that may have constituted the edge of an ancient volcano. Mauritius as an island enjoys a tropical maritime climate (Fig. 21.1).

Economic and Socio-cultural Context

Historically, the island of Mauritius has been a colony of the United Kingdom and France. It was also visited by sailors from different parts of the world. Mauritius became independent in 1968 and was proclaimed a Republic in 1992. It is worth noting that the island had no indigenous inhabitants, and despite its proximity to large land masses, it had no native mammalian population other than bats when the first European colonists arrived (Mauritius Facts, Geography, & History, 2022). To

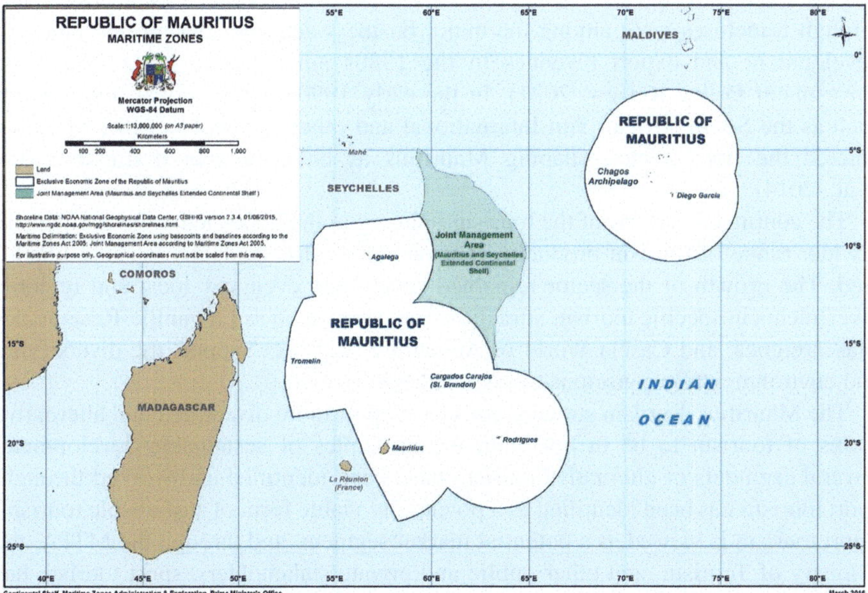

Fig. 21.1 Territorial Island Typologies (Mauritius). (Source: Republic of Mauritius)

date, the island has a total population of 1.3 million, and Mauritius is defined by its constitution as a secular country.

Due to its political and social stability, Mauritius can focus on economic investments and related infrastructural development in various sectors including tourism. In addition to the economic contributions, Oladeji I. Kolawole et al. (2017) argues that some positive contributions of tourism range from "*positive socio-cultural impacts on the community people's quality of life, positive cultural identity and community pride and encouraged local involvement in tourism enterprises*" *to name a few*. The multi-ethnicity of Mauritius has been among the cultural and sociological factors that have added to the local tourism experience over and above the development and diversification of the local tourism product. Tourism experiences are often enhanced with the support of the local population, as it fosters emotional solidarity and sympathy (Hasani et al., 2016).

Tourism Development

Tourism emerged as an industry in Mauritius when there was a need to diversify from the mono-crop and traditional sectors such as sugarcane plantation and manufacturing (Industrial Policy and Strategic Plan, 2020–2025). The developments in tourism from the 1980s onward were financed mainly from domestic capital being injected from the sugarcane industry (Tandrayen-Ragoobur & Kasseeah, 2018). The government of Mauritius played a major role in the marketing of tourism at that time and also invested in a national carrier. Car rental companies and restaurants through franchises were among the major business activities during that phase of development and export revenues in the 1980s amounted to USD 42 million (Investment Policy Review, 2021). In the early 1990s, international hotel chains such as the South African Sun International and other stakeholders like Meridien entered the hotel sector, shaping Mauritius as an up-market destination (Ian et al., 2014).

The continued success of the tourism industry in the 2000s led to investments in a wider range of tourism products. Tour operators also gradually became liberalized. The growth of the sector remained steady and even saw local and regional investments in specific tourism attractions and sites, such as La Vanille-Reserve des Mascareignes, and Casela World of Adventure, to better support the diversifying and environmentally conscious tourism market.

The Mauritius Tourism strategy aims to support more diversified and alternative forms of tourism to be in line with the principles of sustainable development. Several segments of alternative tourism have been identified and worked through; sport tourism has been identified as a potentially viable form of sustainable tourism. Sport tourism is viewed as a potential market segment, and through the MTPA, the Ministry of Tourism, and other public and private stakeholders, sport tourism has expanded as an alternative form of sustainable tourism opportunity for the local economy.

Socio-economic Contributions of Tourism

The economic contributions of the tourism industry were constant from the end of the 1970s to 2019. The evolution of the tourism industry in Mauritius was gradual throughout the late 1980s and through the 1990s and "*international tourism receipts grew steadily from as low as US$9.7 million in the 1970s to US$2.16 billion in 2018, constituting 34 percent of the country's total exports,*" making tourism one of the most important economic drivers (UNWTO, 2020). The value added from the tourism sector has not only positively influenced investments and contributed to the local GDP, but it has also contributed to reducing unemployment on the island. Statistics Mauritius (AHRIM, 2019) confirms that as of 2019, employment within the Mauritian tourism industry accounted for 12.8% of total employment in Mauritius.

The UNWTO (2015) reports that despite the economic contributions of tourism, there needs to be a degree of caution when it comes to the design, planning, and implementation of tourism-related activities, products, and services, more so for small island developing states (SIDS) due to their multiple vulnerable characteristics. The financial contributions of the tourism sector will never outweigh its potentially negative economic, socio-cultural, and environmental impacts. Stainton (2019) asserts that some common negative effects of tourism are economic leakages, seasonal jobs, increased inflation, and infrastructural costs, among other deleterious effects.

Faced with real challenges, such as "connectivity, scarce and fragile natural resources, economic leakages, and climate change," as reported by UNWTO (2015), SIDS cannot carry on the business as usual scenario. That is, the development of alternative forms of tourism is no longer a choice but a priority. Accordingly, the government of Mauritius has over the past years devised strategies for sustainable development in tourism, as well as adopted several strategies toward meeting and successfully implementing the goals of sustainable development while simultaneously meeting the Sustainable Development Goals (SDGs) (WTO, 2019).

Environmental Contributions of Tourism

Proper planning, management, and monitoring of tourism activities are needed in island destinations like Mauritius. Accordingly, such planning allows for better management of scarce resources, making optimum use of these resources, while responding to travelers' needs. Understanding the potential environmental impacts of tourism activities allows for the assessment of the needs and effects arising from land, water/marine, and air-based activities. The tourism sector's interaction with the environment is complicated, and many activities can have negative environmental consequences if proper planning and management are not pursued with intention (Stainton, 2019). Uncontrolled tourism-related activities and services threaten numerous natural regions and other limited resources. Overuse or mismanagement of resources that occur due to tourism activities may lead to negative environmental impacts, such as soil erosion, increased pollution, discharges into the sea, natural habitat loss, greater pressure on endangered species, and increased vulnerability to forest fires. With increased pressure on limited natural resources, other social disturbances may likewise arise (Stainton, 2019). Despite limited positive environmental impacts, it is worth noting that awareness and sensitization about the protection of biodiversity may happen through tourism.

Sport Tourism Activities

Organization of the Sport Tourism Offer

Tourism-related products and services are known to play a significant role in the development and growth of Island States, as well as larger countries worldwide (Roche et al., 2013). The number of international sport tourists has witnessed a growth due to the popularity of professional sports (Malchrowicz-Mosko & Munsters, 2018). Mauritius has noticed an increase in its visitor numbers, as it has expanded its horizon on travel and tourism by exploring different aspects of sport tourism (Ministry of Tourism, 2019). Historically, diving and horse racing have been the two most popular sports activities on the island. Horse racing, for example, dates to 1800 when the Champ De Mars racecourse was opened to the public. Mauritius even holds the oldest racecourse in the southern hemisphere and the world's second-oldest hippodrome.

As an island state, Mauritius also proposes a range of water/marine-based sports activities. For instance, the Mauritius Scuba Diving Association reports that Mauritius has over 30 dive registered centers throughout the island. Many of these diving centers are PADI (Professional Association of Diving Instructors) and CMAS (Confédération Mondiale des Activités Subaquatiques) certified, providing a quality and safe experience to both locals and visitors. Through the hotels and resorts around the island, most tourists visit these diving centers which offer experienced divers, safe and sophisticated equipment, and qualified English-speaking instructors.

Mauritius offers many other marine/water-based sports activities, as well as other land and air-based sports activities both for locals and tourists. These range from athletics, badminton, basketball, boxing, taekwondo, volleyball, weightlifting, golf tournaments to trails, and many more. Despite being an island destination, Mauritius has been able to align itself to offer some air-based sports activities. Though limited in variety, the current air-based activities range from tandem skydiving, helicopter rides, microlight scenic flights, seaplane rides, paragliding, and parasailing (Mauritius Attractions, 2022), which are often practiced by tourists.

Additionally, to ensure the continued success of the sports and tourism industry, the MTPA launched its "Mauritius 365" campaign in 2016. The campaign is designed to promote Mauritius year-round as a destination with a variety of activities. These varied market segments aim to boost tourism in Mauritius during the low season. Some of the proposed activities include events such as the Mauritius Open Golf Event, Rugby 10 competition, international chef event, and cycling events.

Baldacchino (2012) notes that given the specific characteristics and assets of island states, the importance of understanding and offering the island's sociocultural, experiential, mystical, and eco-biological spheres is important. According to Kurtzman (2005), sports activities can be planned and categorized into five main pillars, namely:

1. Sport tourism events, ranging from golf championship tournaments, trails, and hiking events, as well as scuba diving events.
2. Sport tourism attractions refer to the attractions and activities such as sports museums, amusement parks, and skiing. For instance, these include visits to the Photography Museum in Port-Louis which has many photos and information on outdoor activities over time. Sports attractions in Mauritius range from natural assets such as the Le Morne Kitesurfing beach and other man-made attractions, such as the amusement park at Belle Mare. Additionally, the Splash and Fun Leisure parks are common local sports attractions.
3. Sport tourism tours enhance all outdoor activities such as the Superbowl, which can be found in commercial buildings at various locations around Mauritius, such as Bagatelle and Super U—Grand Bay.
4. Sport tourism resorts may include golf resorts, ski resorts, and spa resorts that enhance tourist experiences related to sports activities. Common resorts that offer such activities in Mauritius belong to large hotel groups, such as Constance Group of hotels, Beachcomber Hotels, and Lux Resorts, among others.
5. Sport tourism cruises include the activities of yachting charters, as well as health, and fitness cruises. Yacht charters are primarily available in the region of Grand-Baie, Flic-en-Flac, Tamarin, and the East and South-East regions of Mauritius.

Hallmark Events

To develop sustainable sport tourism and further meet the sustainable development goals, sports policies, plans, and strategies have been designed with the aim of making unique use of the resources and reshaping the future of the tourism industry (Ministry of Tourism, 2019). A strategic outlook has been put forward and the local government has opted to focus on hosting hallmark events in the sporting field, with a specific purpose in mind. For instance, such events were expected to increase the visibility, attractiveness, and business opportunities of the local venue in the short and long term. Furthermore, through such events being held in Mauritius, the island can refocus on the importance of traditional customs and ameliorate the international public image as an island destination (Clark & Misener, 2015). Some examples offered in Mauritius under such events range from: golf events such as the SMB

Golf Open and the MCB Tour Championship, which have successfully attracted 80,000 golfers in 2018, an important achievement (Hamuth, 2019).

Outdoor Recreation

Offering outdoor recreation has always been common among island developing states, as well as other tourism destinations. Islands have made outdoor recreation a core component of their tourism product, as they focus on maximizing nature-based activities (Melly & Hanrahan, 2020). Given the specific geographical locations, topologies, and natural assets, nature-based products and activities can take varying forms while they can also ensure environmental conservation and preservation (Marasinghe et al., 2021). Within the African region, Mauritius is a reference destination and is reputed to have developed and enhanced its variety of outdoor activities (Desai et al., 2014). The varied water sports-related activities, such as kayaks, parasailing, kite surfing, pedal boats, and diving, among others, are known to have boosted the level of the sports industry in Mauritius (Mymauritius.travel). Blessed with crystal clear waters, coral-sea lagoons, and beautiful beaches, the island offers experiences such as breath-taking snorkeling and diving. Audley Travel (2020) reports that the activities offered in Mauritius over the past decade have positively contributed to attracting sports professionals and visitors wishing to experience the best adventures that sport tourism in Mauritius has to offer.

As Outdoor Recreation encompasses human adventures, the Mauritius Tourism Promotion Authority (MTPA, 2022) has collaborated with various partners to promote sport tourism, capitalizing on the greenery, scenery, and vibrancy of the island. Trails have been a very successful segment of outdoor recreation activities, having added value to the tourism product (Ministry of Tourism, 2022). During trails and other nature-based activities, the sumptuous landscapes and exceptional routes are discovered by hikers, who become acquainted with the local biodiversity. Trails such as the Standard Bank Royal Raid and Ultra-Trail Raid Light Beachcomber (UTRB), where both international and local runners participate, have been highly successful and reputed given the international publicity offered by the organizers. They have added increased value and recognition to the sport tourism industry in Mauritius. Other equally successful international trails include the Dodo trail, offering three tracks to explore the southern and western coasts of the island; the Ciel Ferney Trail which is held annually at La Vallée de Ferney, allowing all participants of the competition to reconnect with nature and explore the native forests of Mauritius (About CIEL Ferney Trail—Ferney, 2022); the Kestrel Trail; and Heritage Trail Festival. All of the above-mentioned activities have contributed massively to the success of reuniting visitors around the globe to participate and share the awareness of different dimensions of sustainability, green tourism, and sociocultural tourism (MTPA, 2022). Other outdoor activities that are of interest and currently being marketed as part of the tourism product are golf tournaments and cricket competitions, to name but a few.

Among the various variables that constitute sustainable sport tourism products and activities, it is important to refer to what researchers describe as entertainment. Loi (2009) refers to entertainment as a form of activity added to the traditional tourism product to look more appealing. Furthermore, it allows for greater spending, whereby tourists are eager to spend on entertainment to have the feeling of satisfaction, excitement, and fun while indulging in tourism activities. Island states like Mauritius have very specific characteristics and should focus on offering sports activities as defined by Antar-Ecotono (2004), who focuses more heavily on "active tourism." Antar-Ecotono (2004) states that "active tourism is defined as travels inspired by participating in sporting activities of various physical rigors while utilizing natural resources without depleting them." Active tourism is forecasted to be valued at 7 billion euros annually by 2023, as the appeal for a more health-conscious lifestyle is promoted among consumers interested in active sports (Karantzavelou, 2022), such as swimming, tennis, running, and cycling.

Tourism, sports, and entertainment are interlinked within the tourism product, as they are often combined to offer attractive packages and hence increase tourism arrivals to small island states. SIDS are, however, often challenged with their specific typologies, characteristics, and vulnerabilities, and therefore, Mauritius offers its tourism and sports activities with caution. It matches its vision with what Kirillova et al. (2014) describe as "the environmental qualities of a destination impact on overall tourist experience and re-visitation." To this effect, the main types of sport tourism activities being offered in Mauritius range, namely, from land-based activities, marine-based activities, and air-based activities.

Land-Based Activities

Land-based activities are very common in island developing states, given their specific typologies and characteristics. Given the volcanic origins of Mauritius Island, for instance, very distinct land-based activities are practiced and promoted as sport tourism activities. National parks are among the most common features used for land-based activities.

Water- and Marine-Based Activities

The Republic of Mauritius is a well-reputed holiday destination. Statistics Mauritius (2019) reports that a high percentage of visiting tourists come to the island to discover the beach, related beach entertainments, and of course, water-based/marine sports activities. When interviewed about their purpose of visit to Mauritius, it was very common to find the following statements among Europeans and South African tourists, "*We are only here for the beach and water sports." There is no sea like in Mauritius anywhere else." "All I want to do in Mauritius is deep sea fishing, snorkeling, and scuba diving. If I have time, I look at other things.*" Mauritius offers a variety of water-based activities, ranging from windsurfing, fishing, rafting, swimming, diving, snorkeling, underwater scuba-diving activities, whale watching, kayaking, cruising, yachting, speed-boating, and lake activities.

Li et al. (2011) describe entertainment as one of the resources that may affect a visiting tourist's choice when opting for a destination. They further add that entertainment is the wealth and the diversity, geography, topology, and natural features of locations contribute to the local tourism industry's success. Mauritius has the advantage of having beautiful sites throughout the island; some common beaches are Pereybere, La Cuvette, Cap Malheureux, Belle Mare, Flic en Flac, Albion, Tamarin, Blue Bay, La Cambuse, and Grand Baie, to name a few that offer most of the water/marine-based activities either through hotel organizations or private-owned businesses.

Air-Based Activities

Air sports offer a vast range of activities and are often described as "Extreme sports." The main sports activities offered in the air vary from activities such as paragliding, parachuting, aerobatics, and skydiving among others. Given the level of security, infrastructural investments required, and expertise, not all destinations may offer the same types of air sports activities.

The above-mentioned land-based, water/marine-based, and air activities are accessible around the island. To further provide a better insight into their potential location, together with other common activities available around the island, this work has considered mapping the available activities. The map below has been designed to illustrate some of these activities, as well as other natural features, protected areas, and major infrastructures (Fig. 21.2).

Governance of Sport Tourism

The government of Mauritius has adopted these guiding principles in designing, planning, and offering sport tourism to ensure that the sports activities and facilities respect the three spheres of sustainable development (Tomislav, 2018). In the tourism sector, the concept of sustainable development has been used as a guiding principle to create a variety of tourism products (Ruhanen, 2013). Sport tourism in Mauritius has been considered a sustainable alternative form of tourism, given the close and significant relationship between sustainable development and tourism (Pan et al., 2018).

Local policies and strategies devised for sport tourism in Mauritius have been designed to offer the three major sports categories as defined by Bjeljac et al. (2017); major sporting events; medium competitions at the regional and national levels; and minor competitions at the local level. Coordinated efforts and planning are favored, as opposed to informal planning measures that exacerbate existing issues rather than promote the destination's sustainable growth (Amir et al., 2015). The success of

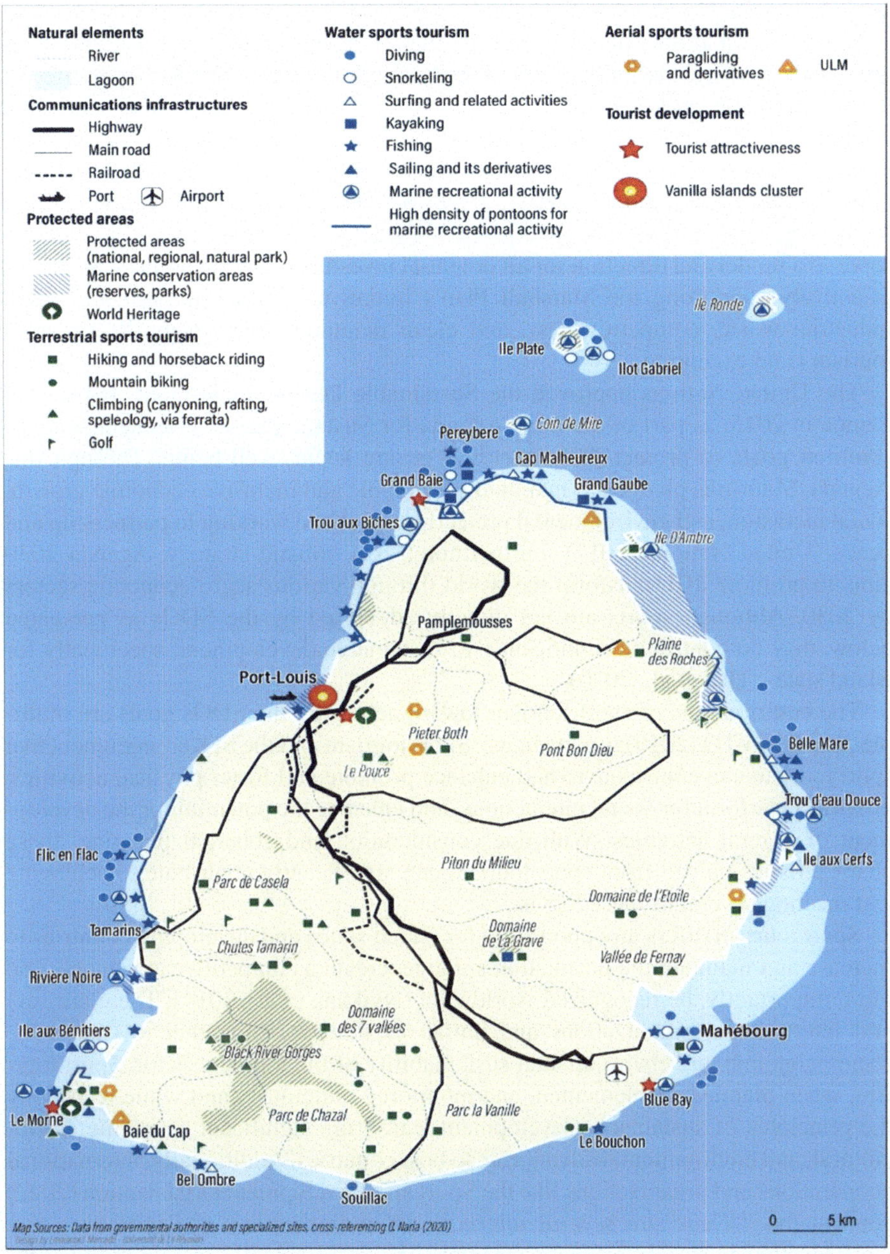

Fig. 21.2 Map of Mauritius designed to locate some of the mentioned activities as well as other natural features, protected areas and major infrastructures

sport tourism in Mauritius will heavily depend on the participation of all concerned stakeholders, the proper planning and conduct of the events and activities, and stringent analysis of related impacts (Hall, 2012; Vujko & Gajic, 2014). To ensure the long-term sustainability of sporting events and activities, the management and arrangement of activities should align with the new national program of the government of Mauritius. The UN reports that the recently unveiled 2020–2024 national plan, entitled "*Toward an Inclusive, High Income and Green Mauritius, Forging Ahead Together*" outlines the government's goals for the next 4 years. The plan will serve as a model and direction for all potential investors and stakeholders, as it aims at actively supporting the Marshall Plan's initiatives on investments, job skills, industrial waste, competitiveness, and clean manufacturing (Page, 2021); sport tourism is no exception.

The United Nations approved the Sustainable Development Goals (SDGs) in Samoa in 2015, as part of the 2030 Agenda for Sustainable Development, a set of common goals to protect the planet and ensure social well-being. Through this Agenda, Mauritius pledged to maintain sustainable and inclusive economic growth, social inclusion, and environmental protection, all while working in partnership and peace (United Nations, 2017). Furthermore, as a holistic strategy, Agenda 2030 aims to promote and transform the world through various socio-economic sectors by 2030. Although sports are not directly addressed by the SDGs as presented below, they do contribute significantly to sustainable development, specifically for island states (Hsu et al., 2020).

The contributions of sport tourism toward achieving the SDGs goals are multiple. The UNWTO (2019) reports in its "Sport tourism and the SDGs" document that sport tourism can contribute to and enhance promote and foster physical activities, provide a platform for social interactions, and enhance the potential for the development of several activities. With due consideration and coherent planning, these characteristics gained from sport tourism can significantly contribute to achieving and meeting several of these SDGs.

Sport tourism offers an opportunity for island states and developing countries to increase and maintain the destination's image, create a competitive advantage, and more importantly, be recognized worldwide (Nyikana et al., 2014). There are several international organizations and initiatives that work together to ensure the socio-economic and environmental sustainability of the tourism sector. Sport tourism, when planned, implemented, and monitored judiciously and while respecting the principles of sustainable development, can bring significant economic, socio-cultural, and environmental advantages to host countries. Additionally, international cooperations and organizations like the Sport and Development Organization (2022) confirm that "*There are several international organizations and initiatives that work together to ensure the socio economic and environmental sustainability of the tourism sector.*" Additionally, the same organization reiterates that the tourism sector is a decisive pillar that can contribute to the 2030 Agenda for sustainable development, as well as meeting the Sustainable Development Goals (SDGs).

The UNWTO (2020) has posited that sport tourism contributes to meeting sustainable development goals, as it is designed and offered through very distinct characteristics. Sport tourism has hence gained significant recognition confirming that it can respond to the three spheres of sustainable development. Sport tourism contributes to economic growth as it encourages investments. It has been reported that via tourism activities, Goal 8—"*Decent work and economic growth*"—for Mauritius is confirmed as this sector only provided 14% of the total employment. SDG 1—"*No poverty*"—again this particular goal is achievable and the economic contributions arising from local employment such as tourist guides, local entrepreneurs such as food/fruit sellers, and handicraft goods are some examples of how Mauritius is working toward this goal through the tourism sector's contributions. This, in turn, stimulates the local economic activities of the destination (Duignan et al., 2020; Gursoy et al., 2017; Melo et al., 2021). Sports events also have the opportunity to boost the local tourism economy and the service industry for months, even years, before and after any sports event, thereby supporting the goals of sustainable development. The horse racing events in Mauritius confirm the possible contributions of sport tourism to sustainable development as they generated Rs 6.2 billion in GDP in 2018, supported 2170 employees with an annual labor income of Rs 2 billion, and brought in Rs 679M for the government (Statistics Mauritius, 2020).

Furthermore, sports meet the social criteria of sustainable development, as it allows and encourages tourists and residents to meet and engage in social exchanges through modern sports facilities and improved infrastructure. Burnett (2019) argues that sport tourism can increase community solidarity and pride. Furthermore, sport tourism-related activities often led and sponsored by local municipal or district councils around Mauritius often provide opportunities to engage with people from different cultures and are a source of education, social exchange, fun, leisure, and recreation for both locals and tourists (Gursoy et al., 2011, 2017). Ministries and hotel providers, as well as private and non-governmental organizations in Mauritius, often regroup and host regional sea-based and land-based national tournaments such as cycling and trail competitions as a means to foster socio-cultural cohesion.

These contributions from sport tourism confirm the positive contributions of this sector toward meeting other SDGs, such as SDG 3, "*good health and well-being,*" reinforced by the commitment of the local Ministry of Youth Empowerment, Sports, and Recreation to promote Mauritius as an "active sports island." The initiative of the Ministry is to encourage tourists to participate in sports activities, either actively or passively to participate in sports events, whether land-, air-, and water-based activities. Moreover, to align with the promotion of sport tourism, the National Physical Activity policy focuses on sports and tourism activities that will help support and achieve sustainable development goals. The national sports and physical activity policy, as stated by the Ministry of Youth Empowerment, Sports, and Recreation (2022), has as its objective to enhance economic development, reduce unemployment, encourage the practice of sports activities, and enhance community well-being.

Sport tourism activities additionally foster social harmony and harmony with the natural and man-made environment (Hsu et al., 2020). Sport tourism has been researched as having the potential to contribute to environmental sustainability as well. Sport tourism encourages locals and tourists to pay more attention to environmental protection and conservation (Liu et al., 2011). The local ecology is further sustained through the proper use and management of natural parks and recreational spaces. Often, roads and public amenities are preserved in good condition, while natural resources are conserved through sport tourism (Masa'deh et al., 2017).

Socio-economic Impacts of Sport Tourism

The contribution of tourism to economic well-being depends on the quality and the revenues of the tourism offer (Makoondlall-Chadee et al., 2017). The government of Mauritius recognizes the economic potential of sport tourism and its related socio-cultural benefits; it has therefore invested massively in infrastructural developments and local athletes. For example, financial assistance is provided to athletes competing at high levels to further encourage them and positively promote sport tourism. The SDGs table sums up the monthly amount provided as financial support to high-level athletes per sports category. The government of Mauritius has provided incentives for developmental projects in the accommodation and catering sectors, and the Ministry of Sports, Youth Empowerment, and Recreation has also announced several regional projects to promote and enhance the contributions of sports and sport tourism, such as the recent construction of the Cote D' Or multi-sports complex. Several investments have been earmarked for enhancing sport tourism infrastructure in collaboration with the private sector, such as leading hotel groups like the Beachcomber and the Constance hotels, among others.

Sports events, namely, mega events, can create positive economic and socio-cultural impacts, as it enhances a destination's image (Getz & Page, 2016). Locals take pride when mega events occur in their area and such events also help target developing tourism markets. For instance, mega events such as the *international Mauritius Ocean Classic-Surf Ski Week (2008), the Lux Mauritius Marathon (2009), the Freeride Attitude Challenge Surf (2012), or more recently the Mauritius Tour Beachcomber in surfing and the Shangri-La Golf Le Tousserok Mauritius (2015), to name a few,* contribute to the island's enhanced image via these recognized sports events. The inter-relatedness of economic and socio-cultural benefits of sport tourism cannot be overlooked (Mair & Smith, 2021).

Sports activities are generally known to enhance well-being and help communities and societies enjoy health benefits and increase their personal development while enhancing social cohesion. Taks et al. (2015) describes the social impacts of sport tourism as "*the way that an attitude or manner is changing because a sporting event is being held in a specific country*." Sport tourism positively has a dissuasive effect on youth; through their interest in sports, they can avoid becoming the prey of social evils such as alcoholism, prostitution, gambling, and drugs. Hiking activities and trails organized throughout the island of Mauritius, such as the Dodo Ferney Trail, attract the youngsters who spend their free time in training.

The MTPA is also working toward the promotion of more green and sport tourism as it develops and markets more trails and adventure-based activities. The policies and strategies implemented by local authorities aim to attract and engage both locals and visitors in the practice of sports. This positive behavioral change in residents will lead to a more satisfied and better-off society, thus creating a more welcoming environment for visitors (Uysal et al., 2020).

Environmental Impacts of Sport Tourism

Sport tourism has become more popular in many tourist destinations, but especially in small island developing states like Mauritius. The related socio-economic positive and negative impacts have been addressed; as such, it is important to identify and understand the related environmental positive and negative impacts. As Beeharry et al. (2014) state, "*Stakeholders are those who know exactly what prevails in for instance coastal zones and know exactly what will the local population benefit from.*" To ensure the environmental sustainability of sport tourism, it is crucial to establish environmental enablers and variables. Beeharry et al. (2020) further support that it is crucial to inform and educate users of coastal areas who are practicing marine/water-based activities in lagoons around Mauritius about the associated risks, as well as potential adaptation and mitigation strategies, that can foster environmental sustainability. Such practices and collaborations are of prime importance when integrating the sustainability concept into sport tourism activities and events.

Sport tourism activities are known to contribute to several positive environmental impacts. For instance, conservation and preservation-related information and education are very much present in the practice of land-, water-, and air-related sports activities. The local government works collaboratively with regional NGOs like the Mauritius Wildlife Foundation, international organizations such as the World Bank, Global Equity Fund, and UNEP, and individual donors to educate, sensitize, and promote environmental conservation while practicing sports activities. The government of Mauritius has included in its National Biological Strategy (2017) the concepts and principles to promote forest, terrestrial, freshwater, coastal, and marine aquatic biodiversity, biotechnology, and sustainable ecotourism. The MMCO, an international whale-watching commission observer, is present in Mauritian waters to maintain the sustainability of marine-based activities such as yachting, pleasure boats, whale watching, catamaran tours, and scuba diving. The organization monitors marine species such as whales and turtles. It aims to raise awareness of the importance of safeguarding and appreciating such creatures among local fishermen, maritime operators, local communities, and tourists (Marine Megafauna Conservation Organization, MMCO, 2022).

Fishing, diving, kayaking water-skiing, wakeboarding, windsurfing, and snorkeling are other common marine-based activities. To ensure the sustainable use and conservation of marine resources during these practices, several stakeholders, including the government and non-governmental organizations such as Reef Conservation, are committed to working together to maintain and preserve the limited natural resources (Reef Conservation, 2022). Constant monitoring and sustainable policies and practices are being implemented to limit the degradation and loss of marine biodiversity, especially when snorkel and dive boats drag their anchors, install shark nets, or when snorkelers accidentally hit corals or disrupt sand and sediments (Margaret, 2021).

Golf clubs tend to adopt sustainable practices and this is in turn translated to players and visitors. In Mauritius, the concept of preservation and conservation is promoted by most golf courses that offer quiet, natural spaces without predators or invasive plants. Similarly, Heritage Resorts and Hotels, through the Heritage Golf Club, adopt strong sustainability concepts—Heritage Golf Club (Veranda group). This creates numerous benefits for birds, animals, and wildlife. Other popular land-based activities offered in Mauritius are hiking and trekking. These activities allow for significant environmental awareness as hikers and trekkers are taught about the natural environments they visit. They may also develop a sense of appreciation, respect, and conservation that will in turn help protect the natural habitat of native animals and plants (Hernandez, 2017). Hiking and trails are popular sustainable sport tourism activities that drastically eliminate the consumption of fossil fuel for vehicles as people walk around to discover destinations (Puzyreva et al., 2022).

Despite the numerous positive environmental impacts that sport tourism may foster, it cannot eliminate all possible negative impacts. All if not most of the proposed sports activities, whether land-, water/marine-, or air-based, have negative environmental impacts. Coral depletion and bleaching, sand erosion, invasive or extinct species, deforestation, land reclamation and restructuring, and air and water pollution are common negative environmental impacts of sport tourism. Some other common negative impacts of land-based activities, such as hiking and trekking on summits, create air pollution through campfires, littering, and plastic waste among others. These unfortunately become a nuisance as they disturb and interfere with nearby plants and animals. Dr. Tatayah, Conservation Director at the MWF, reports in an interview that visited islets and natural reserves are in an indescribable mess after visits and trails of tourists and locals (Lemauricien, 2022).

Conclusion

The success of the local tourism industry of Mauritius has been steady over time and has brought along both positive and negative impacts. The evolution of sport tourism over time has been significant for many countries, and Mauritius is a developed tourism destination aimed at enhancing its sport tourism sector. Small island states like Mauritius have taken this international interest as an opportunity and have re-engineered their national strategies. As the diversification of local tourism products and activities seemed a feasible option for the long-term sustainability of the sector, alternative forms of tourism, such as sport tourism, appeared as the game changer.

Small islands like Mauritius aim to develop local enterprises for coherent economic progress and stability; this goal can be promoted via sport tourism and the many medium- and small-sized companies that offer many sport tourism activities.

Njoroge et al. (2017) also note that the social exchanges and interactions occurring from sports activities and tournaments are of benefit to both host countries and visitors. Mauritius aims to increase its destination image by enhancing its sport tourism sector, and Chueng et al. (2016) support this concept as a potentially positive social benefit. To enjoy a sustainable, socially sound development, the local community must always be recognized and confirmed as a key player in the process of sport tourism planning and implementation. For Mauritius, with a panoply of varied sport tourism activities, ranging from land-, water/marine-, and air-based sports activities, education remains crucial to avoid over-use and over-exploitation.

References

Amir, A. F., Ghapar, A. A., Jamal, S. A., & Ahmad, K. N. (2015). Sustainable tourism development: A study on community resilience for rural tourism in Malaysia. *Procedia – Social and Behavioral Sciences, [online], 168*, 116–122. https://doi.org/10.1016/j.sbspro.2014.10.217

Antar-Ecotono. (2004). *El Turismo de Naturaleza en España y su Plan de Impulso*. (Nature Tourism in Spain and Its Momentum Plan) Ministerio de Industria, Turismo y Comercio, Madrid, Spain. (In Spanish) [Google Scholar].

Audley, Travel. (2020). Indian Ocean Travel Guides. Available from: https://www.audleytravel.com/indian-ocean/region-guides. Accessed June 21, 2022.

Baldacchino, G. (2012). Governmentality is all the rage: The strategy games of small jurisdictions. *The Round Table, 101*(3), 235–251. https://doi.org/10.1080/00358533.2012.690964

Beeharry, Y., Makoondlall-Chadee, T., & Bokhoree, C. (2014). Policy analysis for performance assessment of integrated coastal zone management initiatives for coastal sustainability. *APCBEE Procedia, 9*, 30–35. https://doi.org/10.1016/j.apcbee.2014.01.006

Beeharry, Y., Bekaroo, G., Bussoopun, D., Bokhoree, C., & Phillips, M. R. (2020). Perspectives of leisure operators and tourists on the environmental impacts of coastal tourism activities: A case study of Mauritius. *Environment, Development and Sustainability*. https://doi.org/10.1007/s10668-020-01080-7

Bjeljac, Z., Curcic, N., & Ivolga, A. (2017). Tourismological classification of sporting events. *Journal of the Geographical Institute Jovan Cvijic, SASA, 67*(1), 53–67. https://doi.org/10.2298/ijgi1701053b

Burnett, C. (2019). Framing a 21st century case for the social value of sport in South Africa. *Sport in Society*, 1–16. https://doi.org/10.1080/17430437.2019.1672153

Cheung, S. Y., Mak, J. Y., & Dixon, A. W. (2016). Elite active sport tourists: Economic impacts and perceptions of destination image. *Event Management, 20*, 99–108.

Clark, R., & Misener, L. (2015). Understanding urban development through a sport events portfolio: A case study of London, Ontario. *Journal of Sport Management, 29*(1), 11–26. https://doi.org/10.1123/jsm.2013-0259

Desai, et al. (2014). Grey-based model for forecasting Mauritius international tourism from different regions. *Grey Systems: Theory and Application, 7*(2). [Online]. Available from: https://www.emerald.com/insight/content/doi/10.1108/GS-04-2017-0008/full/html. Accessed: June 13, 2022.

Duignan, M. B., Down, S., & O'Brien, D. (2020). Entrepreneurial leveraging in liminoidal olympic transit zones. *Annals of Tourism Research, 80*, 102774. https://doi.org/10.1016/j.annals.2019.102774

Dunn, R. (2009). The Sun Yet Warms His Native Ground. *American Scientist, 97*(2), 167–170.

Getz, D., & Page, S. J. (2016). Progress and prospects for event tourism research. *Tourism Management, [online], 52*(0261-5177), 593–631. https://doi.org/10.1016/j.tourman.2015.03.007

Gursoy, D., Chi, C. G., Ai, J., & Chen, B. T. (2011). Temporal change in resident perceptions of a mega-event: The Beijing 2008 Olympic games. *Tourism Geographies, 13*(2), 299–324. https://doi.org/10.1080/14616688.2010.529935

Gursoy, D., Yolal, M., Ribeiro, M. A., & Panosso Netto, A. (2017). Impact of trust on local residents' mega-event perceptions and their support. *Journal of Travel Research, [online], 56*(3), 393–406. https://doi.org/10.1177/0047287516643415

Hall, C. M. (2012). SuStainable Mega-eventS: Beyond the Myth of balanced approacheS to Mega-event SuStainability. *Event Management, 16*(2), 119–131. https://doi.org/10.3727/152599512x13343565268294

Hamuth. (2019). *When sports boost the economy*. [online] Le Defi Media Group. Available at: http://defimedia.info/when-sports-boost-economy. Accessed September 5, 2022.

Hasani, A., Moghavvemi, S., & Hamzah, A. (2016). The impact of emotional solidarity on residents' attitude and tourism development. *PLOS ONE, [online], 11*(6), e0157624. https://doi.org/10.1371/journal.pone.0157624

Hernandez, J. (2017). Hiking, the truth about environmentally safe and healthy hiking. [online] Environment South Africa | Green Blog | Eco Bulletin. Available at: https://www.environment.co.za/food-health-and-the-environment/environmentally-safehealthy-hiking.html. Accessed June 15, 2022.

Hsu, B. C.-Y., Wu, Y.-F., Chen, H.-W., & Cheung, M.-L. (2020). How sport tourism event image fit enhances residents' perceptions of place image and their quality of life. *Sustainability, 12*(19), 8227. https://doi.org/10.3390/su12198227

Ian, C., Enaida, F., Meser, H., & Tiwning-Ward, L. (2014). *Tourism in Africa: Harnessing tourism for growth and improved livelihoods* (p. 20433). Co-publication-International Bank for Reconstruction and Development/The World Bank.

Industrial Policy and Strategic Plan for Mauritius 2020–2025, "*Revitalising Mauritius' economic growth*", Ministry of Industrial Development, SMEs and Cooperatives (Industrial Development Division) & UNCTAD.

Investment Policy Review for Mauritius. (2021). *United Nations Conference on Trade and Development (UNCTAD), Division on Investment, Technology and Enterprise Development*, UN New York, Geneva.

Karantzavelou, V. (2022). *New study assesses active sport tourism facilities across 50 countries*. [online] TravelDailyNews International. Available at: https://www.traveldailynews.com/post/new-study-assesses-active-sports-tourism-facilities-across-50-countries. Accessed July 20, 2022.

Kirillova, K., Fu, X., Lehto, X., & Cai, L. (2014). What makes a destination beautiful? Dimensions of tourist aesthetic judgment. *Tourism Management, 42*, 282–293. https://doi.org/10.1016/j.tourman.2013.12.006

Kolawole, O. I., Mbaiwa, J. E., Mmopelwa, G., & Kgathi, D. L. (2017). The socio-cultural impacts of tourism on people's quality of life in Maun, Botswana. *Special Issue on Environment, Tourism and Comtemporary Socio-Economic Issues in the Okwango Delta and Other Ecosystems, 49*(Special Issue), 73–86.

Kurtzman, J. (2005). Sports tourism categories. *Journal of Sport & Tourism, 10*(1), 15–20. https://doi.org/10.1080/14775080500101502

Lemauricien. (2022). *Environnement: Campings nocturnes en montagne interdits*. [online] Available at: https://www.lemauricien.com/actualites/societe/environnement-campingsnocturnes-en-montagne-interdits/483898/. Accessed June 15, 2022.

Li, S., Blake, A., & Cooper, C. (2011). Modelling the economic impact of international tourism on the Chinese economy: A CGE analysis of the Beijing 2008 Olympics. *Tourism Economics, 17*(2), 279–303. https://doi.org/10.5367/te.2011.0025

Liu, C., Chang, C., Chen, C., Liu, Y., & Ni, K. (2011). A study of sport tourism's environmental impact a case study of kenting traveling destination in Taiwan, research paper of Meiho University Institutional Repository http://ir.meiho.edu.tw/handle/987654321/1663.

Loi, K. I. (2009). Entertainment as a tourism development tool in Macao. Unpublished Doctoral Dissertation, James Cook University.

Mair, J., & Smith, A. (2021). Events and sustainability: Why making events more sustainable is not enough. *Journal of Sustainable Tourism, 29*(11–12), 1–17. https://doi.org/10.1080/09669582.2021.1942480

Makoondlall-Chadee, T., Bokhoree, C., & Sumputh, D. (2017). A Systemsthinking approach towards promoting sustainable tourism in Mauritius. In *Conference proceedings of BEST EN Think Tank XVII: Innovation and progress in sustainable tourism*. James Cook University. ISBN: 978-0-9954470-7-3.

Malchrowicz-Mosko, E., & Munsters, W. (2018). Sport tourism: A growth market considered from a cultural perspective. *Ido Movement for Culture. Journal of Martial Arts Anthropology, 18*(4), 25–38.

Marasinghe, S., Perera, P., Simpson, G. D., & Newsome, D. (2021). Nature-based tourism development in coastal wetlands of Sri Lanka: An importance–performance analysis at Maduganga Mangrove Estuary. *Journal of Outdoor Recreation and Tourism, 33*, 100345. https://doi.org/10.1016/j.jort.2020.100345

Margaret, P. (2021). 5 Islands in Mauritius that will take your breath away "island in Mauritius". [online] Tripatini. Available at: https://tripatini.com/profiles/blogs/5-islands-in-mauritius-thatwill-take-your-breath-away-island-i-3. Accessed July 14, 2022.

marinemegafaunaconservation.org. (n.d.). *Home*. [online] Available at: https://marinemegafauna-conservation.org/?fbclid=IwAR1g-. Accessed September 5, 2022.

Masa'deh, R., Nasseef, M. A., Alshayeb, H., Ojilat, J., & Alshafiee, M. (2017). The effect of sport tourism management on support for tourism development. *Journal of Management and Strategy, 8*(3), 20. https://doi.org/10.5430/jms.v8n3p20

Mauritius Attractions (2022), https://mauritiusattractions.com/

Mauritius Facts, Geography and History. (2022). Mauritius introduction and quick facts. Available at: https://www.britannica.com/place/Mauritius. Accessed June 6, 2022.

Melly, D., & Hanrahan, J. (2020). Tourist biosecurity awareness and risk mitigation for outdoor recreation: Management implications for Ireland. *Journal of Outdoor Recreation and Tourism, 31*, 100313. https://doi.org/10.1016/j.jort.2020.100313

Melo, R., Sobry, C., & Van Rheenen, D. (Eds.). (2021). *Small scale sport tourism events and local sustainable development. A cross-national comparative perspective*. Springer. https://doi.org/10.1007/978-3-030-62919-9

Ministry of Tourism (2019, 2022), Ministry of Tourism. (2018). Strategic plan 2018–2021 – "Adapting to the changing global environment"

Ministry of Tourism. (2018). Strategic plan 2018–2021 – *"Adapting to the changing global environment"*. Republic of Mauritius_ chrome-extension://efaidnbmnnnibpcajpcglclefindmkaj/https://govmu.org/EN/infoservices/travelandtourism/Documents/Web_Final_STRATEGIC_Plan_2018-2021.pdf

MTPA. (2022). 'Sport tourism in Mauritius'. [Online]. Available from: https://www.mymauritius.travel/articles/sports-tourism-mauritius. Accessed: June 15, 2022.

National Biodiversity Strategy. (2017). *National Biodiversity Strategy and Action Plan (NBSAP) 2017–2025*. Republic of Mauritius (RoM).

Nigel, R., Rughooputh, S.D., & Boojhawon, R. (2010). Land cover of Mauritius Island. *Journal of Maps, 11*, 217–224.

Njoroge, J. M., Atieno, L., & Do Nascimento, D. V. (2017). Sport tourism and perceived socio-economic impact in Kenya: The case of Machakos county. *Tourism and Hospitality Management, 23*, 195–217.

Nyikana, S., Tichaawa, T., & Swart, K. (2014). Sport, tourism and mega-event impacts on host cities: A case study of the 2010 FIFA World Cup in Port Elizabeth. *African Journal for Physical, Health Education, Recreation and Dance, 20*(2:1), 548–556.

PAGE Strategy 2021–2030 | Un-page.org https://www.un-page.org/about/governance-strategy/clone-governance-strategy

Pan, S.-Y., Gao, M., Kim, H., Shah, K. J., Pei, S.-L., & Chiang, P.-C. (2018). Advances and challenges in sustainable tourism toward a green economy. *Science of The Total Environment, [online], 635*, 452–469. https://doi.org/10.1016/j.scitotenv.2018.04.134

Puzyreva, M., Gunn, G., & Simoes, J. (2022). A Value on the Priceless: Ecological goods and services generated in the Seal River Watershed, IISD. Canada. Retrieved from https://policycommons.net/artifacts/2680418/a-value-on-the-priceless/3703792/ on 18 May 2024. CID: 20.500.12592/nm6sxb.

Reef Conservation. (2022). *Environmental profile – Reef conservation.* [online]. Available at: https://www.reefconservation.mu/about-mauritius/environmental-profile/. Accessed July 11, 2022.

Roche, S., Spake, D. F., & Joseph, M. (2013). A model of sporting event tourism as economic development. *Sport, Business and Management: An International Journal, 3*(2), 147–157. https://doi.org/10.1108/20426781311325078

Ruhanen, L. (2013). Local government: Facilitator or inhibitor of sustainable tourism development? *Journal of Sustainable Tourism, 21*, 80–98.

Saddul, P. (1995). *Mauritius: A geomorphological analysis*. Mahatma Gandhi Institute.

Sports and Development Org. (2022). *How does sports contributes to sport tourism*. Available at: https://www.sportanddev.org/en/article/news/how-does-sport-contribute-tourism. Accessed: September 23, 2022.

Stainton, H. (2019). *Social impacts of tourism.* [online] Tourism Teacher. Available at: https://tourismteacher.com/social-impacts-of-tourism/

Statistics Mauritius. (2020). International travel and tourism. [Online]. Available from: https://statsmauritius.govmu.org/Documents/Statistics/ESI/2021/EI1567/Tourism_Yr20.pdf. Accessed: June 17, 2022.

Statistics Mauritius, AHRIM. (2019). AHRIM Annual Report 2018/2019.

Taks, M., Chalip, L., & Green, B. C. (2015). Impacts and strategic outcomes from non-mega sport events for local communities. *European Sport Management Quarterly, 15*(1), 1–6. https://doi.org/10.1080/16184742.2014.995116

Tandrayen-Ragoobur, V., & Kasseeah, H. (2018). Mauritius' economic success uncovered in Ramtohul, Ramola & Eriksen, Thomas. (Eds). The Mauritian Paradox: Fifty Years of Development, Diversity and Democracy, Réduit : University of Mauritius Press.

Tomislav, K. (2018). The concept of sustainable development: From its beginning to the contemporary issues. *Zagreb International Review of Economics and Business, 21*(1), 67–94. https://doi.org/10.2478/zireb-2018-0005

United Nations. (2017). *The 2030 agenda for sustainable development.* [online] Un.org. Available at: https://www.un.org/development/desa/jpo/wp-content/uploads/sites/55/2017/02/2030-Agenda-for-Sustainable-Development-KCSD-Primer-new.pdf. Accessed July 15, 2022.

UNWTO. (2015). Sustainable tourism for development guidebook enhancing capacities for sustainable tourism for development in developing countries. Available at https://www.e-unwto.org/doi/book/10.18111/9789284415496. Accessed June 15, 2022.

UNWTO. (2019). *International tourism highlights* (2019 Edition). United Nations World Tourism Organisation.

UNWTO. (2020). *Yearbook of tourism statistics*. UNWTO.

Uysal, M., Berbekova, A., & Kim, H. (2020). Designing for quality of life. *Annals of Tourism Research, 83*, 102944. https://doi.org/10.1016/j.annals.2020.102944

Vujko, A., & Gajic, T. (2014). Opportunities for tourism development and cooperation in the region by improving the quality of tourism services – The 'Danube Cycle Route' case study. *Economic Research-Ekonomska Istraživanja, 27*(1), 847–860. https://doi.org/10.1080/1331677x.2014.975517

World Tourism Organization. (2019). *Sport tourism and the sustainable development goals (SDGs)*. UNWTO. https://doi.org/10.18111/9789284419661

Chapter 22
Towards the Development of a Sustainable Sport Tourism Destination in Madagascar: The Key Role of Inhabitants as Stakeholders

Lilia Khelifi

Introduction

This chapter examines the state of sport tourism in Madagascar, the fifth largest island in the world. Its size is both an advantage and a disadvantage for developing tourism. Indeed, the island offers a diversity of climates and landscapes. It is renowned worldwide for its endemic biodiversity, and for its niche crops, such as vanilla. But in a recent publication (Razafindrakoto et al., 2017), three researchers demonstrate the paradox of Madagascar's economic development since its independence in 1960. According to their study, the country is getting poorer even though it has many assets that other states with similar profiles have managed to improve the economic situation of their inhabitants.

The data collected for this chapter were obtained during a five-week research stay in June 2022 to the following localities: Antananarivo, Tulear, Anakao, and Mahambo (see Fig. 22.1). Another field survey for a master's thesis on the emergence of surfing in Madagascar allowed us to investigate Antananarivo, Mahambo, and Fort-Dauphin in 2017 for a month (see Fig. 22.1). A wide range of stakeholders in the sport and/or tourism sector were interviewed, ranging from institutional actors to informally involved residents. The scientific literature on sport in Madagascar is quite limited and focused on the colonial period (Combeau-Mari, 2004, 2007, 2009; Nativel, 2016; Bordelès, 2021). Academics and sport journalists interviewed during the field study reported difficulties in accessing information and statistics. Besides, social science students at the University of Antananarivo do not pursue topics related to sport, according to the interviews conducted with lecturers and students.

L. Khelifi (✉)
Sorbonne University—Faculty of Letters, Research Unit Mediations, Paris, France

D. Van Rheenen et al. (eds.), *Sport Tourism, Island Territories and Sustainable Development*, Sports Economics, Management and Policy 24,
https://doi.org/10.1007/978-3-031-51705-1_22

Fig. 22.1 Situation map of Madagascar

Regarding tourism, a report by the World Tourism Organisation (UNWTO) in 2019, entitled *International Tourism Highlights*, states that "international tourism growth remains faster than the global economy"[1] (OMT, 2019: 2). Considering the low proportion of international tourists in Madagascar and the Covid crisis, the issue of domestic tourism has recently been identified. This research also examines the definition of sport tourism since "in many articles the expression 'sport tourism' is used without the author taking care to specify what is implied in its representation"[2] (Pigeassou, 2006: 147). Is it tourism or sport that first drives the individual to practise a physical activity when travelling? Lastly, it is necessary to question what sustainable development means (Jollivet, 2002), as well as the notion of development was criticised and historicised (Rist, 2013). "By adding an adjective to the concept of development, it is not really a matter of questioning capitalist accumulation; at most, it is a matter of adding a social component to economic growth, as it

[1] Translated by the author. In the original language in the text: "La croissance du tourisme international reste supérieure à celle de l'économie mondiale."

[2] Translated by the author. In the original language in the text: "Dans de nombreux articles l'expression 'tourisme sportif' est utilisée sans que l'auteur prenne le soin de préciser ce qui est sous-entendue dans sa représentation."

was once possible to add a cultural dimension, and today an ecological component"[3] (Latouche, 2003: 23). In a book supervised by the geographer Peyvel (2019), the authors examine how tourism relates to an "education for travel."[4] The contributions analyse how tourism practices are learned and transmitted and question the issues of circulation and re-appropriation of knowledge.

In light of the current state of knowledge, this chapter aims to question the processes at work in Madagascar at the junction of sport tourism and sustainable development, examining the practices of tourists and the inhabitants who work with/for tourists. To this end, an overview of the physical and socio-economic situation will be described before discussing the current issues at stake in sport tourism on the Big Island.

Physical and Socio-economic Characteristics of Madagascar

Physical Characteristics of the Big Island

Madagascar is a big island in the south-west Indian Ocean located off the coast of Africa between Mozambique and Réunion Island. The fifth largest island in the world, it measures approximately 591,896 km^2, which is equivalent to the size of France. This indication reminds us of the stakes involved in the choice of cartographic projection, which shapes a subjective vision of the representation of the world.

Five climatic zones are identified in Madagascar, which reinforces the faunistic and floristic diversity of the island, but also the possibilities of economic activities for the inhabitants. At present, the state accounts for five wilderness areas, 24 national parks—four of which are marine—24 special reserves, 12 new protected areas, 5 biosphere reserves, and 20 Ramsar sites. Three sites are listed under UNESCO World Heritage. In 2019, there were reported to be 126 protected areas in the country. The protected areas benefit from various statuses. Madagascar is endowed with endemic resources due to its biogeographic isolation and its variety of climates and landforms (Sarrasin, 2007). The protection of nature by enclosing it and its supposed importance, according to subjective criteria (Mougenot, 2003), feed the endless debate of nature conservation ideology, as defined in the Western world for the last 150 years (Depraz, 2008). This vision imposed from a small part of the world to the whole planet reduces the possibilities of actions and representations in the relations between human societies and nature. However, several so-called developing states are seizing on this argument, which has been labelled "greenwashing," to obtain funds from the international community. The geographer C. Fournet-Guérin points out that "in order to obtain subsidies today, in Africa as

[3] Translated by the author. In the original language in the text: "En accolant un adjectif au concept de développement, il ne s'agit pas vraiment de remettre en question l'accumulation capitaliste, tout au plus songe-t-on à adjoindre à la croissance économique un volet social comme on a pu naguère lui ajouter une dimension culturelle, et aujourd'hui une composante écologique."

[4] To quote the title of the book.

elsewhere, it is necessary to display sustainable city themes, even if they are totally artificial and not a priority for the city and its inhabitants: this is the 'green washing' of urban policy in Africa"[5] (Sénat, 2011: 65).

Socio-economic Data

A whole national statistical system[6] has been established in order to meet the challenges set by the United Nations on the Sustainable Development Goals (Desrosières, 2010). A population census was conducted in 2018 and the results were published on the official website in 2021. The resident population was estimated at 28,177,762 in 2021; more than 80% live in rural areas. The World Bank provided a more recent assessment: there were about 27,691,019 inhabitants in 2020, while it estimated the population at 26,262,013 in 2018. The number of inhabitants in Madagascar increased by 457.5% between 1960 and 2021, according to the data of the World Bank (see Fig. 22.2). Little more than half of Madagascar's resident population is concentrated in the central highlands. The World Bank estimated Madagascar's GDP at US$13,721 billion in 2020. Gross National Income per capita in current international purchasing power parity was estimated at $1540 in 2020. In 2021, Madagascar was ranked the 10th poorest country in the world, according to the United Nations. Agriculture remains the main sector of activity since three quarters of the population are employed in the agricultural sector, according to the figures produced by the World Bank in 2016. Almost 17% of the population works in the tertiary sector. Most employment in Madagascar occurs in the informal sector. The population is fairly young since more than half of them are under 20 years old

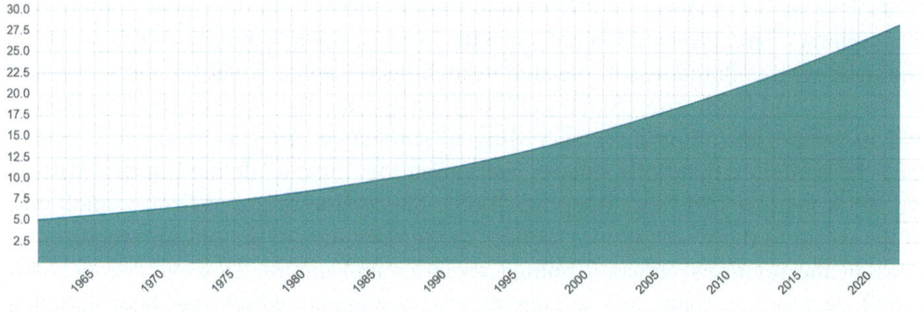

Fig. 22.2 Population trends

[5] Translated by the author. In the original language in the text: "Pour glaner des subventions aujourd'hui, en Afrique comme partout ailleurs, il faut afficher des thématiques de ville durable, fussent-elles totalement artificielles et en rien prioritaires pour la ville et ses habitants: c'est le '*green washing*' de la politique urbaine en Afrique."

[6] INSTAT: https://www.instat.mg/

and less than 3% are over 65. Although a young population can be perceived as an asset, the respondents most often considered this to be problematic, particularly because of the limited access to education.

The Paradox of a Rich State with a Poor Population

Madagascar has been an independent state since 1960. It was officially a French colony from 1897 to 1946, before becoming an overseas territory until 1958. Francophonie is still in effect, but the currency of Franc was replaced by the Ariary in 2003. When it was introduced, the ariary had an exchange rate of EUR 1 for MGA 1400. Since then, the ariary has steadily lost value. On 25 August 2018, the exchange rate was EUR 1 for MGA 3830.58, while on 20 January 2019, the exchange rate was EUR 1 for MGA 4100.

In a study carried out by two consultants for *The Africa Region Working Paper Series*[7] about the state of the tourism sector in Madagascar, the authors assert that "Madagascar's extraordinary natural capital, both terrestrial and marine, makes it a target destination for a diverse range of tourists"[8] (Christie & Crompton, 2003: 1) and that "Madagascar enjoys a world-class seaside capital"[9] (ibid.: 2). Served mainly by air—79% of tourists arrived by this means of transport in 2019, according to the World Tourism Organisation—tourism is relatively undeveloped in Madagascar. Although the 2019 tourism statistics produced by Madagascar's Ministry of Transport, Tourism and Meteorology show a steady rise in the number of visitors over the past 10 years, the country is struggling to reach the target of 500,000 visitors per year by 2023. The number of tourists in 2019 is estimated at 348,660 by the Ministry, while the World Bank obtains 486,000.

The pandemic situation and the 2-year closure of the borders between March 2020 and March 2022 have definitely undermined this objective. Now served by

[7] Supported by the World Bank among others, "The Africa Region Working Paper Series promotes the dissemination of applied research and policy studies that have the potential to improve the economic and social environment in sub-Saharan Africa. The Series publishes early-stage works to stimulate timely discussion within the region and among our client countries, donors, and the research community" (Christie & Crompton, 2003: 1). Translated by the author. In the original language in the text: "La Série de documents de travail de la Région Afrique active la diffusion de la recherche appliquée et des études de politique qui ont un potentiel pour améliorer l'environnement économique et social de l'Afrique subsaharienne. La Série publie des œuvres en étape préliminaire afin de stimuler la discussion en temps opportun dans la région et parmi nos pays clients, les bailleurs de fonds et la communauté de chercheurs."

[8] Translated by the author. In the original language in the text: "L'extraordinaire capital naturel de Madagascar, à la fois terrestre et marin, en fait une destination cible pour une gamme variée de touristes."

[9] Translated by the author. In the original language in the text: "Madagascar possède un capital balnéaire de classe mondiale."

nine airlines and welcoming international tourists in a brand-new airport in Antananarivo, the State aims to develop the country's touristic image. The field survey revealed a strong competition with the other islands of the Indian Ocean. Political instability, the prevailing insecurity, the lack and low quality of accommodations and the conditions and cost of travel within the country are all arguments which lead international tourists to choose another destination in the Indian Ocean. La Réunion and Mauritius, in particular, account for a significant number of tourists each year, unlike Madagascar. Stafford et al. (2002) provide an overview of tourism in Madagascar from 1990 to 2000. But the director of the regional office of tourism of Analamanga[10] emphasises that the State does not accurately quantify the exact number of tourists. Only the registration of arrivals is applied. "The information is not reliable: number of real tourists versus foreign visitors, their spending, average length of stay, distribution in the country, and their socioeconomic characteristics"[11] (Christie & Crompton, 2003: 2).

Although Malagasy public policies do not encourage sport tourism development projects, it is the self-organisation of a few individuals that enables some activities to be established and sustained in identified territories. "Some contributions develop and deepen the idea that sports leisure is a vector of tourism development provided that the supporting activity becomes a product"[12] (Pigeassou, 2006: 146). "It [sport] has become a social norm, a vector of economic development and an asset for the enhancement of territories. But at the same time, it is the bearer of a certain number of financial (commercialism, exploitation, imbalances, etc.), moral (eugenics, cheating, doping), and environmental (disfigurement of sites, pollution, waste of energy) abuses that tarnish its image and call into question its original virtues of personal development and improvement of living conditions"[13] (Bessy, 2006: 180).

[10] One of the 22 regions of Madagascar, it is located in the province of Antananarivo.

[11] Translated by the author. In the original language in the text: "Les informations ne sont pas fiables: nombre de vrais touristes par opposition aux visiteurs étrangers, leurs dépenses, la durée moyenne de séjour, la répartition dans le pays, et leurs caractéristiques socio-économiques."

[12] Translated by the author. In the original language in the text: "Quelques contributions développent et approfondissent l'idée que les loisirs sportifs sont un vecteur du développement touristique à condition que l'activité support devienne un produit."

[13] Translated by the author. In the original language in the text: "Il [le sport] est devenu une norme sociale, un vecteur de développement économique et un atout de valorisation des territoires. Mais il est en même temps porteur d'un certain nombre de dérives financières (mercantilisme, exploitation, déséquilibres...), morales (eugénisme, tricherie, dopage) et environnementales (défiguration de sites, pollution, gaspillage d'énergie) qui écornent son image et remettent en question ses vertus originelles d'épanouissement de la personne et d'amélioration des conditions de vie."

An Uneven Sport Tourism Sector

The Legacy of Colonial Sports

"Sport was established in the French colonies towards the end of the nineteenth century, shortly after its birth in metropolitan France […]. At the turn of the twentieth century, it was to accompany colonial and post-colonial nationalism"[14] (Abbassi, 2009: 5). In its colonial enterprise, the French authorities imported the sports practised in order to help discipline the colonised population. This begins with the banning of local fighting practices and the spread of gymnastics (Combeau-Mari, 2007). "He [Pierre de Coubertin] sees this as a way of establishing the power of the metropolis, its practices and its values in a more effective and positive way"[15] (Bouvier, 2010: 2). The historian Combeau-Mari (2004, 2007, 2009) has widely worked on the issues of colonial sports in Madagascar and in Réunion Island. Although football remains the most popular sport on a global and national scale, rugby was the first sport introduced in the Big Island by French settlers in the early 1900s. Rugby is deeply associated with politics. It contributed to the anti-colonialist struggle of the Malagasy (Combeau-Mari, 2007; Bordelès, 2021). "Surprisingly, they place rugby at the centre of their activities. This is in stark contrast to the widespread support for football in Africa"[16] (Combeau-Mari, 2007: 130).

Rugby matches—among others—were the beginnings of sport tourism, with teams touring nationwide and between the metropolis and the colonies. Rugby remains a popular sport in Madagascar today, especially in the highlands. Associations, such as *Terres en Mêlées*, seize this sport to promote values and help meet the sustainable development goals set by the UN. "In complex procedures that can only be analysed over a long period of time, the terms 'appropriation' or 'use' do not mean 'interiorisation'"[17] (Combeau-Mari, 2007: 134). Despite independence, the links maintained with the former colonising country are manifested in several ways. As a member state of the International Organisation of the Francophonie, Madagascar organised the 16th edition in 2016. Francophonie Summits are held every 2 years since 1986. The Games of the Francophonie have been organised since 1989 under the aegis of the International Committee of the Games of the Francophonie. Antananarivo was the host city in 1997.

[14] Translated by the author. In the original language in the text: "Le sport s'implante dans les colonies françaises vers la fin du XIXe siècle, peu de temps après sa naissance en métropole. Il va, au tournant du XXe siècle, accompagner les nationalismes coloniaux et postcoloniaux."

[15] Translated by the author. In the original language in the text: "Il y voit une manière d'asseoir de manière plus effective et positive le pouvoir de la métropole, de ses pratiques et de ses valeurs."

[16] Translated by the author. In the original language in the text: "Fait surprenant, ils placent le rugby au centre de leurs activités. Ce choix se démarque singulièrement de l'adhésion généralisée pour le football en Afrique."

[17] Translated by the author. In the original language in the text: "Dans des procédures complexes qui ne peuvent être analysées que sur la longue durée, les termes 'appropriation' ou 'utilisation' ne signifient pas 'intériorisation'."

The Current Sporting Landscape

Even when a State provides statistics and data, it is never enough to understand a local or national context and the way people express their sporting activity and their mobility. The weeks spent in the country between September and October 2017 and between May and June 2022 and meeting with diverse local actors as part of a research project on the emergence of surfing are essential to understand the current issues identified by the inhabitants themselves. It is difficult to assess the number of athletes in the Big Island for at least two reasons. There is no data collected by public authorities and very limited academic research, at least available. The production of statistical data is currently required by supranational organisations: the UN and international sports federations. Besides, even when public statistics exist about the number of license holders in a sport, it is not enough to understand the number of practitioners, their relationship to the practice, and so on. As the most globalised sport, football is most widely practised in Madagascar. It was following its independence that a Malagasy football federation was created in 1961. In a paper entitled "Football: Esperanto of Culture," H. Gicquel indicates that "it is and it makes the event by at least five characters: its globalisation, its media impact, the financial sums invested in it, the excesses it provokes and its ritualisation"[18] (Gicquel, 1995: 269). The respondents confirmed the popularity of football in Madagascar. Petanque, rugby, and athletics are also prevalent among the Malagasy. Nevertheless, sport stakeholders interviewed lament that sport ranks the lowest in the state budget.

From now on, nature tourism has become the current state priority in the highlands' region, according to the director of the tourism office of Analamanga. Inspired by the Réunion model, hikes and trails have been developing in the Malagasy highlands for about 10 years. The ambition is to stimulate domestic and regional tourism on the scale of the Indian Ocean. If one of the stakeholders of the *Ultra-Trail des Ô Plateaux* (UTOP) affirms that hiking is the most accessible and least expensive sport, other stakeholders in the sports world consider it to be a sport for the upper class. Beyond the question of the cost of equipment, it is above all the physical expression of walking for pleasure that questions the possibilities of individuals according to their social class. Indeed, walking is most often a matter of not having the choice to move around, and the exercise of a physical and exhausting job explains the low level of interest in hiking and trail walking. Sport in general is not very popular in Madagascar, according to the sports journalists interviewed, because the priorities of the individual are pyramidal, and housing and food are well ahead of integration into a sporting community. It is important to remind that the "right to sport" is the result of public policies in favour of its democratisation since the last century in France (Jamet, 1995) and in other Western countries. It is now promoted by the United Nations through the Sustainable Development Goals.

---

[18] Translated by the author. In the original language in the text: "Il est et il fait l'événement par au moins cinq caractères: sa mondialisation, son impact médiatique, les sommes financières qu'on y investit, les débordements qu'il provoque et sa ritualisation."

Sport tourism in Madagascar can be understood by its geographical situation and its cultural proximity with Réunion Island. The Réunionese, the South Africans, as well as the French from the metropolis are the main vectors of sport tourism, especially on the coasts. Indeed, the field survey revealed that the majority of hotels and sports activities offered are owned by foreigners from the three regions previously mentioned who have decided to settle in Madagascar. Most of the interviewees asserted that they were looking for places that were still "wild" and spared from the Western model. The respondents often spent a few years in another Indian Ocean Island before moving to Madagascar and were disappointed with their experience because of the development of these regions, especially in Réunion Island where buildings predominate according to the interviewed. However, security in terms of political stability but also in the physical activities offered is one of the main obstacles to the growing development of sport tourism. The lack of training, on the one hand, and the lack of transport and accommodation infrastructures, on the other hand, were also mentioned. Lastly, Madagascar will be the next host country for the 2023 Indian Ocean Island Games. Surfing will be on the programme for the first time. These Games embody a way of emancipation from the Western normative framework, while reproducing this model in the Indian Ocean space.

A Low Coastal Tourism

Since Madagascar is not a small island, inhabitants do not all have a direct link with the ocean. Even people living near the ocean have not necessarily included it in their living space, as conceptualised by the geographer Frémont (2009). The ocean is regarded as a taboo in Malagasy society in most cases because it is full of beliefs. It produces a repulsive effect, that is why few people immerse themselves and know how to swim. Named "fady," one of the cultural prohibitions consists in not eating pork before entering the water.

A distinction must be made between the potential and the resource (Sarrasin & Ramahatra, 2013). The latter is made possible by the development of actors who depend on their social and cultural capital. The development of sport is driven by local actors with limited economic capital. It is not primarily the result of institutional actors. The legacy of colonisation can be a way to understand the current sport scene (Naria, 2010), but recent migrations, whether short or long term, from France, Réunion Island, and South Africa, shed light on local contexts, especially on the coasts. Indeed, there is a split between the highlands—the capital Antananarivo and its surroundings—and the coasts (Fournet-Guérin, 2009). Public money is mostly invested in the highlands, to the detriment of the coasts. Entrepreneurial individuals participate in the diffusion of maritime sports, such as surfing, kitesurfing, diving, and so on. The impression of exploring virgin lands attracts tourists, some of whom decide to settle down for a long time.

Some inhabitants with no economic capital were able to grasp the practice of surfing and its related opportunities. It is the inhabitants themselves who play a key role in the offer and development of sport tourism. Initially appropriated for their personal enjoyment, they soon realised the economic opportunity offered by these activities. The Malagasy surfing federation was created in 2019 on the initiative of a Malagasy resident of Fort-Dauphin with the help of the Réunionese association *Indian Ocean Surfers*. But insecurity, lack of infrastructure and facilities, and the high cost of airfare represent the main obstacles to the development of tourism. Road conditions and the high cost of domestic flights strengthen the difficulties for tourists to move around the country. Paradoxically, this limited access makes it an unspoiled area of the world sought after by some tourist profiles. A certain amount of economic capital is required to travel around the country, as a driver is needed who will also act as a guide and ensure the safety of travellers. The Covid pandemic has resulted in the closure of Madagascar's borders between March 2020 and March 2022, increasing the difficulties for tourism actors. Hotel owners had to adapt their rates to local demand to avoid bankruptcy. The aim of developing sport tourism for local stakeholders is not to deprive the inhabitants of their own living areas or to exclude them from this economic opportunity (Amirou, 1995).

Mougenot (2003) emphasises that nature is a plural notion and questions the hierarchy of powers and knowledge on the basis of a survey conducted among various actors involved in the Communal Development Plans for Nature in Wallonia (Belgium). Several coral reefs surround Madagascar. These are assets for water activities, such as diving, surfing, stand-up paddling, etc., but these ecosystems have been endangered for the last few decades (Vasseur, 1997). In addition to the anthropogenic pressure exerted by the fast-growing population and fishing, it should not be neglected that "the environmental impact of nature sports is therefore very difficult to characterise from a scientific point of view"[19] (Mounet, 2007: 164). The current ambition of the Malagasy government is to develop sport tourism and coastal tourism. Thus, the president of the republic declared that he intends to transform the beach of Tamatave into the beach of Miami. The construction work has begun. Also, the Ministry of Tourism launched a call for tenders in 2021 to evaluate the kitesurfing spots in the country and propose lines of development for this activity. The results were presented internally in May 2022. Furthermore, "with the support of donors, in particular the World Bank and numerous non-governmental organisations (NGOs), the Malagasy government has prepared a 15-year National Environmental Action Plan (NEAP).

The implementation of this NEAP began in 1990 with the adoption of the Environmental Charter"[20] (Froger, 2010: 17). Meanwhile, the *QIT Madagascar Minerals* (QMM) company set up in Fort-Dauphin (Sarrasin, 2006) and caused great tension among the local population. Feasibility studies begun in 1986 resulted

[19] Translated by the author. In the original language in the text: "L'impact environnemental des sports de nature est donc très difficile à caractériser d'un point de vue scientifique."

[20] Translated by the author: "Soutenu par les bailleurs de fonds, notamment la Banque Mondiale et de nombreuses Organisations Non Gouvernementales (ONG), le Gouvernement malgache a élaboré un Plan National d'Action Environnemental (PNAE) d'une durée de 15 ans. L'exécution de ce PNAE a débuté en 1990 avec l'adoption de la Charte de l'Environnement."

in the creation of the company—only 20% of which is owned by the Malagasy government—and a port in Fort-Dauphin. The other 80% belongs to *Rio Tinto Fer et Titane*, a Quebec-based subsidiary of the Anglo-Australian multinational mining group *Rio Tinto*. B. Sarrasin (ibid.: 11) reminds that "although no village is located on the ilmenite deposit that QMM intends to exploit, nearly 6,000 people live within a few kilometres of the Mandena sector, the first phase of the mining project [...]. In a context where poverty is even more marked in rural areas than in urban areas, agriculture represents the main source of income for 80% of the population of this region."[21] Tensions were recently rekindled in early 2022 due to the dumping of sewage into rivers and the presence of dead fish, which triggered roadblocks by local residents.

Lastly, Madagascar possesses a few offshore islands, but only Nosy Be and Sainte-Marie are inhabited and geared towards receiving tourists. These islands provide major tourism benefits. Indeed, Nosy Be is renowned for being a hot spot for international tourism in the country. The Scattered Islands in the Indian Ocean are small French islands located around Madagascar. Currently uninhabited, these are the subject of diplomatic tensions between the two states.

Conclusion

Currently not well developed for tourism, Madagascar has many assets with regard to sport tourism. The government, in conjunction with UN guidelines, is attempting to meet the sustainable development goals through sport, while tourism represents a flourishing economic sector to be developed. The Malagasy sports landscape has inherited the colonial past as well as globalisation. Conceptions of nature, tourism, and sport are extensions of colonial issues, which are nowadays described as post-colonial. The state apparatuses set up in the former colonies reproduce the model imposed by the former colonising country. Mobility issues, as well as environmental problems and health risks, are among the problems that the State must tackle in order to sustain the development of sport tourism for all.

References

Abbassi, D. (2009). Le sport dans l'empire français: un instrument de domination ? *Outre-mers, 96*(364–365), 5–15. https://doi.org/10.3406/outre.2009.4411

Amirou, R. (1995). *Imaginaire touristique et sociabilités du voyage*. Les Presses universitaires de France.

[21] Translated by the author: "Bien qu'aucun village ne soit situé sur le gisement d'ilménite que QMM compte exploiter, près de 6000 personnes vivent à quelques kilomètres du secteur de Mandena, la première phase du projet minier [...]. Dans un contexte où la pauvreté est encore plus marquée en milieu rural qu'en milieu urbain, l'agriculture représente la principale source de revenus de 80% de la population de cette région."

Bessy, O. (2006). Le développement durable: un défi pour le sport ? Approche française des connaissances et perspectives de recherche. In D. P. Bouchet, & C. Pigeassou (dirs.), *Management du sport: actualités, développements et orientations de la recherche* (pp. 179–207). AFRAPS.

Bordelès, F. (dir.). (2021). *Histoires du rugby malagasy, de ses origines aux tournées des Makis en France: joueurs sacrés et sacrées joueuses, 1900–2021*. Catalogue de l'exposition rugby gasy.

Bouvier, (2010). Colonisation et pratiques sportives, Journal des anthropologues [En ligne], 120–121 I 2010, mis en ligne le 16 juillet 2014, consulté le 18 mai 2024. https://journals.openedition.org/jda/4208; https://doi.org/10.4000/jda.4208

Christie, I., & Crompton, E. (2003). *République de Madagascar: Etude du Secteur Tourisme*. Africa Region Working Paper Series, 63, 149 p.

Combeau-Mari, E. (2004). *Sports et loisirs dans les colonies XIXe–XXe siècle*. Le Publieur.

Combeau-Mari, E. (2007). Le sport colonial à Madagascar (1896–1939). *French Colonial History, 8*, 123–138. https://doi.org/10.1353/fch.2007.0005

Combeau-Mari, E. (2009). *Le sport colonial à Madagascar: 1896–1960*. Société française d'histoire d'outre-mer.

Depraz, S. (2008). *Géographie des espaces naturels protégés. Genèse, principes et enjeux territoriaux*. Armand Colin.

Desrosières, A. (2010). *La politique des grands nombres. Histoire de la raison statistique*. La Découverte.

Fournet-Guérin, C. (2009). La dimension spatiale du fait national à Madagascar: un État-nation en Afrique ? *L'Espace Politique, 7/1*. https://doi.org/10.4000/espacepolitique.1244

Frémont, A. (2009). *La région, espace vécu*. Flammarion.

Froger, G. (dir.). (2010). La mesure des effets des formes de tourisme responsable. *Cahier du CEMOTEV, 3*, 154 p. https://hal.archives-ouvertes.fr/hal-00845223/document

Gicquel, H. (1995). Le football: espéranto de la culture. In D. J.-P. Augustin, & J.-P. Callède (dirs.), *Sport, relations sociales et action collective* (pp. 269–272). MSHA.

Jamet, M. (1995). Sport et politique entre *Welfare* et laisser faire. In D. J. -P. Augustin, & J.-P. Callède (dirs.), *Sport, relations sociales et action collective* (pp. 347–363). MSHA.

Jollivet, M. (2002). *Le développement durable, de l'utopie au concept. De nouveaux chantiers pour la recherche*. Elsevier Masson.

Latouche, S. (2003). L'imposture du développement durable ou les habits neufs du développement. *Mondes en développement, 1*(121), 23–30. https://doi.org/10.3917/med.121.0023

Mougenot, C. (2003). *Prendre soin de la nature ordinaire*. Éditions de la Maison des sciences de l'homme, Institut national de la recherche agronomique.

Mounet, J. (2007). Sports de nature, développement durable et controverse environnementale. *Natures Sciences Sociétés, 15*, 162–166. https://www.cairn.info/revue%2D%2D2007-2-page-162.htm

Naria, O. (2010). Les pratiques sportives. *Les Cahiers d'Outre-Mer, 250*, 275–290.

Nativel, D. (2016). Loisirs et productions spatio-temporelles citadines à l'époque coloniale au Mozambique et à Madagascar (1900–1970). *Cadernos de Estudos Africanos, 32*, 33–52.

OMT (organisation mondiale du tourisme). (2019). *Faits saillants du tourisme*. Madrid, 24 p. https://doi.org/10.18111/9789284421251

Peyvel, E. (dir.). (2019). *L'éducation au voyage: pratiques touristiques et circulation des savoirs*. Presses Universitaires de Rennes.

Pigeassou, C. (2006). Le tourisme sportif: Identification et caractérisation d'un nouvel objet d'étude dans les sciences du sport. In D. P. Bouchet, & C. Pigeassou (dirs.), *Management du sport: actualités, développements et orientations de la recherche* (pp. 143–176). AFRAPS.

Razafindrakoto, M., Roubaud, F., & Wachsberger, J.-M. (2017). *L'énigme et le paradoxe. Economie politique de Madagascar*. IRD Editions.

Rist, G. (2013). *Le développement. Histoire d'une croyance occidentale*. SciencesPo. Les Presses.

Sarrasin, B. (2006). Economie politique du développement minier à Madagascar: l'analyse du projet QMM à Tolagnaro (Fort-Dauphin). *VertigO, 7*(2), 1–14.

Sarrasin, B. (2007). Géopolitique du tourisme à Madagascar: de la protection de l'environnement au développement de l'économie. *Hérodote, 127*, 124–150. https://doi.org/10.3917/her.127.0124

Sarrasin, B., & Ramahatra, H. (2013). *Environnement et développement à Madagascar. Du plan d'action environnemental à la mise en valeur touristique des ressources naturelles*. Karthala.

Sénat. (2011). *Rapport d'information fait au nom de la délégation sénatoriale à la prospective (1) sur les villes du futur: Villes du futur, futur des villes: quel avenir pour les villes du monde ?* https://www.senat.fr/notice-rapport/2010/r10-594-1-notice.html

Stafford, J., Sarrasin, B., & Ramahatra, H. (2002). Portrait du tourisme à Madagascar: entre nature et éloignement. *Téoros, 21*(2), 70–72.

Vasseur, P. (1997). Écosystèmes côtiers en danger dans la région de Tuléar. Analyse des agressions humaines et problèmes de gestion. In D. J.-M. Lebigre (dir.), *Milieux et sociétés dans le sud-ouest de Madagascar* (pp. 97–120). Centre de Recherches sur les Espaces Tropicaux (Université Michel de Montaigne, Bordeaux 3).

Chapter 23
Sport Tourism in Mayotte: A Sector of Activity Between Confidentiality and Potential

Dahlin Mikidachi Mondroha

Introduction

Mayotte, an island of the Comoros archipelago located halfway between Africa and Madagascar, is a territory with an area of 374 km^2, composed of two main islands: Grande-Terre (325 km^2) and Petite-Terre (18 km^2) and about 30 uninhabited islets. According to the panorama of Mayotte (INSEE, 2020), Mayotte has 212,600 inhabitants, a population that is increasing rapidly. According to INSEE (2022), the island's demographic situation has been catastrophic since 2012, with particularly dynamic population growth (+3.8% per year on average). The huge challenge revolves around the youth under the age of 21 years, who represent 60% of the population. Thousands of young people are left to their own devices, in the absence of appropriate means and a real political will (Fontaine, 1995).

Mayotte and its population are attractive due to its socioeconomic context (Math, 2012). The population is growing steadily, as there is a large immigrant population from neighboring islands in the territory. The territory is thus considerably behind in terms of economic development, education and public facilities (Blanchy, 1990) despite its status as a department. This situation is also complemented by a strong mix of cultures that leads to multilingualism on the island (Barbier et al., 2005). Mayotte is currently experiencing a very difficult situation by way of a social crisis, a worrying situation (Barbier et al., 2005; Mayotte, 2015; Math, 2012). By acceding to the status of department, Mayotte is now experiencing social, economic, cultural development and a profound transformation occasioned by this process of political integration (Roinsard, 2014).

D. M. Mondroha (✉)
Department of Social Cohesion, Youth and Sports of the Departmental Council of Mayotte, Mayotte, France

D. Van Rheenen et al. (eds.), *Sport Tourism, Island Territories and Sustainable Development*, Sports Economics, Management and Policy 24,
https://doi.org/10.1007/978-3-031-51705-1_23

Within this framework of profound societal changes, Mayotte has experienced the development of recreational leisure, tourism, and sport tourism for two decades (Saffaride, 2018). These activities can be carried out in an exceptional natural environment, previously preserved from mass tourism. Indeed, an exceptional fauna and flora with more than 650 native plant species and 20,000 ha of forests cover 53% of the island (Barthelat, 2019). This forest fauna is very varied and composed of terrestrial mammals and forest birds. The lagoon of Mayotte is a world-class marine heritage site that is being studied for UNESCO heritage listing, with more than 1100 km^2 of surface and 270 km of reefs (Guézel et al., 2009).

The objective of this chapter is to study the role of the local population in the process and sectoral development of sport tourism. More specifically, it is to understand the contribution of these sport tourism activities in the organization of sustainable development policies of the Mahoran territory. The chapter will analyze more specifically the Sustainable Development Goals 1, 3, 8, and 10 of the World Tourism Organization (2019) (Map 23.1).

Elements of Context on the Island

The Weight of Tradition

As permitted by article 75 of the Constitution[1], local law inspired by Muslim law and African and Malagasy customs applies only to Mahorais who have retained their personal status. The law of July 11, 2001, on Mayotte maintains the existence of this local law civil status and specifies the possibilities of renouncing it in favor of common law civil status. Ninety-five percent of the Mahorais are Muslims of Sunni faith, but their practice of Islam is moderate. The areas of application of local law civil status are specified by the Overseas Program Law (LOPOM) of July 21, 2003: the status and capacity of persons, matrimonial regimes, inheritance, and gifts. The law on immigration and integration of July 24, 2006 entrusted the celebration of civil marriages to the civil registrar alone, whereas previously a marriage celebrated before the cadi under local law could be validated under common law by the simultaneous presence of a civil registrar. Thus, in Mayotte, society functions with the weight of important traditions, marked by rurality (Barthes, 2003), which give rhythm to the daily life of the populations and slow down the opening toward new activities such as tourism and sport tourism (Fontaine, 1987).

[1]Constitution of October 4, 1958 *(Updated January 2015)* https://www.assemblee-nationale.fr/connaissance/constitution.asp

Map 23.1 Presentation of Mayotte. (Source: http://afriquemonde.org)

The Growth of a Very Young Population

According to Candau and Rey (2015, pp. 1159–1172), Mayotte's population has increased sharply from just over 90,000 in 1990 to over 200,000 in 2010 to just over 300,000 in 2019. Mayotte has become the first maternity hospital in Europe with an average of 10,000 births per year. INSEE states that in 2017, 256,500 people live in Mayotte. The population remains much younger than elsewhere in France: the average age is 23 years (compared to 41 years in metropolitan France) (INSEE, 2020). As a result, there are significant migratory flows and a large part of the population lives with very few resources: 77% of the inhabitants live below the national poverty line, five times more than in metropolitan France (INSEE, 2019). In 10 years, this population has grown by 42%, Mayotte the second most populous island in the southwest Indian Ocean, after Mauritius. The capital, Mamoudzou, absorbs more than 53,000 inhabitants (28% of the total population). The birth rate and immigration, essentially illegal, from neighboring islands are increasingly high and are also the cause of this very significant population growth: the average annual growth rate recorded between the last two censuses is 3.1%. The population of Mauritius is young: 55% of the total population is under 20 years old, the highest percentage of all French territories.

Significant Delays in the Economic and Employment Fields

According to Édouard Fabre and Jannick Rivière (INSEE, 2014), three out of four Mahorais who have left the school system or who have never worked are without a diploma. This is a clear improvement since 2009, but an important figure that sums up the Mahoran situation. In 2014, 36% of the working-age population of Mahoran had never been to school, whereas in metropolitan France, the figure was only 2.6%. We have seen a slight improvement since 2009 in this phenomenon, but Mayotte has twice as many people without diplomas as in metropolitan France. It is important to note that the percentage of people who went to school without a diploma in Mayotte is 7% higher than in metropolitan France. For people with a qualifying diploma, the difference is very significant since the percentage in Mayotte is 27% in 2014, while in metropolitan France, it is practically at 69%, if we take the example of female education graduates.

According to INSEE (2019), the market sector is underdeveloped in Mayotte, particularly industry and the market tertiary sector, which concentrate far fewer jobs than at the national level. *Conversely*, the non-market tertiary sector concentrates more jobs than elsewhere in France, as in the other DOMs: 55% of jobs, compared to 30% at the national level.

Apparent labor productivity (GDP per job) is significantly lower than other French departments, including overseas: 51,686 euros per job, compared to 83,204 euros at the national level. In 2018, the GDP per capita is nearly four times lower than the national level… The economic fabric, still very informal, partly explains this low productivity.

In summary, a dramatic situation characterizes the island from a social and economic point of view, exacerbated by enormous demographic pressure from a traditional society and illegal immigration (Mayotte, 2015; Askandari, 2009). The development of Mayotte today rests on "heavy challenges" (Sueur et al., 2012).

An Attractive Natural Heritage

A Natural Setting Conducive to Leisure Activities and Sport Tourism on Land and in the Air

Mayotte has a very rich and preserved climate and land atmosphere, both for its fauna and flora. A tropical island, specific, rich, and abundant, Mayotte is nicknamed the perfume island because it hosts countless rare and striking plant species (baobabs, vanilla, cocoa, jasmine, ylang-ylang…) as well as a varied fauna (makis, dogfish…) (Barthelat, 2019).

This rich natural heritage has contributed to an emerging development of tourism in Mauritius (Gay, 1999). This development can be based on multiple leisure activities and terrestrial sport tourism, such as hiking, horseback riding, mountain biking, canyoning, climbing, or golf, as well as the emerging air sport tourism, with paragliding and microlight.

As an illustration, in the middle of one of the largest lagoons in the world, Mayotte is a playground and adventure for hiking enthusiasts. The island is one of those rare places that have managed to preserve an authentic environment. The numerous trails that run along the corners of the island of Grande-Terre as well as Petite-Terre allow you to discover the dense bush and forest of Mayotte. The 13 trails of Grande Randonnée (GR) of Mayotte are not all regularly maintained, but efforts are underway to restore them. On the island, there are almost no accommodations on the trails, but it is relatively easy to reach nearby villages in order to eat or stay (MKG, 2018) (Table 23.1).

On land, there are other possibilities with mountain biking, especially in Grande-Terre. The ONF has already listed nearly one hundred kilometers of trails for guided

Table 23.1 The 13 GR hiking stages

Numbers	Locations	Distance
Step 1	Mamoudzou to Koungou	10.5 km
Step 2	Koungou to Dzoumogné	18.8 km
Step 3	Dzoumogné to M'tsamboro	11.9 km
Step 4	M'tsamboro to Mtsangamouji	10.8 km
Step 5	Mtsangamouji in Chiconi	15.7 km
Step 6	Chiconi to Chirongui	17 km
Step 7	Chirongui to Dapani	8.73 km
Step 8	Dapani to Mtsamoudou	10.28 km
Step 9	Mtsamoudou in Bandrélé	8.32 km
Step 10	Bandrélé to Tsaranoro	14.8 km
Step 11	Tsaranoro to Vahibéni	11 km
Step 12	Vahibéni in Mamouzdou	10.8 km
Step 13	Dzaoudzi to Pamandzi (Petite-Terre)	14 km
Total		162.63 km

Source: Conseil Général de Mayotte (2019)

outings (HTL, 2019). Faced with the growth of sports and tourist leisure activities, in 2021, a National Nature Reserve of the Forests of Mayotte (2800 ha) has been formalized in a perspective of sustainable development. As a result, sports leisure and tourist activities can be practiced in this tropical island environment marked by the variety of landscapes, reliefs and coasts, and its preservation.

A Strong Attraction for Ocean Sport Tourism Activities

Due to its insular geomorphology, Mayotte has a vast lagoon of 1100 km^2, one of the largest closed lagoons in the world, with its 195 km long coral reef and its second internal barrier, a rare geological epiphenomenon (Guilcher, 1965). Essentially, it covers a very rich biodiversity with its 7.3 km^2 of mangrove swamps and as many grass beds. In total, 25 species of marine mammals are present, but also 2300 marine species, including 760 species of fish and two species of turtles (HTL, 2019).

The island is a leading destination for diving. The water is never below 25 °C and the waves are very limited thanks to the protection by the coral reef and a weak wind. The currents, except in the passes, are also very weak and thus meet all the conditions for a safe start.

Kitesurfing and wakeboarding are practiced on the island's jagged coastline, which does not allow for stable winds that are conducive to kitesurfing. However, Petite-Terre, which is less disturbed, is better suited to it, as well as other spots on Grande-Terre, especially on the east coast and in the south. Kayaking and paddle boarding take place thanks to the presence of the coral reef, which makes the lagoon ideal for admiring the sea floor through a transparent lens. Most coastal hotels rent them. Day or half-day kayak excursions are organized and recommended to reach the islets bordering Grande-Terre. The most popular excursion is to Bandrélé islet, accessible from Sakouli beach.

Big game fishing attracts many tourists. The sport fishing offer is slowly but surely developing. More accessible, jacks, barracudas, kingfish, groupers, and wahoos are found in the lagoon. Big game fishers, who fish outside the lagoon, will find something to do in Mayotte, but it is important to know that there are very few service providers who organize these trips.

Around the numerous islets of the lagoon, the activity of pleasure boating has developed progressively, while sailing enthusiasts take advantage of the "southern winter" currents between June and October. Equally important but more touristic, cruises to the bays and coves of the neighboring islands are increasingly successful, allowing snorkeling, diving, or activities derived from sliding and surfing.

Thus, in 2010, the Natural Park was created in Mayotte, under the direct authority of the Prefect and the Directorate of Food, Agriculture and Forestry, which aims in particular to discover the marine environment and biodiversity through the organization of recreational activities and the professionalization of tourism stakeholders (Legoff, 2010).

Sport Tourism, Issues, and Perspectives in Terms of Sustainable Development

The Difficulties of Governance of Local Actors

All the actors of the territory understand the potential of the exceptional natural heritage which can contribute to the increase of the requests for the activities of the natives and especially which could take part in the attractiveness of sport tourism. Moreover, there is a real opportunity for all project leaders who can be financed by the European Regional Development Fund (ERDF) and the European Social Fund (ESF) of the European Union to develop projects around tourism and sports activities.

However, there is a lack of real coordination of these activities around the heads of networks in the territory (CROS Mayotte, Prefecture of Mayotte, Departmental Council, 2018). For several reasons, these activities are often not financially accessible to the vast majority of the indigenous population since they are often paid for in the territory. Two problems that have arisen are the lack and state of dilapidated infrastructure dedicated to sport tourism activities, as well as the lack of training for qualified professionals, such as in the field of climbing or even canoeing. These aspects can be linked to the structural problems of the last 15 years of regional development (Roinsard, 2014) related to insecurity, poor waste management, and an institutional landscape not well suited to the development of tourism, and therefore sport tourism. The territory is also penalized by the price of airline tickets and the lack of accommodation.

However, there are planning documents to remove these obstacles. Thus, the Schéma de Développement et d'Aménagement du Tourisme et des Loisirs de Mayotte (HTL, 2019) expresses the levers for the implementation of strategic actions that would favorably impact the development of tourism and leisure. An ambitious tourism development scheme that also derives from the principles of balanced development and long-term planning. In addition, for the Forestry Guidelines of the Department of Mayotte (OFDM),[2] the island forest constitutes spaces of valorization but fragile and potentially threatened by the population explosion and economic development of the island.

At the level of the sports movement, Mayotte is one of the 22 departments considered by the "Pôle de ressources nationales des sports de nature" that has a policy of deploying nature sports, but the departmental council has not committed to the CDESI-PDESI approach due to the budgetary context. It can be noted that during the assises du sport (CROS Mayotte, Prefecture of Mayotte, Departmental Council, 2018), a roundtable devoted to the issue of nature sports was an opportunity to emphasize the beneficial contributions of sports practices in nature in terms of

[2](LAW No. 2014-1170 of October 13, 2014 on the future of agriculture, food and forestry) Forestry Guidelines for the Department of Mayotte Prefiguring the Forest and Wood Program of the Department of Mayotte.

environmental education and personal development. The state periodically implements actions in favor of nature sports. Indeed, since 2018, the DJSCS of Mayotte and the local authority organize the Meetings of nature sports. This event takes place over a weekend, in a different town each edition. More than 2000 Mahorais discover various activities such as stand-up paddle, water climbing, hiking, snorkeling, swimming, kayaking and pirogue mahoraise, or traditional beach activities.[3] In terms of training courses for nature sports, the actors have a desire to use nature sports to enhance, educate, and preserve the environment for the population.

Ultimately, the actors of the territory are aware of the stake they represent and initiatives are underway in "green sports" through mountain biking, hiking, climbing; and in blue sports, via kayaking, canoeing, swimming, and diving. There is a common desire on the part of the various stakeholders in the territory to use sport tourism and nature sports to enhance, educate, and embellish the local environment (Saffaride, 2018). However, it is clear that Mayotte has not yet fully exploited the island's potential. In order to discern a better organization of sport tourism activities on the territory, a joint consultation between stakeholders in the sports, land use, tourism, agriculture, and environment sectors becomes urgent and necessary.

Emerging Sport Tourism: Levers of Sustainable Potential

Contrary to many other sports activities on the island, we note that sport tourism activities are practiced in an intense way, but outside of an organized sports association framework, sometimes in natural sites that are not necessarily specifically developed for the practice. The inhabitants are confronted with economic, social, and environmental difficulties, but they could benefit from a controlled sustainable development of this sector of activity on the whole territory.

Since the COVID-19 crisis, this sector is booming, with many people engaging in such activities. Sport tourism is one of the tourist sections that is expanding rapidly since sports events of all types and sizes are increasing more and more and attracting many tourists. We can mention the Ylang race (2010), the run and bike (2010), the turtle run (2010), the lagoon canoe race (2014), the trail in the city (2015), and even the turtle round (2016).

The islanders combine the balance of the socio-educational, cultural, and sporting activities, on the one hand, by local, cultural activities carried by natives for the most part, and on the other hand by activities related to sport tourism, leisure, outdoor activities, and hiking. It is still difficult to integrate the local actors in all this step since they are recent elements in the territory. Nevertheless, the population is progressing and integrating considerably all its sport tourism organizations in the local landscape.

---

[3] https://www.sportsdenature.gouv.fr/developper-les-sports-de-nature-en-territoire-atypique-exemple-a-mayotte accessed on March 5, 2023.

Sport tourism can teach values such as fairness, teamwork, equality, discipline, inclusion, perseverance, and respect. It can make healthy sport very beneficial. Since the COVID-19 crisis, the relationship to health has become very serious throughout the territory (Senate, 2022). Mayotte has increased in recent years the number of practitioners, walkers, hikers, following the confinement and a questioning of eating habits favorable to the practice of tourism and sport.

Mayotte welcomes cruise ships that stop for a few hours at the perfume island; the cruise passengers have the choice to visit a part of the territory with guided and targeted tours. Hundreds of artisans benefit from these few specific moments in the year. Thus, new forms of social and territorial integration for local populations exist. The territory is developing these infrastructures to give more autonomy to the indigenous population through training in the field of sport tourism animation. For example, development work is being studied: for example, bicycle paths in the Dembeni-Mamoudzou urban community (Cadema) or in the 3CO intercommunity. Kilometers of long-distance footpaths are being renovated and cleaned. In recent years, the number of accommodations has increased significantly and is complemented by urban transport works to facilitate the movement of the population toward economic and tourist activities.

Sport Tourism Issues for the Local Population

For the people of Mauritius, the practice of sport tourism represents various stakes and interests. We can mention environmental awareness for schools, immersion in an environment for some volunteers, and the transmission of values for the different actors of the territory. For the collective reception of minors, specific conditions are required to the different teachers and educators. Nature education through sport tourism is a simple lever to encourage and activate from a very young age in educational structures.

In the island, the sports equipment, sites, and arranged places allow the practice. It is essential to know the facilities and equipment that are necessary for the practice of these activities. This will allow the development of natural areas, the local cultural heritage, the interpretation trails, but also the activities that are conducive to the discovery of the island. This will allow to really engage in a sustainable development approach. Sport tourism, as the name indicates, is closely connected to the landscapes and natural sites on the island, and this is very positive in the sense that the numerous facilities created (trails, parking, lodges …) delimit the areas of practice, accommodation, and parking and channel the public.

Mayotte suffers from an imbalance of the territory, between the countryside and the city (Fontaine, 1995). Nature sport tourism can play an important role in rebalancing and using the natural space that offers diversity and richness of these landscapes. The increased growth and democratization of tourism and sports implies economic effects on the territory. But it is especially the tourist offer that will have impacts on the island. These structures are important assets for the communes and inter-communities. Their image is enhanced. The territories thus appear more dynamic and necessarily attract more sport tourists. The enhancement of the island through sport tourism has contagious consequences on the other sectors of public policy.

The evolution of ecological means of transportation becomes an essential need in our territory. The representation of routes adapted to soft mobility must be a territorial stake in Mamoudzou as in the rest of the territory. To think in this direction is to support the practice of sport tourism in a connection with other ecological means of transport.

Sport tourism can greatly contribute to the development and reinforcement of sports excellence on the island. Sports events that promote certain athletes can serve as a model. Events for all can integrate high level athletes, who are gateways to regular practice. Mayotte must seize the opportunity of these activities, which are in full expansion to prepare our future champions and make the island known.

Even if the lagoon and its natural terrestrial heritage constitute economic resources with considerable potential for exploitation in the tourism sector, it is nevertheless a fragile heritage, whose conditions of reasoned exploitation must be accompanied by protection and conservation measures, which have been initiated for some years now (Bernardie-Tahir & El-Mahaboubi, 2001).

Conclusion

The actors in Mauritania have realized that in order to respond favorably to the different challenges of sport tourism in the territory, it is important to implement a strategy that relies on different levers, particularly through the mobilization of the network and partners. The implementation of the approach that focuses on the deployment of sport tourism activities is at the crossroads of tourism, sport, and recreation development logics. The aim is to make sport tourism a tool for the sustainable development of the island.

References

Askandari, A. (2009). *Logiques politiques et mahorité dans la postcolonie de Mayotte* (p. 241). L'Harmattan.

Barbier, G., Campion, C.-L., Desmarescaux, S., Fischer, G., Gournac, A., Juilhard, J.-M., et al. (2005). *Rapport d'information sur la situation sociale à La Réunion et à Mayotte, Commission des affaires sociales du Sénat*, n° 140, séance du 19 décembre.

Barthelat, F. (2019). The illustrated Flora of Mayotte. In *Muséum national d'Histoire naturelle* (Inventaires & biodiversité; 15, 688p). Biotope, Mèze.

Barthes, C. (2003). *L'état et le monde rural à Mayotte*. Coll. Economie et développement, CIRAD-Karthala,

Bernardie-Tahir, N., & El-Mahaboubi, O. (2001). Mayotte: des parfums au tourisme.les nouveaux enjeux du littoral. *Les Cahiers d'Outre-Mer, 216*, 369–396.

Blanchy, S. (1990). *La vie quotidienne à Mayotte* (240 p). L'Harmattan.
Candau, F., & Rey, S. (2015). *Economic Review, 66*, 1159–1172,
Conseil Général de Mayotte. (2019). *Schéma de Développement et d'Aménagement du Tourisme et des Loisirs de Mayotte.*
CROS Mayotte, Prefecture of Mayotte, Departmental Council. (2018). *Assises du sport à Mayotte.* https://www.cros-mayotte.fr/non-classe/les-assises-du-sport-a-mayotte/
Fabre, É., & Rivière, J. (2014). *Training and employment.* Mayotte. INSEE. Flash, No 21. Published on 11 Nov 2015.
Fontaine, G. (1987). *Société et économie de Mayotte* (450 p). PhD thesis, University of Nice.
Fontaine, G. (1995). *Mayotte* (250 p). Karthala.
Gay, J.-C. (1999). Les balbutiements du tourisme mahorais. Travaux & documents. *Propos géographiques sur le Sud-Ouest de l'océan Indien, 11*, 137–152. ffhal-02175041fr.
Guezel, R., Quartararo, A., Aboutoihi, L., Saindou, K., Salaun, P., Ybrahim, B., & Aranaud, J. P. (2009). *Richesses de Mayotte—Parc naturel marin de Mayotte—Les hommes et l'océan—Mission d'étude pour la création d'un parc naturel marin à Mayotte Agence des aires marines protégées* (72 p).
Guilcher, A. (1965). *Les récifs coralliens et le lagon de l'île de Mayotte* (Mémoire de l'ORSTOM, n°11, 210 p). ORSTOM.
HTL. (2019). *Schéma Régional de Développement du Tourisme et des Loisirs de Mayotte* (361 p).
INSEE. (2020, July). https://www.insee.fr/fr/statistiques/4210656—INSEE Analyses. *Inequalities in living standards have increased*, No. 25. INSEE. https://www.insee.fr/fr/statistiques/4622454
Insee, (2022), Estimation de la population au 1er janvier 2022 – Séries par région, département, sexe et âge de 1975 à 2022, janvier.
INSEE Analyses. (2019, September). *L'insertion sur le marché du travail à Mayotte*, N°21.
INSEE Première. (2019, February). *À Mayotte, près d'un habitant sur deux est de nationalité étrangère*, n°1737. INSEE. https://www.insee.fr/fr/statistiques/3713016
Legoff, N. (2010). *La rapide création du Parc Naturel marin de Mayotte.* EchoGéo [En ligne], Sur le Vif, uploaded on 06 Apr 2010, accessed on 02 Mar 2023. http://journals.openedition.org/echogeo/11808
Math, A. (2012). Mayotte—La situation économique et sociale. *Chronique internationale de l'IRES*, n° 132, 41–54.
Mayotte 2025, une ambition pour la République. (2015). *Préfecture de Mayotte.* http://www.mayotte.gouv.fr/Politiques-publiques/Amenagement-du-territoire-Politique-de-la-Ville-et-Cohesion-Sociale/Mayotte-2025-une-ambition-pour-la-RepubliqueV1
MKG. (2018). *Etude sur l'hébergement marchand à Mayotte* (118 p). AFD, Département de Mayotte, ADIM, Observatoire économique de Mayotte.
Roinsard, N. (2014). Living conditions, poverty and social protection in Mayotte: A multidimensional approach to inequalities. *Revue française des affaires sociales, 28–49*. https://doi.org/10.3917/rfas.144.0028
Saffaride, A. (2018). *Tourism and heritages* (410 p). Valorisation et développement de territoires insulaires dans l'océan indien: Mayotte, doctoral dissertation in the framework of École doctorale Normandie Humanités (Mont-Saint-Aignan, Seine-Maritime).
Senate Report, Cohesion, Performance, Rayonnement, Quels tremplins pour le sport en outre-mer, 2018–2019, Senate, Report No. 140. http://www.senat.fr/fileadmin/Fichiers/Images/delegation/outre_mer/Delegation2017-2020/DSOM_synthese_rapport_sport_final.pdf. Information report by Ms. Catherine Deroche, Mr. Jean-Luc Fichet, Mr. Dominique Théophile and Ms. Laurence Cohen, on behalf of the Social Affairs Commission, 2022, Mayotte: A hypertensive health care system, No. 833 (2021–2022)—27 July 2022.
Sueur, J.-P., Cointat, C., & Desplan, F. (2012, July 18). *Mayotte: un nouveau département confronté à de lourds défis.* Rapport d'information, n° 675, Sénat.
World Tourism Organization. (2019). *Sport tourism and the Sustainable Development Goals (SDGs).* UNWTO, Madrid. https://doi.org/10.18111/9789284419661

Chapter 24
Implication of Tourism Development in Rodrigues: Ecotourism and Sport Tourism in a Sustainable Perspective

Henri Agathe

Introduction

Rodrigues is the smallest island of the Mascarene archipelago, composed also of Mauritius and Réunion Island, located on the Mascarene Continental shelf in the south-west of the Indian Ocean.

Rodrigues is part of the Republic of Mauritius (ROM). The ROM comprises the islands of Mauritius, Rodrigues, Agalega, Saint Brandon, and other islands each quite remote from one another (see Fig. 24.1). The ROM has a total land area of 2040 km^2 and a coastal line of 322 km that extends over a vast oceanic area with an equivalent Exclusive Economic Zone (EEZ) of 2.3 million km^2. Rodrigues, the second largest island of the ROM, lies about 560 km north-east of the main island Mauritius, has a surface area of 108 km^2, and in 2019, had a population of 44,000 inhabitants and has received 77,831 tourist visitors (Report, *Assise du tourisme de Rodrigues*, 2021). Since 2002, Rodrigues enjoys a large degree of autonomy within the ROM provided by Rodrigues Regional Assembly (RRA) Act, 2002. The governing body of the island is the Rodrigues Regional Assembly (RRA), which has a minimum of 17 elected members and an Executive Council headed by a Chief Commissioner. There are six other Commissioners who are responsible for different domains, such as tourism, also responsible for other subjects like employment and Information and Communication Technology (ICT), as well as sport, inclusive of a portfolio with responsibilities for culture, art, and youth (RRA Act, 2002). Tourism is considered the most important pillar of Rodrigues' economy and its main engine for growth.

Rodrigues is quite different from Mauritius, resulting from its long history as a predominantly peasant and rural community. Until the 1970s, agriculture was its main economic activity, which was essentially of a peasant and subsistence nature,

H. Agathe (✉)
Regional Assembly of Rodrigues, Port Mathurin, Mauritius

D. Van Rheenen et al. (eds.), *Sport Tourism, Island Territories and Sustainable Development*, Sports Economics, Management and Policy 24,
https://doi.org/10.1007/978-3-031-51705-1_24

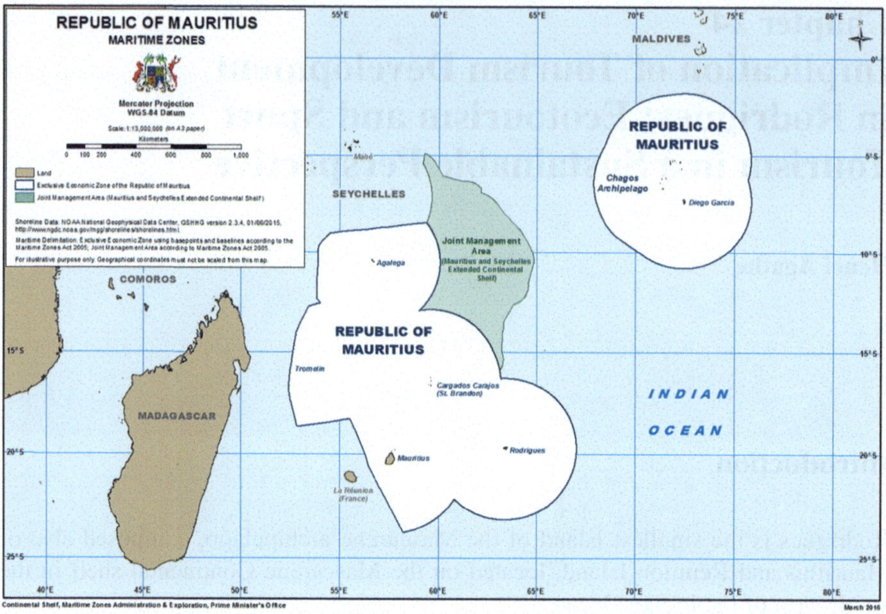

Fig. 24.1 Location of the Republic of Mauritius with its immense maritime zones, as encircled, of approximately 2.3 million km^2

whereby most of the production was aimed at satisfying the needs of the local population, while the remainder was exported towards the Mauritian market.

In the 1970s and 1980s, Rodrigues experienced a series of changes, both political and climatic, due to a severe drought, which led to a transformation of the socio-economic fabric of the island. Tourism emerged as a key development sector. Rodrigues is now moving towards a predominantly market economy where the needs of the population is for the most part satisfied through markets and small businesses.

The lifestyle of the population, generated by the agrarian society and the unique natural assets developed during a long history marked by geographical isolation, is considered as the main driver of tourism development of Rodrigues. The people of Rodrigues have a friendly and open nature and enjoy a relaxed lifestyle, all of which is very attractive to visitors from more staid and hurried lifestyles. The cultural heritage is characterised by a simple way of life with its own culinary art, traditional folk music and dances, a Creole language, and a rich history. The natural heritage assets are a beautiful hilly landscape surrounded by a magnificent large lagoon which is more than twice the size of the island, extensive beaches, limestone plains, and a diverse, endemic, and indigenous flora and fauna.

The history of tourism development in Rodrigues is closely linked with the development of air travel. Tourism was almost non-existent in the 1960s. The small airport of Rodrigues inaugurated its first flight of a six seater *Piper Navajo* in July 1972 (airmauritius.com, 12 Oct 2022). However, it is only in the early 1980s that tourists started to visit Rodrigues for recreational purposes. Tourism development really took off in December 1987, with the entry into operation of an ATR 42-500 by Air Mauritius when it started daily commercial flights between Rodrigues and

Mauritius. The growth has been progressive since then, except for a dip in 2005 and 2006 linked very likely with the outbreak of the chikungunya epidemic on the island; then in 2008, related to slowdown in international tourism due to the global financial crisis and also recently in 2020 and 2021 due to the Covid pandemic.

Ecotourism has been and is still significant in Rodrigues. Ecotourism defines an activity or a particular type of tourism product in natural areas that conserves the natural environment and sustains the well-being of local people. The largest accommodation capacity and related activities are provided by small enterprises spread throughout the island. Sport tourism is a major contributor to ecotourism in Rodrigues. The natural landscape offers a perfect setting for sport activities and adventure (Naria et al., 2022). Sport tourism in Rodrigues is mainly about participating in activities or hosting events in relation to its natural landscape. There are no visitors coming as passive or active participants to big events or tournaments. Most of the visitors are active participants to sport activities and are accompanied by friends or family members. It has significantly contributed in making Rodrigues known as a destination and contributes substantially to the development of tourism in Rodrigues in terms of the visitor numbers, income generation, and the emergence of new entrepreneurs. In reference to the issue addressed by the World Tourism Organization (2019), the analysis of my chapter on Rodrigues will deal specifically with SDGs 1, 8, 10, and 14 of sport tourism.

Presentation and Characteristics of Rodrigues Islands

Geographic and Landscape Profile

Rodrigues is a subtropical volcanic island situated approximately 560 km to the east of Mauritius at around longitude 63° 25′ East, and latitude 19° 43′ South (see Fig. 24.1). Its maximum length East–West is 16.5 km and maximum width North–South is 7.5 km. The landmass is surrounded by a fully developed asymmetrical peripheral fringing reef system, forming an almost uninterrupted band around a lagoon and measuring approximately 90 km in circumference.

The landscape characteristic of the island is significantly determined by its geology and topography (Check & Hume, 2009). Two principal regions are identified. The 'higher lands' consists of a central ridge of low mountains running from east to west forming the backbone of the island. Mont Limon is the highest point 396 m above mean sea level. The higher lands are of volcanic origin, with basalt creating the main rock structures. The higher lands flatten towards the southwest of the island into a 'plain' of limestone (Aeolian Calcerentic deposits) at Petite Butte-Plaine Corail, (See Fig. 24.2). Smaller deposits are located at Graviers, Trou d'Argent, and Pointe Coton. Rodrigues is the only island that has a vast expanse of limestone throughout the Mascarenes Islands. The limestone plain of Petite-Butte-Plaine Corail hosts an underground system of caves. Among the nine caves, the most famous and visited ones are Caverne Patate the longest of a length of 1 km and la Grande Caverne the largest having a length of 500 m.

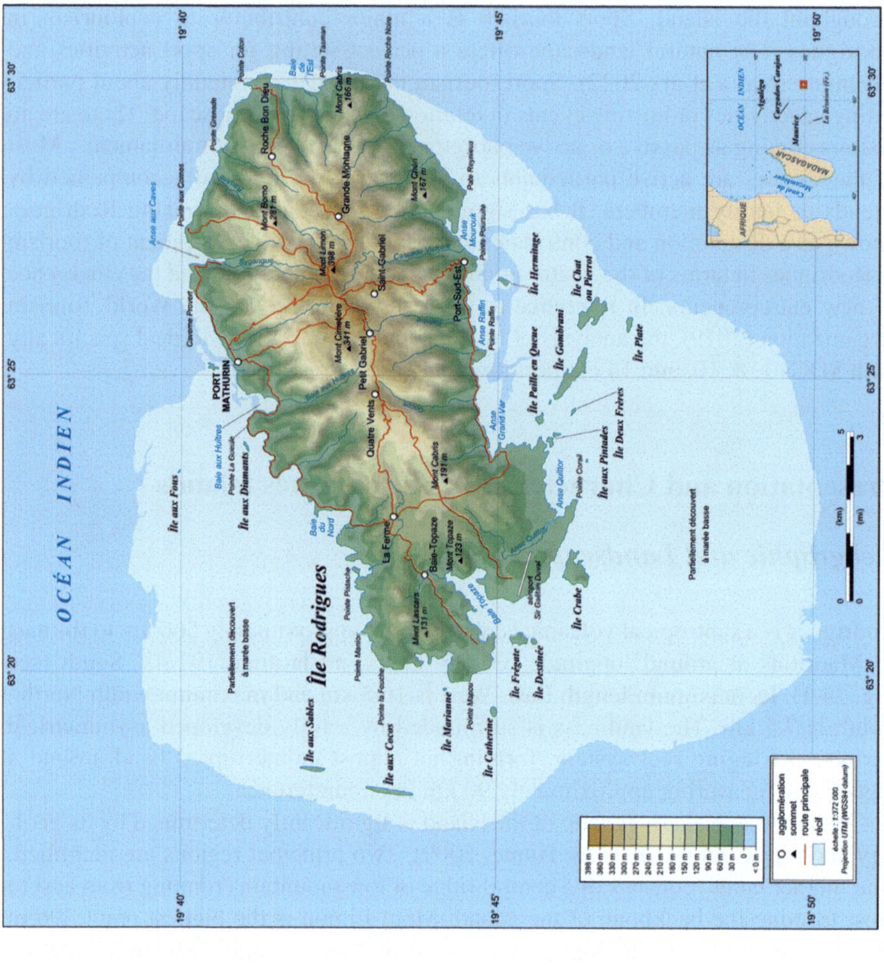

Fig. 24.2 Topographic map for Rodrigues. (Source: Wikimedia Commons)

Numerous rivulets radiate from the high ridge towards the sea which have created a series of deep ravines that open out into valleys of 300 m width that cut the coastal topography. These valleys have broad flat land formed by alluvial soils, which provide opportunity for market gardening activities.

These topographical features of Rodrigues constitutes an invaluable heritage and offer a strong infrastructure for hiking. The numerous hills and slopes offer breathtaking panoramic viewpoints of the island and of its lagoon.

The distance between the coast and reef barrier varies considerably, being greatest in the North-West and South-West sectors at 7 km offshore. In the west and in the south of the island, the reef is located about 4 km offshore. On the north-east, east and south-east, the reef is considerably closer to land, with a distance rarely exceeding 1 km. The reef encloses an approximately 240 km^2 lagoon which is more than twice the surface area of the island of Rodrigues (108 km^2) (KPMG/UNDP, 2009). The lagoon comprises many islets, many are of tiny size, and some are attractions for tourists, namely, Ile aux Cocos, Hermitage, Gombrani, Ile aux Crabes, and Ile aux Chats. The lagoon is quite shallow with an average depth of less than 2 m. The maximum tidal range is around 1.90 m. Most areas are exposed at the low spring tides during winter, when the lowest tides are experienced. The lagoon is connected to the open sea via three major channels cutting the reef. One was dredged for Port Mathurin harbour activity in the north and two natural channels: one in the south adjacent to Port Sud-Est and one in the East in the St Francois Bay area. There are many other smaller reef passes which are scattered along the 90 km coral reef, most of which are used by fishermen equipped with small fishing boats to access open sea fishing ground. This set of conditions makes the lagoon and the reef quite accessible for a variety of recreational, fishing, and sport tourism activities.

The south-east trade winds influence the island climate, providing warm dry winters from April to October and hot wet summers from November to April. The average mean temperature ranges between 22.1 °C in August and 27.2 °C in February in the lowlands from 1991 to 2020 according to the figures of the Mauritius Meteorological Office. There is, however, a variety of micro climates across the island, owing to its hilly topography; however, weather data are unfortunately not recorded in a variety of places on a consistent basis to document these micro climates. Cyclones prevail during summer months from November to April. The historic average annual rainfall for the island is 1200 mm; however, this figure tends to vary considerably during the recent past years.

Cultural Context

In 1735, after several failed attempts earlier in the eighteenth century, a permanent French settlement was established. Some white settlers arrived in Rodrigues bringing African slaves to work their stock and farms. Today, most of the inhabitants are of African descent. There is a minority of mixed-race people, descendants of the first European settlers, and a minority of people of Asian origin from India and China mainly.

The main religion is Christianity, dominated by Roman Catholicism, with other smaller Christian churches, and there are small communities of Hindus, Muslims, and Buddhists. The main language is Rodriguan Creole, which is now taught at schools on an optional basis. English and French are the official languages and the medium of teaching. Rodriguan Creole is close to Mauritian Creole, but is recognised as a different language in the Republic of Mauritius, given that many words are pronounced differently and displays original idioms. There exists a dictionary of Rodriguan Creole since 2021. People born in Rodrigues Island are called Rodriguans regardless of their parents' ethnic or national origins. The cultural awakening occurred at the end of the 1970s, prompted by the construction and consolidation of the cultural identity of Rodrigues through the elements of food, music and dance, crafts, and language.

Diversification of the Tourist Economy: From Ecotourism to the Contribution of Sport Tourism

Owing to its very small size and its geographical location away from the other islands of the Mascarene archipelago, Rodrigues has received less attention and initiatives from the colonial powers compared to Mauritius and Réunion during the colonial periods. The combination of its isolation and the small size of its territory has inhibited its development.

The export-base theory highlights the vulnerability associated with the small size of island territories. The smaller the economy, the more it needs to specialise in a small number of export-oriented products in order to optimise scarce resources. This results in economic structures that are not very diversified, highly dependent on foreign trade, not very autonomous, and vulnerable to external factors (Agathe, 2006). As an illustration, specialisation in sugar cane production has allowed Mauritius to develop an efficient economy integrated to the global economy, achieved through the concentration of sugarcane plantations and factories, thus allowing economies of scale and gains in productivity and competitiveness. Rodrigues, on the other hand, has not been able to achieve such a specialisation of its economy. Nevertheless, its diversified agrarian economy, aimed at satisfying its own needs while exporting surpluses and some specific products to Mauritius, has allowed the development of a form of economic efficiency giving rise to an original autonomous economic system.

Three main periods can be distinguished in the recent history of the Rodriguan economy:

Transition from a Peasant Society to a Market-Oriented Economy: 1950–1982

The first period, beginning in 1950 and ending in 1982, is characterised by an increasing monetisation of the economy, thanks to the injection of money from outside while remaining a self-subsistence economy. In the aftermath of the Second World

War, the injection of money from outside and the improvement of the maritime link between Rodrigues and Mauritius enabled a transformation of the local economy. The injection of money took place through the transfer of pioneer and soldier allowances and through improved public investment. The first major development initiative on the island was the land restoration programme carried out from 1955 to 1968 under the responsibility of the British agronomist Philip Hotchin. Many people of working age were recruited to build terraces on an area of 2800 hectares. The 1971–1975 Four Year Development Plan of Mauritius then allocated a budget of 32 million rupees to Rodrigues. In response to the great drought of the 1970s, the state hired a large number of general workers in the civil service to mitigate the effects.

The improvement of the maritime link broke the marked isolation of Rodrigues from 1937 onwards, as it allowed the development of trade and the acceleration of the movement of people with the regular service to Rodrigues by the cargo ship Zambezia from 1937 to 1947, and the link provided by the ships of the Dutch company KPM (Koninklijke Paketvaart Maatschapij) on the Batavia–Mauritius route. From 20 August 1955 until 1990, the MV Mauritius served the Rodrigues–Mauritius route (North Coombes, 1971). The Mauritius Pride then took over from the MV Mauritius and marked a break in the island's maritime history as it was a 92-m-long ship carrying passengers and all types of cargo.

The monetisation of the economy and the increase in the frequency of maritime links led to changes in the lifestyle of the population, which made greater use of the market and imported products. Items such as rice, flour, oil, and soap became commonplace. The period is also characterised by the development of dependence on imported foodstuffs and the emergence of a trade deficit, as the growth of exports was less than that of imports. However, the population did not change its way of life, continuing to engage in self-subsistence farming. This period was therefore a transition from the traditional agrarian society to a market-oriented economic mode of operation.

The Rodriguanisation Shift: 1983–2002

The second period, beginning in 1983 and ending in 2002, is characterised by the development of markets, the "rodriguanisation" of the administration and the improvement of social infrastructure (roads, schools, health, etc.). The emergence of the tourism sector has had a significant knock-on effect on agriculture, handicrafts, catering, transport, etc. This second period was marked by a major political transformation that had important economic impacts. The coming into power of the progressive Mauritian Left was reflected in a better consideration of Rodrigues by the central government. Thus, from 1982 onwards, an elected representative of Rodrigues was systematically appointed Minister of Rodrigues, a system which continued until the island was granted political autonomy in 2002. These political changes allowed Rodrigues to move away from its image as the *Cinderella of the Mascarenes* (Barat et al., 1985; Jauze & Yeung Ching Yung, 1998).

The first major change was a marked improvement in the economic and social infrastructure with the extension of the road network, the reinforcement of services such as electricity, telephone, and water distribution and the creation of new health centres and schools. This allowed for a clear improvement in the living conditions of the population and conditions favourable for local entrepreneurship. This led to an increased monetisation of the economy, the development of markets, an increase in the number of businesses and the production of services.

Improved air links and road networks have also led to the development of the tourism sector. The entry into service of an ATR 42 aircraft, providing regular service between Mauritius and Rodrigues in December 1987, was a major event as it constituted the starting point of the tourist activity. Better connected to the rest of the island and particularly to Port-Mathurin, new hotel resorts were developed at Pointe Coton and Mourouk in the east and south-east of the island, respectively. The available data show a continuous and strong growth in tourist arrivals for the period 1995 to 2019, except for 2005 and 2006, indicating a still buoyant sector, employing 1400 people in 2018 that comprises 7.4% of total employment (Fig. 24.3).

The active discrimination policy of *rodriguanisation* by the government implemented from 1982 onwards, consisted of favouring the recruitment of Rodriguans into the civil service in Rodrigues at all levels, resulting in an increase in the purchasing power of the population, a progressive change in consumption patterns and the progression of urbanisation. The *rodriguanisation* of the administration also enabled the creation of a "middle class," a professional cadre with confidence to promote new initiatives and investments. In a more economic term, it has allowed the progressive constitution of physical, social, and human capital that will serve as a lever for the island's subsequent economic development.

The economy of Rodrigues has gradually diversified and the private sector has become increasingly important. A growth dynamic was established with the acceleration of the transformation of the local economy during the 1990s and 2000s. It

Fig. 24.3 Foreign tourists' arrival. (Compiled from data from several Editions of Digest of Statistics for Rodrigues, Statistics Mauritius)

has moved away from the stationary state in which it was confined to become a more diversified economy.

There has been a continuous decline in the number of employees in the public administration from 3745 in 1994 to 3258 in 2000 and 2752 employees in 2012. On the other hand, total employment in the private sector has continued to grow. Historical sectors, such as agriculture and fisheries, contribute to the importance of the private sector. However, the dynamics are particularly driven by small-scale manufacturing of goods (handicrafts, furniture, metalwork, and sewing), construction, trade, hotels, restaurants, and transport. The Rodriguan economy is diversifying, with a private sector that is becoming more important and revealing a real growth dynamic. For these sectors, there has been an increase in the number of companies, production volumes and values created as well as in the number of jobs (Fig. 24.4).

Towards an Autonomous Region and the Structuring of a Tourist Economy: From 2002 Onwards

The last period begins in 2002, with the accession of the island to the status of administrative and political autonomy. Rodrigues is granted the possibility to manage its local development in the context of a market economy marked, however, by a strong deficit of economic activity. Despite the economic diversification and growth, Rodrigues is still characterised by a lack of employment opportunities with a relatively high unemployment rate and a large trade deficit, with a dependence on

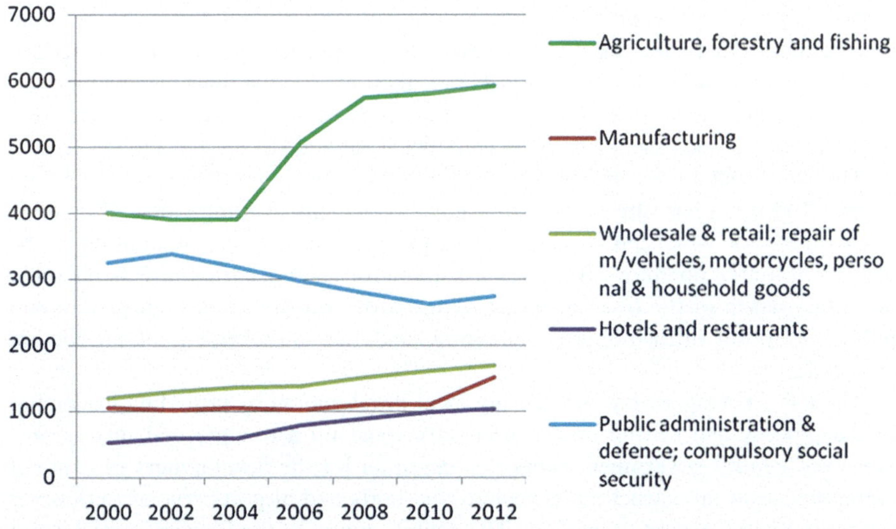

Fig. 24.4 The changing structure of the economy as reflected by employment. (Compiled from data from several Editions of Digest of Statistics for Rodrigues, Statistics Mauritius)

the outside world to meet its needs, particularly for food. This is partly due to the constraints of limitations on capital mobilisation, the island's agrarian origin, small size, and isolation affecting the economy. These challenges highlight the need for a development approach that addresses these structural imbalances.

The strategy adopted by the RRA is deployed at several levels. The vision looks to Rodrigues being a green and sustainable island through an integrated approach that recognises the necessity to develop the economy while restoring the marine and terrestrial ecosystems (RRA Five-Year Government Programmes, 2012, 2017). To this end, several projects like Marine and Terrestrial Reserves, the banning of plastic carrier bags and periodic closures of octopus fisheries, have been effectively implemented. The RRA defines the promotion of an entrepreneurial culture as a cornerstone to success, while its economic development places an emphasis on the capabilities of its people and on the improvement of the investment climate for the development of micro-, small-, and medium-sized enterprises. Professional training schools, such as the Mauritius Institute for Training and Development (MITD) and Polytechnics, provide national general vocational training schools; Vatel for the hospitality sector; additionally, a Lycée Agricole and a Business School have been set up to promote capacity building. Job opportunities in the tourism sector are attracting the interest of young people who seek to study courses appropriate to the sector. The management and operation of many tourism businesses require varied and advanced competences.

The strategy, first of all, is to strengthen the traditional sectors of agriculture, livestock, and fisheries, for which there is a demand in Mauritius due to the competitiveness of these products. This reinforcement also aims at import substitution in order to reinforce food security. This will involve the modernisation of traditional sectors in order to respond to market changes and also to create greater added value. Instead of limiting itself to the production of raw materials, it is important to add an industrial component that will make it possible, by transforming this production to generate more activities and value locally. In addition to the transformation of raw materials, whether in agriculture, livestock, or fisheries, their production must also be rationalised to become competitive. Fisherpeople are more and more getting involved in recreational fishing and other ecotourism activities in and outside of the lagoon.

The second strategy consists in promoting new sectors. Tourism is one such sector and Rodrigues has some competitive edge to face global competition. Tourism is a driving sector which pulls other sectors in its wake because, in addition to the activity of tourist operators, it creates a demand for fishing, agriculture, handcrafts, etc. The growth of the tourism sector depends on investment in both private and public reception infrastructure, marketing, and the improvement of production standards.

The RRA is responsible for steering local development by providing leadership in the process and setting up the necessary local infrastructure and institutions. However, central government support is essential for the development of external communication infrastructure, especially the ports and airports crucial to develop connectivity and access to markets. The actual runway at the Plaine Corail Airport is 1800 m long. The central government is looking for funding to develop a new runway of 2100 m length to enable light passenger aircrafts of the type A321 Neo to land in Rodrigues. There is also a Master Plan to develop the port which now has a wharf that is120 m in length. Major plans are underway to improve both the airport with a new runway and the port.

The overall strategy is therefore to strengthen the production of products in which Rodrigues has comparative advantages, such as garlic, onions, sheep, goats, cattle, fish products, lemon-based condiments, vegetables, etc. This will make it possible to increase the export of products (honey, coffee, ham, lemon) with a high degree of uniqueness due to the existence of microclimates and to promote import substitution, particularly in the food sector. Developing access to niche markets should make it possible to generate significant added value and thus support the development of the traditional sectors of agriculture, livestock, and fisheries.

Sport Tourism in Rodrigues

From Ecotourism to Sport Tourism

The beginning of tourism in Rodrigues is very recent and is the initiative of some locals. Its mode of emergence is therefore closely linked to the itinerary and experience of these people. The course of their lives marked by events and encounters led them to take the decision to open an *Auberge* (a form of tourism accommodation) or to initiate an activity as a tourism operator. As mentioned earlier, tourism really took off in Rodrigues in 1987, with the start of operation of ATR 42 aircraft by Air Mauritius, offering daily flights between Mauritius and Rodrigues. In addition to the small Pte Venus Hotel owned and operated by Air Mauritius in the 1970s, a few locals, namely, Ty Wye Ching Siew and Dorisson Bégué, created *Auberge Ciel d'Eté* and *L'Auberge les Filaos* located at Port Mathurin and Anse aux Anglais. Henri Meunier started as a tourism operator providing diving services and discovery tours across the island to address the demands formulated by tourists visiting the island. There was thus a demand for sport tourism from the beginning.

Tourism then spread to the interior of the island on the higher lands with the breakthrough of ecotourism in the 1990s (Gay, 1998) and the local Catholic Church was instrumental in that. The aim of the Catholic Church was to provide the population with a complementary source of income to their usual activities and to promote the autonomy of families in terms of availability of food. In 1988–1989, it approached the Catholic Church in Réunion Island to learn from the initiatives the setting up as *Gîtes*, in other words, tourism *chez l'habitant* (in the homes of the locals). The idea at the start was that tourism should benefit the people and not just a few large businesses and also to offer tourists the possibility of discovering Rodrigues from a different perspective with local people and culture rather than the ubiquitous hotel resorts. It was a question of preserving the environment and the social and cultural authenticity of the island in order not to exert additional pressure on fragile ecosystems through a massive arrival of tourists. This was the first time where there have been a clear initiative to develop a type of tourism product. The whole concept corresponds point to point to eco-tourism. This initiative led to the setting up of some *gîtes* in the rural areas. During the 1980s, the tourism sector was concentrated in the north-eastern part of the island. From the beginning of the 1990s, it spread through the inner land and also along the east and south-east coasts of the island, which have the most pristine beaches, with the construction of the first international class beach hotels, the Cotton Bay Hotel (48 rooms) and the Mourouk Ebony Hotel (30 rooms) with the initiative of Willy Auguste, in association with a Swiss investor (Agathe, 2016). Ecotourism seems to be well suited to the island, given the limited investment of capital and the existence of various natural heritage assets.

The Organisation of the Offer and the Actors for the Development of Sport Tourism

Progressively, local actors are becoming aware that the potential of the natural heritage can be enhanced by sport tourism in a perspective of environmental preservation. The two hotels have played a pioneering role in the development of sport tourism by hosting two clubs, Scuba Diving at Cotton Bay and Windsurfing at Mourouk, which have greatly contributed to the promotion of Rodrigues as a tourism destination. This first successful step in sport tourism was made possible thanks to the vision of two locals. One of them, Mr Willy Auguste, has worked in the hotel sector in South Africa and Mayotte before becoming the promoter of the Mourouk-Ebony Hotel. He became aware of the potential of scuba diving in terms of tourist arrivals and income generation in Mayotte. The other, Mr Percy Gontran, had worked in the travel industry and travelled the world as a flight attendant on Air Mauritius and was an avid windsurfer. The information and knowledge they had, the opportunity to participate in international tourism fairs, and their ability to network made it possible for these two clubs to become a reality. Other key players who contributed to this success included personalities in the field of diving, windsurfing, journalism, tour operators, and travel agencies. Most of these players strongly believed in the authenticity of the destination. According to Mr. W. Auguste, "*it is truly thanks to a deep conviction, a real will, and a frank collaboration of the partners that allowed the realisation of a diving club at Cotton Bay Hotel and a windsurfing club at Mourouk Ebony. Mr. Percy Gontran as CEO of Mourouk Ebony and Sydney Caré of Cotton Bay have been decisive in this joint effort. Jacky Dégrémont, a diving instructor from France, was the very first instructor of the diving club. The joint promotional work helped bring in the customers.*"

The Main Sport Tourism Activities

Hiking and trailing are activities which were demanded right at the start of tourism in Rodrigues. Because of the variety of landscapes within short distances due to topography and size, the island offers plentiful opportunities for hiking. As the topography is not very high and the differences in altitude are not enormous, the walks are suitable for almost all age groups. The island has an extensive network of footpaths used by former generations for their movement around the island to fulfil their different needs before motor transport became commonplace. Even though the footpaths are unfortunately not well marked, thus limiting somewhat the undertaking of this activity, the availability of services provided by competent hiking guides partly address this problem. Since 2017, these guides are grouped in a professional Association called *Association des Guides de Randonnée de Rodrigues (AGRR).* In addition to the extensive network of footpaths, the island has a considerable number of heritage assets associated with fisheries, agriculture, handicraft, and water resources that can only be reached by foot.

Taking trails in Rodrigues is more often a walk than a hike. It is a whole experience of immersion in landscapes between sky, land, and sea and immersion in the history of the island. According to Marie Paule Pierre Louis, the initiator of the AGRR, "*There are a considerable number of frequent visitors who come to Rodrigues to perform this activity from Réunion Island and France but also more and more from Mauritius to enjoy the richness of the history and the diversity of landscape.*"

The network of footpaths for hiking runs across the whole island. The most popular routes follow the eastern coast paths between Mourouk and Point Coton via Graviers, Trou d'Argent, and Saint Francois.

There are two limitations for the full expansion of hiking. Firstly, there is a lack of quality maps at appropriate scales. Secondly, very few footpaths are adequately marked and equipped with signage. This is very unfortunate given that hiking as an activity has existed for more than 25 years and the elementary conditions for the good functioning of the activity have not yet been put in place. All paths need to be clearly signposted and interpretive information and adequate maps developed.

The network of footpaths handed down by past generations is a cultural heritage in part being abandoned given that it is no longer being widely used. The promotion of hiking, together with maintenance, marking, and mapping, can be an effective way to protect them. These paths are part of the island's heritage assets and are an interesting way to discover the island. Interpretation centres can be added to this infrastructure to add to the experience of hikers.

Trail running has emerged as an activity in 2009 with the *Trail de Rodrigues* (Trail of Rodrigues). This has now become very popular for tourists and the local population. The Trail de Rodrigues is the largest annual event; it takes place in November and has four different courses, namely, Tortu of 6 km, Greko of 15 km, Solitaire of 27 km, and Perroquets of 50 km. A fifth course, Hibou of 73 km, was added to the programme in 2022. This ultra course starts in the east of Rodrigues at Point Coton and arrives at the Francois Leguat Turtle Reserve at Anse Quitor in the West. In 2019, 1472 participants attended the event, but during the following 2 years, the number of participants decreased due to border closures linked to the COVID pandemic.

There are now three other trails: the Golden Bat Trail which takes place at night; the MCB Port Mathurin Trail; and the Rodrigues Championship UBP Welcome Trail attended by 800 participants for the first two and 500 for the last race, respectively. These four trails are now part of the International League of Trails, meaning that the performance of participants attending these trails in Rodrigues is assessed according to international standards.

Trail competitions benefit from the same infrastructure as for hiking and take advantage of the variety of island landscape and topography. The courses alternate between coastline and beach edges, rocky coasts and creeks, grassy slopes, and rocky areas.

There are four diving centres in Rodrigues: Eco Evasion and Alex Salamenders in the North, Bouba Diving Centre in the South, and the Jacky Diving Centre in the East. Six qualified diving instructors provide coaching services in these centres. There are excellent dive sites in Rodrigues, all around the island but especially outside the lagoon, where this activity has a good reputation. Diving can be undertaken throughout the year but is sometimes restricted by weather conditions; depending on the site, the diving trip may depend on tidal conditions as well. The centre based in Port Mathurin offers outings at any time of the day, while those located elsewhere can undertake only two diving sessions per day at high tides. It can reasonably be estimated that diving attracts between 2500 and 3000[1] visitors in Rodrigues a year, translating to between 600 and 800 per centre. The bulk of visitors come from Réunion Island and France. The length of stay of the visitors from Réunion Island is between 5 and 8 days, whereas for visitors from France, the length of stay is more than 10 days. They are mainly middle-class and professional people, primarily in the 25–35 age group. Youth between 8 and 18-year-olds also undertake this activity, often with their parents.

Youths from Rodrigues are quite engaged in this activity as either professionals at the level of diving centres or the Marine NGO Shoal's Rodrigues or as amateurs at the level of the network Rod Dive Lovers, comprising some 40 participants.

Diving areas in Rodrigues are numerous and can be grouped into three categories: inner lagoon areas, usually confined in passes given the shallow depth; reef slopes and on the shelf after the reef slope. The *Grande Passe* of Port Sud Est provides a wide range of diving spots suited for all categories of divers and at depths varying between 4 and 18 m. The sites are famous for the diversity of fauna. It is reported that no less than 140 types of species can be found in these waters, comprising fish, dolphins, barracudas, turtles, sharks, with the exception of whales. The Passe of Saint François is reported as a fantastic diving area on its own, hosting canyons with wonderful corals and fish.

The reef slopes in the outer lagoon are famous for their architecture with caves, canyons, and holes, rich in the diversity of flora, especially corals. Sites are found around the island but display different features in the west where the depth varies between 5 and 35 m. Sites in the north-east in front of Grenade and Rivière Banane display comparable features. In some areas after the reef slopes, there are pinnacles

[1] According to figures provided by Diving Centres.

of corals at 16–18 m deep, displaying great varieties of corals. These areas are common in the north but can be found also in the south.

According to the diving centres, there are about 15 diving areas in the north, the east, and the south, respectively. One major characteristic of diving sites in Rodrigues is that many wonderful dives can be made at depths between 15 and 20 m. Given there is still no decompression chamber available on the island, this condition enables divers to enjoy the richness of the experience of diving in Rodrigues without having to take additional risks with dives of greater depths. For the centres located in the north, dive trips can be made at any time through the harbour channel and around the whole island. In the south, the trips are organised according to the tide and weather conditions; at low tides, navigation in the lagoon's outside channels is very often not possible.

There are at least four operators in Big Game fishing in Rodrigues: Rod fishing club; *Oiseau des Iles*; Why Not; and the independent Gezim that organise fishing trip outside of the lagoon. They are all located in the north in the compound of Port Mathurin. The operators usually have two boats which they rent with all the attached services on a day rate. Fishing outside of the lagoon is limited by weather conditions. The main season for the activity is during summer between the end of September through May. The stock of fish in the vicinity of Rodrigues and even further on the Hawking Banks and Eastern Banks is good and the quality of fishing is also considered good. The main catches are Blue Marlin and Skipjack Tuna. The operators practise the method of catch and release as much as possible. The clients are usually professional people, CEOs and politicians from all over the world and have high spending power. The active sport tourists are often accompanied by their family; while they undertake fishing, their families enjoy other activities on land, especially hiking. The duration of stay is a least 5 days and may extend up to 3 weeks, according to local tourism surveys.

There are seven kitesurfing businesses on the island, all located between Mourouk and Graviers. Osmosis, the oldest Club, employs four instructors. There is a certain degree of informality in the provision of services in the sector and it is estimated that there are about 20 to 25 instructors in Rodrigues, who offer coaching in the lagoon. Kitesurfers arrive from all over the world, but more generally from France and Réunion Island. Mauritian tourists very often perform the activity on their own without having recourse to the services of local clubs. Given that Rodrigues is quite a windy island, the activity can be undertaken all year round. This activity enjoys a natural feel even though its performance requires the use of equipment made from petrochemicals. Every year there is an international festival of kitesurfing over 1 week, with an international, well-known surfer invited with the aim to promote surfing in Rodrigues as a destination. The annual number of visitors is estimated to be 2000–2500[2] annually.

The most common sites for kitesurfing are located in the south between Graviers and Mourouk. While the limited depth of the lagoon may restrict the time

[2] Estimated by the author on the basis of figures provided by the Clubs.

kitesurfing can be undertaken, the calmness of the lagoon water and the omnipresent wind in the south in particular, make the area a surfer's paradise. Long courses can be undertaken all around the island when the tide and weather conditions allow. The presence of islets in the south allows visitors to combine kite surfing with picnic trips to the islets in the lagoon.

Sailing regattas are historically very popular, as are traditional fishing boats (named *pirogue)*. Given the importance of fisheries in the lagoon, sailing offers an opportunity for fishers to display their knowledge of the lagoon and skill at sailing. North-Coombes (1971b) underlined the excellence of Rodriguans at sea at least in the lagoon. More and more tourists are attracted by regatta events, especially tourists from Mauritius, who come mainly to watch. Sailing has the potential to reinforce the image of Rodrigues as a sustainable island destination and to encourage the traditional ship carpentry of Rodrigues that has all but disappeared.

Over the last two decades, the number of sports has grown due to the passion and commitment of pioneers and the contribution of several players in the travel and tourism sector. There is a tendency towards the formalisation and greater professionalisation of every activity at all levels. Firstly, there is an increase in the number of service providers for every activity. There is an increase in the number of sport tourism activities as well as nature-based activities in general. The advent of the Trail de Rodrigues in 2009 and the growth of this activity is living proof of this development. The activities are being better regulated by authorities or self-regulated by the service providers themselves, who regroup themselves in associations. There is an improvement in the quality of services provided, with the availability of better trained and competent instructors. There is a greater ability for the service providers in the sector to work with other service providers and distributors locally and elsewhere.

It has to be mentioned that 'Expedition Africa' 2019 took place in Rodrigues. It is a sport adventure event, all in one contest, consisting of diving, swimming, racing, rowing, trail, cycling, climbing, and orienteering taking place over 5 days for a total distance of 400 km. Expedition Africa 2019 demonstrated Rodrigues goes beyond having the physical features/heritage for such events but also shows it can adapt itself in welcoming new events and organising complex international competitions. Rodrigues has hosted some events of the 10th Indian Ocean Island Games held in Mauritius in July 2019 (officially known as *Jeux des îles de l'océan Indien Maurice 2019* or simply *JIOI Maurice 2019*). The Indian Ocean Island Games is a multi-sport event for athletes representing the National Olympic Committees of Indian Ocean island nations.

The Importance of the Local Population Within the Organisation of the Sport Tourism Sector

Given that ecotourism is the predominant form of tourism in Rodrigues, the organisation of the sector is very much embedded within the society. Tourism is also a private led sector, meaning the civil society takes the lead in the undertaking of various activities at various levels. On top of that, it has to be noted that there has never been a large public administration for the sector, which entails a mode of governance giving due consideration to collaboration among stakeholders.

Except for big game fishing, where the two main operators are foreigners, since it is an activity where it is necessary to combine a relatively high capital investment capacity and also a high level of technical training of fishermen, and for kitesurfing and diving, where the first instructors to operate were foreigners, all of the operators in sport tourism are locals. The foreign operators are long established on the island and can be considered as locals. Among the service providers, most of the *gîtes* and guest houses are owned by locals. The three main beach hotels on the eastern coast are owned by Mauritian investors. Tour operators and specialised travel agents selling the destination of Rodrigues and sport activities in Rodrigues are located outside of the island and are not of Rodriguan origin.

Rodriguans participate in sporting activities as well. They are a powerful means of providing leisure, healthy, and adventurous activities to the population. There exist several trail and regatta events which are very popular among the population. Kitesurfing has also been adopted by local people, with the clubs having adopted policies to provide their services free of charge to Rodriguans. As a result, the island has produced some renowned champions.

Sustainable Development Policies on the Island, Especially Those Related to Sport Tourism and How They Are and Have Been Implemented

The Governance of Actors

Two important planning documents have defined development approaches and targets for sustainable tourism development in Rodrigues: the 2002 'Tourism Development Plan for Rodrigues' (TDPR) and the 2009 'Sustainable Integrated Development Plan for Rodrigues' (SIDPR). The five-year government programmes for each mandate, since the advent of autonomy in 2002, outline medium term strategic targets for sustainable development in Rodrigues. Both of these documents recognise the importance of tourism as the main pillar of the Rodriguan economy, acknowledging the industry's linkages with other sectors and its importance in employment creation and revenue generation. The Tourism Development Plan for Rodrigues (TDPR) identified a carrying capacity of 100,000 tourists to be achieved through a steady tourism development by 2015. In 2009, the Sustainable Integrated Development Plan for Rodrigues (SIDPR) recognises that many of the recommendations of the TDPR have already been met. The carrying capacity may evolve with improvement in infrastructure and management of environmental assets and effective product development.

The various policies conducted since the advent of autonomous government include the preservation of the island's cultural heritage (formalisation and teaching of Creole at school, listing of the Sega Tambour, the local folk dance, by the

UNESCO as a Universal Heritage, …), improvement in project management for ecosystem conservation (closure of octopus fishery, operationalisation of marine and terrestrial natural reserves), development of infrastructure, and the promotion of entrepreneurship and vocational training. All of these policies have contributed to leverage the carrying capacity.

The impact of tourism on the environment is limited. The beaches and seaside are protected. The same environmental laws of the ROM also apply to Rodrigues. A setback of 30 metres from the highest watermark is mandatory for any development along the coast. Construction on sand dunes is not permitted. Rodrigues is also included in the Sensitive Environmental Areas Plan; although it is not legally binding, it serves as guidelines for major development. There is no legally binding land use plans in Rodrigues, but most of the land is state-owned. This allows the RRA to have some control on land use development. Unfortunately, land sometimes acts as a barrier to entrepreneurial development due to the lengthy process for the granting of land leases.

The Economic, Environmental, and Socio-Cultural Impacts of the Sport Tourism Activities

The positive impacts of sport tourism are considerable. Given the positive reputation enjoyed by the main sport tourism activities in Rodrigues, this segment of the tourism sector accounts for a large number of visitors. However, this success should not be attributed solely to the quality of delivery for these sports but rather to the destination as a whole. The operators of the main activities assert that a larger number of sport tourists are accommodated in *gîtes* and rural guest houses than in hotels. This is possible because of the availability of *gîtes* and rural guest houses across the island and this in return enables the distribution of income to the local population. This allows tourists to encounter the local population and to enjoy the local cuisine, which derives its quality from the use of fresh food, especially seafood.

According to our investigations, sport tourists are reported to be high spenders, as compared to the average tourist, thus generating high revenue for the local economy. A tourist who comes for a main activity, whether it is kitesurfing or big game fishing, for example, often undertakes another sport activity like diving or hiking during his stay. These active sport tourists are often accompanied by their families or friends. The family or friends usually undertake softer activities like hiking or adapted short courses in Trail competitions or will visit and enjoy other features and products of the island. Most of the operators in sport tourism are locals. Sport tourism is a major contributor to economic growth and economic development, as it allows the emergence of new entrepreneurs in various fields and the distribution of income across the country and the population.

One of the strengths of Rodrigues is the presence of well-trained and effective local NGOs in the field of environment, whether in marine or terrestrial conservation. These NGOs have very significant contribution in natural reserves management. There is a well-established practice of collaboration between the RRA, tourism operators, and these NGOs to manage the environment, monitor the status of natural resources, and undertake conservation (Jhangeer-Khan et al., 2015). There is partnership between the diving operators, the NGO Shoals Rodrigues and the South-East Marine Protected Area (SEMPA) to clean passes (removal of basket trap and other wastes) and to plant coral. *L'Association des Guides de Randonnée* works with the public and the RRA for clean-up campaigns of trails across the island. Sport tourism serves as leverage for environment protection and conservation. It serves as a leverage for the maintenance of cultural heritage as well, such as the network of footpaths, our traditional cuisine, and so forth.

Conclusion

Sport tourism has a very prominent role in the development of tourism in Rodrigues. There is a very strong linkage between sport tourism and the environment. Sport tourism relies on the outstanding natural landscapes of the island, while it also creates awareness of the importance of these resources and supports initiatives to protect and conserve the natural environment. However, the success of sport tourism goes far beyond this strong linkage with the natural environment. Other elements are also instrumental, such as the presence of intangible cultural heritage, the safety and friendliness of people, the existence of vocational training facilities, the presence of small guest houses, the presence of dedicated and effective local environmental NGO, adequate transport, and other services. Sport tourism operates within the wider context of the island as a tourist destination. The linkages are strong with the environment but also with the whole community. Sport tourism acts as the cornerstone of ecotourism and enables the island to promote many of its assets, which had been stymied in the constraints of isolation and economic backwardness. Isolation itself is no longer a constraint but is the condition that generates uniqueness through the generation of endemicity of fauna and flora and unique cultural, social, and economic traits. If it is true that this sector is promising, it remains that it is still perfectible from the point of view of the signposting of spaces, sites, and practices, of the communication, of the synergy of the actors in a perspective of sustainability. To draw tourists to sporting activities and events in Rodrigues, travelling to Rodrigues needs to be made easier. The limitations on luggage allowances on the ATR72, the limited number of seats, and the high prices for seats and excess luggage are issues that have to be addressed. While it is recognised that the carbon footprint of air travel is quite high, the overall carbon footprint of sport tourism activities in Rodrigues remains relatively low.

References

Agathe, H. (2006). *La gestion de l'espace des territoires insulaires: contraintes et perspectives. Une étude comparative des marchés fonciers des Mascareignes (Maurice, La Réunion et Rodrigues)*, Thèse de doctorat, Ecole des Hautes Etudes en Sciences Sociales, Paris, 341 p.

Agathe, H. (2016). Histoire économique et perspectives de développement à Rodrigues, Unpublished paper, 19 p.

Barat, C., Carayol, M., & Chaudenson, R. (1985). *Rodrigues, La Cendrillon des Mascareignes, Institut de Linguistique et d'Anthropologie de La Réunion*, Université de La Réunion, 116 p.

Check, A., & Hume, J. P. (2009). *Lost land of the dodo: The ecological history of Mauritius, Réunion and Rodrigues*. Bloomsbury Publishing, 480 p.

Gay, C. (1998). *La dynamique touristique de l'île Rodrigues (Maurice)*, Actes des 7es Journées de géographie tropicale de Brest, Nantes, Ouest éditions/Presses académiques, 708 p., pp. 205–212.

https://www.airmauritius.com/news/air-mauritius-resumes-commercial-flights-to-madagascar1, 12 October 2022.

Jauze, J. M., & Yeung Ching Yung, J. (1998). *Rodrigues – La troisième île des Mascareignes*. L'Harmattan, 268 p.

Jhangeer-Khan, R., Agathe, H., & Yvergnaux, Y. (2015). *The Rodrigues Octopus Fishery: A case study from the Island of Rodrigues*. Blue Ventures, 47 p. https://blueventures.org

KPMG/UNDP. (2009). *Sustainable integrated development plan for Rodrigues*. RRA, 453 p.

Naria, O., Van Rheenen, D., Melo, R., & Sobry, C. (2022). *Sports tourism, island territories and sustainable development: approach to a methodological framework for research based on the insular context of the Indian Ocean in Sport Tourism and Its Territorial Development and Opportunities*, Mazza B. Sobry C, Cambridge Scholars Publishing, 19–40.

North-Combes, A. (1971a, Reprint in 2002). *The Island of Rodrigues*. Book Printing Services Ltd, 337 p.

North-Combes, A. (1971b). *The Island of Rodrigues*, Port Louis, Mauritius, 276 p.

Rodrigues Regional Assembly. (2012). Discours Programme 2012–2017, "*Consolider et réussir l'autonomie*", 100 p.

Rodrigues Regional Assembly. (2017). Discours Programme 2017–2022, "*Consolider et réussir l'autonomie*", 100 p.

World Tourism Organization. (2019). *Sport tourism and the sustainable development goals (SDGs)*. UNWTO. https://doi.org/10.18111/9789284419661

Chapter 25
Sport Tourism: An Undeniable Asset for Sustainable and Local Development in Comoros

Ouledi Ahmed

Introduction

The Comoros is an archipelago of islands and volcanic islets, aligned over 225 km, along a southeast/northwest submarine plateau. The total area of the archipelago is 2236 km^2. The exclusive economic zone is estimated at over 160,000 km^2. It covers 900 km^2 of continental shelf and 427 km of coastline. The climate is tropical and humid with an oceanic influence and two seasons. It is subject to three successive wind regimes: (a) the monsoon or kashkazi; (b) local southwesterly winds originating from the southern high pressures and channeled between Madagascar and Africa; and (c) the austral trade wind or kusi.

Due to its geographical location at the northern entrance to the Mozambique Channel, its climatic characteristics and the possibility of practicing a wide variety of tourism and sports activities throughout the year, Comoros has many natural and historical attractions and undeniable assets to reach the immense global tourism market through the development of sport tourism. Indeed, the country benefits from an important tourism potential due to its rich landscapes, its remarkable natural environments and its rich historical and cultural heritage. The environment of the Comoros Islands is characterized by a diversity of habitats, a great wealth of specific flora and fauna, and a high rate of endemism with elements of global importance such as the coelacanth and the Livingstone's dogfish. More than 2000 species of terrestrial plants have been recorded. The fauna is very diverse and balanced with all zoological species present.

The marine and coastal environment is also very rich, characterized by mangrove swamps, coral reefs, coastal forests, and sea grass beds with a remarkable and diverse marine fauna. These coastal areas also contain the most beautiful

O. Ahmed (✉)
University of Comoros, Moroni, Comoros

D. Van Rheenen et al. (eds.), *Sport Tourism, Island Territories and Sustainable Development*, Sports Economics, Management and Policy 24,
https://doi.org/10.1007/978-3-031-51705-1_25

landscapes, such as the white sand atolls, the underwater peak of Vailleux Bank, but also ports and historic towns that bore witness to the many waves of migration and cultural landscapes shaped by several centuries.

Sport tourism is a form of tourism that respects people and the environment and can contribute to improving the living conditions of local communities. It allows the sport tourist to enjoy an active holiday in a healthy and preserved natural environment. It gives them the opportunity to come into contact with the natural elements, to discover new places, and to challenge themselves physically. It offers them the opportunity to combine pleasure and discovery by meeting communities and indulging in a variety of activities: hiking, climbing the Karthala volcano, diving, fishing, exploring the regions on foot, and tasting the good local dishes.

The promotion of sport tourism will increase the economic income of the local population and help reduce current practices that affect the marine and terrestrial ecosystem through the active participation of all stakeholders, including the private sector, local communities, and government. The aim of this study is to identify and promote the place of nature sports activities in the tourism offer of Comoros and to see to what extent they could contribute to sustainable and local development, notably by strengthening the country's economic activity and community development. Sport tourism is thus an undeniable asset for sustainable and local development in Comoros. We have several years of experience in the promotion of ecotourism and sustainable tourism in relation to the implementation of SDGs 1, 8, 13, 14, and 15.

General Presentation of the Comoros

Geographical Data

The Comoros Archipelago is located at the northern entrance of the Mozambique Channel, halfway between the East Coast of Africa and the NorthWest of Madagascar. It occupies a total land area of 2236 km^2, unequally divided into four islands which are, from North to South, Ngazidja (Grande Comore), Mwali (Moheli), Ndzuwani (Anjouan), and Maoré (Mayotte) which cover, respectively, 1147 km^2, 290 km^2, 424 km^2, and 375 km^2. The island of Mayotte is under French administration.

The country was created by major volcanic events at the end of the Tertiary era and is characterized by a rugged terrain, with peaks and cones that leave little room for plains. The exclusive economic zone is estimated at more than 160,000 km^2. It covers 900 km^2 of continental shelf and 427 km of coastline. The climate is tropical humid with an oceanic influence and two seasons. It is subject to three successive wind regimes: (a) the monsoon or kashkazi, (b) local southwestern winds coming from high pressure in the south and channeled between Madagascar and Africa, and (c) the southern trade wind or kusi.

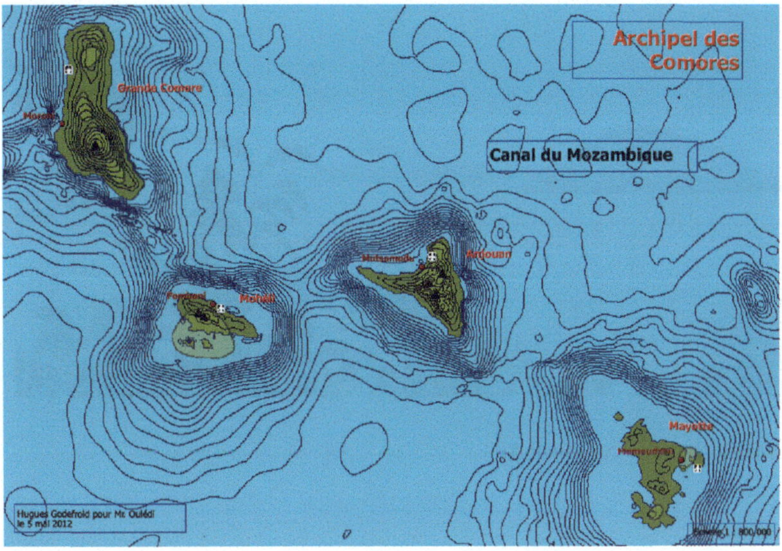

Presentation map of Comoros archipelago. (Source Ouledi A.)

Geological Data

The islands of the archipelago are built from a hot spot, with a progression of the age of the volcanic formations from west to east. Although all of them are of volcanic origin, the Comoros Islands present varied morphological and pedological characteristics, depending on the degree of evolution of volcanism. The island of Grande Comore, consisting of two shield volcanoes, not dissected by erosion, one of which has undergone several phases of activity during the twentieth century (Karthala), presents morphological characteristics in relation to its very young age. On the other hand, Mohéli and Anjouan have reached a more advanced stage of maturation and are deeply eroded. Mayotte presents an even more advanced stage of evolution, characterized by a very intense erosion and alteration, and by a coastal morphology showing a very indented coast, signs of an active subsidence. The development of coral formations around these islands supports the pattern of a progression of the age of the volcanic formations from west to east: the perimeter of Grande Comore is bordered only by a few embryonic fringing reefs, whereas a more developed fringing reef almost completely surrounds Moheli and Anjouan, and there is a continuous reef barrier, delimiting a lagoon of more than 12 km in its greatest width, around Mayotte. Geochronological studies confirm the morphological evidence and the degrees of evolution of volcanism: the rock samples studied show an age of between 130 and 200,000 years for the island of Grande Comore, 5 million years (Ma) for Mohéli, 3.9 Ma for Anjouan, and 7.7 Ma for Mayotte.

Moheli National Park. (Source Ouledi A.)

Demographic Data

The resident population of the Comoros is estimated at 758,316 inhabitants (Union of the Comoros, INSEED, 2017). It is evolving relatively rapidly. It has, in fact, increased from 335,150 inhabitants in 1980 to 758,316 inhabitants in 2017. It remains marked by the weight of young people, with 52.4% under 15 years age. The average annual growth rate rose from 1.9% between 1980 and 1991 to 2% between 2003 and 2017. Although the Union of the Comoros has not achieved its demographic transition, the birth rate is declining at the same time as the mortality rate is falling. Indeed, the crude birth rate fell from 35.6% in 2003 to 31.8% in 2017. The average number of children per woman (Total Fertility Rate) also tells us a downward trend. Estimated at 7.1 children per woman in 1980, the 2017 census tells us 4.2 children per woman. Mortality was 7.9% in 2003 but decreases relatively to 7.1% in 2017. As for maternal mortality, according to the DHS-MICS2012, there were 172 deaths per 100,000 women, unfortunately, in 2017, it is estimated to be 195 deaths per 100,000 women.

The islands are therefore densely populated with 407 inhabitants/km^2, with this density even reaching 772 inhabitants/km^2 on the island of Anjouan. The population of the Comoros is the result of successive migrations which have led to a mixture of different races and cultures (Portuguese, Indonesians, Malays, Persians, Arabs, and Africans have immigrated to the archipelago during the many invasions that the archipelago has undergone in its history or simply brought back as slaves).

Socioeconomic Data

Comoros is a Small Island Developing State (SIDS) with a GNP per capita of US$450. It ranks 132nd out of 177 countries with a Human Development Index (HDI) of 0.547 (Human Development Report 2005). From 2012 to 2017, the Human Capital Index (HCI) of the Union of Comoros declined from 0.44 to 0.4, placing the country 123rd out of 157 countries.

Since 2005, a strategic framework for growth and poverty reduction (SCRP) has been in place in Comoros. The Government of the Union of the Comoros has adopted this instrument to organize and harmonize its efforts and those of its partners in the fight against poverty. The DSCRP was used as a resource mobilization tool at the Donors' Conference of December 8, 2005 in Mauritius. The DSCRP was then replaced by the adoption of a Strategy for Accelerated Growth and Sustainable Development (SCA2D) 2015–2019, which was revised in 2017 and extended until 2021. The government then developed and adopted in 2019 the Plan Comores Emergent (PCE) to 2030 to translate this ambition into a reference framework for development planning.

The development of the Emerging Comoros Plan was preceded by a long process of domestication and prioritization of the Sustainable Development Goals (SDGs), which helped build a broad national consensus around the 2030 Agenda for Sustainable Development and to agree on the SDG targets to be prioritized at the country level. Thus, the Union of Comoros has prioritized 68 targets, including 32 in the social field, 20 in the economic field, 12 in the environment, and 4 in governance.

The Comoros Emergent Plan is therefore designed around the dual objective of contributing to the achievement of both the SDGs and the country's ambition to emerge, while integrating the implementation needs of the African Union's Agenda 2063 for the fulfillment of the Pan-African Vision for an "integrated, prosperous and peaceful Africa, led by its own citizens, and representing a dynamic force on the world stage." The ECP is built on five major key pillars which are the key areas identified as being likely to generate strong and sustainable growth that will foster the country's emergence, namely, (1) tourism and handicrafts, (2) the blue economy, (3) financial and logistical services, (4) agriculture, and (5) industrial niches.

Sociopolitical Data

A former French colony, Comoros declared its independence on July 6, 1975. The Comoros experienced a long and turbulent post-independence period, marked by chronic political and institutional instability. This long period of crisis culminated in the secession of the island of Anjouan. Following the reconciliation agreement known as the Fomboni agreement, a new constitution was adopted on December 23, 2001, establishing the Union of the Comoros. This constitution granted the islands broad internal autonomy and also established the principle of a rotating presidency among the islands of the country.

The Fomboni Agreement enabled the country to enter an era of relative institutional and political stability, marked in particular by three democratic changes of leadership. The country has taken advantage of this lull to focus on economic and social development issues and has regularly adopted national reference strategies for this purpose since 2001. Since 2018, constitutional reforms have been carried out that reduce the broad autonomy of the islands in favor of a certain centralization in the functioning of the State.

General Situation of the Tourism Sector

Comorian tourism is still in its infancy, both in terms of infrastructure and tourist numbers. While nearly two million tourists are attracted each year to the islands of the western Indian Ocean and this figure is growing each year, in the case of Comoros, only 35,865 tourist arrivals were recorded for the year 2018. Europeans numbering 22,554, including mainly French (21,534) visitors, occupy the first place at the level of tourists coming to Comoros followed by Africans (10,979) and the rest of Europe (2332). This tourism is limited to business tourism (29%) and affinity tourism (46%) from the numerous Comorian communities living abroad.

For 10 years, the number of tourists has been increasing steadily (except in 2009 and 2013) from 14,753 in 2008 to 35,865 in 2018 with a jump between the years 2017 and 2018. This increase is confirmed by the increase in hotel arrivals from 6641 in 2013 to 19,513 in 2018. With an average stay of 7 days, overnight stays increased by 194% between 2013 and 2018 from 46,487 nights in 2013 to 136,561 nights in 2018. Capacity has increased significantly due to renovations and expansions of some establishments.

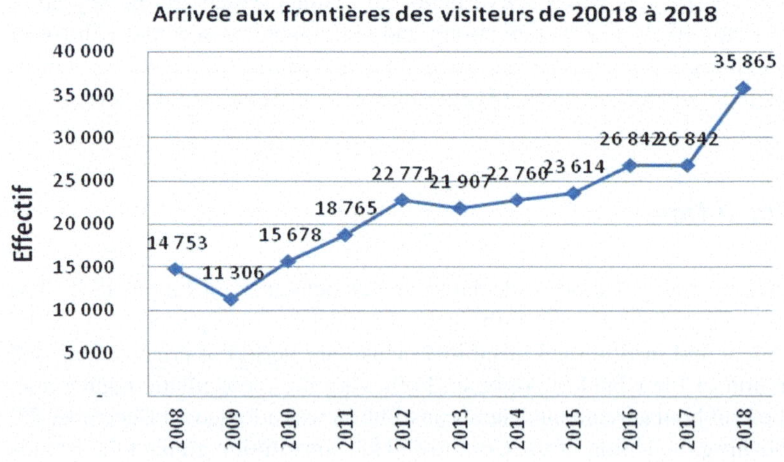

Evolution and structure of tourist arrivals. (Source: Tourism Statistical Yearbook 2018)

Tourism experienced some expansion during the 1990s with the activities of the Galawa Beach Hotel. The Comoros destination then declined due to the lack of a real promotion policy and hotel infrastructure. The hotel offer is largely below capacity. The quality of the accommodation on offer is very uneven between the islands of the archipelago. Tourism activity is currently limited to business tourism, missions, and stays by Comorians from the diaspora. Tourism for pleasure is very limited. A few tourist relays have been created but several important constraints hinder the development of tourist activity.

Most of the monuments and cultural sites are little known by the general public and are in a state of degradation, due to the lack of protection, maintenance, and development. The inter-island links, especially maritime ones, are not adapted to the tourist clientele and the prices of the air links with the emitting countries are prohibitive. The government has chosen, in consultation with the stakeholders, to revitalize the tourism sector and make it one of the main growth sectors in the coming years. A tourism strategy framework document has been developed and serves as a reference framework for the promotion of tourism.

Traditional culture. (Source Ouledi A.)

The natural and cultural heritage, the important potentialities of discovery and sports activities, and the village activities form a very attractive tourist offer on the whole national territory. More than 45 sites of tourist interest have been identified (flora and fauna, archaeological sites, religious buildings, buildings related to the former sultanates, fortifications and burial grounds). The country also has a knowledge in handicrafts, notably in carved woodwork, traditional jewelry and especially, cabinet making and sculpture, basketry and embroidery. The cultural richness also extends to other events such as the ceremonies of the big weddings, the songs, the dances, the Comorian cooking, the museum of the National Centre of Documentation and Research Sciences (CNDRS), the traditional medicine, the games, and the village activities. Sports activities such as scuba diving, deep-sea fishing, and mountain hiking are important components that can help launch this sector.

Ylang-ylang, perfume plant, in Anjouan. (Source Ouledi A.)

Employment generated by the tourism sector in general is low. Many of these jobs are in the hotel industry and tourism revenues have been steadily increasing over the past 5 years. Tourism's share of GDP is only 10% and the sector suffers from a lack of infrastructure.

Beach of Bouni in Grande Comore. (Source Ouledi A.)

A few international airlines fly to Comoros, but with an operational accommodation capacity of 763 rooms and 1526 beds in 2018, Comoros' tourism supply remains insufficient and has little impact on the economy in terms of employment and economic spin-offs. Nevertheless, tourism revenues have increased significantly over the past 5 years.

The Sport Tourism Sector

The practice of sport tourism in Comoros dates back to the early 1960s with the adoption of the status of internal autonomy and French Overseas Territory. At that time, there was a development of individual and collective sports. Metropolitan French people spent most of their afternoons practicing water skiing on the coast. A small number of them devoted themselves to trolling or long-line fishing. At independence, and especially under the revolutionary regime of Ali Soilihi, the practice of sport tourism had clearly decreased. It gradually resumed with the construction of the Galawa Hotel, a Sun International Group establishment. Since then, this type of tourism remains limited and only concerns a few people practicing trekking, scuba diving, and deep-sea fishing.

Comoros has a potential for sport tourism, but the exploitation remains modest. This potential can be used as a driving force for all segments of tourism products, including ecotourism, beach tourism, cultural tourism, and agrotourism, through the discovery of an exceptional coastline, protected endemic species, and a preserved natural habitat. Tourism development would contribute to the reduction or elimination of poverty and thus respond to SDGs 1, 8, 13, 14, and 15.

	Number of rooms	Number of beds	Number of travel agencies	Number of restaurants
Grande Comore	329	658	51	23
Anjouan	312	624	7	17
Mohéli	122	244	ND	7
TOTAL	763	1526	58	47

Source: Counting from the tourism statistical yearbook, 2018

This can be seen as a way to improve the quality of life of the population. The product is almost exclusively focused on the ascent of Mount Karthala and does not generate real economic benefits for the country. The country's tourism policy, as set out in the 2030 Emerging Comoros Plan, is "to make the Comoros a leading tourist destination in the Indian Ocean region." The diversification of the tourism offer with the creation of a tourism hub around the islands, according to their specificity, will enable Comoros to position itself on a dynamic tourism market in the Indian Ocean.

The tourism offer is designed to develop a seaside resort in Ngazidja on the island of Grande Comores, a tourism zone on the island of Anjouan with the development of agro-tourism, and an ecotourism zone around the National Park of Moheli on the island of Moheli. In addition to these specialized areas, there is the development of a tourism of discovery of the Karthala volcano and the area of presence of the coelacanth. Sport tourism is envisaged through sport fishing and scuba diving. The enhancement of the natural and cultural heritage associated with the development of sport tourism will allow Comoros to take better advantage of its sustainable development.

The Main Sites and Areas for Sport Tourism

Comoros is a great adventure ground for hiking, offering the opportunity to discover the richness of the fauna and flora. The potential of the archipelago in terms of trekking is undeniable. Comoros offers a large number of hiking possibilities. The main sport hike is the ascent of the Karthala volcano (2361 m), which can be compared to that of the Piton de la Fournaise in Réunion Island. The Karthala area can be a strong structuring element for the development of a "trekking" product. Apart from the volcano, it is possible to practice trekking in a variety of natural spaces (islet of Selle, high altitude crater lakes, interior plains of the islands, coastal areas, etc.), allowing tourists to meet the local population and to discover the richness of the fauna and flora.

Summit of the Karthala volcano. (Source Ouledi A.)

The sea, or more particularly the coral reef, also offers important possibilities for sporting activities, such as:

- Big game fishing.
- Scuba diving: The sites of Comoros are qualified by professionals as being remarkable for their reefs and the biodiversity of fauna and flora.

On Grande Comore, diving sites are concentrated in the north of the island and in the south. There are also sites frequented by expatriates and nationals like Itsandra, Iconi, Itzoundzu, and Chindini. In Anjouan, the main dive sites are in front of the Al-Amal hotel, in Moya, Pomoni and around the islet of La Selle. In Moheli, the dive sites are mainly located around the islets of Nioumachoua.

At present, there are not enough hotels to welcome a tourist clientele for sports. However, for this type of product, the question of accommodation and comfort is central. It is thus a question of increasing the hotel capacity of certain zones in the country with the establishment of accommodations with the international standards and with capacity of reception for the sport tourists.

Trekking in the region of Diboini. (Source Ouledi A.)

The Economic Weight of Sport Tourism

Sport tourism is in its infancy but is clearly growing. The demand is increasingly strong and its promotion must be aimed preferentially at hiking–trekking, scuba diving, and deep-sea fishing, which provide alternative income-generating activities. This evolution should be considered with a participatory approach involving the village communities and targeting maximum benefits for the local population. This is a way of responding to SDGs 1 in particular.

This approach also has the major advantage of promoting the financing of the management of protected areas currently being set up and equity in the distribution of benefits related to the development of biodiversity. The sustainability of sport tourism will depend on the effectiveness of its financing and the role of sponsors. The organization of regional and international sports events should contribute to attracting sport tourism enthusiasts who could at the same time discover the cultural diversity of the country.

The sport tourism offer should be coupled with cultural tourism, which is the entry point for the involvement of local communities in its organization and popularization. The Comoros archipelago has a large number of historical sites, elements of tangible cultural heritage, but also events and traditions, or intangible cultural heritage, which can be the object of tourism. The tangible heritage is mainly concentrated in the cities of the archipelago (medinas, mosques, palaces), while the elements of intangible heritage are found everywhere in the archipelago (festivals, religious processions, traditional events, etc.). Thus, the Comorian culture, result of the meeting of many influences, offers a tremendous wealth and can be the object of a tourist valorization, and in particular in the current tourist context where globalization pushes each destination to distinguish itself from its competitors. A large number of jobs could be generated and contribute to the sustainable development of the archipelago.

Local Issues and Positioning of Stakeholders with Regard to Sport Tourism and Sustainable Development

The Comoros is a country with a high potential for tourism, but exploitation remains modest. This potential encompasses all segments of tourism products: ecotourism, beach tourism, sport tourism, cultural tourism, and agrotourism, through an exceptional coastline, protected endemic species, and a preserved natural habitat.

Dzialandze Lake on Anjouan Island. (Source Ouledi A)

The tourist sector in general is still weakly developed. The product is almost exclusively focused on beach tourism and does not generate real economic benefits for the country, unlike neighboring countries in the southwest Indian Ocean.

The overall tourism offer is designed to develop a seaside resort in Ngazidja on the island of Grande Comores, a tourism zone on the island of Anjouan with the development of agro-tourism, and an ecotourism zone around the National Park of Moheli on the island of Moheli. In addition to these specialized areas, there is the development of a tourism of discovery of the Karthala volcano and the area of presence of the coelacanth. Sport tourism is envisaged through sport fishing and scuba diving. The enhancement of the natural and cultural heritage associated with the development of sport tourism will allow Comoros to take better advantage of its sustainable development

Sustainable Development Perspectives for the Future

The questions of how to serve the Comoros, the establishment of structures dedicated to sport tourism, and the training of personnel will also have to be answered. Indeed, there are many opportunities for the development of sustainable and profitable tourism for the population. The Plan Comores Emergent by 2030, adopted in 2019, is part of this perspective. The Plan details the key ideas and large-scale projects that will drive the structural transformation of the country's economy. It constitutes the mobilization tool of the archipelago to succeed in its development while ensuring our full regional integration, honoring our international commitments, and contributing to the well-being of humanity. It integrates the achievement of the Sustainable Development Goals by 2030 and the African Union's Agenda 2063. In this Ten-Year Plan, tourism is considered as one of the key sectors to promote the sustainable development of the country. In the framework of this, reforms are conducted in other sectors (finance, investment) and must necessarily impact on the tourism sector. Through the substantial increase in tourist arrivals (more than 200,000 by 2030), revenues linked to the tourism sector will have to increase significantly and have an impact on the Gross Domestic Product and job creation, and ultimately contribute to the fight against poverty. The cross-cutting nature of tourism will create positive externalities on the sectors directly linked to its activity. Comoros could benefit from the aura of its national soccer team, obtained in 2021 during the African Cup of Nations, to attract sport tourism enthusiasts and professionals and boost this sector in the short and medium terms.

Human Capital to Promote Sport Tourism in the Islands

Human capital is one of the major catalysts identified by the Emerging Comoros Plan to promote the sustainable development of the country in general and tourism promotion in particular. Significant efforts are underway to build up a high and growing level of skills and experience to meet the country's present and future issues and challenges. Investments are focused on education, health, youth, social welfare, employment, sports, culture, and heritage. Training is provided at the University of the Comoros to build capacity to increase the quantity and quality of the tourism offer. The current orientation is based on:

- The consolidation of existing courses in tourism and management at the University Institute of Technology and respond to SDGs 4
- The opening of a higher school for training in tourism, hotel management, and tourism agency management
- The opening of bachelor's and master's degrees targeted at the preservation, conservation, and enhancement of natural, cultural, and historical heritage
- The development of a research pole on the development of tourist sites, local products, or the sciences of the sea and the earth

Conclusion

Sport tourism in Comoros is still in its infancy, both in terms of infrastructure and tourist numbers. Sport tourism will allow active stays in a rich and varied natural environment of the islands of the archipelago. It will give the opportunity to the sport tourists to come in contact with the natural environment, to discover new places by exceeding themselves physically. It offers them the possibility to combine pleasure and time of discovery by meeting the communities and by indulging in multiple discoveries: hiking, climbing the Karthala volcano, diving, fishing, exploring the regions on foot, and tasting the good local dishes. The promotion of this form of tourism will increase the economic income of the local population and will contribute to reducing the current practices that affect the marine and terrestrial ecosystem through the active participation of all stakeholders, including the private sector, local communities, and public authorities. Sport tourism is thus an undeniable asset for sustainable and local developments in Comoros. We have several years of experience in the promotion of ecotourism and sustainable tourism in relation to the implementation of SDGs 1, 8, 13, 14, and 15.

References

Ouledi, A., & Abdérémane, H. (2013). Comoros Islands. Ecotourism, a sustainable source of economic income. Institut de la Francophonie pour le Développement Durable. Issue 95. 4th Quarter 2013.

Union of Comoros. (2005). Tourism strategy framework document, "Comoros a destination in the making for tourism". Conference of donors in favor of the Comoros, 8 December 2005.

Union of Comoros. (2012). Tourism Policy Note, Report 1: Assessment and Diagnosis (final version), ABGE Project.

Union of Comoros. (2018). Annuaire statistique tourisme.

Union of Comoros. (2019). Comoros Emergent Plan 2030, Official document.

Union of Comoros, CPAD. (2019). Presentation note of the Tourism sector.

Union of Comoros, Ministry of Tourism, 2022. Report of the 8th edition of the International Assizes of Sustainable and Responsible Tourism co-organized by the Ministry in charge of Tourism and Tourism without Borders, Moroni, 10 and 11 May 2022.

Union of the Comoros, INSEED. (2017). Report of the General Census of Population and Housing. 2017 (RGPH, 2017).

World Tourism Organization. (2019). *Sport tourism and the sustainable development goals (SDGs)*. UNWTO. https://doi.org/10.18111/9789284419661

Chapter 26
Sport on Maldivian Islands. At the Intersection of Tourism, Development, and Power Relations

Claudio Melli, Stefano Malatesta, and Simone Tosi

Introduction

Sport is a pivotal interpretative key to understanding current tourism policy in the Maldives. Experts have investigated the social importance of sport and the fact that it is far more than a leisure activity. Sport plays a central role in the production of culture: it is intertwined with power relations, processes of social integration, and the economic sector. It is of great interest to analyse these dimensions in local places such as small islands, because they contribute to our understanding of manifold contemporary processes. In the Maldivian archipelago, sport tourism is one of the main new sectors of development being pushed for expansion by the government. Such expansion demands the "creation" or "invention" of a physical territory if it is to materialise. This chapter examines the role of sport as a driver of development in the case of an insular state, and the interconnectedness of tourism, economy, and soft power.

The chapter opens with a brief outline of the physical and political geography of the Maldives, which is key to our analysis given its distinctive characteristics (Sections "Physical Geography and the Political Administration of the Maldives: Insularity and Decentralisation" and "The Maldives: Human Geography Between Polarisation and Hyper-centrism"). Next, it addresses the topic of land reclamation,

C. Melli (✉)
Department of Sociology and Social Research, University of Milano-Bicocca, MaRHE Center, Magoodhoo, Faafu, Maldives
e-mail: c.melli@campus.unimib.it

S. Malatesta
Department of Human Sciences and Education "Riccardo Massa", University of Milano-Bicocca, MaRHE Center, Magoodhoo, Faafu, Maldives

S. Tosi
Department of Sociology and Social Research, University of Milano-Bicocca, Milan, Italy

D. Van Rheenen et al. (eds.), *Sport Tourism, Island Territories and Sustainable Development*, Sports Economics, Management and Policy 24,
https://doi.org/10.1007/978-3-031-51705-1_26

currently the main strategy being deployed in the country to reduce the islands' exposure to the effects of climate change and to facilitate urban development projects (Section " Land Reclamation Processes as Urban Resilience"). In the following section, sports are analysed in relation to Maldivian power dynamics and land reclamation strategies (Section "Sport in the Maldives"). Section "The Case Study: Magoodhoo, Faaf Atoll" presents the case study of the Rajjee Atoll Volleyball Championship 2022 and examines sports on the islands more generally. The concluding section suggests that the chapter's two main focuses, land reclamation and sport are closely interconnected. The topics involved in this chapter raise several dimensions of relevance for SDGs Agenda. On the one hand, they contribute to a better understanding of gender relations and the dynamics of women's empowerment (SDG 5) and the importance of inclusive quality education (SDG 4). On the other hand, the relationships between territorial transformation policies and climate change (SDGs 11 and 13) are highlighted.

Physical Geography and the Political Administration of the Maldives: Insularity and Decentralisation

The Republic of Maldives is an archipelagic state[1] in the Indian Ocean that consists of 1192 coral islands (Shaig, 2006), located 600 km South of India, and South-West of Sri Lanka. The territory extends approximately 900 km from North to South across the 15th Parallel, and 120 km from East to West along the 73rd Meridian. Together with two other Indian Ocean archipelagos, Chagos and the Laccadives, the Maldives form the ecoregion with the largest coralligenous atoll system on the planet.

Maritime space constitutes the majority of Maldivian national territory. The country's official name, "Dhivehi Raajjeyge Jumhooriyya" (Maloney, 2013, p.X), literally means "Kingdom of Islands," as well as "Kingdom of the Dhivehi people" (Romero-Frias, 2012, p.XXI). Thus, insularity is a fundamental category for understanding the geography of the country. Only nine islands have a natural area of over two square kilometres (Ghina, 2003), although many of them are implementing reclamation projects to increase their surface area.

The Maldivian government divided the 1192 islands into 26 natural atolls and 19 administrative districts labelled using the letters of the Maldivian alphabet. This approach was further confirmed in 2010 via a game-changing legislative decree. More specifically, in 2010, the government of Mohamed Nasheed enacted the *Act on Decentralization of the Administrative Divisions of the Maldives*,[2] a law that

[1] In this chapter, we adopt the United Nations Convention on the Law of the Sea (UNCLOS) definition of an archipelagic state as a group of islands forming a state and representing a single unit. Furthermore, the waters around, in between or connecting the islands, even beyond three miles from the coast, are part of the state's internal waters and therefore sovereign.

[2] Local Government Authority, 2010. Act on Decentralization of the Administrative Divisions of the Maldives 2010 (Maldives) Law No. 7/2010 (Decentralisation Act).

enabled the decentralisation of state power by making each island a separate administrative division. In addition to the redistribution of the administrative geography, the law provided for the election of island chiefs by the inhabitants and greater autonomy from the central state. Indeed, the islands now independently manage several areas of governance, including land policies and waste management.

The Maldives: Human Geography Between Polarisation and Hyper-centrism
Although the Maldives are formed by a vast number of small islands, the population speaks a language of its own (Dhivehi), uniformly practises Sunni Islam, and has a shared national history.

Definitive data for the 2022 census is not yet available, but the Maldives Bureau of Statistics has released interim results stating that 452,495 is the current population of the Maldives. The official data will be published in the coming months, but the upward trend seen in the 2014 census (which officially set the resident population at 407,660) seems set to continue.[3] However, the reliability of the census data is highly controversial in light of documents produced by various international agencies. For example, the United Nations Development Programme (UNDP) posits a significant discrepancy between the official data and the actual situation.

Apart from the use of different demographic data retrieval systems, it is plausible that the gap between estimated and actual data is underpinned by a pattern of significant growth in the Maldivian population. From 2006 to 2014, the demographic data show that, in less than a decade, the total population grew by over 94,000 inhabitants, or 31% in percentage terms (Malatesta et al., 2014), representing a reversal from the previous pattern of population decrease to one of population increase.

The archipelago's demographic data reveals a dual trend that is highly relevant to this chapter, because it conveys the special role played by the capital on even the most far flung of the peripheral islands. The capital-periphery relationship plays a key part in the social life of Maldivians, particularly with respect to the socio-economic sector, the religious sphere, and the construction of a national identity.

According to the 2015 census data, 157,935 inhabitants resided in the capital of Malé in an area of only 5.7 sq km, constituting about 38% of the total population. Numerous social groups tend to settle in or around this city, given its greater concentration of employment opportunities. Among the 1192 islands, therefore, about one-third of the population resides in the capital Malé, while the remainder (249,725) is settled on the 187 inhabited islands or resides temporarily in the approximately 159 resort islands (Ministry of Tourism, 2021).

In the outlying islands, the distribution of the population is extremely uneven, as few of the inhabited centres have more than 1000 residents, and only two islands besides the capital have a population of over 10,000, namely, Addu (32,000) and Fuvahmulah (12,000).

[3] Maldives Bureau of Statistics, Ministry of National Planning Housing & Infrastructure. In the 2015 census, the distinction between the resident population and the broader Maldivian population was implemented for the first time, unlike in previous censuses. https://statisticsmaldives.gov.mv/nbs/wp-content/uploads/2015/10/Census-Summary-Tables1.pdf

As stated by Malatesta and Schmidt di Friedberg (2017), "the human geography of the Maldives is characterised by the spatial antithesis between concentration and dispersion, and by centre/periphery relations and imbalances. Such spatial patterns act, and interact, within the geography of the archipelago, in relation to population distribution, capital accumulation, resource management, and accessibility to services" (p. 17). This description evokes a nation that is obliged to manage a high concentration of inhabitants and services in its capital; at the same time, the peripheral islands still account for almost half the population (44%), thus retaining their local importance, while remaining challenging and uneconomic to serve due to their geographical characteristics. This configuration of hypercentrism in the capital on the one hand, and controlled but increasingly autonomous suburbs on the other, persists to this day.

These distinctive characteristics of the Maldivian archipelago also necessarily bear consequences in the political sphere, from both large-scale and local perspectives. From a political point of view, it is clear that the capital Malé carries much more weight than all the other islands in the archipelago. Not only that, but the city acts as a "supercentre," where "everything happens" and the most important developments take place. This is so much the case that exiting the national political arena often goes hand in hand with moving away from Malé (Colton, 1995). Yet, although politics enjoys greater prominence in the everyday life of the capital, where the focus is often on international issues, the political weight of the peripheral islands is often underestimated at the national level. An example of this is the transition from Abdulla Yameen to Ibrahim Solih Precedencies, where much of the political and consensus-building effort was concentrated in the capital, but nevertheless the gap that secured victory for Ibrahim Solih's Maldivian Democratic Party (MDP) came from the votes of the peripheral islands. Without further dwelling on the topic, we might conclude that this political dualism of capital and periphery implies that the *res publica* on the outlying islands is managed via small-scale resources and policies that are often completely unrelated to the mechanisms that play out in the capital.

Land Reclamation Processes as Urban Resilience

In this chapter, we define "land reclamation" as an engineering process through which an area with natural characteristics that would not otherwise be exploitable for such purposes is made fit for agricultural, industrial, or urban use. Through physical intervention such as the draining of a marine area and the construction of various kinds of infrastructures, it is possible to create new spaces or enable new uses.

Land reclamation is a common practice in many archipelagic regions (Zhu et al., 2022; Wittezaele, 2021). In the Maldives, this operation consists of dredging sand from shallow lagoons. Although making new land available facilitates new development and infrastructural projects and represents an attempt to ease the pressure of immigration to Malé, it can also significantly damage coral reefs or create vast plumes of sediment that travel to other reef platforms.

The Maldives is also one of the lowest states in the world, making it extremely prone to climate change effects. Indeed, at these latitudes, land space is also claimed back from marine space in response to rising sea levels, a factor that features prominently in the political discourse. "We are one of the most vulnerable countries on Earth and therefore need to adapt," Maldives Vice President Mohamed Waheed Hassan stated in 2010 in the World Bank Report,[4] warning the world that around 200 islands in the Maldives are projected to be submerged by 2100. Thus, land reclamation fulfils a primary function as an adaptive response to climate change, but also bears prominent social, political, and economic implications. In the next section, we briefly present three cases that illustrate different outcomes and implications of land reclamation processes in the Maldives: Addu, Hulumalé, and Meedho.

The first case is Addu, the second largest city in the country. Maldivian President Ibrahim Solih recently approved an investment of $150 million for a land reclamation project on the island. The intention is to create a land area of approximately 228 hectares for residential, commercial, and industrial development.[5] The proposal has provoked ambivalent reactions: on the one hand, the inhabitants have welcomed the initiative with enthusiasm, as the project is expected to drive development and thus create better working and living conditions. On the other hand, a great debate has arisen, centred mainly on the significant and irreversible damage that the project will wield on the lagoon and surrounding reefs.

The second example concerns Hulumalé, the largest land reclamation area in the Maldives. The island has been entirely reclaimed from the sea, obtaining 4.3 sq. km of land intended to reduce the demographic pressure on Malé. The project was launched in 1997 and is still ongoing. The rapid population growth that has affected Malé in recent decades, making it one of the most densely populated cities on the planet, first triggered an upward expansion in terms of the construction of skyscrapers. When even this "upper" space was running out, in order to meet existing and future demands for housing and productive development in the Malé region and as a response to the threat posed by rising sea levels, the government began to seek additional "outer" space via extensive land reclamation interventions (Naylor, 2015).

The island of Hulumalé represents an extraordinary case of land reclamation that has enabled a large-scale extension of the city. As a result, Malé has remained the administrative centre of the country, while relying on the neighbouring reclaimed island for its additional space requirements. The first phase of the reclamation of Hulumalé, leading to the recovery of 188 hectares, began in 1997 and was completed in 2002. Two years later, the island celebrated the arrival of its first 1000 residents. Reclamation of a further 244 hectares of land was completed in 2015, and by late 2019, over 50,000 people were living on Hulumalé.

[4] https://www.worldbank.org/en/news/feature/2010/04/06/climate-change-in-the-maldives

[5] More specifically, the areas concerned are Hithadhoo, Hulhudhoo/Meedhoo (38 hectares including two island resorts of 5 hectares each), and Maradhoo/Maradhoo-Feydhoo (75 hectares).

Hulumalé was reclaimed by pumping in millions of cubic metres of sand, bringing the island to sit about 2 metres above sea level as an initial solution to the imminent rise in sea level. Areen Ahmed, director of business development at the Housing Development Corporation (HDC) that oversees the entire Hulumalé project, told BBC Travel that "After the 2004 tsunami, a programme for enhancing resiliency through safer islands was introduced. Hulumalé is being developed through careful considerations of climate change in its architecture and communities. Buildings are oriented north–south to reduce heat gain and improve thermal comfort. Streets are designed to optimise wind penetration, reducing reliance on air conditioning. And schools, mosques and neighbourhood parks are within 100–200 m walking distance of residential developments, reducing car use."

The artificial island of Hulumalé, built using millions of cubic metres of sand pumped from the seabed. (Credit: Hassan Mohamed)

The third case is the island of Meedhoo in the Dhaalu Atoll, where the small land surface was seen as the leading cause of vulnerability following the damage wreaked by the 2004 tsunami. As a result, artificial expansion of the area became a political necessity and a large-scale defensive land reclamation project was proposed as an adaptive climate change strategy (Schmidt di Friedberg et al., 2020). As noted, the restricted size of the settled area was the primary factor underpinning its vulnerability to rising sea-levels and massive events like a tsunami, but there are additional reasons for the development: namely, to reduce the risk of overpopulation by creating sites for future settlements, and to provide safe shelter for the local fleet of fishing vessels, one of the largest in the region. The harbour, which is protected by areas of reclaimed land, serves as both a shipyard and a storage area (Malatesta & Schmidt di Friedberg, 2017).

Land reclamation in the Maldives is a remarkable process that combines economic, social, engineering, urban planning, and political processes, a total social fact (Mauss, 2016). The demographic polarisation outlined in the previous section, in addition to the rapidly rising population growth rate, has put constant pressure on recent governments to identify a solution for the increasing scarcity of living space, such that land reclamation has become an integral part of Maldivian life. Reclamation statistics from the Ministry of National Planning, Housing and Infrastructure estimate that over 14 sq. km of land had been reclaimed up to 2015, representing a 9.5% increase in the total area of the inhabited islands, a percentage that is set to rise in conjunction with new economic developments and climate change. Coral reef reclamation poses a real threat to the marine biodiversity of the Maldives, in that it can lead to a significant loss of ecological habitat in the affected areas. Although individual impact and risk assessments have been conducted for each coral reef reclamation project, there has been little assessment of the cumulative risk to the overall biodiversity of the country and the health of the coral reef due to increasing development (Stevens & Froman, 2019).

Sport in the Maldives

In the last section, we outlined the scope and importance of land reclamation in the Maldives, which is driven by the need to promote sustainability first and foremost (in terms of protecting against the adverse effects of climate change) and economic development. The reclaimed sections of the islands move the core of settlements away from the sea (thus making them safer during events such as tsunamis, strong winds, and violent monsoons) and raise the island above sea level, in preparation for projected future rises. These factors, together with the growing need for housing and global challenges such as tourism, underpin the drive to reclaim land.

As mentioned at the beginning of this chapter, sport is a significant ingredient in the political and communications dynamics related to land reclamation programmes. The reading offered here focuses on one of the main implications of this practice, namely, the creation of new spaces for sporting activities and sports facilities, which are built as a result of the new urban landscape and the politics of consensus that is implemented in it. Sport is at the centre of multiple interweaving dynamics, including tourism policy, identity-related and social dimensions, local branding and territorial development strategies, and political relations. Sport lends support for land reclamation by contributing to promoting local areas and offers a mechanism whereby local political leaders and councils can enhance their relations with the superordinate territorial scales.

Sports for Tourists

The Maldives are represented in the imaginary of the average tourist as a natural paradise, and tourism here is profoundly related to sporting activities. As early as the 1970s, the first tourists in the country began to combine leisure activities with a specific type of sport that won immediate success in the archipelago, spearfishing. Such a connection between travel destinations and practising sports gives rise to what we call sport tourism: sport fishing, in fact, is a complement to the traditional "3S" (sun, sea, and sand) holiday, and this makes these islands a favourite place in which to practise various forms of the sport in a unique setting. Although fishing was one of the first attractions of the Maldives, the resort islands now offer a wide array of recreational water-based activities for tourists. Tourists are generally interested in any activity that can fill their time, transforming it from boredom to active relaxation. Table tennis, volleyball, football, even golf thus become important tourist resources that cannot be missed in any tourism offer package.

The latest statistics also confirm this trend: according to the latest *Maldives Visitor Survey* (2021), sport is a key-attraction for 38% of visitors. Although fishing was one of the first attractions associated with the Maldives, today the country boasts several water sports among its tourist offerings: snorkelling, windsurfing, kitesurfing, sailing, and water-skiing. Since the 1970s, resorts have been the most iconic and successful tourism facilities in the Maldives (dell'Agnese, 2018) and, from the early beginning, sport has played a significant role. Within their marketing, the proposed healthy lifestyle is characterised by various sports activities that can be practised both on water and on land. For the tourist, combining relaxation with the practice of one or more sports conditions the choice of destination, and also shapes the form of the accommodation facilities; sport plays a central role in the mechanisms of spatial and territorial planning and development as well, being "measurable." The role of sports in the current development of tourism in the Maldives appears particularly relevant. Historically, in the period when Maldivian tourism was confined to resorts, sports were just another component of the overall tourism product (particularly water sports), which served above all to increase the variety of leisure activities.

The tourism scene in the Maldives has been shifting quite rapidly for a few years now: the 1999 Tourism Act authorised the opening of guest houses in the country, a market that was prohibited until a few decades ago (MTCA, 2007; MOT, 1999). Due to the peculiar characteristics of the country and its tourist sector, sport has a key part to play in the emerging guest house market, as it does in the everyday spaces and lives of Maldivians. Sports spaces on the local islands increasingly offer novel meeting places. Unlike sports spaces in resorts, the impact of sports facilities on guest houses has not been studied yet. However, sport creates opportunities to encounter and talk with tourists, local people, and staff at sports facilities, in contrast with the "environmental bubble" (Cohen, 1972) generated by resorts (dell'Agnese, 2018).[6] Sport, more than ever before, is an aggregator of ideologies.

[6] dell'Agnese provided an illustrative spatial analysis of Maldivian resorts (dell'Agnese, 2018, 2019)

It is an opportunity for dialogue and hybridisation while playing football in a holiday-paradise is also a socio-cultural encounter, bringing into play different philosophies, and ways of understanding religion, ritual, ethics, and other components of identity. Indeed, sport is one of the most used yet least communicated drivers of socialisation practices on the guest house islands (Melli, 2022). With the "opening/exposition" of inhabited islands with the growing guest house market in the Maldives, land-based sports may be new venues for social encounters that can relate to meeting grounds, which have created new spaces for cultural interpretation.

Sports for Locals

There is no national sport in the Maldives, but sea sports, such as snorkelling, fishing, and surfing, are widespread. On the islands inhabited by local communities, sport practised on land is of particular scholarly interest due to its various interactions with the political establishment.

In the past, the most widely practised games were the "traditional" sports of bashi, thinmugoali, vaadhemun, and baibala; nowadays globally inspired, competitive sports such as football and volleyball are becoming increasingly popular. (Photo by Ahmed Ibrahim: Bashi match, Kalaidhoo Vs Dhanbidhoo, 30th November 2012)

"Structured" sport in the Maldives was introduced in the nineteenth century following the arrival of the British. This led to the gradual spread and uptake of different sports on various islands across the archipelago. Initially, sports and sports clubs were started by voluntary associations that aspired to more systematic management of facilities and training. According to Musthafa (2016), cricket was the first imported sport practised in the Maldives and football followed soon thereafter. The introduction of organised sports in the 1940s led to the creation of football teams and the first national tournament was held in 1948. With the growing interest in sports, an administrative office was inaugurated in 1950 to co-ordinate sports in the capital Malé, leading to a gradual introduction of other sports such as tennis, volleyball, basketball, athletics, and table tennis.

The Maldivian Olympic team at Athens 2004. https://olympics.com/ioc/maldives

The organisation of sports in the Maldives is centrally controlled by the government through the Ministry of Sports Youth and Development, which is under the aegis of the President. This is the government organisation responsible for overseeing all sporting activities in the Maldives and promoting a culture of sport in Maldivian society. All sports-related associations and federations come under this Ministry, which seeks to support their activities on the ground, internationally, nationally, and locally on the islands. Under the Ministry of Youth and Development are the National Sports Council, the Maldives National Sports Institute and the Maldives National Olympics Committee. The first of these bodies is the main governmental and funding authority for the national sports associations. The council's main role is to guide its member organisations in the development of sports. The Maldives National Institute of Sport was officially established in 2012 and is responsible for the training of sports personnel by providing high-quality education, training, and research to enhance the competence of those who work in the sports sector. Its stated objectives include the development of specialised sports services to ensure that athletes receive professional treatment for injuries and obtain scientific support throughout their development so that they can achieve peak performance. The mission is to improve sport in the Maldives by attaining sporting excellence, enhancing the national image, and fostering the development of sports professionals through sports education and scientific approaches.

The National Olympic Committee was established in 1985 and was recognised by the International Olympic Committee in the same year. This organisation played an important role in raising the sporting profile of the Maldives to a higher level, as it facilitated the Maldives' participation in sporting events and competitions outside the country, such as the Indian Ocean Games in 1979, where a first international prize (a silver medal) was obtained in badminton (Musthafa, 2016). In 1988, a Ministry of Sports was established in Malé with the aim of incorporating the practice of sports as an integral part of the nation's development, and in the same year, the Maldives took part in the Olympics for the first time at Seoul.

Maldives Olympics committee. (Official website, 2022)

After 1981, the associations slowly increased in number; modern sports were also introduced in schools and consolidated over the 1990s. Children were introduced to athletics, table tennis, badminton, football, netball, volleyball, and basketball and local and inter-school tournaments were organised.

Although some of the new sports have become highly popular, such as football for men and netball among women, they continue to receive little recognition at national and international levels today. The Government has recently moved to address these gaps in its policy agenda. Through multiple projects and the investment of a significant amount of funds in sports, the Minister of Youth, Sports, and Community Empowerment, Ahmed Mahloof, is attempting to lead the Maldives onto the international chessboard, by wielding soft power in the domain of sports. Some of the main strategies have included the recruitment of Francesco Moriero and Simao Costa, a pair of internationally recognised coaches for the Maldivian National Football Team; the purchase of footballers with foreign experience at a high level; the creation and expansion of sports spaces in the capital and on many of the outlying islands, including the goal of having a football pitch on every inhabited island and a projector in bars throughout the country for sports broadcasting; the organisation of the Beach Volleyball Pro Tour Challenge 2022 at the Sun Island Resort; the Maldives Sports Awards at CrossRoads resort with an audience of eminent sportspeople such as Roberto Carlos, Charles Puyol, Giba, in an effort to showcase the country's present and aspired-to future sporting achievements.

Volleyball World Beach Pro Tour—Maldives Challenge—2022. (Photo Credit: Volleyball World)

The Case Study: Magoodhoo, Faaf Atoll

Sport is an activity that, above and beyond its recreational and playful dimension, performs complex and crucial functions within societies. In this section, we home in on a specific case that illustrates how sport can function as an effective vehicle for soft power on a Maldivian Island, Faaf-Magoodhoo. Scrutiny of sport and its infrastructures can shed light on social ties, political and economic relations and narratives, thus enhancing our understanding of local contexts. Tournaments and sporting events help to define the status of specific places/islands within the atolls where they are situated, as well as vis-à-vis the central government. Soft power mechanisms and relationships in the Maldives are shaped by insularity given the country's archipelagic geographical conformation. Indeed, political relationships from the local to the international level are often the outcome of the interaction between sport, business, and power. The case study of Faaf-Magoodhoo we present here is particularly emblematic. First, because it is an example of a Maldivian island that has been focusing on a certain type of urban development. It has also been the main location of the authors' fieldwork for the past several years.

Faaf-Magoodhoo Council has invested in the development of the island mainly via sport and land reclamation projects. The island hosts the MaRHE Centre (Marine Research and High Education), which was officially inaugurated on 28 January 2009 in the presence of former President Mohamed Nasheed, a research centre whose mission is to conduct fieldwork in the areas of tourism, human geography, and the environmental, marine, and biological sciences.[7]

Combining the approaches of cultural, island, and sports studies, our research entailed the integration of several techniques: participant observation; analysis of visual materials (images, video, and maps) and government documents; semi-structured interviews with local community actors on the island of Magoodhoo and with local/national actors involved in sports (politicians, players, coaches, ministry officials); analysis of secondary data made available by the Maldives government and local authorities; and direct involvement on the part of one of the authors as an assistant coach of the Magoodhoo men's volleyball team in 2022.

Why in Magoodhoo?

"During the past decade, the Maldives have changed. The implementation of the Decentralization Act in 2010 that afforded more powers to local island councils, allowing guest house tourism in inhabited islands (something that was banned under the 1978–2008 regime), as well as improving ferry services between islands, has transformed hitherto remote and often impoverished islands into much wealthier and attractive places to live. No longer are Maldivians forced to move to Malé to seek work: many can now earn a good income working in their home islands" (Mohamed Nasheed, in Malatesta et al., 2021). This small portion of a speech by former President of the Republic, Mohamed Nasheed, emphasises that the recent opening up of the guest house market and the Decentralisation Act have given an unprecedented boost to local development on the islands.

The island of Magoodhoo is located in the Faaf Atoll, 135 km South-West of Malé. The name of the atoll may be translated as "North Nilandhoo Atoll." It is over 30 km long and 25 km wide, and is the second least populous atoll in the country after Vaavu Atoll, with about 6000 inhabitants (2021).

---

[7] https://marhe.unimib.it

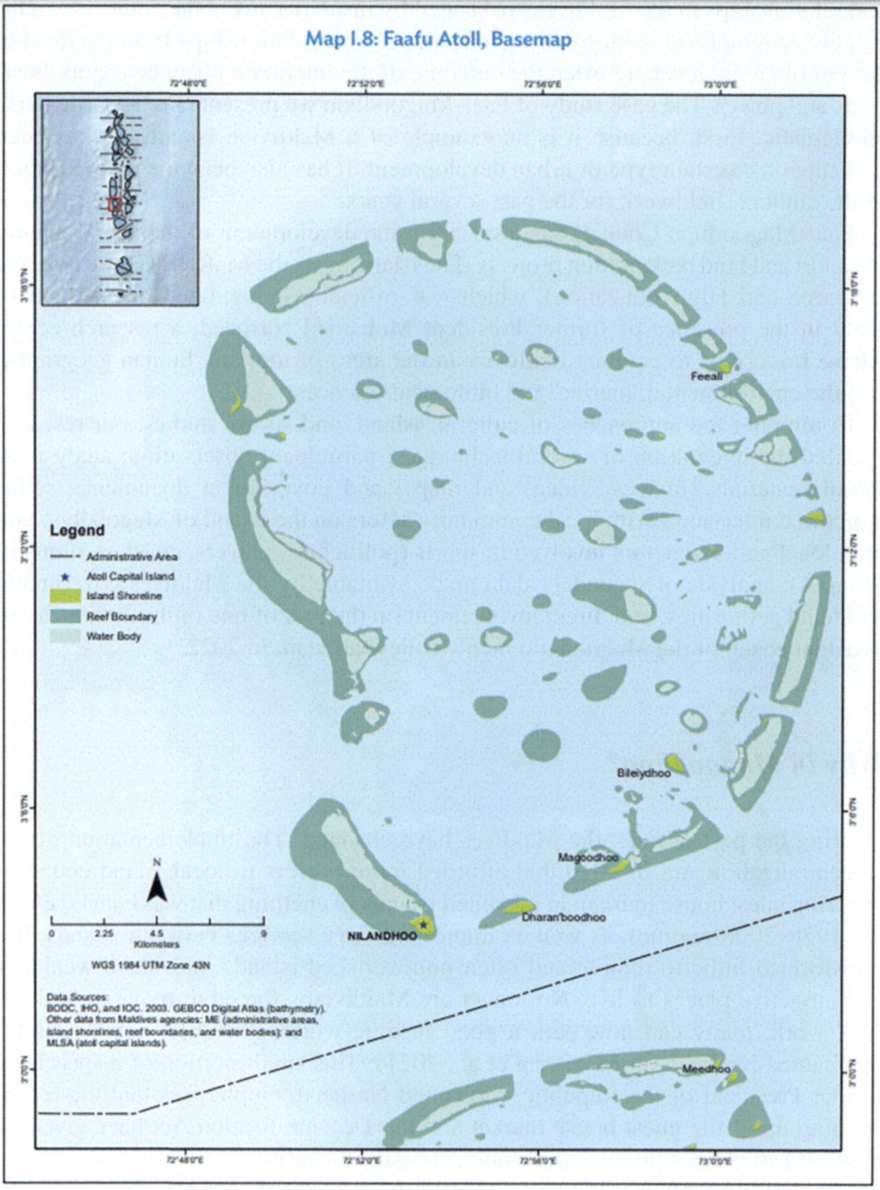

Map of Faaf Atoll. (Maldives Atlas Multi-hazard Risk 2020)

A first glance at the map reveals that inhabited islands are concentrated along the "half-moon" of the southern border, in Nilandhoo, Faaf-Magoodhoo, and Bilehdhoo, thus facilitating the flow of people and goods in this area. These three islands comprise the majority of the atoll's population, along with its schools, healthcare centres, main moorings for deep-sea fishing fleets, local transport network, and, of course, sports facilities. In light of this basic consideration, we can already deduce that Faaf-Magoodhoo occupies a hierarchically significant position within the political and social life of the atoll.

A further consideration is that until 1999 Faaf-Magoodhoo was the administrative capital of the atoll and the seat of the supra-local government. This brought about the construction of infrastructure to house officials and the celebration of public functions, the main hospital for the entire atoll, the largest school in the region, and the main docking point for commercial traffic. Faaf-Magoodhoo underwent significant phases of transformation during the final decades of the twentieth century, which allowed it to establish a hegemonic position underpinned by its function as the capital and visibility in the construction of public spaces (Malatesta et al., 2016).

It is evident from the photograph below that the island's institutional spaces, such as its office and administrative complexes, the research centre, and sports and religious facilities, occupy a significant portion of the developed area.

A photograph of Faaf-Magoodhoo island illustrating that the institutional buildings and football pitch take up a disproportionate share of the inhabited space. (Photo by Luca Fallati)

The island in 2023, following the most recent land reclamation interventions. (Photo by Ammadaj Siraj, 2022)

This position allowed Faaf-Magoodhoo to benefit for a number of years from privileged channels for selling local products and accessing public transport, conferring it with advantages over neighbouring islands: "in this sense, the island … found itself for years at the centre of a network of very relevant relations on a supra-local and regional scale" (Malatesta et al., 2016, p. 14). Consideration of the island's administrative status is particularly key to our analysis here because Faaf-Magoodhoo has a peculiar relationship with Nilandhoo, its main counterpart within the social, political, and economic life of the atoll. In 1999, as part of a broader transfer of power, Nilandhoo was appointed capital of the atoll instead of Faaf-Magoodhoo, a move that clearly led to the enhancement of the two islands' administrative and political status.

Indeed, as a result of the change, the central hospital, the atoll's new government buildings, the leading school hubs, the main fishing docks, public transport terminals, and fish processing facilities were all relocated to Nilandhoo. The economic and political downgrade imposed by the central government has shaped the current political status of the island.

Sport in Faaf-Magoodhoo

Let us now examine the use of sport by leaders and councils to promote development projects, soft power in consensus building, and sport as a bargaining tool in Faaf-Magoodhoo. Faaf-Magoodhoo Councils have launched several projects involving sport. In keeping with the nation as a whole, the men in the island community have displayed a keen interest in football since the 1970s, thanks to the British influence, while the women continued to play the traditional sport of *bashi*, although the latter has died out recently.

Faaf-Magoodhoo sports grounds. (Ammaday Siraj, 2022)

The football pitch is the main landmark in the area, which is of great importance in terms of urban planning. Furthermore, it is due to be renovated over the next few months (early 2023) with a view to better "showcasing" Magoodhoo to its neighbouring islands, both as a key island within the atoll.

Several other facilities have recently been upgraded along with the football pitch: work on a new volleyball court was initiated towards the end of 2022, and a new guest house was opened close to the football grounds. In addition, in spring 2023, the school's kindergarten building is due to be reduced in size, making way for a multi-sports field for the pupils. Over the past two years, volleyball has become one of the main sports practised on the island. This is due to an initial project to promote and invest in the local team as a means of boosting the local area, with the full support of the Island Council. There is also renewed interest in the upkeep of the sports fields, the expansion of the outdoor fitness area, and the daily practice of volleyball, which has largely replaced football. This shift in sports activity is also reflected in the construction of a beach volleyball court just off the MarHE Centre campus, with a view to providing a softer surface to play on and facilitating matches between residents and students.

The recent construction of sports facilities feeds into an ongoing relevance of sports in the local agenda. Magoodhoo Island has exploited a range of soft power dynamics within the domain of sport. This has included the co-opting of international level actors. An example is the hiring in 2018 of one the authors of this chapter as the coach of the women's volleyball team and subsequently as coach to the men's soccer team in preparation for the qualifying round of the *Minivan Championship 2018 cup*. In light of the new enthusiasm for volleyball, the island council realised that this sport offers a potential means of challenging Nilandhoo, which has a stronger tradition of soccer. This led in the following years to the setting up of various volleyball associations, and especially a men's volleyball team that trains intensively, thus moving into a space that was previously predominantly female.

Upcoming political and administrative elections are at the heart of consensus-building projects surrounding sport, especially over the past six years, borne out by the lively sports scene in Faaf-Magoodhoo.[8] Recent projects have seen new initiatives, including the expansion of five local sports associations (both male and female), as well as the further development of the Island's Youth Centre.

Rajjee Atoll Volleyball Championship 2022

The Rajjee Atoll Volleyball Championship 2022 was a tournament that took place over almost a month across the different atolls of the Maldives. The subtitle of the tournament, "a tournament for all Islands," emphasised the government's will to involve the outlying islands. This competition proceeded by administrative-geographical levels, with teams initially playing off against the others from their own atoll, and proceeding to higher territorial levels, with the final being played at the national level.

[8] The reading offered here should be viewed in light of the various electoral contests that are held in the country. The next general elections are due to be held in September 2023.

Event Poster

In Magoodhoo, the Council did remarkable work, bidding successfully against Nilandhoo, the capital of the atoll, to host the local stage of the tournament. The local matches began on 11 March 2022 and the preparatory work for the event saw strong involvement on the part of the entire island population. First, the sports facilities were significantly modified: spaces were enlarged nearby the court, access for security and emergency personnel was enhanced, and a broadcasting stage and official gallery with executive seats for the national and local authorities were built.

Subsequently, work was done on the surface of the court, while the nets, balls, lines, and other technical equipment were replaced, using financial support from the Government. In addition to the preparation of the space and infrastructural interventions, key emphasis was also laid on the sporting outcome. Magoodhoo Island Council invested relevant energies and adopted several strategies to get the highest possible standard. First, the council set out to optimise the preparation of the team from a technical point of view. Its primary move was hiring a coach of national calibre to lead the men's team, Mohamed Nadheem, a former national player of both volleyball and soccer in the 1990s, with one of the authors of this chapter as assistant coach. For the women's team, the council engaged former player Shaira Adam, who had also played volleyball at the national level in the Maldives.

Players who usually reside in Malé were also invited to participate. For the occasion, they negotiated time off work to be able to train with the local team at least twice a day. In some cases, the services of semi-professional players were purchased. Thanks to sponsorship and state funding, event T-shirts were also provided to all the players and their entourages. Interestingly, a former player on the National

Youth Team, despite residing in the nearby (and competing) island of Feeali, preferred to play with the Magoodhoo colours. This raises the theme of relational tensions between the islands, and the role of sport as a key means of communication. In the context of sporting competitions, the residents of different islands invite one other over, compete, fight, unite or fall out, as in any "derby" among local sports teams. Players residing in given islands of the atoll but registered in others were the object of debate, whistling during matches, and gossip, which in the Maldives functions as a true social regulator (Colton, 1995). In some cases, matches were suspended or a team withdrew, not showing up to the final prize-giving ceremony, as in the case of the aforementioned Feeali.

All the teams from the atoll's islands attended the Volley Rajjee Atoll Tournament on Magoodhoo: Magoodhoo, Feeali, Byledhoo, Nilandhoo, and Dharaboodhoo. The Island Council also inaugurated a male Football Academy and a female Handball Academy at the opening ceremony of the volleyball tournament. Underlying this move is the fact that Nilandhoo already had its own Football Academy. Furthermore, the ceremonies were in keeping with the agenda of the Minister of Sport: to encourage the practice of sport on every inhabited island.

The event also triggered multiple other related dynamics. In the first place, as already outlined, it prompted the improvement of the sports facilities and the reorganisation of urban areas with relevance to the tournament. A small number of accommodation facilities were placed under severe stress in attempting to host, often with little notice, entire teams and their back-up squads. Undoubtedly, such an important tournament also influenced the local tourist sector, or the hosting of Maldivians by other Maldivians. Much of the accommodation was in guest houses paid for by the Island Council, but these could not cope with the number of guests generated by a tournament of this size. The visit of the Minister of Sport, the Presidents of the Volleyball Federations, national referees and their entourage of almost 60 people, alone absorbed all the accommodation laid on by the council. Consequently, various alternative strategies were deployed to accommodate the large influx of people. Support was requested from "external" institutions, such as the MaRHE Center, or overnight stays were set up informally with friends residing on the Island. In addition, public transport was intensified to allow the teams and their supporters to move easily to and from Magoodhoo, with the provision of an almost daily service instead of the usual weekly one. In sum, embedded in a sporting fixture, we can observe differential relationships among islands, infrastructural development, and political agenda.

Conclusions

Describing land reclamation processes in the Maldives, sport, and one key event such as the Atoll Rajje Tournament, allows us to observe several connections:

The Relationships Between Islands and Local Actors

A sports tournament such as the Rajjee Atoll Volleyball Championship 2022 points to the political relations inherent in sport. For example, we may note the case of Feeali, which throughout this tournament expressed internal tensions with Magoodhoo or some of its players. Nilandhoo, on the other hand, although it is the current capital of the atoll, did not host the tournament. Sport, in this sense, is a tool that can re-shape the balance among islands, giving, even for a short lapse of time, Magoodhoo a new centrality. The relationships at different scales that were on display during this tournament convey the notion that the local islands, albeit peripheral and distant from Malé, can have considerable impact on national politics, as borne out by the publicity received by the event and the visit of the Sports Minister to the Island. In addition, intra-archipelagic relations and political negotiations show that relationality, connections, and local alliances are what truly define island politics. In the case of Faafu Atoll, one of the least populated atolls with the smallest number of inhabited islands, it is clear that the relationship between Magoodhoo and the capital Nilandhoo is emblematic of this complex network of connections.

The sports tournaments that are organised in the atoll with variable frequency contribute significantly to maintaining more constructive political relations that would otherwise be compromised by frictions that are still very much alive. Tournaments between these two islands and exchanges of players may be seen as largely reflecting a policy of pacification (Melli, 2020). Sports tournaments are also one of the few, if not the only, reason to travel to a neighbouring island to support one's local team. This leads to the renewal of family ties, increased uptake of car hires and accommodation services, and a general boost to the local economy. However, it can also serve the purposes of claiming or defending one's local territory or reaffirming established political hierarchies.

The Infrastructure

Thanks to both national and local policies, sports spaces were built or renovated and a short-term boost was provided to the construction of other spaces. Possible knock-on benefits of this emphasis may relate to future infrastructural development on Magoodhoo. Furthermore, sports policies for young people such as setting up of dedicated academies have great impact in a country like the Maldives, constituting one of the cornerstones of building consensus in the local islands. Moreover, the Rajjee Atoll Volleyball Championship 2022 has been, for Faaf-Magoodhoo, one of the leading national-level events of recent years. Indeed, the visits to the island of "legends of Maldivian Sport" and the Minister of Sport Ahmed Mahloof boosted the competitiveness of the tournament and the participation, as well as offering an opportunity to players seeking more lucrative contracts in Malé to show off their talent. Some of the players that took part in the tournament have since participated in other events and are attempting to become pros on the main island.

A sporting event of this type also intertwines with the dynamics of tourism. The accommodation and food hospitality offered (a bar and a bakery/pizza restaurant were opened during the tournament), the additional visitor services, the press and social media coverage, the enhancement of local transport services, and the embellishment of the sport all illustrate the potential to foster the growth of both local and international tourism.

While it is not possible here to empirically document the trends in domestic tourism, we think it is remarkable to consider the possible future evolution of domestic tourism in the Maldives. Considering the possible future evolution of it, it can be emphasised that the growth of sport facilities in land reclamation areas can be a driver for local facilities, with the possible enrichment of alternatives for the sport marketing and development of the Maldives, for years focused on water sports. In terms of its medium-term spin-offs, the case of the Atoll Rajje Tournament is even more interesting. The strategies implemented by the Faaf-Magoodhoo Council leverage sport to attain more ambitious aims than the mere "promoting of sport on the Island." Advocating for a large land reclamation project that will involve the whole island, the Council is seeking a solution, as in Hulumalé, to respond to local people's needs for more stable job opportunities and a better housing situation. The tournament was an opportunity to show the Minister of Sport the island of Magoodhoo needs a new and large-scale reclamation project to follow the development of the new port in 2017. Therefore, this tournament served as a valuable showcase not only for the players, but above all for the island, pointing up the fact that its living space and facilities are currently at the limit of their capacity, but that it offers potential centrality in the atoll. After the tournament, the reclamation project was approved with great enthusiasm by the local administration. The reclamation intervention that began in September 2022 involves the pumping of sand to clear 17 hectares of land to be used for housing or public facilities that will generate employment.

Magoodhoo's land reclamation project, however, does not end there. Indeed, not only the area adjacent to the port will be enlarged, but also the far end of the island. Here the Magoodhoo Island Council, again soon after the volleyball tournament, won a public tender for the construction of a domestic airport. Under a recent scenario, every atoll must have an airport, and Faafu was one of the last to get one. Again, with great political acumen, the Council, guided by Abdulla Waheed, got the approval for an even more extensive reclamation project: airport reclamation operations over 30.8 hectares, construction of a 3339 sq. m revetment, and a 90 sq. m cave. The value of the project is MVR 206.88 million (source: Maldives Transport and Contracting Company Plc). A project of this type will place Faaf-Magoodhoo and the nearby areas on a new communications level, as well as enhancing local accommodation facilities and transport. Furthermore, these policies have brought visibility to the island, which has recovered a political role that it had not filled for a long time. The importance of Faaf-Magoodhoo within the insular dynamics of the atoll and on the national political chessboard could be re-visited.

An "In Progress" Reclamation Project and Some Future Ideas

Sport offers a lens for learning about multiple aspects of a society. The innovative approach of this study lies in its in-depth analysis and focus on the peculiar geographical characteristics of the Maldives. More specifically, our case study shows how sports tourism interacts with soft power at both the local and national levels. It is also of value to study sport tourism in the Maldivian archipelago in light of the country's distinctive characteristics; its fragmentation, centre-periphery paradigm, and isolated image among those who do not live there all make it a privileged territory for observing relational dynamics.

Sport is a resource for local development, largely thanks to its high visibility and attractiveness to tourists. Alongside this "manifest function," it fulfils various "latent functions." First, it drives employment, mutual alliances, and political manoeuvring; second, it offers a focus for economic strategies (land reclamation, urban development, land use); finally, as a key factor in the contemporary geopolitics of the Maldives, it contributes to the country's internal cohesion policies (at different scalar levels: island, atoll, archipelago), as well as its international positioning.

The case study of the Faaf-Magoodhoo stage of the Rajjee Atoll Volleyball Championship 2022 produced several outcomes. First, it enhanced the island's image within the atoll and contributed to a rearrangement of the local chessboards. Also, at the end of the event, the Magoodhoo women's team won their category (and got into the next stage) while the men's team ranked second, with the (potential) effect of temporarily quelling enthusiasm for the sport. While the football and handball academies undoubtedly generated an initial boost in interest among young people, this has waned over time, as the teenage population typically moves to Nilandhoo or Malé to pursue their education, given the total absence of a management structure in terms of coaches, staff, and players.

An interesting ongoing process is that the strategies deployed to date in conjunction with sport have both incentivised sports spaces (such as the beach) and at the same time left them empty. A clear reflection of this is Faaf-Magoodhoo's preclusion, due to a lack of players, from the inter-zonal football tournament, which is relatively important at the national level.

Finally, we should report an interesting turn in the Faaf–Magoodhoo reclamation project: sports spaces share their previous central role with the project of the new airport. These latter considerations show how land reclamation is becoming increasingly central to the political agendas of the local islands and how it should be understood not only as an engineering strategy, but also as one of the constantly evolving forces that define the geography of the archipelago.

References

Cohen, E. (1972). Toward a sociology of international tourism. *Social Research, 39*, 164–182.

Colton, E. O. (1995). *The elite of the Maldives: Sociopolitical organisation and change*. London School of Economics and Political Science.

dell'Agnese, E. (2018). "One Island, one resort". Il turismo enclave alle Maldive come eterotopia pianificata. *Bollettino della Società Geografica Italiana serie, 14, 1*(1), 27–39.

dell'Agnese, E. (2019). 'Islands within Islands?' The Maldivian resort, between segregation and integration. *Tourism Geographies, 21*(5), 749–765.

Ghina, F. (2003). Sustainable development in small Island developing states. The case of the Maldives. *Environment, Development and Sustainability, 5*(1), 39–165.

Malatesta, S., & Schmidt di Friedberg, M. (2017). Environmental policy and climate change vulnerability in the Maldives: From the 'lexicon of risk' to social response to change. *Island Studies Journal, 12*(1), 53–70.

Malatesta, S., Schmidt di Friedberg, M., Mohamed, M., Zubair, S., & Bowen, D. (Eds.). (2021). *Atolls of the Maldives. Nissology and geography*. Rowman & Littlefield ("Rethinking the Island").

Malatesta, S., Muller, S., di Friedberg, M., Squarcina, E., Cajiao, A., & Di Pietro, A. (2014). Stato centrale e centri periferici: la geografia politica delle Maldive tra omologazione e segregazione. In *ACTAS, XIII Coloquio Internacional de Geocrítica. El control del espacio y los espacios de control*. Universidad de Barcelona.

Maloney, C. (2013). *People of the Maldive Islands*. Orient BlackSwan.

Mauss, M. (2016). *Saggio sul dono*. Giulio Einaudi Editore.

Melli, C. (2020). *Islam e Leadership Locale alle Maldive. Uno Studio Etnografico*. Unpublished, MA Thesis. University of Milano-Bicocca.

Melli, C. (2022). The role of sport in the relationship between tourists and locals: Case study of the Maldivian guesthouses. In B. Mazza & C. Sobry (a cura di) (Eds.), *Sport tourism and its territorial development and opportunities* (pp. 41–64). Cambridge Scholars Publishing.

Ministry of Tourism. (1999). *The Maldives Tourism Act*, Law n.2/99, https://www.tourism.gov.mv, Last Access 1 Jan 2023.

Ministry of Tourism. (2021). *Tourism yearbook 2021*. Malé, Republic of Maldives.

Ministry of Tourism and Civil Aviation. (2007). *Maldives third tourism master plan*. https://www.tourism.gov.mv, Last Access 2 Jan 2023.

Musthafa, A. (2016). *Sports history of Maldives*. [online] Createdbymustho.blogspot.kr. Retrieved from: http://createdbymustho.blogspot.kr/2016/01/sports-history-ofmaldives_24.html.

Naylor, A. K. (2015). Island morphology, reef resources, and development paths in the Maldives. *Progress in Physical Geography, 39*(6), 728–749.

Romero-Frias, X. (2012). *Folk Tales of the Maldives*. NIAS Press.

Schmidt di Friedberg, M., Malatesta, S., & dell'Agnese, E. (2020). Hazard, resilience and development: The case of two Maldivian Islands. *Bollettino Della Società Geografica Italiana, 3*(2), 11–24.

Shaig, A. (2006). *Climate change vulnerability and adaptation assessment of the Maldives land and beaches*. James Cook University, Centre for Disaster Studies.

Stevens, G., & Froman, N. (2019). The Maldives archipelago. In C. In Sheppard & W. Seas (Eds.), *An environmental evaluation volume II: The Indian Ocean to the Pacific*. Academic Press.

Wittesaele, C. (2021). Art, land reclamation and green governmentality in Indonesia: Teja Astawa's Dewa Murka and Tita Salina's 1001st Island — The most sustainable Island in Archipelago. *Journal of Southeast Asian Studies, 52*(2), 309–335.

Zhu, J., Yao, Y., Li, L., & Guan, W. (2022). *Impacts of land reclamation on hydrodynamics in Zhoushan Archipelago*. Third international conference on geology, mapping and remote sensing (ICGMRS), Zhoushan, China, 2022, pp. 74–77.

Chapter 27
The Role of Sport Tourism in Diversifying Tourism Activities in Zanzibar: The Need for Institutional Support for Sustainability

Makame Omar Makame and Said Shaib Said

Introduction

Sport tourism refers as 'all forms of active and passive involvement in sporting activity, participated in casually or in an organized way for non-commercial or business/commercial reasons that necessitate travel away from home and work locality' (Standeven & De Knop, 1999: 12; also see Downward et al., 2020). Sports play an important role in destination marketing, as they can promote and help to brand destinations (Dimanche & Sodja, 2006). Sport tourism is growing rapidly all over the world (Whitehead & Wicker, 2018; Pedauga et al., 2020) and sports activities linked with tourism have become significant economic opportunities globally. Sports-related travel such as travelling to watch sport, travelling to play sports, and travelling to sports attractions contribute to economic development of the host destination, bolstering business for various industries, such as hotels and aviation.

Sport and tourism are concerned with the important economic activities in the developed and developing countries (Swart & Bob, 2007: Mazza & Sobry, 2022). The United Nations rates this type of tourism as the fastest growing sector in recent years (WTO, 2001; Li et al., 2012) Based upon the World Tourism Organization (WTO) estimates, 43% of the world jobs are associated with the tourism industry in 2010. According to Fernandez (2014), participants, guides, spectators, and

M. O. Makame (✉)
School of natural and Social Science, The State University of Zanzibar, Zanzibar, Tanzania

Department of Marine Conservation, Ministry of Blue Economy and Fisheries, Zanzibar, Tanzania

S. S. Said
Department of Marine Conservation, Ministry of Blue Economy and Fisheries, Zanzibar, Tanzania

D. Van Rheenen et al. (eds.), *Sport Tourism, Island Territories and Sustainable Development*, Sports Economics, Management and Policy 24,
https://doi.org/10.1007/978-3-031-51705-1_27

organizers of the sports events generate expenditure and employment opportunities. WTO has estimated average contribution to gross domestic product of sport tourism in 2001 to 1–2% of GDP, while tourism's contribution was 4–6% in 2001. According to the World Tourism Organization (WTO), sport tourism generates annual revenues of $USD 800 billion globally. Furthermore, according to Kasimati and Dawson (2009), from 1997 to 2005, sport tourism brought about an annual increase in gross domestic growth (1.3%), resulting from economic activities with regard to holding sport competitions. Unemployment declined by 1.9% on an annual basis. Therefore, sport can considerably influence sport-based services and products in terms of functional and visual dimensions. Hence, it is believed to be one of the most effective factors influencing national growth and revenue in the twenty-first century (Brown & Nagel, 2002).

Tourism brings economic benefits to regions, states, and countries, but there are usually substantial environmental and socio-economic costs associated with such growth (Davenport & Davenport, 2006: Apostolopoulos & Gayle 2002; Royle, 2001). Small Islands nations, such as Zanzibar, are vulnerable to various stresses, including those related to mass tourism; however, sport tourism has recently been regarded as a vehicle towards sustainable economic growth (Lim & Patterson, 2008). This is because sport tourism offers an opportunity that may help to overcome some of the negative impacts of mass tourism. According to Lim and Patterson (2008), islands like Bonaire in the Dutch Caribbean and Malta in the Mediterranean, offer successful case studies of sports market development through diversification of investment sport tourism. According to the WTO (2019), sport tourism can help in achieving SDGs 1, reducing poverty, SDGs 2, promoting sustained, inclusive, and sustainable economic growth and SGDs 14, and conserving and sustainably using the oceans, seas, and marine resources for sustainable development. These opportunities may likewise help to achieve SDGs 3, health and well-being, and SDGs 5, ensuring gender equity and empowerment.

In Zanzibar, sport tourism is slowly being recognized as an important niche market within the tourism sector. However, this subdivision of tourism is haphazardly managed, which lacks vision and strategic planning from actors. Currently, tourism in Zanzibar is mainly based on recreational boating, sun (sunbathing), coastal walks, sand, coast- and marine-based ecotourism, swimming, snorkelling, diving, cruises, recreational fishing, and excursions within the MCAs. However, in recent years, we have witnessed diversification in activities that fall under sport tourism. These include kite surfing, Jet Skiing, yachting, bike riding on the beaches, and sport tournaments. With an intentional emphasis on promoting a blue economy, Zanzibar strives to increase the number of tourists per year. The diversification of tourism activities is considered to be one of the many strategies to reach the goal (Carius & Job, 2019). However, the current sports-related activities observed, such as kite and jet ski among others, are loosely managed and their potential environmental impact within the MCAs are not known.

For example, there is an increase in the number of complaints from hotel owners and tourists on jet skiing that operate in shallow water close to the beach or sandbanks where tourists and locals normally swim. Injuries caused by Jet Skis to these swimmers have also been reported over time. This means there is conflict between hoteliers and the Jet Ski rental business. This highlights the need of coordination and management of the sport tourism activities between different actors within the MCAs in Zanzibar. This chapter analyses sport tourism activities practised in the study areas and the opportunities they offer for future development. It also examines the conflicts between these activities and other marine users. Finally, the chapter addresses the institutional support needed in Zanzibar to help with the management and operations of sport tourism in MCAs. As such, this chapter illustrates how sport tourism can contribute to these goals and what needs to be considered when developing sport tourism to realize its contribution to sustainable development.

Table 27.1 below presents various characteristics of Zanzibar, which include social, economic, and the geography of Zanzibar. Indicators of insular determinism and others highlight the potential of Zanzibar as a sport tourism destination.

Study Area

Zanzibar is a semi-autonomous state that, together with mainland Tanzania, forms the United Republic of Tanzania (URT). It is composed of the two main islands of Unguja and Pemba, which both lie about 40 km off the mainland shores. There are also a large number of smaller islands, of which the larger ones are almost all inhabited. The coastline of Zanzibar stretches over a distance of 370 km, and is influenced by monsoon winds and local storms, especially on the eastern side. This study was conducted in four Marine Conservation Area (MCAs) in Unguja Island, namely, Menai Bay Conservation area (MBCA), Mnemba Island and Chwaka bay Marine Conservation area (MIMCA), Changuu–Bawe Marine Conservation area (CHABAMCA), and Tumbatu Marine Conservation Area (TUMCA) (Fig. 27.1). These areas host more than 90% of the tourism hotels and resorts and where most of the sport tourism activities are practised.

Data Collection

The data from this research was collected through various qualitative methods such as interviews, key informant interviews (KII), focus group discussion (FGDs), questionnaires, and a review of secondary data. Interviews, including KII, were conducted with owners of businesses related to sport tourism, such as tour operators, hoteliers, the tourism commission, experts from the department of the environment, the department of marine conservation, and managers of all (four) MCAs, the department of fisheries, and some members from the Shehia fisheries committee. These methods were utilized to obtain qualitative and quantitative information to collect adequate and reliable information from stakeholders that would help to achieve the objectives of this study (Table 27.2).

Table 27.1 Characteristic of Zanzibar

Islands (facts in 2019)	**Zanzibar**
Social economic characteristics	
Capital	Zanzibar Town
Geographical size (Km2)	2656
Population (thousands)	1,671,598
Share of population under 20 years old (%)	49
Share of population above 65 years old (%)	2.8
Unemployment rate (%)	14.3
Unemployment rate of population under 20 years old (%)	27
GDP per capita (US$)	1099
Average annual income (US$)	NA
Sport participation index (%)	NA
Tourism direct contribution to GDP (% of total GDP)	29.20
Geography of the island	**Zanzibar**
Island typology (mono-island, multi-islands, archipelagic)	**Archipelagic**
Number of islands	54
Geographical size of the whole island	2656
Political status of the island (independent, autonomous region, islands under the authority of a metropolis...)	Autonomous region
Name of the main colonizing country	Britain, Arab
Indicators of insular determinism	**Zanzibar**
Location	Indian ocean
Type of geomorphology of islands (oceanic island, pre-continental island)	Oceanic island
Number of islands	54
Geographical size (Km2)	2656
Number of inhabited islands	12
Main means of accessibility	Water, air
Number of seasonality	4
Number of conservation area (national park, regional park, protected area)	6
Number of marine conservation area	5
Number of World Heritage Sites	2
Indicators globalizing the structuring context of the islands	**Zanzibar**
GDP per capita (US$)	1099
GNP/GDP (US$)	NA
Gini Index (0 to 1)	0.3
Type of principal vulnerability (natural hazard, health, security)	Malaria, Covid 19
Tourism in islands	**Zanzibar**
Number of tourists	394,185 (2021)
Number of cruise passengers	94 (2021)
Change in attendance	260,633 (2020)
Length of stay (in days)	9
Total revenue (millions US$)	380.1
Share in GDP (%)	29.2
Number of beds	20,991 (2021)

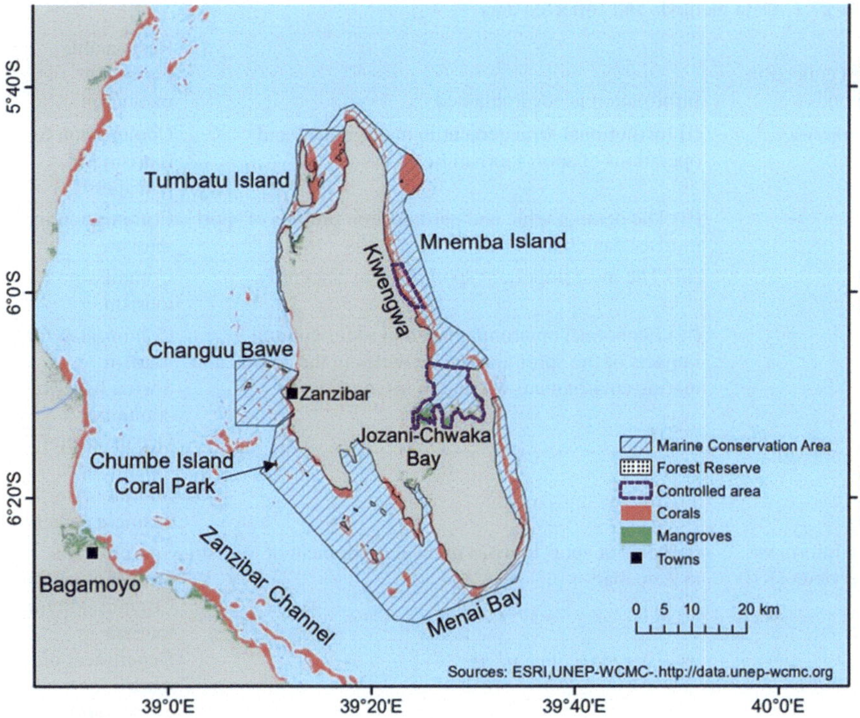

Fig. 27.1 Map showing location of study sites (Zanzibar Marine Protected Areas). (Source: DMC, 2022)

Results and Discussion

Tourism in Zanzibar

Tourism was identified in the late 1980s as a sector with major potential for driving economic development in Zanzibar. The industry has since grown from about 18,000 visitors in 1985 to 520,809 in 2018. These visitors reached a record-breaking 538,300 thousand in 2019, and nearly 189 thousand tourists in the first half of 2021, despite the Covid-19 pandemic (see Table 27.1). The number of tourism hotels increased from two hotels during the revolution in 1964 to over 600 in 2023 (The Citizen, 2023). The tourism sector in Zanzibar accounts for 27% of GDP. These growth indicators of the tourism industry in Zanzibar elucidate not only an attractive environment for investment but also innovations and development of services, including sport tourism. This has been one of the important innovative strategic features to ensure the growth and sustainable economic development in line with SDGs in and across the islands. However, increasing tourism investment in Zanzibar has exerted pressure on suitable land for this activity, which is already a scarce resource. Because of this, the Government of Zanzibar has recently announced

Table 27.2 Data methods and collected data

Data collection methods	Information needed/obtained	Responsible personnel/ institution
Interviews	(i) Institutional arrangement in management and operations of sport tourism in MCAs	Commission for tourism MCAs' managers
	(ii) The demographic and participation profiles of sport tourists Zanzibar	Commission for tourism
	(iii) The most important sport tourism categories	Commission for tourism
	(iv) Economic, environmental, and socio-cultural impacts of the sport tourism activities in the coastal and marine environment within the MCAs/MPAs	Commission for tourism Shehia fisheries committee Department of environment Sport tourism business owners
Key Informant Interviews (KII)	(i) Roles of sport tourism in the development of tourism in Zanzibar	Sport tourism business owners Commission for tourism Department of marine conservation
	(ii) The main sport tourism (natural) sites and spaces	Department of fisheries Sport tourism business owners Commission for tourism
Focus group discussion (FGDs)	(i) Impacts of sport tourism in the coastal and marine environment within the MCAs/MPAs	Shehia fisheries committee Staffs from DMC, DFAD Commission for tourism
Questionnaires	(ii) Impacts of sport tourism in the coastal and marine environment within the MCAs/MPAs	Staffs from DMC, DFAD Department of environment
Secondary data review	(i) The characteristics of the island, especially their typology, geographical size, population, GDP, annual net earning average, main professional sectors, political system, colonial past, and future trajectories (ii) The geographic and natural indicators, such as, location, climate, type of geomorphology, accessibility (iii) Present the ways in which the sport tourism sector has changed during the new millennium (e.g. over the past two decades) on the island and future trajectories for sustainable development	Commission for tourism Department of environment

plans to allow investment on more than 50 islets surrounding the two main islands (Unguja and Pemba) of Zanzibar. Following this announcement, over ten islands are already leased out for investment mainly related to high-end eco lodges. This plan offers diversification in tourism investment, which previously had been focused on the mass tourism in only certain areas of development.

Apart from white sandy beaches and colourful coral reefs and seas, Zanzibar offers various diverse tourism products, such as beach, nature/ecotourism, culture and tradition, history and archaeology, water sports, honeymoon opportunities, performing arts, conferences, culinary, and medical tourism (Gössling et al., 2006). Globally, the tourism industry has become a significant sector in both developing and developed countries, as it holds tremendous potential in economic, livelihood, and socio-cultural development (Shah & Gupta, 2000; Mutayoba & Mbwete, 2013). In Zanzibar, tourism has brought many benefits to the local economy in terms of direct and indirect employment, foreign exchange income, and infrastructure development (Gössling et al., 2006; Mutayoba & Mbwete, 2013). The tourism industry has stimulated the growth of regional and international transport services by air and sea and has provided training and transfer of technology, management and technical skills (Mutayoba & Mbwete, 2013). It has also contributed to the development of small businesses linked to the tourism industry, such as tour operators and food suppliers (Mutayoba & Mbwete, 2013). A study by Mutayoba and Mbwete (2013), for example, revealed that the hotel industry in Nungwi village, located South of Unguja town, employs a good number of local people; however, most of the employed villagers were male. This was also reported in other villages that are within the marine conservation areas.

Sport Tourism in Zanzibar

Sport tourism as a segment of tourism can contribute to sustainable development and helps to achieve 17 Sustainable Development Goals (SDGs), if properly developed and managed. The comparative strength of sport tourism lies in (1) engagement in physical activities, (2) opportunities for interactions, and (3) high development potential almost anywhere (small events or light activities like walking in particular). Thanks to these characteristics, sport tourism can play an important role in achieving various SDGs if developed with consideration of these goals. Zanzibar is not the home of large or mega sports events such as football, athletics, swimming, cycling, golf, marathon, tennis that could draw competitors and spectators from around the world. In the past, Zanzibar used to host an International Triathlon every November, which was proven successful as it drew competitors and spectators form other parts of Africa and Europe, particularly from Italy. Despite its potential in hosting these big events, poor infrastructure and lack of proper planning has impeded Zanzibar towards diversification of its tourism industry. Only a few resorts owned golf courses, while infrastructure for tennis, swimming, cycling, and football are also scarce, making it very difficult to attract big events.

On the other hand, Zanzibar has experienced increased investment in small-scale sports activities for recreation. Although these activities may not be solely responsible for influencing decision of the tourists for destination selection, the diversity of such recreational activities increases the value of tourist experience in Zanzibar. In return, these activities add employment for the youth living within the Marine Conservation Areas (MCAs) where most of these activities take place. This study has observed various marine tourism sports activities across the MCAs in Zanzibar. These include Jet Ski, kitesurfing, kayaking, sport fishing, sunset boat cruising, yachting, beach bike riding, parasailing, skydiving, and beach soccer/volleyball. Most of these activities are new in these areas that complement traditional activities such as swimming, snorkelling, diving (Melo et al., 2021 also see Table 27.3). The introduction of Jet Ski is quite new in Zanzibar; it has been used as a recreational activity in various destinations. Unfortunately, this activity has been linked with various environmental impact and injuries.

The popularity of these marine-related sports activities varies across the MCAs. Jet Ski and skydiving are most popular in the TUMCA, compared with other areas. The study found that apart from individual owners of Jet Ski businesses, some dive centres within hotels and resorts are also found to operate Jet Ski for their guests. The number of Jet Ski operators has increased tremendously over time to the extent of causing conflicts between Jet Ski operators and users of intertidal zones such as swimmers and fishers. Jet Ski tourism has also spread to other MCAs, such as CHABAMCA and MBCA. Kitesurfing is very common in the MBCA, where it was minimally observed in TUMCA, especially Kendwa village. Kitesurfing is a wind powered surface water sport using a kite and a board to move across the water. Kitesurfing harnesses the power of the wind through a large parachute type kite to propel a rider across the water on a small surfboard or a kiteboard. Kayak is mostly operated within the mangrove ecosystem in the MBCA, although in other conservation areas, it is also practised. Good coverage of the mangrove forest in these conservation areas offers good opportunities for the development of this sport. However, there is a report of mangrove deforestation associated with the kayak development as an emerging sport tourism activity, especially in Unguja Ukuu village within the MBCA.

The study found that the majority of the marine-related sport tourism activities were practised and owned by locals, highlighting the diversification in the local livelihood portfolio. Similarly, the conservation fees paid by tourists who participate in these tourism activities contribute to community development and in financing conservation activities. This means sport tourism has a potential to address the SDGs in various ways. If these activities are well planned, they can assist in achieving poverty reduction (SDGs 1), sustainable economic growth (SDGs 2), life below water (SDGs 14), and gender equity and empowerment (SDGs 5) within the MCAs and beyond.

Table 27.3 Sport tourism activities and their characteristics

Sport tourism in islands	Zanzibar
Most important sport tourism categories:	
Island sport tourism of opportunity (choice of an activity without planning to carry it out)	Marathon, golf, Ngalawa race, skydiving
Island sport tourism of action (choice of an activity with plans to carry it out)	Diving, parasailing, snorkelling, beach soccer, kitesurfing
Island sport tourism for competition (choice of an activity linked to the preparation or practice of the competition)	Jambiani Beach soccer, Pemba Tourist Bonanza, Ngalawa race
Event-based island sport tourism (choice of an activity linked to a sports, work, conference, or cultural event, etc.)	Jambiani Beach soccer, Pemba Tourist Bonanza, Ngalawa race
Number of tourists who participate in sport tourism activities. (in national or local tourism survey)	10–15%
Number of habitants of sport tourism activities	2%
Number of tourist of sport tourism activities	10–15%
Number of associative structures	150
Number of structure merchants	80
Number of sites and spaces	25
Number of jobs in totality sport tourism sector	6%
Number of jobs in a sport tourism sector dynamics	6%
Promising sport tourism activities	(Air)parasailing, Diving
Sport tourism projects (location, equipment, events, new activities...)	Along the coast of Islands
Number of specialized stores	150
Number of specialized media	12
Number of merchants who maintain sites and spaces	7
Number of incoming travel agencies	10
Sport tourism events	Zanzibar
Number of total events	17
Number of international events	8
Number of national events	12
Number of regional events	5
Number of local events	17
Promising sport tourism activities to be organized into events	Bike

Sustainability of the Sport Tourism Activities

This study also shows so that not all tourism's effects are positive (Jiménez-García et al., 2020). According to Jiménez-García et al., (2020), the rapid growth of sport tourism and its lack of planning have made some tourist destinations victims of its success, testing their carrying capacity. These challenges reduce the pace towards sustainability and achievement of these goals (SDGs 1, 2, 5, 8, 12, and 14) linked with tourism (Jiménez-García et al., 2020). The majority of the tourism sports activities take place within the marine and coastal environment, areas already under stress from various issues such as climate change. To achieve sustainability and realize SDG goals, careful planning is needed, as well as coordination between institutions responsible for the sport tourism business. This is because some of these sports activities are known to cause negative environmental impacts and conflicts in areas where there are multiple users. The majority of the individuals interviewed for this study believed that marine sport tourism activities affect the marine environment within the Marine Conservation areas. In the MBCA, there is a report that kayak business owners have cut down mangrove forests to create a path to allow movement of the kayaks within the forest.

The linkage between Jet Ski use and environmental impacts are obvious, as it has been established in various destinations. The environmental impacts associated with Jet Ski include the risk of waste oil spills, contamination of reefs, sand banks, and other adjacent habitats. Potential degradation of shallow zone marine life forms and habitats, including inducing beach erosion was also noted. Moreover, these vessels can cause impacts on marine biodiversity such as dolphins, whales, sea turtles, fish, crabs, shore birds, and other reef life, such as corals and sea grasses. The underwater noise levels emitted by PWCs are harmful to marine fauna (whales and dolphins) and shore birds, potentially causing disruption to the ecosystems within the MCAs. This means while there remains a need to diversify sport tourism activities to create more jobs and inclusion, these should be carefully planned to mitigate against deleterious impacts on coastal and marine resources within these vulnerable ecosystems.

Jet Ski vehicles are potentially dangerous in the marine zones just as conventional land motor vehicles (Cars or motorcycles) are potentially risky on the roads. Therefore, without proper training skills and an adequate level of knowledge in operating and managing them, Jet Skis have increasingly become a safety liability. This has had the effect of potentially endangering the lives of other water and beach users through potential collisions, strikes, or running over unsuspecting swimmers, snorkellers, scuba divers, seaweed farmers, and fishers. Jet Ski and kitesurfing in Zanzibar generally operates within 100–300 meters from the high water mark, meaning that they are operating in areas with high congestion of multiple uses.

This area is also used by female seaweed farmers to increase their income. It is also used by guests in the hotels for swimming along the shoreline. The conflicts between seaweed farmers and hotel owners in both MIMCA and MBCA are very common. In Paje and Jambiani, the introduction of kitesurfing has had a significant impact on seaweed beds, as they are not compatible. In most cases, hotel owners and kitesurf businesses regularly seek to push seaweed farmers away from these areas. Seaweed farming contains a network of pegs/sticks that have the potential of injuring both swimmers and kitesurfers. But perhaps the biggest conflict arises within the tourism industry itself. There is an ongoing conflict between swimmers and kitesurfers, on the one hand, and between swimmers and Jet Ski users, on the other. Many cases of injuries and fights related to Jet Ski and kitesurfing have been reported in Nungwi within MIMCA, Paje and Jambiani within MBCA, and in Kendwa within TUMCA.

The following explanation from a tourist in Jambiani highlights the extent of this conflict:

> Having noted that many kite surfers were out, my husband and I chose to bathe out of their routes. However, one of the surfers, clearly inexperienced, came quite close to where we were. In order to make my self-seen I stood up in the water, my head and chest over the surface. The kite surfer did not seem to be paying attention to what was going on in the water and came rushing towards me. I tried to dash out of his way, and was literally inches from being hit at full speed.

Similar incidents have been reported for Jet Skis Most of the reported injuries were associated with Jet Ski because they operate in areas which are heavily used by other users. Swimmers are more vulnerable to Jet Skis compared with other users. Although the tourists themselves are the one who hire these vessels, they threaten the safety of other tourists who are engaging in other activities such as swimming in the same area. The study found that some of the Jet Ski or kitesurf renters have no knowledge on how to operate the vessels and business owners did not bother to give enough training to the renters to avoid accidents. The intensity of the Jet Ski problem can be summed up by the following testimony from one interviewed stakeholder in Kendwa and Nungwi villages:

> We have our guests swimming, we have guests who are snorkeling down there in the shallow reef close to the beach and you have fiber boats and Jet Skis coming with very high speed, one of our guest was nearly injured the other day.

In fact, there are no guidelines that govern these sports activities towards sustainability. Furthermore, the majority of these businesses have no licenses or permits and, thus, operate illegally. Sport tourism businesses, especially Jet Ski, kitesurfing, and kayaks, are owned by local providers and therefore offer additional employment and offer diversification in the livelihood. However, if these activities are not exercised safely, it will destroy the tourism brand, especially in areas where these activities are common. International traveller's selection of a destination is influenced by various factors, including safety and positive feedback from the peers. Thus, negative comments provided by tourists as a results of injuries associated with Jet Ski or any other sports are not good for the industry. The study found that there were no regulatory policy frameworks guiding the use of Jet Ski; neither were there any efficient institutional coordination between key stakeholders, such as the local government, Zanzibar Environmental Management Authority (ZEMA), Zanzibar Maritime Authority (ZMA), and others to govern the operations of Jet Ski in a safe and environmentally friendly manner. To address multiple users' conflicts within the inter-tidal zone, local governments, in collaboration with Department of Marine Conservation, ZMA and ZEMA, conducted a series of meetings with stakeholders and came up with a spatial planning framework. This plan demarcates areas for swimming, which are 100 m from the high water mark; only beyond 100 m, can Jet Ski and other speeding boats and yachts operate (Fig. 27.2). This zoning regulation is expected to reduce conflicts between multiple users, with an aim to reduce injuries to fisherwomen and tourists alike.

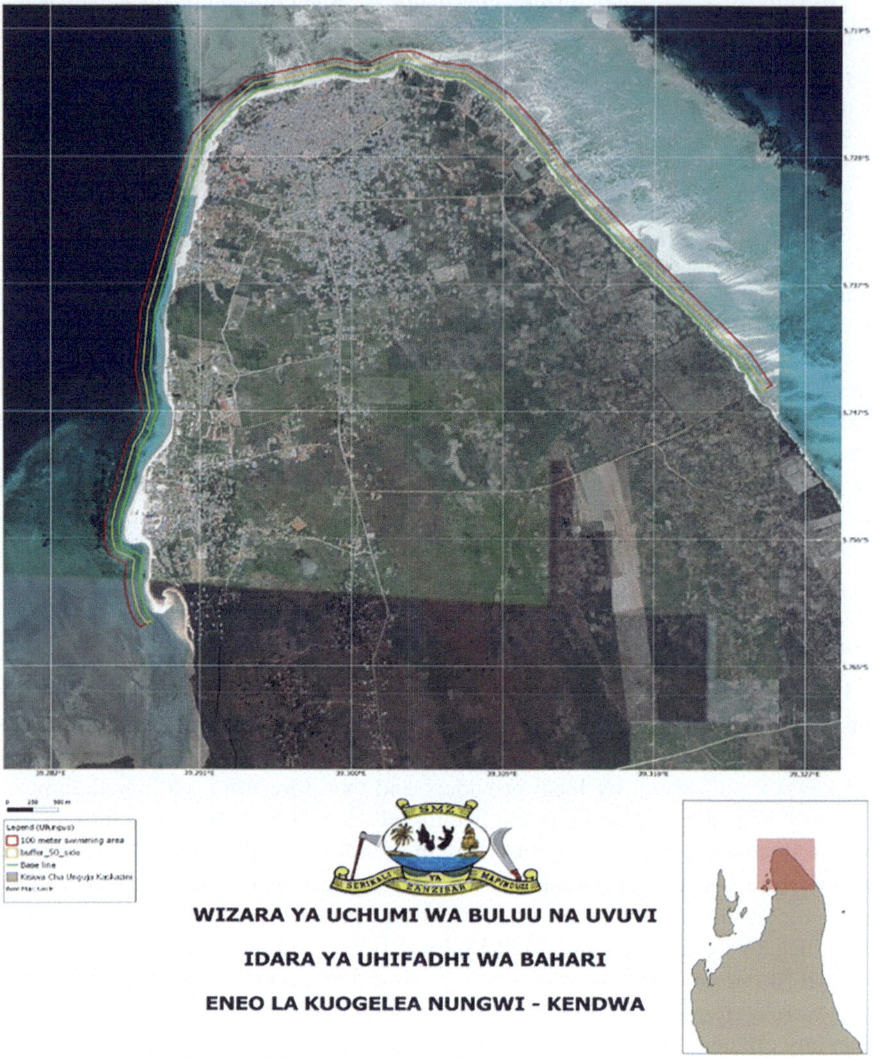

Fig. 27.2 Demarcation plan that regulates Jet Ski use and Swimming at Kendwa village in TUMCA and Nungwi Village at MIMCA. (Source, DMC)

Opportunities for Future Growth

The results of this study highlight the fact that tourism experienced in coastal and marine areas has become diversified over the past decade from predominantly snorkelling and diving to new sports activities such as Jet Ski, kitesurfing, kayaking, and skydiving. However, these activities are not well planned and coordinated to maximize profit, user experience, and sustainability. Apart from marine-related sports activities, Zanzibar could benefit from various sports activities such as football, athletics, golf, tennis, and cycling. That only thing which is needed is investment in various sectors linked with these specific sport activities. Kurtzman and Zauhar (2003) have identified various general characteristics that influence various sport tourism activities in every destination. These include state-of-the-art sport facilities and/or unique sports facilities, which generally house sports events, such as stadiums, arenas, bowls, domes, etc. For the development in sports within the resorts, Kurtzman and Zauhar (2003) believed that the availability of teachers, trainers, coaches with a great deal of expertise and personal visibility was very important. The study also highlighted the need of high-tech instructional apparatuses for practice and game play, as well as opportunities to compete and practice fundamentals, fine-tuning, and comprehensive strategy.

Conclusion and Recommendations

The findings of this chapter suggest that there are direct linkages between sport tourism and sustainability. In fact, a large number of jobs have been created through the observed diversification towards sport tourism activities. However, more needs to be done to reduce environmental footprints of the sports activities within the marine conservation areas. Jet Skis, in particular, pose various degrees of threats on human safety and marine biodiversity, exacerbating conflicts between Jet Ski operators and other users (e.g. swimmers, snorkellers, scuba divers, fishermen, seaweed farmers, etc.), which have the potential to increase in the future because of the growing development of shoreline areas for both tourism, fisheries, and urbanization activities. To achieve sustainability, the study recommends several steps that include provision of licenses and registration for all sport tourism activities, while increasing training and facilitating enforcement. Demarcation plans that have been proposed should be attached with the guidelines so as to increase coordination between actors.

References

Apostolopoulos, Y. & Gayle, D. (2002). *Island Tourism and Sustainable Development: Caribbean, Pacific, and Mediterranean Experiences*. Retrieved from http://publisher.abc-clio.com/9780313013645

Brown, M., & Nagel, M. (2002, Sept). The size of the sport industry in the United States. In *The 10th European sport management congress* (Vol. 4, No. 7).

Carius, F., & Job, H. (2019). Community involvement and tourism revenue sharing as contributing factors to the UN Sustainable Development Goals in Jozani–Chwaka Bay National Park and Biosphere Reserve, Zanzibar. *Journal of Sustainable Tourism, 27*(6), 826–846.

Davenport, J., & Davenport, J. L. (2006). The impact of tourism and personal leisure transport on coastal environments: a review. *Estuarine, coastal and shelf science, 67*(1–2), 280–292.

Dimache, F., & Sodja, M. (2006). A sports portfolio for the development and image of a tourism destination: The case of the French Riviera. In *Illes belears Forum 2006 pre-conference academic seminar* (p. 119–129). Illes Balears Forum, Publishing.

Downward, P., Rasciute, S., & Muniz, C. (2020). Exploring the contribution of activity sports tourism to same-day visit expenditure and duration. *Journal of Sport & Tourism, 24*(2), 111–126.

Fernandez, M. (2014). El impacto tur ´ ´ıstico de los eventos deportivos: un estudio de caso. *Cuadernos de turismo, 33*, 59–76.

Gössling, S., Bredberg, M., Randow, A., Sandström, E., & Svensson, P. (2006). Tourist perceptions of climate change: A study of international tourists in Zanzibar. *Current Issues in Tourism, 9*(4–5), 419–435.

Jiménez-García, M., Ruiz-Chico, J., Peña-Sánchez, A. R., López-Sánchez, J. A. (2020). A Bibliometric Analysis of Sports Tourism and Sustainability (2002–2019). *Sustainability, 12*, 2840. https://doi.org/10.3390/su12072840

Kasimati, E., & Dawson, P. (2009). Assessing the impact of the 2004 Olympic Games on the Greek economy: A small macroeconometric model. *Economic Modelling, 26*(1), 139–146.

Kurtzman, J., & Zauhar, J. (2003). A wave in time-the sports tourism phenomena. *Journal of Sport tourism, 8*(1), 35–47.

Li, M., MacIntosh, E. W., & Bravo, G. A. (2012). *International sport management*. Human Kinetics.

Lim, C., Patterson, I. (2008) Sport Tourism on the Islands: The Impact of an International Mega Golf Event, *Journal of Sport & Tourism*, 13:2, 115–133, https://doi.org/10.1080/14775080802170346

Mazza, B., & Sobry, C. (2022). *Sport tourism and its territorial development and opportunities*. Cambridge Scholar Publishing.

Melo, R., Sobry, C., & Rheenen, V. (2021). *Small scale sport tourism events and local sustainable development: A cross-national comparative perspective*. Springer.

Mutayoba, V., & Mbwete, R. (2013). Is booming tourism in Zanzibar pro-poor? A micro-economic impact analysis. *KIVUKONI Journal, 1*(2), 104–120.

Pedauga, L. E., Pardo-Fanjul, A., Redondo, J. C., & Izquierdo, J. M. (2020). Assessing the economic contribution of sports tourism events: A regional social accounting matrix analysis approach. *Tourism Economics, 28*, 1354816620975656.

Royle, S. (2001). *A Geography of Islands: Small Island Insularity*, Routledge, 237 p.

Shah, K., & Gupta, V. (2000). *Tourism, the poor and other stakeholders: Asian experience*. ODI Fair Trade in Tourism paper.

Standeven, J., & De Knop, P. (1999). *Sport tourism*. Human Kinetics.

Swart, K., & Bob, U. (2007). The eluding link: Toward developing a national sport tourism strategy in South Africa beyond 2010. *Politikon, 34*(3), 373–391.

The Citizen. (2023). *Zanzibar urges preparedness as investment projects rise*. Tuesday, January 3, 2023. https://www.thecitizen.co.tz/tanzania/zanzibar/zanzibar-urges-preparedness-as-investment-projects-rise-4073980

Whitehead, J. C., & Wicker, P. (2018). Estimating willingness to pay for a cycling event using a willingness to travel approach. *Tourism Management, 65*, 160–169. https://doi.org/10.1016/j.tourman.2017.09.023

World Tourism Organisation (UNWTO). (2001). *Sport & Tourism—Deporte y Turismo*. Available online: https://www.e-unwto.org/doi/pdf/10.18111/9789284404681. Accessed on 7 Dec 2022.

World Tourism Organization. (2019). *Sport tourism and the Sustainable Development Goals (SDGs)*. UNWTO. https://doi.org/10.18111/9789284419661

Part III
Comparative Analyses of Islands and Island Archipelagos

Chapter 28
Sport Tourism, a Catalyst for Sustainable Development in Island Territories: Dynamics, Stakeholders and Challenges in the Indian Ocean

Olivier Naria, Derek Van Rheenen, Ricardo Melo, and Claude Sobry

Introduction

International tourism has grown rapidly in recent decades and has been seen as a vehicle for economic development in many island developing states. However, these islands face many imperatives as mentioned by UN Secretary General Ban Ki-Moon on World Environment Day (5 June 2014): "*Small island developing states face many challenges. For many, their remoteness affects their ability to be part of the global supply chain, increases import costs—particularly for energy—and limits their competitiveness in the tourism industry. Many are increasingly vulnerable to the impacts of climate change.*" Despite these challenges, many of these islands have embarked on a path of tourism development due to the paradisiacal and exotic nature of the destinations (Butler, 2006; Silva et al., 2013). The mass tourism development model has nevertheless produced mixed results for the islands (Dehoorne & Saffache, 2008). Yet, the islands offer undeniable assets with

O. Naria (✉)
Department STAPS, Université de La Réunion, Le Tampon, île de la Réunion, France
e-mail: olivier.naria@univ-reunion.fr

D. Van Rheenen
Cultural Studies of Sport in Education, University of California Berkeley, Berkeley, CA, USA
e-mail: dvr@berkeley.edu

R. Melo
Coimbra Education School, Polytechnic of Coimbra, Coimbra, Portugal

C. Sobry
Faculty of Sport and Physical Education, Université de Lille, Lille, France
e-mail: Claude.sobry@univ-lille.frxt

D. Van Rheenen et al. (eds.), *Sport Tourism, Island Territories and Sustainable Development*, Sports Economics, Management and Policy 24,
https://doi.org/10.1007/978-3-031-51705-1_28

beaches, coral reefs, remarkable heritage (UNESCO, 2013), exceptional biodiversity (Teelucksingh & Watson, 2013) or multi-ethnic populations. They attract far more tourists than their permanent population and they can consume resource use beyond sustainable levels (Baldachino, 2014). Islands are therefore directly exposed to the impact of global warming, pressure on natural environments (Klöck & Fink, 2019) as well as social and economic challenges. Sustainable development therefore takes on its full meaning in these territories. Its implementation is vital to combine the well-being of populations with the preservation of natural resources. In the wake of the Millennium Development Goals (MDGs), the United Nations has set up 17 Sustainable Development Goals (SDGs) to respond to the challenges facing the world and the islands within the framework of the 2030 Agenda (UN, 2015).

A string of islands and islets in the Indian Ocean, the whole region received 4.1 million tourists in 2019. Due to their geographical configurations and climatic particularities, these islands offer the possibility of practising a wide variety of sport tourism activities in optimal conditions. Previously reserved for a few initiates, more and more tourists are interested in sports activities during their trips. Within this context, various stakeholders are looking into the economic, socio-cultural and environmental issues of these activities, anxious to make the most of their impact on local life.

After having participated in the process of developing leisure activities on the islands since the 1980s, the objective of this chapter is to study the role played by the actors in the development of tourism in the sports areas and territories of the Indian Ocean, which today constitute a significant economic market. More precisely, the objective is to analyse the incentives and regulations of these activities at the intersection of island tourism and sport territories contributing to the sustainable development of this sector of activity. We will then study the models of initiatives that have been currently enacted within the islands, as well as the perspectives of sustainable development of sport tourism for these territories.

Firstly, we propose to present the structuring of the Indian Ocean market by underlining the evolution of the public and commercial offer, as well as the new environmental stakes of this sector. In the second section, we will analyse the sport tourism policies of the islands by studying the logics of actors in the SDGs. Our reflection will be based on incisive examples at the intersection of the island contexts and the sustainable development actions undertaken.

Research Methodology

Theoretical Framework

The theoretical framework on sports leisure is also part of an approach on the contribution of leisure and sport tourism to the sustainable development of territories, based in particular on the work of Higham and Hinch (2006), Bouchet and Sobry

(2019) and Melo et al. (2021). It is part of the renewal of the issue of sport tourism and sustainable development, which today uses a broader and more qualitative framework of analysis of the territories where the geographical, social, cultural and environmental dimensions are taken into account. It aims to strike a balance between the economic wealth generated by sport tourism while respecting the footprint left on the environment and cultural heritage, tending towards a more sustainable practice of its activities. Finally, the proposed framework is in line with the theoretical perspective opened up by the work of Naria et al. (2022) on the elements of island determinism, which, combined with the factors of the structuring context of the islands, make it possible to define a common methodology for analysing these territories in order to achieve the SDGs (cf. Fig. 28.1).

The Methodological Approach

The methodological approach is based on both a qualitative and quantitative approach. This chapter studies the islands of Mayotte, Réunion, Mauritius and the autonomous region of Rodrigues, Madagascar, as well as the archipelagos of the

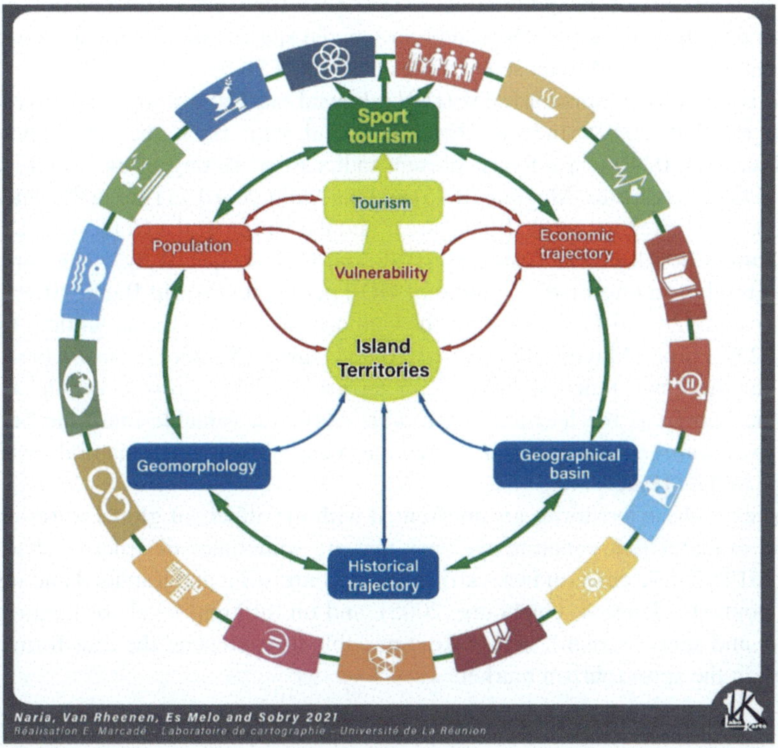

Fig. 28.1 Conceptualisation for the analysis of sport tourism and Sustainable Development Goals in islands' territories. (Source: Naria et al., 2022)

Indian Ocean, the Comoros, the Maldives and the Seychelles. We were also able to collaborate on our study subject with a network of academic actors in each island.

Qualitatively, it is based on the comparative analysis of existing documentation, reports from professional institutions of tourism and sports practices as well as public documents from local actors of the islands (see Table 28.1). It is also completed by semi-structured interviews with political, tourism, sports, economic and environmental actors, carried out during the period of 2011–2015. The data is also supplemented by an analysis of press articles. Finally, empirical observations of hiking and diving will be used.

Quantitatively, it is based firstly on a survey of the institutional and economic players in the sector, and then on the analysis of tourism and sport tourism planning documents and websites specialising in these activities in each of the islands (see Table 28.2). The objective was to determine the creation of associations, companies and events in the sector, as well as the location of sites and areas for the practice of sport tourism. Despite the fragmented and non-diachronic data, this inventory allowed us to build a database from the cross-referencing of different information.

Populated today by 30 million inhabitants, these islands have as many similarities as differences (Jauze, 2019). They have in common a relative geographical proximity, island issues, a historical legacy with rich cultural elements (Bertile, 2013). They have experienced the same French or British colonisation, and have seen the admixture of like populations. At the same time, each island or archipelago represents a particular case. The islands are mixed spaces where African, Asian and European intersect within these territories (see Table 28.3).

The Human Development Index (HDI) showed marked differences between the islands (UNDP, 2019). Indeed, Réunion Island with 0.88, the Seychelles and Mauritius with 0.80, respectively, present indices that belong to the "high human development" countries. Mayotte (0.75) and the Maldives (0.71) are in the "medium development" category, while Madagascar and the Union of the Comoros are in the "low human development" category (0.48 and 0.56, respectively). This situation also highlights an uneven distribution of GDP per capita (World Bank, 2019). It is, respectively, 729 and 1258 euros for the Comoros and Madagascar, while it ranges from 8232 euros (Mayotte/France) to 20,834 euros (Réunion/France) (see Map 28.1). In the environmental field, these territories offer a resource-rich oceanic coastline known as the leeward coast, which offers a valuable maritime heritage (lagoons, coral reefs, underwater seabed, etc.) and a diversity of natural resources specific to the "Hauts" area.

However, these territories are confronted with the effects of global warming and tropical climates that generate cyclones that are sometimes destructive (Klöck & Fink, 2019). These case studies, carried out in contrasting geographical and institutional contexts (Jauze & Guébourg, 2005), and on different levels of maturity (of tourism and sport tourism), will make it possible to apprehend the transformations at work in the sport tourism market.

Table 28.1 Sources of studies related to sport tourism in the islands

Islands	Data
Réunion (France)	*"Les loisirs de nature dans le secteur du tourisme,"* SYPRAL (2014); *"Schéma territorial de développement du sport de La Réunion,"* DDJSC, CROS (2016); Combeau-Mari E., (1998), *"Sport et décolonisation, La Réunion de 1946 à la fin des années 1960,"* Paris, L'Harmattan; Bessy O., Naria O., (2003), *"Loisirs et tourisme sportif de nature à la Réunion"* Report to the CCEE, 150 p; *"Actualisation du schéma de développement et d'aménagement touristique de la Réunion"* (SDATR) (2018), *"La Charte du parc national de La Réunion,"* Parc national de la Réunion, (2014), 202 p.; Bouchet P., Gay J-C. (1998) *"Les Hauts de La Réunion conquis par les loisirs,"* in Mappemonde, n°58, pp. 31–37.
Madagascar	*"Tourism Sector Study"* (2003), World Bank; *"Madagascar Vision for Tourism 2023,"* Ministry of Transport, Tourism and Meteorology (2018); *"Madagascar Emergence Initiative,"* (2020) *"Tourism Marketing Plan for Madagascar,"* Paris, (1995); World Bank, *"Tourism Sector Study,"* Africa Region Working Paper Series No. 63 155 pp. (2013)
The Seychelles	*"The Seychelles Tourism Master Plan: Facts and Findings From 2000 to 2020,"* Republic of Seychelles (2019); Seychelles Tourism Master Plan (2018); *"The Republic of Seychelles, systematic country diagnostic,"* World Bank (2017), Cazes-Duvat, *"Atlas de l'environnement côtier des îles granitiques de l'archipel des Seychelles,"* Université de la Réunion, 88 p; *"Seychelles tourism value chain analysis,"* Commonwealth Secretariat, Seychelles Tourism Board (2010).
Mauritius	*"Assises du Tourisme Committee on Visibility"* (2017) Mauritius Tourism Promotion Authority; *"Adapting to the changing global environment, strategic plan 2018–2021*, Ministry of Tourism (2018), *"Maurice Ile Durable (MID)"* (2013); *"Entertainment of leisure tourists in island destinations: Evidence from the Island of Mauritius"* V Gowreesunkar, M Sotiriades African journal of hospitality, tourism and leisure 4 (2015), 1–19
Rodrigues	Jauze. J. M., Yeung Ching Yung. J. (1998), *"Rodrigues - La troisième île des Mascareignes,"* L'Harmattan, Paris, 268 p.; Deloitte, Touch (2001), *"Tourism Development Plan for Rodrigues"* Volume I, 133 p; KPMG/UNDP (2009), *"Sustainable Integrated Development Plan for Rodrigues,"* RRA, 453 p.
The Maldives	*"Strategy Action Plan"* (SAP), Republic of Maldives (2019–2023); *First Tourism Master Plan* (1984–1993); *Second Tourism Master Plan* (1996–2005); *Third Tourism Master Plan* (2007–2011), *Fourth Tourism Master Plan* (2013–2017), Republic of Maldives, Ministry of Tourism, *Sport tourism strategy* (2020) Ministry of Tourism.
The Comoros	*"Support to the Commissariat Général au Plan for the elaboration of a land strategy, for the formulation of a tourism development strategy and the elaboration of a diagnostic study and a private sector development strategy"* (2019–2035) Republic of the Union of the Comoros; *"Plan d'Aménagement et de Développement Touristique de l'Union des Comores,"* Republic of the Comoros (2019); *"Schéma de développement touristique de l'Ile d'Anjouan,"* NGO, Tourisme sans Frontières (2019); *"Plan d'Aménagement et de Gestion du Parc National Mont Ntringui (2017–2021),"* Republic of the Comoros.
Mayotte (France)	*"Schéma Régional de Développement du Tourisme et des Loisirs de Mayotte,"* Département de Mayotte, HTL (2019); *"Schéma d'Aménagement Régional de Mayotte"* (2012); *"Etude sur la randonnée à Mayotte (2017)"* ONF, Conseil Général; Quartararo A., (2009). *"Tourism and leisure in Mayotte."* Master 2 thesis for the Agence des aires marines protégées. 89 pp; *"Synthèse des enquêtes réalisées auprès des structures professionnelles de sport, de loisirs et balades en mer,"* Parc Marin, l'Agence des aires marines protégées, (2013).

Table 28.2 Sources of the database on practice sites and spaces

Islands	Sources
Réunion (France)	Bessy and Naria (2004); Sypral (2014); Pôle de ressources national sports de nature [online]. French Ministry of Sports [accessed 26 April 2020]. Available from: http://www.sportsdenature.gouv.fr/; Marine reserve [online]. Réserve naturelle marine de la Réunion [accessed February 2, 2020]. Available at: http://www.reservemarinereunion.fr/; Ile de la Réunion Tourisme [online]. Ile de la Réunion Tourisme [consulted on 26 April 2020]. Available from: https://www.reunion.fr/.
Madagascar	Ministry of Transport, Tourism and Meteorology, [online]. Republic of Madagascar [accessed 19 April 2020]. Available from: http://www.mttm.gov.mg/circuits-et-aventures/, Office national du tourisme [online]. Republic of Madagascar [accessed 19 April 2020]. Available at: http://madagascar-tourisme.com/fr/; Office du tourisme de Nosy-Bé [online]. Republic of Madagascar [Accessed 19 April 2020]. Available at: https://nosybe-tourisme.com/fr/, Office du tourisme de Diego-Suarez [online]. Republic of Madagascar [Accessed 19 April 2020]. Available at: https://diegosuarez-tourisme.com/fr/, Office du tourisme de Tulear [online]. Republic of Madagascar [Accessed 19 April 2020]. Available at: http://www.tourisme-tulear.com/fr/, Office du tourisme de Fort Dauphin [online]. Republic of Madagascar [Accessed 19 April 2020]. Available at: https://fortdauphin-tourisme.com/fr/, Office du tourisme de Vatovavyfitovinany [online]. Republic of Madagascar [Accessed 19 April 2020]. Available from: https://www.vatovavyfitovinany-tourisme.com/fr/, Surf spots [online]. Wannasurf.com ltd [accessed 19 April 2020]. Available from: https://fr.wannasurf.com/spot/Africa/Madagascar/index.html Hiking trails [online]. vtt.mg|bike, motorbike, walking trails, 2020. VTT.mg company [accessed 19 April 2020]. Available from: https://www.vtt.mg/circuits-vtt-moto-marche-a-pied/circuits-madagascar/ Walking trails [online]. Trace Directe enterprise [accessed 19 April 2020]. Available at: https://www.tracedirecte.com/59-randonnees-madagascar.htm, World Bank (2013).
The Seychelles	Ministry of Tourism, [online]. Ministry of Tourism [accessed 20 April 2020]. Available from: https://www.seychelles.travel/fr/ surf spots website [online]. Wannasurf.com ltd [accessed 20 April 2020]. Available from: https://fr.wannasurf.com/spot/Africa/Seychelles/ diving spots [online]. Okeanos Cruise company [accessed 20 April 2020]. Available at: https://www.okeanos-cruise.com/full-gallery-atoll-granitique-carte-des-spots.php hiking trails [online]. Randopiton.re [accessed 20 April 2020]. Available at: https://randopitons.re/mot-cle/digue-seychelles-1297, "Hiking in Seychelles, 15 routes to discover" R. Ravon, R. Latournerie, 2017, Seychelles tourism board, SNPA, 42 p.
Maurice	Ministry of Tourism, [online]. Republic of Mauritius [accessed 8 April 2020] available at: https://www.tourism-mauritius.mu/fr; surf spots website [online]. Wannasurf.com ltd [accessed 8 April 2020]. Available from: https://fr.wannasurf.com/spot/Africa/Mauritius/index.html, hiking trails [online]. Randopiton.re [accessed 8 April 2020]. Marsala, V., Galli, A., Paglia, G., & Miccadei, E. (2019). Landslide Susceptibility Assessment of Mauritius Island (Indian Ocean), Geosciences 2019, 9, 493. Available at: https://randopitons.re/randonnees/region/ailleurs/maurice

(continued)

Table 28.2 (continued)

Islands	Sources
Rodrigues	KPMG, Rodrigues Regional Assembly, Final SIDPR, 2009; Rodrigues Autonomous Region Link, [online]. Republic of Mauritius [accessed 8 April 2020]. Available at: http://mdr.govmu.org/English/Rodrigues/Pages/Rodrigues-Administration.aspx, hiking trails [online]. Randopiton.re [accessed 8 April 2020]. Available from: https://randopitons.re/randonnees/region/ailleurs/rodrigues
The Maldives	Ministry of Tourism, [online]. Republic of Maldives, [accessed 16 April 2020]. Available from: https://maldives-a-la-carte.com/, surf spots [online]. Wannasurf.com ltd [accessed 16 April 2020]. Available from: https://fr.wannasurf.com/spot/Asia/Maldives/, "The Maldives," Le petit futé, 2020–201.
The Comoros	Site office national du tourisme, [online]. République Union des Comores [accessed 16 April 2020]. Available at: https://www.comores-online.com/portal/index.php/fr/tourism, World Bank (2013). Republic Union of Comoros, (2019), Institutional Capacity Building Project (PCRI II)—Comoros tourism sector strategic plan (2019–2035), AIC Group, 2 IEC. Union of Comoros, (2017). RNAP PROJECT "Réseau National des Aires Protégées". Union of Comoros, 2019. CPADC, Plan Comores Émergent.
Mayotte	Pôle de ressources national sports de nature [online]. French Ministry of Sports [accessed 2 April 2020]. Available at: http://www.sportsdenature.gouv.fr/; Topo "cascades M'Tsapere ": Collectif Mayotte, 2eme Edition (2014); Agence des aires marines protégées, *Parc marin de Mayotte, les hommes et l'océan*, Enquêtes sur les pratiques professionnelles sportives, de loisir et de ballades en mer (2013); pistes de randonnées pédestres [online]. RandoGPS.net [accessed 2 April 2020]. Available at: https://www.randogps.net/randonnee-pedestre-gps-mayotte-976.php (2020); surf spots [online]. Wannasurf.com ltd [accessed 2 April 2020]. Available at: https://fr.wannasurf.com/spot/Africa/Mayotte/index.html, Guide de la plongée de Mayotte, Département de Mayotte (2021)

A Dynamic Process of Sport Tourism Development and a Significant Spatial Heterogeneity of Practice Areas in the Islands

According to Bouchet and Bouhaouala (2009), several phases can be identified concerning the changes that occur in the supply and consumption behaviour of sport tourism. Accordingly, we have analysed the marketing of sport tourism on the Indian Ocean islands in three main phases (see Graph 28.1).

The Evolution of the Structure of the Sport Tourism Market (1960–2022)

The Initial Phase of Sport Tourism (1960–1980)

After a colonial period marked by the emergence of leisure activities, sport tourism appeared in the 1960s with the improvement of accessibility to the Indian Ocean (Gay, 2017).

Table 28.3 Characteristics of Indian Ocean islands

	Réunion (France)	Mayotte (France)	Madagascar	Mauritius-Rodrigues	The Seychelles	The Comoros	The Maldives
Location	Western Indian Ocean	Mozambique Channel, between Africa and Madagascar	Indian Ocean, 400 kilometres off the African coast, at the level of Tanzania and Mozambique	It is located in the western part of the Indian Ocean, between Réunion Island to the west and Rodrigues Island to the east	1100 km northeast of Madagascar	North-western Madagascar	South-West Indian subcontinent
Area (km^2)	2 512	374	587 041	1 865	455	1 862	298
Climate	Tropical	Maritime tropical with two seasons (rainy and dry)	Tropical	Tropical two seasons: the hot and humid southern summer and the mild and dry southern winter	Equatorial.	Tropical	Hot and humid

Islands (number)	1	2 islands, Grande-Terre and Petite-Terre (11 km^2), about thirty scattered islets	1 main island, 11 islets	2 islands main, Maurice, Rodrigues (108 km^2), 2 outer island groups Saint-Bradon (3 km^2) and Agalega (21 km^2)	Archipelago of 115 islands, two distinct island groups; the Inner Island Group of which 3 main ones, Mahé (157 km^2), Praslin (38 km^2) and La Digue (10 km^2) and the Outer Island Group of which Amirantes, Alphonse Group, Farquhar Group and Aldabra...	Archipelago of 3 islands, Grande-Comore (1148 km^2), Anjouan (424 km^2) and Mohéli (290 km^2)	26 atolls, 1190 coral islands, 201 inhabited islands
Capital	Saint-Denis	Mamoudzou	Tananarive	Port-Louis	Mahé	Moroni	Malé
Population (2019)	860 000	256 500	26 262 000	1 265 000	94 676	828 000	533 000
Natural environment	Active volcanism, exceptional biodiversity, pitons, cirques and ramparts listed as UNESCO heritage sites	185 km of indented coastline, with bays and inlets, 1100 km^2 of lagoon, 270 km of reefs, 53% forest cover	2 sets of highlands and coastal sedimentary basins, 5000 km of coastline, 5% of the world's biodiversity	243 km^2 of lagoon, 150 km of coral reefs	Islands of granite or of coral origin	Grande Comore Island active volcanic	More than 80% less than one metre above sea level; marine biodiversity exceptional
GDP/capita (in euros) (2019)	20 834	8 232	1 258	19 738	14 626	729	9 099
HDI (2019)	0.88	0.75	0.48	0.80	0.80	0.56	0.71

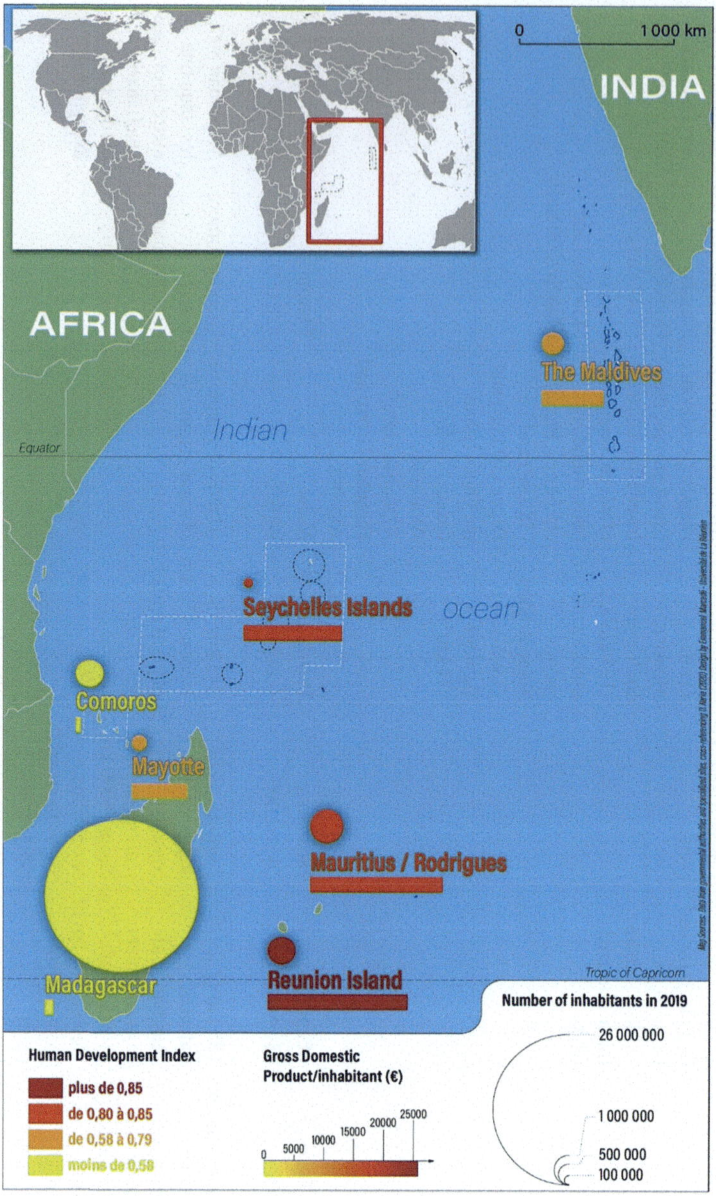

Map 28.1 Socio-economic disparities in the Indian Ocean islands. (Sources: UNDP, 2019; World Bank, 2019; INSEE Réunion, INSEE Analyses, 2019)

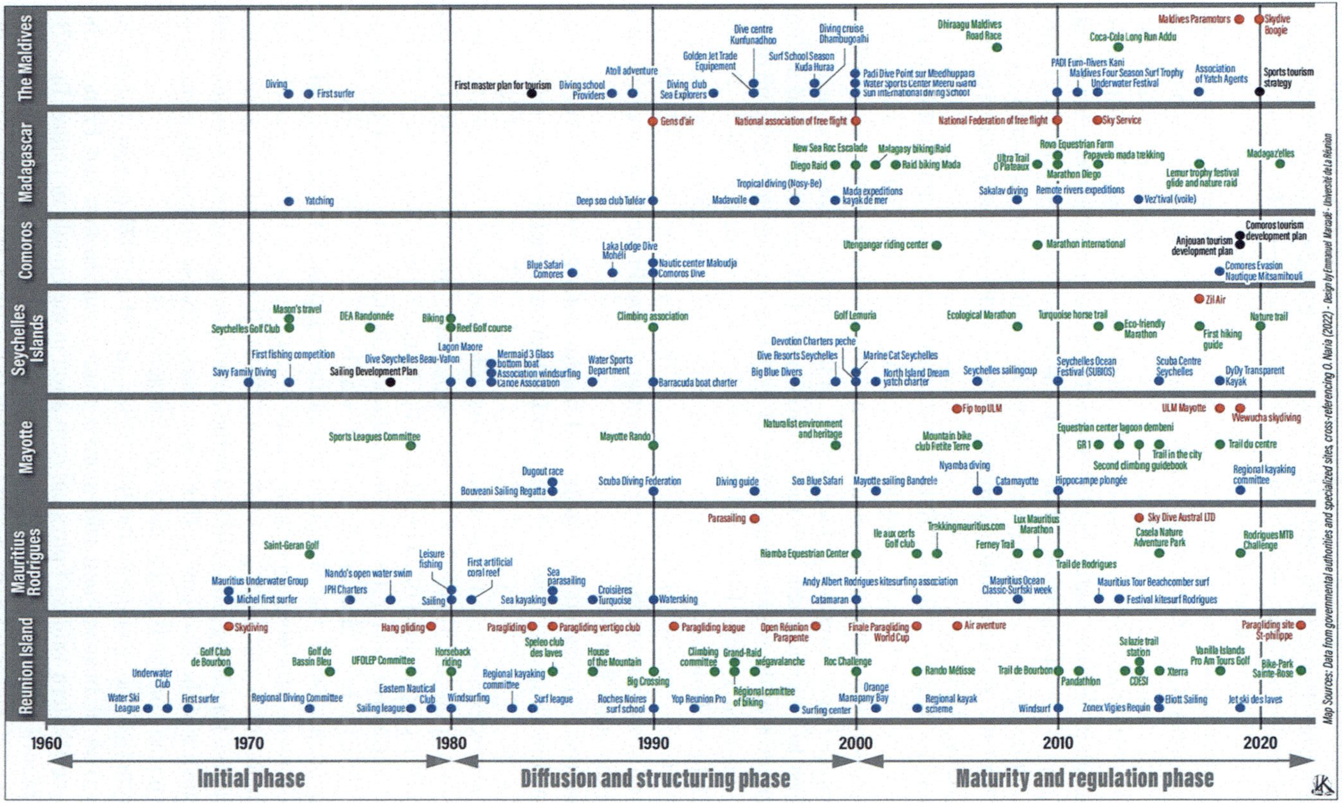

Graph 28.1 References on the development of sport tourism on the Indian Ocean islands. (Source: data from institutional documents and specialised sites, cross-referenced with O. Naria in 2022)

During the 1970s, the Maldives was a world away from global tourist centres (Gay, 2001). Surfing arrived in the Maldives in 1973 through Tony Hinde, an Australian surfer (Pluss, 1986, 180). He is said to have taken a surfing break in Sri Lanka and landed in Malé. The development of tourism really began in the early 1980s with the liberalisation of the tourism sector by the local authorities. Twenty-seven accommodation facilities were opened between 1979 and 1982, with a capacity of 1992 rooms (Gay, 2001). Pluss (1986) indicates that diving has great potential: "*it is the major attraction in the Maldives and its colourful and detailed presentation illustrates that paradise is given an extra dimension here*" (p. 228). This activity will develop with the "safari boats" which facilitate travel between the islands near the airport for diving trips. Diving and surfing activities became the driving force behind holidays in the Maldivian islands. From then on, Kaafu Atoll concentrated 90% of the accommodation capacity as well as the majority of recreational activities (Gay, 2001).

On Réunion Island, it can be observed that the activities offered during this period were diversified thanks to its geomorphological assets. In the aerial sector, the first parachuting (1969) and hang-gliding (1979) flights were organised by expert metropolitans (Bouchet & Gay, 1998). In the aquatic sector, the coastal strip was the first to be concerned with the development of diving, surfing and sailing with the creation of the water-skiing league (1965), followed by the setting up of the Réunion underwater club (1966) and the arrival of the first surfer (1967). Finally, the golf clubs of Bourbon (1969) and Bassin Bleu (1974) reflect the popularity of these activities for tourists and local notables. At the same time, the first hiking and walking trails were promoted by the committee of the French Union of Secular Physical Education Works.

In Mauritius, it is essentially aquatic and seaside activities that were developed on the coastline (Singaravelou, 1989). Mauritian tourism benefited from the capital released by the sugar industry and the impetus of the local Rogers group (Serviable, 1983). Services exclusively for the tourist clientele were set up with the Mauritius Underwater Group (MUG) (1970), as well as pleasure cruises with JPH. Charters (1975) and the organisation of an event with the Nando's Open Water Swim (1977). Golf is an attractive sector for the international clientele discovering the Indian Ocean, and this is how the Saint-Geran Golf was born (1973). In Rodrigues, at the beginning of the 1980s, the Mauritian authorities supported the start-up of guest houses, following a training opportunity based on an exchange between the Catholic Church in Réunion and Rodrigues (Agathe, 2016). It was from this point that a training course for hiking guides linked to the development of gites was established. The structuring of the Rodriguan offer is also marked by the progressive demand of tourists who stay on the island to discover new activities.

In Seychelles, the opening of the Seychelles airport in 1971 corresponds to the gradual arrival of businessmen and tourists from Great Britain and France who exported the Western values of physical recreation in the middle of nature. In 1970, the Savy family organised the first diving trips off Mahé, then to Praslin on the boat "Stella Polaris" (Naria & Sherwin, 2011). Founded in 1972, Mason's Travel was the first tour operator on the island to organise expeditions within the archipelago. In 1976, Clerk Louis Gallant (FIC) and members of the Duke of Edinburgh Awards

Scheme, in collaboration with the Seychelles Forestry Department, developed the first hiking trails on Mahé. In 1977, the Seychelles Ministry of Sports put forward the Sailing Development Plan. The objective of this plan was to raise awareness among students so that the ocean would no longer be an anxiety-provoking element and to encourage vocations in the management of the emerging tourist and sports offers.

Other islands of the Indian Ocean at this time were not as concerned by the phenomenon of sport tourism because of the urgency required from the public authorities to eradicate poverty and food insecurity (Serviable, 1983).

Diffusion and Structuring Phase of Sport Tourism (1980–2000)

During this second period of development, sport tourism activities developed at a particularly sustained rate, marked by the structuring of a constantly dynamic market, but also the increase in the number of tourists and locals. The role of the media in promoting island destinations, tour operators and hotel groups, as well as enthusiasts, helped to diversify the islands' market supply.

Réunion continued the development of the sector with the spread of windsurfing (1980) as well as the creation of the first surfing club, Maddoï's team (1980), following the Boucan surf club (1983) as well as the surfing league (1984). The coastline has likewise seen the rapid development of scuba diving. This activity is mainly promoted by the Regional Agency for Leisure Tourism (ARLT), created by the Regional Council in 1982. The sites of Saint-Gilles-les-Bains and Saint-Leu are popular spots for visitors from mainland France, as well as for a small number of locals interested in this type of leisure sport (Bessy & Naria, 2005). A nautical plan was promoted by local actors to structure the 1984 sector. In another sector, it was the turn of the "Hauts" to be developed thanks to the initial and combined action of the Commissariat de l'Aménagement des Hauts (CAH) and the Office National des Forêts (ONF) to implement the Plan d'Aménagement des Hauts proposed in 1978. The creation of the Maison de la Montagne in Cilaos (in 1987) and the hiking committee (1988) were part of this important dynamic. Paragliding was established on the Trois-Bassins site (1984). An international event developed with the organisation of the Grand-Raid (1994) which put the island on the world map (Bessy & Naria, 2004).

In the Seychelles, many private companies developed and proposed new offers, notably the Lagon Maore, the Mermaid 3 Glass Bottom Boat, as well as the structuring of the windsurfing association and sea canoes. From 1991 onwards, the supply of pleasure boats accelerated. It consisted of rental companies that sold yachts to owners, usually foreign. From there, companies then managed the yachts on behalf of the owners and made them available for charter to tourists who would use them for leisure. These included Barracuda Boat Charter and Devotion Charters Pêche. The Seychelles Ministry of Sports and Recreation then set up the Department of Water Sports, which included sailing, windsurfing and kayaking, attached to the National Sports Council in order to professionalise the clubs and private providers.

In the early 1990s, the Seychelles Climbing Association opened an equipped trail with 50 distinct routes at Boulder Anse d'Argent on La Digue Island. In the same year, J. Barreau organised the first trekking tours on Mahé.

In Mayotte, the coastline represents a privileged space with the establishment of private structures such as the Maore Lagoon (1981), the Bovéani Sailing Regatta (1985), as well as pirogue races (1985). Gradually, diving was identified by institutional actors as a flagship product that marked the destination. Moreover, in 1996, Mayotte's slogan became "the lagoon island." Board sports became more prevalent, with activities such as paddle, sea kayak, etc. At the same time, Mayotte Rando was set up with the ambition of discovering the green assets of the seahorse island. However, the poor development of roads and public facilities inhibited the development of tourism.

In Mauritius, the government began to accelerate the "beach hotel" model (Magnan, 2005) to boost its tourism offer. At this time, the number of public beaches accessible to the local population decreased as hotels and resorts took over the coastline for their development projects, creating tensions between the local population and tourists (Pebarthe, 2001). From 1980 onwards, in order to attract divers from around the world, some 20 ships were sunk to create artificial reefs mainly off the west coast of the island by the Mauritius Marine Conservation Society.[1] The Mauritian coastline diversified with these activities as shown by the installation of Turquoise Cruises (1987), Solar Sea Walk and the Mauritius Scuba Diving Association the same year (1988). The establishment of Coral Diving (1992), Diving World LTD (1993), the arrival of watersking (1990), kayaking (1990) and the organisation of an event with the Marlin World Cup (1998) all occurred during this period. The introduction of parasailing (1995) marked the beginning of aerial conquest. In Rodrigues, efforts were made for the modernisation of the island and the programming of social infrastructures (Agathe, 2016) allow the development of an authentic tourism with its culinary art, traditional dances and diving, snorkelling and boating activities.

In the Maldives, tourism accelerated from 1981 onwards with the lengthening of the airport runway, which allowed long-haul aircraft to land (Gay, 2001). In addition, the industry progressed with the introduction of the first two tourism master plans from 1983 to 1992, then from 1996 to 2005. These planning documents quickly transformed the Maldives into a world-renowned holiday destination (Kundur & Murthy, 2013). Magnan (2005) stated that "*the challenges of developing archipelagic territories, perfectly suited to very small islands, generates relatively large and rapid revenues*" (p. 98). The Maldivian government implemented the island-hotel concept (Gay, 2001). As such, each island would have a resort that might offer a varied range of water sport tourism activities. Surfing, windsurfing, water skiing, parasailing, diving, deep-sea fishing and catamaran sailing were the most popular activities. In this concert, Magnan maintained that "*it is strong spatial constraints and an economic logic of profitability that have favoured a development in 'leaps and bounds', relayed locally by a 'spreading of oil'*" (p. 138). The new

[1] https://www.lexpress.mu/node/247670, accessed on 5 December 2021.

modes of transport with speedboats and seaplanes accelerated the spread of sport tourism in the atolls (Gay, 2001).

In Madagascar and the Comoros, sport tourism developed with a time lag. On the big island, European adventurers came to indulge in land-based activities such as hiking, trekking, quad biking or other 4 × 4 safaris through authentic spaces in the valleys or forests. The islands of Sainte-Marie and Nosy-Be were the most attractive areas. In the Comoros, diving and sailing activities were mainly present as emerging opportunities. In these two islands, the public authorities do not play their role as leaders in tourism (Christie & Crompton, 2003; Rajeriarison, 2013). Some actions by tour operators and hotel groups will sell the destination. It is mainly adventurers and small private structures that will accompany the development of the sites and create a sport tourism offer. This situation is explained by the fact that these two islands are among the poorest states in the world (UN, 2019).

Maturity and Regulation Phase of Sport Tourism (2000–2022)

The new millenium has been characterised by an accentuation of the phenomenon which sees the development of associations and companies providing services, but also new activities proposed to tourists and the development of several international events which will have promoted a modern image of the islands. This phase of maturity has been progressively marked by the enactment of laws and regulations, aimed at protecting and preserving the environment in a sustainable way.

In Réunion, several events have ensured that the island is promoted in the international media, notably the Roc Challenge (2000) in climbing, the Orange Manapany Bay in surfing (2001), the final of the Paragliding World Cup (2003), as well as the Trail de Bourbon and the Mascareignes (2010 and 2011), which precede the organisation of the Grand-Raid during these 2 years. The Tourism Development Plan for Réunion Island (SDATR, 2018) states that sport tourism is a diversified sector that needs to be strongly developed.

In the Seychelles, events also enliven the territory with the first Ecological Marathon, the Turquoise Horse Trail, the Eco-friendly Marathon or the Nature Trail. In 2000, the Lemuria course brought together international golfers in Praslin, in partnership with the Constance Resort (32). From 2005, off the east coast of Mahé, a South African company developed the Eden Island project, with 500 high-standing villas, delivered with their own mooring for yachting activities,[2] on a 56 hectare polder. In 2006, the Ferney Trail was rehabilitated for hiking and 4 × 4 activities. In 2008, the CIEL Ferney Trail took advantage of the event dynamic in the Indian Ocean, gathering thousands of competitors. The government's parks and gardens authority were responsible for the design and maintenance of the hiking trails. In 2009, this oversight was taken over by the Seychelles National Parks Authority, which maintains the trail system in connection with the management of the national parks. Building on these dynamics, the Seychelles Tourism Authority published the

[2] https://www.mansionglobal.com/articles/eden-island-where-the-famous-flee-for-freedom-205769 accessed on 8 June 2022.

first official hiking guide in 2017. These activities will lead to the development of small accommodation units near these hiking trail heads.

In the Maldives, the government promotes luxury tourism with five-star resorts whose postcard images of rooms on stilts set in the lagoon are propagated by the slogan "*Maldives, always natural*" (2011). These resorts operate on a self-sufficient basis, as there is limited interaction with neighbouring island communities at this time. Owners are large international companies, and to a lesser extent, local entrepreneurs. Resorts train their own employees on site. In 2001, 97% of island hotels offered a diving club for tourists (Magnan, 2005). Safari ships offer short cruises through the atolls as multi-service spaces for these activities. They operate with floating beds on which to live while sailing to various sport tourism sites. They accounted for 6% of the total bed capacity for 153 safari ships (Ministry of Tourism, 2019). According to the authorities, there are 3000 divers per day. Carrying capacity is reached at the sites closest to the hotel islands. This situation generates pressure on the environment and ecosystems, accentuated by the rise in sea level. In order to diversify its tourism offer, since 2010, Maldivian law allows tourists to stay on local islands. At the end of 2018, there were 167 island hotels, 218 diving centres, including 48 PADI diving resorts, and 613 travel agencies (Ministry of Tourism, 2019). Finally, the realisation of a network of 17 international and domestic airports favours the spread of sport tourism in the archipelago.

In Mauritius, Gowreesunkar and Sotiriadis (2015) noted that nautical activities are among the "favourite entertainment" of tourists and locals. The seaside coastline welcomes new activities with the development of catamaran, snorkelling and windsurfing from 2000. The Mauritian resorts associate the names of their establishments with sport tourism events whose activities are the most economically profitable for an international clientele. These include the Mauritius Ocean Classic-Surf Ski Week (2008), the Lux Mauritius Marathon (2009), the Freeride Attitude Challenge Surf (2012), or more recently, the Mauritius Tour Beachcomber in surfing and the Shangri-La Golf Le Tousserok Mauritius (2015). Due to the high number of tourists and sports enthusiasts on the coast, the government has focused on the "*master plan for management and zoning of lagoon for the tourism sector*" (2014) in order to preserve the impact of coastal activities through an assessment of nautical activities and resources.

Rodrigues became an autonomous region of Mauritius in 2001. This year is marked by the implementation of the "*Tourism development plan for Rodrigues*" (2001). As a result, several road improvements allowed the extension of the airport and the construction of hotels. Tourism has been structured with new markets for ecotourism products (Jauze, 1998); the local government favours small accommodation structures so that Rodriguans can benefit from the economic spin-offs. The geological history of the island provides it with natural advantages for diving and fishing, as it is located in the middle of an ancient underwater plateau which abounds in cavities suitable for submarine activities (ORSTOM, 1986). While it is true that several activities are gradually being developed, such as diving, sailing and hiking, one activity has helped to create the identity of the island's sport tourism: kitesurfing. Indeed, Andy Albert, a native of the island, set up the first association (2004), then the "Andy Kitesurf Morne Kite School" (2005) for tourists and young

Rodriguans. The island was then identified as a world kitesurfing hotspot. From there, a dynamic partnership supported by institutional and tourist actors and the population organised a stage of the international championship of the kitesurfing federation two years later. This activity has brought tangible economic benefits to the island.

In Mayotte, the departmental council, the ONF and the association of naturalists seek to promote a hiking strategy (2017) in order to improve the tourist offer in this sector, which is still in its infancy (Gay, 1999). The inventory demonstrates that there are 13 GR stages over 160 km, but several paths need to be rehabilitated (HTL, 2019). Two associations organise regular hikes, the Mayotte naturalists and the Ami Raid Rando, with more than 4000 people per year engaged in this activity (HTL, 2019).

In the other islands, despite the establishment of a Ministry of Tourism in 1994 on the main island of Madagascar and a National Tourism Office in 1992 on the Comoros, the two islands have not managed to develop this sector due to political instability, marked by coups d'état, an uncertain investment climate, despite studies and action plans initiated by international organisations.

During this last period of development, decision-makers and local populations have become aware that the benefits of the sport tourism market are increasingly accompanied by environmental costs. Studies were then carried out to measure the impact of tourist traffic. The results show that carrying capacities have been reached or exceeded at several practice sites. For example, hiking in Salazie and Mafate in Réunion (Robert & Caze, 2000), diving in Beau-Vallon (Mahé), Anse Volbert (Praslin) or Anse Source d'Argent (La Digue) in the Seychelles (Cazes-Duvat, 2001), surfing at Jailbreak (Kaafu Atoll) or Pasta Point (North Malé Atoll) in the Maldives (Kundur & Murthy, 2013) or snorkelling and diving at Trous aux Biches or Grand-Baie in Mauritius (Beeharry et al., 2014).

In order to deal with this vulnerable situation, decision-makers have begun to implement protection strategies. In Réunion, a national park and a maritime reserve have been set up to protect the island's natural heritage and ensure its sustainable development. This approach has been accentuated by the recognition of the Pitons, cirques and ramparts of the island by Small Island Developing States (2013).

In the Maldives, Pluss (1986) shows that "*a whole catalogue of regulations prescribe the behaviour of tourists and resorts*" (p. 159). At the same time, strict legislation regulates the equipment of boats, diving and surfing excursions in order to preserve a protective image of the environment. Underwater hunting is prohibited. In addition, the development of resorts is subject to strict rules because no construction may exceed the tops of coconut trees and they must favour energy production, wastewater treatment and the sorting and recycling of waste.

Since 1999, the Naturalists of Mayotte have been focused on the discovery and protection of heritage sites and the environment. Several regulations have been put in place in order to respect the resources linked to sport tourism activities, such as the regulation of the approach and observation of marine mammals (2004) and the creation of the Natural Marine Park of Mayotte (2010).

In Madagascar, the National Parks authority manages 43 Protected Areas out of the 126 units of the territory, i.e. 10% of the island's surface area, made up of terrestrial and marine national parks or nature reserves.[3] In the Seychelles, the government currently protects 30% of the ocean territory, i.e. 410,000 square kilometres.[4]

These few examples show the stakes involved in preserving the natural heritage, while regulating and increasing the economic benefits of sport tourism, have become a vitally important issue. Each island has a multitude of protected areas with terrestrial protection areas (national, regional and natural parks) and marine conservation areas (reserves and parks) (see Maps 28.2, 28.3, 28.4). The evolution and recognition of the importance of sustainable sport tourism today is visible through legislative provisions, the various environmental protections and the initiatives of stakeholders involved in environmental education.

The Geography of Sport Tourism Sites and Spaces in the Islands

The analysis of the database of 1854 sites and sport tourism areas has made it possible to establish the main structuring of the area (see Maps 28.2, 28.3 and 28.4). It should be noted that the geographical areas of the east coast, which are exposed to the wind, are less well supplied than the west coast.

Large Lagoons, Coral Reefs and Atolls for Water Sport Tourism

The north-west and south-west coast of Mauritius is the preferred area for oceanic activities and cruises with a concentration of diving, surfing, snorkelling and fishing sites. In Rodrigues, there are about twenty diving sites, mainly in the Grand Pass, the outer reefs and the reefs of the deep south. Pleasure fishing is mainly located in the south on Ile aux Chats and at Ermitage inside the lagoon.

On Réunion Island, the National Maritime Reserve covers the water sports of diving, surfing and gliding from Saint-Paul to Etang-Salé. In the Comoros archipelago, snorkelling activities are located in the north of Grande Comore at the site of Galawa or at the trou du prophète, at the salt lake in Moheli, and at the Moheli marine park. Along with the lagoon, there are numerous diving sites on 270 km of reefs and they are very popular at the Passe en S. Surfing is practised during part of the year, particularly at Moya beach. The Choazil Islands and the islet of Saziley are particularly attractive. Among the islands of the Maldives, there are over 150 island-hotels whose sport tourism activities are linked to the accommodation offers of the resorts.

In Madagascar, the bay of Diégo Suarez is a place for marine activities. In the West, tourists can enjoy boating between river cruises and marine excursions through the Tsingy de Bemaraha or the Belo sur mer. In Nosy-Be and Sainte-Marie Island, multi-activities are practised as tourists embark on cruises and fishing trips,

---

[3] https://www.parcs-madagascar.com/apropos.php accessed on 14 October 2021.

[4] https://undp-biodiversity.exposure.co/un-equilibre-entre-croissance-et-durabilite accessed 12 December 2020.

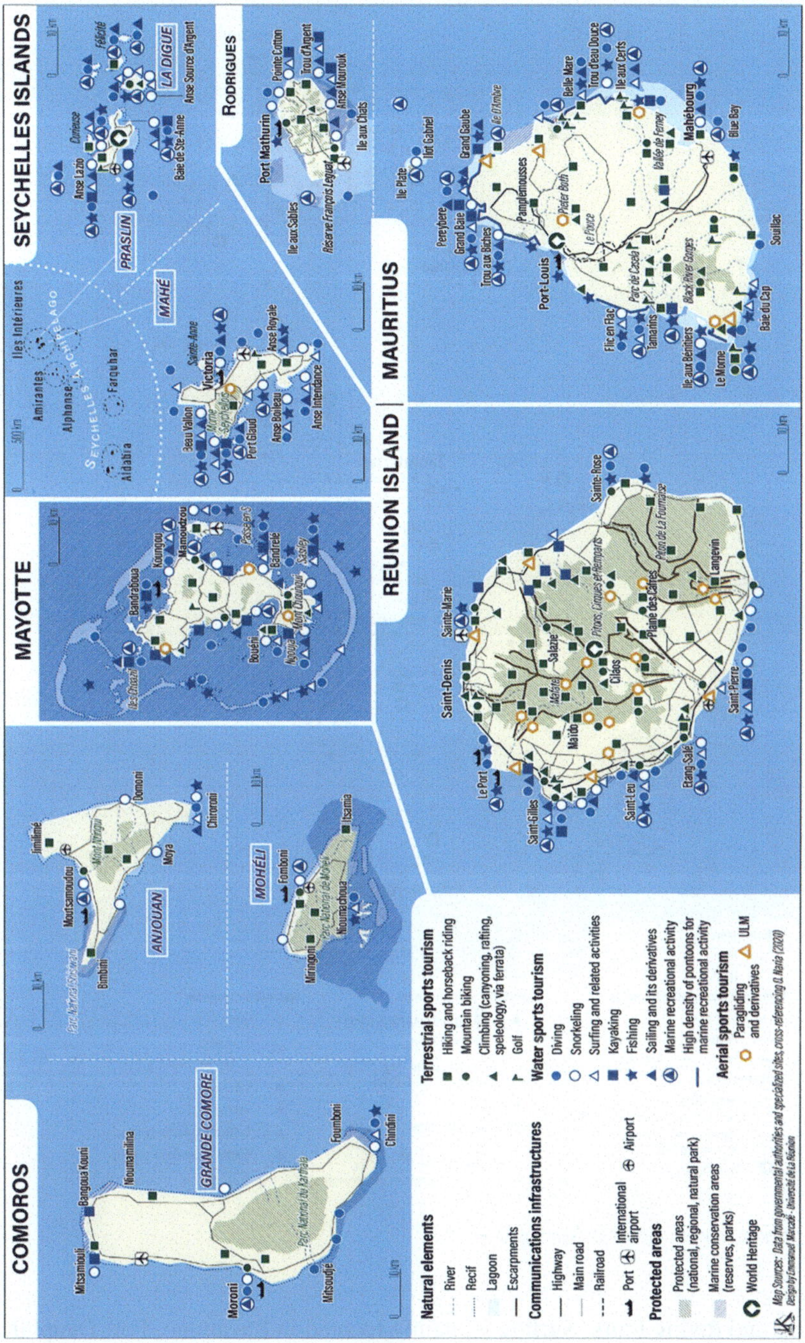

Map 28.2 Sport tourism areas in the Indian Ocean islands

Map 28.3 Sport tourism areas in Madagascar

snorkelling and diving tours. Surfing is mainly located in Lavanono, Fort Dauphin and even Androka. In the Seychelles, the west coast of Beau-Vallon is home to the majority of diving, surfing and snorkelling sites. The islands of Praslins and La Digue offer a range of water activities.

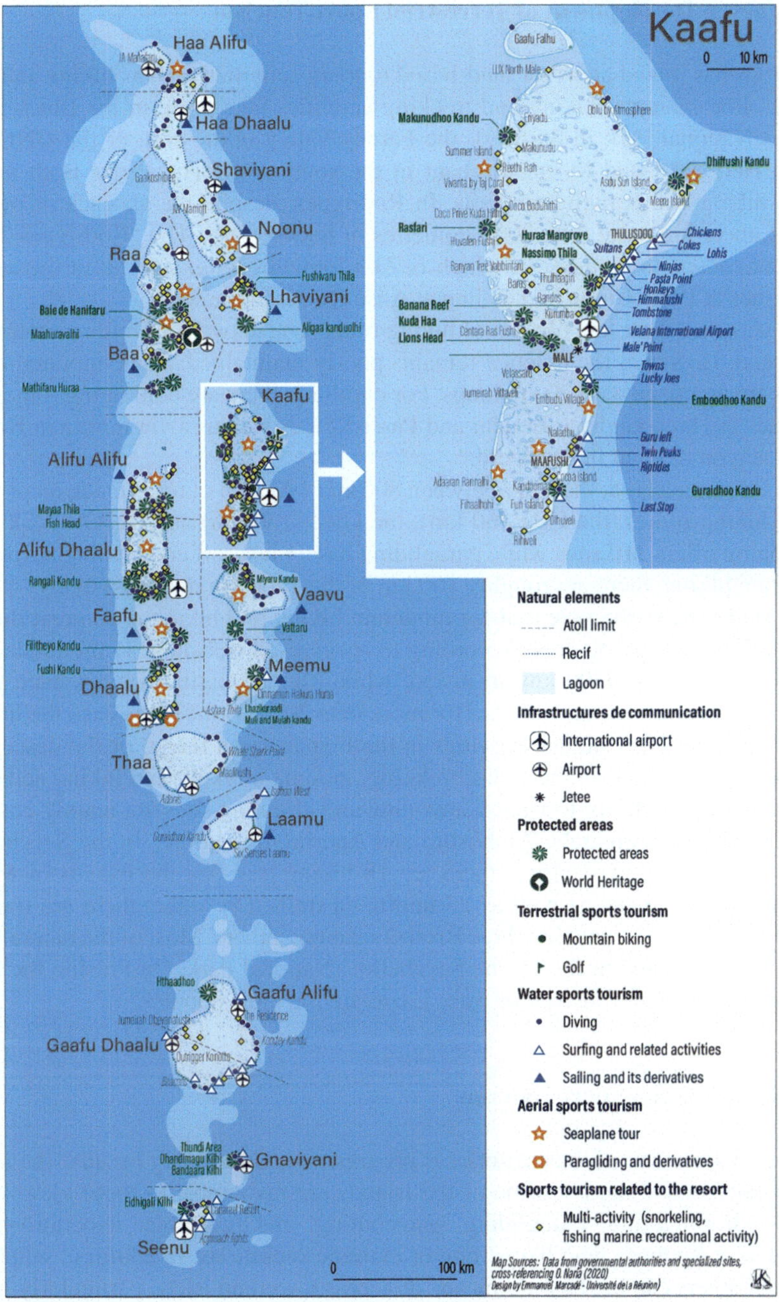

Map 28.4 Sport tourism areas in the Maldives

The Uneven Development of Terrestrial Sport Tourism

Mauritius has inland areas for land-based tourism activities such as hiking and canyoning. For example, hiking and trekking activities take place in the Black River Gorges National Park in the west, the Ferney Valley in the east, or the UNESCO World Heritage-listed Morne Brabant in the south-west. The Pieter Both massif offers hiking and paragliding spots. In Rodrigues, despite the absence of official guides, there are more than 200 kilometres of hiking trails. The "Rodrigues Trail" has four routes, the longest of which is 73 kilometres long, starting in the east of Rodrigues at Point Coton and ending at Anse Quitor in the west.

In Mayotte, there are 146 km of hiking trails. The network of trails circumvents the island. However, the practice remains poorly maintained. Some mountain bike routes are practised with 20 km loops. For climbing, there are bouldering routes and passages in Hagnoundrou, Bouéni and Passi-Kéli as well as a via ferrata in Bouéni and a canyon descent in Mtsapéré.

On Réunion Island, there are between 800 and 1000 km of trails made up of three major hiking routes, the GR1 (60 km), the GR2 (130 km) and the GR3 (38 km). Canyoning offers 210 sites while paragliding has 71 take-off and landing strips. For mountain biking, there are roughly 100 km of the 1400 km that exist today.

In Madagascar, the potential for mountain biking and hiking is so great that no one really knows the figures. According to specialised websites, 55 mountain bike sites for more than 9200 km are listed, while hiking possibilities are even more important with more than 100 sites for more than 20,000 km! In the east, the luxuriant landscapes allow hiking activities in the Fianarantsoa forest, or the descent of the Pangalanes canal. In the south, trekking, mountain biking and hiking activities are carried out in the Isalo Park. Canyoning and climbing sites are mainly concentrated around the Antsirabe-Ambositra and Andringitra regions. In the Seychelles, according to the tourist office, there are 18 hiking trails on the islands of Mahé, Praslin and La Digue, as well as 23 unofficial trails. On Mahé, there are trails at Copolia, the Reserve and the Three Brothers, among others. Most of the nature trails on Mahé are located in the Morne Seychellois National Park. On Praslin, the trails of Glacis Noir, Vallée de Mai or Anse Lazio are the most popular.

The Rise of Aerial Sport Tourism

The aerial discovery of the territories is progressively developing because the relief, the mountains and the coastline of the islands are favourable to these views from above. Microlighting, paragliding, parachuting, and helicopter expeditions are sought after by those who like strong or extreme sensations. In addition, seaplanes and helicopters can be the only means of access to certain islets, particularly in the cirques of Réunion. They allow excursions and parachute jumps near several places of tourist interest in Mayotte, the Seychelles or the Maldives. These different activities structure unusual images in the communication of destinations on social networks or in magazines. Kandima (Dhaalu Atoll) was one of the first resorts to add skydiving to the guest experience.

In conclusion, the supply and demand for sport tourism in the Indian Ocean today represents a particularly buoyant "niche" market by taking advantage of a rich natural heritage and promoting sport tourism products. These activities concern the leisure activities of people from all social backgrounds and in particular the local population, which is increasingly present. The multiple offers of land, air and water activities, the development of numerous natural spots, the growth of a new generation of participative sports events, as well as the increase in the number of people practising them, testify to the dynamism of this market. Sport tourism is also redefining the contours of tourism in the Indian Ocean islands. Wise and Harris (2017) show that sport tourism represents a phenomenon of "regeneration tourism" of the territories. The development of sport tourism in each island illustrates the uniqueness of each territory and culture with its own system of management, convention and regulation, thereby developing the idea of "glocalisation" (Erik, 2004). In this regard, Indian-Oceanic sport tourism is a "hybrid" product combined with folklore, gastronomy and cultural heritage for tourists. Thus, the notion of glocalisation results from interconnections between the global and the local, and allows for a better understanding of this sector of activity. Sport tourism is then mixed with the fertilisation of a set of original activities around multi-practice centres (Nosy-Be, Praslin, La Digue...) or multi-service "safari boats" (for the Maldivian atolls) and an increasingly close relationship with indigenous heritage and the environment. Sport tourism will therefore provide an opportunity for the development and sustainability of socio-economic activities in Indian Ocean destinations in connection with their "local flavours" (Bouchet & Sobry, 2019; UNWTO, 2019), and the intentional promotion of ecotourism (Picard, 2015).

The footprint of sport tourism in the islands is thus increasingly marked. Figure 28.2 evaluates the weight of sport tourism in the islands. More than 4.1 million tourists will be counted in 2019. In the absence of precise data, we can estimate hundreds of thousands of participants in these activities.[5] Our survey includes more than 1854 sites and practice areas in the islands, in relation with the environmental wealth of the territories and the various developments carried out, for 1448 reception structures, associations or small companies. These territories are animated by 115 competitive or participative events. 580 million, i.e. 12% of the total tourist impact. Spin-offs are visible in the thousands of direct jobs in entertainment, management, etc., and indirectly in catering, transport, accommodation, etc. However, this evaluation must be qualified as no data is available for Madagascar, and the data for the Maldives only concerns the diving sector.

From this comparative analysis, we are able to categorise the particularities of sport tourism in these island territories in the following way:

[5] In Réunion, a survey estimated that there were 273,000 sport tourists and 232,000 locals, for a total of 505,000 sports enthusiasts in 2003 (Bessy, Naria, 2003, 61).

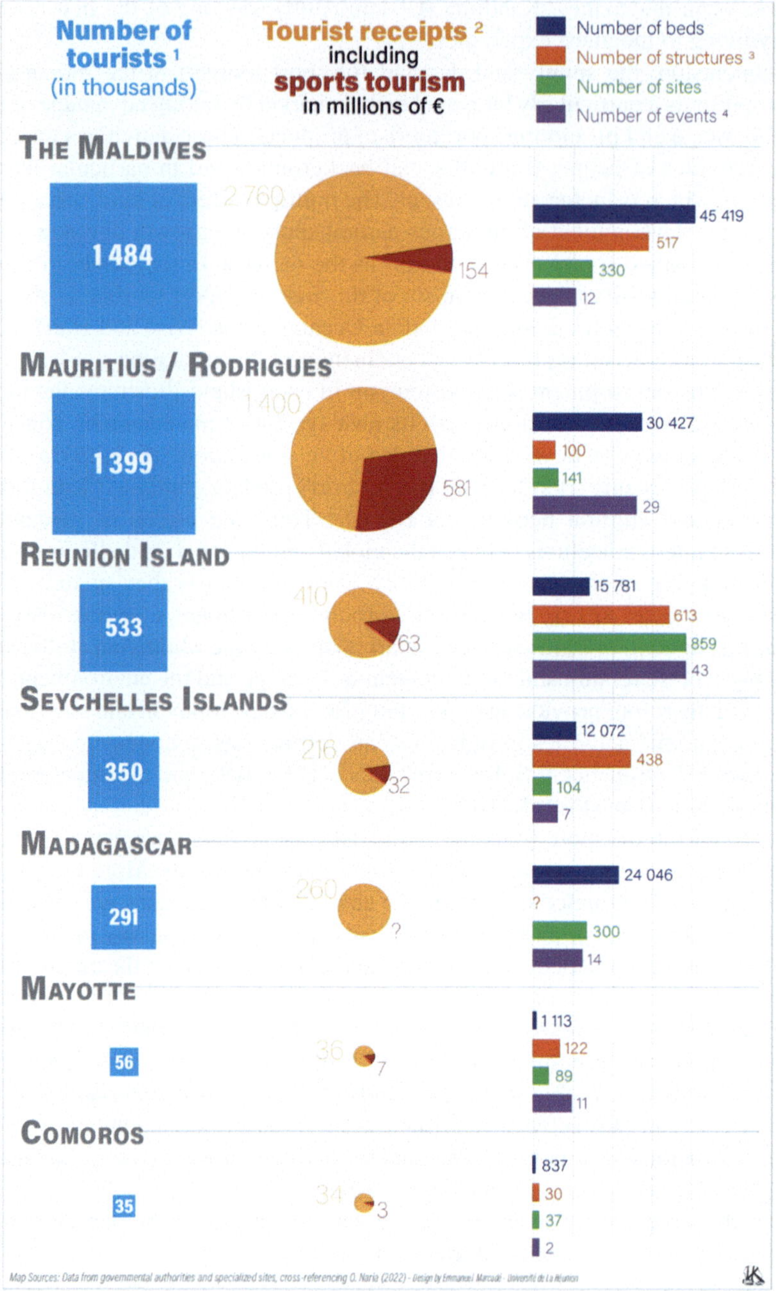

Fig. 28.2 Economic benefits of sport tourism in the islands in 2019. (Sources: data from government authorities and specialised sites, cross-referencing O. Naria, 2022)NB: (1) Tourist numbers are from before the Covid health crisis in 2019(2) Due to lack of available data, the revenue generated from the sector for the Maldives represents only the diving sector, which was worth 154.83 million euros in 2018(3) The structures include associations and companies. For the Maldives, this is the sum of the number of island hotels, diving centres and vessels(4) These are international, national or regional events that are structuring for the islands

(i) Islands where island sport tourism is "specialised." This is the case of the Seychelles and the Maldives, which have adopted a positioning centred on beaches and paradisiacal atolls with aquatic and underwater activities in connection with high-end tourism, which, for a long time, benchmarked their territories before diversifying their clienteles.
(ii) Islands where sport tourism is "diversified." This is the example of Réunion Island, which has a complete sport tourism offer with activities on the coast, in the mountains and in the air.
(iii) Islands where sport tourism is "an opportunity." This is the example of Mauritius with an ocean sport tourism sector that has complemented a mass tourism destination or Rodrigues with an ocean sport tourism sector that has enriched a more ecotourism destination
(iv) Islands where sport tourism is "emerging." This is the case of the Comoros and Madagascar, with very limited external tourism, where sport tourism activities remain marginal. In light of the current levels of development, much remains to be done, such as the development of air access, inter-island links, increased accommodations, tourist products, etc.

Approaches to Sport Tourism Policies and the SDGs in the Indian Ocean

The aim of this chapter has been to understand the levers put in place to develop sport tourism in the islands, by underlining the articulations between political and touristic-sports regulations. These levers offer an understanding of the modes of governance of the islands, which refer to the dynamics of multiple actors in the processes of sustainable development objectives at different stages according to the islands.

Islands and Sport Tourism Policies

In the Indian Ocean, the action of governmental and/or national public authorities is heterogeneous. Nevertheless, several plans and laws reflect the ambition to develop sport tourism territories. All of these documents refer to the consideration of sport tourism as a "niche" sector, as previously mentioned, promising economic development and employment opportunities for the tourism industry at the crossroads of leisure and sport. Political actors are today particularly aware of the stakes that sport tourism presents in the economic dynamics, as well as in the development of territories in a sustainable way.

The mechanisms for implementing public policies on sport tourism depend mainly on the political status of the islands. Indeed, Réunion and Mayotte, regions dependent on France and the European Union, do not function in the same way in the planning and legislative deployment of sport tourism development. This situation leads to a tangle of interactions between the authorities, which often complicates the planning and implementation process.

For its part, the autonomous region of Rodrigues has delegated powers in various fields, including education, fisheries, economic development and tourism. The Rodrigues Tourism Commissioner has the same powers as the Mauritian Tourism Minister. The governments of the other islands implement their tourism development policies in direct link with their tourism ministries, sometimes with the ministries of sports and environment, as well as with the private actors of the territory. Thus, the analysis of the sport tourism offer shows the existence of existing and potential sectors, but unequally structured and distributed according to the islands. It thus reflects a plural level of investment by the public authorities and the private sector, which oscillates, on the one hand, between a game of actors linked to governance and the structuring of a strategic vision for the destination, and on the other hand, a logic of opportunities for rather isolated initiatives, very small companies or associations, which have been able to respond to growing needs in terms of activities and facilities for tourists and locals alike.

The analysis of the planning documents of this sector of activity in the islands reflects the ambivalent interest of the actors on the relationship between sport tourism and the SDGs.

It shows that each territory seeks to bring together public and private actors (see Table 28.4). However, the legislative deployment of sustainable development in sport tourism is disparate due to several factors. The first factor is linked to the managerial culture of tourism through the governance of actors (Taglioni, 2009). The strategies of decision-makers are multiple because they rarely act in the same way because they are linked to the variability of the resources present and the way they are activated by the game of actors (Bessy, 2008). Their positions revolve around interdependence, autonomy or their own interests (Bessy & Naria, 2005). In this context, sport tourism stakeholders must improve their "smart governance" (Giffinger, 2015). This can be understood as intelligent governance of coordination to create a dynamic capable of carrying the economic model necessary for the development of sports territories by answering the questions of the three levels of sustainability (environmental, economic and socio-cultural) in relation to the operationalisation model.

The second factor is explained by the lobbying action of the tourism actors, more particularly the hoteliers, tour-operators, as well as the independent service providers, towards the public decision-makers. This action has been powerful in the development of the sector in the Maldives, the Seychelles and Mauritius. Finally, there is at times a lack of connection between the strategies deployed and the application of regulatory provisions for the development of sport tourism with regard to the issues at stake between environmental, sports and tourism stakeholders and local populations within the same island or on the scale of an archipelago. The examples of the Comoros and Madagascar illustrate our point.

Table 28.4 Islands and sport tourism policies

	Réunion (France)	Mayotte (France)	Madagascar	Mauritius-Rodrigues	The Seychelles	The Comoros	The Maldives
Major areas	Complete offer of land, air and water sport tourism.	Offer around diving	Offer with hiking, mountain biking, fishing and boating	Offer around water sport tourism	Offer around aquatic sport tourism such as diving, surfing, boating...	Hiking and diving offer	Complete offer around water sport tourism
Public actors	State, Regional Council, General Council, EPCI, municipalities, national park, marine park, IRT (associations)	State, Department, CDTM, EPCI, municipalities, GEMTOUR, CCI, Marine Park	State, Ministry of Tourism, Madagascar National Parks...	State, Ministry of Tourism and Leisure, Mauritius Tourism Promotion Authority, Ministry of Environment & Sustainable Development, RRA	State, Ministry of Tourism, Seychelles Tourism Board (STB), Park Authority Seychelles National Parks...	State, Ministry of Tourism, Comoros National Tourist Office, Comorian Tourism Association	State, Ministry of Tourism, Ministry of Atolls Development, Ministry of Planning and National Ministry of Economic Development Development, Ministry of Environment, Energy and Water, NGOs
Private actors	Commercial offers: SYPRAL, companies, tourist-sports guides,	Supply of commercial and public service providers, Groupement des Entreprises Mahoraises du Tourisme, Chamber of Commerce and Industry	Supply of commercial service providers and tour operators National Tourist Office of Madagascar, association of professional tour operators	Association of Inbound Operators Mauritius, Association des hotels de Charme, Association of Tourist Operators	Fishing Boat Owners Association Marine Charter Association/Yacht Operators/Ferry and Schooner Operators, Dive Center Operators, sport tourism operators	Chambers of Commerce, Industry and Agriculture, Comorian Private Sector Federation	Hotel and tour offer (Association of Travel Agents Tour Operators (MATATO), Tourism Liveaboard Association of Maldives (LAM) and Divers Association of Maldives (DAM).

(continued)

Table 28.4 (continued)

	Réunion (France)	Mayotte (France)	Madagascar	Mauritius-Rodrigues	The Seychelles	The Comoros	The Maldives
Main planning documents	"Les loisirs de nature dans le secteur du tourisme," SYPRAL (2014); "Schéma territorial de développement du sport," DDJSC (2016)	"Schéma Régional de Développement du Tourisme et des Loisirs de Mayotte," Department of Mayotte, HTL (2019)	"Madagascar Vision for Tourism 2023," Ministry of Transport, Tourism and Meteorology (2018)	"Adapting to the changing global environment, Strategic Plan" 2018–2021, Ministry of Tourism (2018)	"Tourism Master Plan" (2020–2023); Seychelles Sustainable Tourism Label	"Tourism Law" (2012), "Investment Code" (2017), "Tourism Development Plan" (2019)	"First Tourism Master Plan (1984–1993); Second Tourism Master Plan (1996–2005); Third Tourism Master Plan (2007–2011), Fourth Tourism Master Plan (2013–2017), SAP (2019–2023)
Main legislative deployments of sustainable development	Marine Park Order (2007, 2014) regulating kitesurfing and sailing gear, kayaking and beach gear, swimming and recreational fishing; National Park Charter: regulations for the heart of the park (2012)	Marine Park (2010), lagoon management plan (2002), local action plan of the French Coral Reef Initiative (IFRECOR—2006) Mammal approach (2004) marine mammals by operators tourist (2013) and boater (2014)	Tourism Law, No. 95-017 of 25 August 1995 contains the Tourism Code Tourism; Malagasy Environment Charter (1990), National Environmental Action Plan (PNAE) (1990)	"Mauritius Sustainable Island" (MID) (2013), "Environmental Protection Act 1991" (EPA), "Environment Impact Assessment" (EIA) (1993) "Guidelines for the use of helmets in diving and underwater walking" (2013), "Sustainable Integrated Development Plan for Rodrigues" (2009)	The Environment Protection Act (1994), Policy on Yachting Tourism (2003, 2008), and Boat Charter Policy (2008) for diving activities; Policy for Recreational Diving Business, (2015); Policy for Water Sports Businesses (2017)	Law of 29 December 2011 establishing the Tourism Code, revised in 2021	Environmental Protection and Conservation Regulations (1993, 2006) Marina Regulations, Recreational DivingRegulations

Source: Data collected by the author

The development of sport tourism in the Indian Ocean has brought about economic growth, sustainable environmental management and an innovative approach towards preserving and protecting natural resources. These aspects are only compatible if appropriate measures are taken to improve the efficiency and productivity of resource management and to safeguard the natural heritage features of the islands.

From there, the existing institutional and legislative provisions in the islands for the management and protection of the environment through the use of sport tourism have developed over a relatively long period of time in relation to the increase in tourist numbers in the area. Thus, government recognition of environmental issues is fundamental to the economic development of sport tourism in an SDGs territory. Table 28.4 lists the main measures that intervene in this field with the necessary laws, regulations, charters and best practices in order to achieve the global SDGs.

Finally, the study of strategies shows that decision-makers are interested in respecting the concerns linked to the territories and the place of local populations in the construction of sport tourism offers. Overall, this ambition is not sufficiently emphasised in the documents because the mention of destinations for the benefit of the local population does not necessarily equate to genuine benefits for these populations, who are often on the fringes of the economic benefits of sport tourism. Participatory approaches involving local populations, governments, the tourism industry and environmentalists are particularly disparate. While it is an obligation in the French islands of Réunion and Mayotte, only the planning documents of Mauritius and Seychelles mention the inclusion of local people in such participation structures.

Islands and Initiatives for the Sustainable Development of Sport Tourism

With the ambition to show how much the socio-economic stakes of sport tourism in the islands can be linked to legislative deployments and sustainable development measures, the analysis of two islands provides a potential heuristic.

With the creation of a park in 2007, Réunion Island's tourism development is now part of a sustainable development approach. In view of the biophysical vulnerabilities, the development of the leisure and sport tourism market in the "Hauts" will be confronted with resistance from environmentalists (Naria, 2011). The Charte Réunionnaise de l'Environnement (1992) and the Schéma d'Aménagement Régional (1995) established the principle of a National Park for the sustainable development of the Hauts, combining heritage protection and economic development. This park aims to harmonise the development of activities, of which leisure and sport tourism are part, in a more balanced perspective and to strengthen the policy of the Hauts on 42% of the island. Its core zone is listed as a UNESCO World

Heritage Site under the name "*Pitons, cirques and ramparts of Réunion Island*." At the Park level, the "*category use regulations*" emphasise the "national park spirit" through a guide to good discovery practices to be adopted in the natural environment for all sport tourism activities. Examples include educating sport tourism clients about heritage preservation issues, environmentally friendly discovery practices and the organisation of events subject to authorisation (Réunion National Park, 2013, 149–159).

The case of the Seychelles speaks to the sustainable development measures of ocean sport tourism. Aware that resources are exhaustible, the government introduced the precepts of sustainable development in 1969 with the adoption of the National Parks Ordinance, and later with the adoption of integrated coastal zone management (Cazes-Duvat, 2001, 76) with the "*Indian Ocean Commission Regional Environment Programme*" in 1995. This scheme puts forward an integrated reflection on the sustainable management of beaches. They become playgrounds and issues with the acceleration of aquatic activities. The first "*Policy on Yachting Tourism*" was drawn up in 2003 in order to better manage the growing tourist industry. This regulation limits the number of charter yachts to 200 in order to preserve the environment (Seychelles Tourism Master Plan, 2017). These actions were complemented by the acceleration of the installation of mooring buoys in the busy areas around north-west Mahé, north-east Praslin, and some of the surrounding islands, as well as the requirement for onboard treatment of yacht wastewater. The *Boat Charter Policy* (2008) governs the activities of boat charter operators from a sustainability perspective. According to the "*Policy for Recreational Diving Business*," in 2015 (Seychelles Tourism Master Plan, 2017, 89), six diving spots are allowed in marine national parks such as Ile Coco Marine, Sainte-Anne or Curieuse. Furthermore, ownership or investment in diving businesses is exclusively reserved for Seychellois citizens, with the exception of motorised diving operations and offshore diving operations. For the latter activities, foreigners are allowed to invest provided that the fleet size exceeds three yachts (Government of Seychelles (2017), Tourism Master Plan (2020–2023). In the "Policy for Water Sports Business" (Government of Seychelles (2017), Tourism Master Plan (2020–2023), 198 p), the water sports industry is classified as motorised or non-motorised depending on the craft used for the activity. In order to regulate the cohabitation of swimming and diving tourists, marine recreation areas have been divided with swimming areas, non-motorised areas and motorised areas for environmental and safety reasons. As an indication, Anse à la Mouche and Anse Lazio are for non-motorised only, while Anse Royale, Beau Vallon, Côte d'Or and Perseverance have been designated as motorised areas by the Seychelles Maritime Safety Administration.

At the end of our analysis, the quantitative and qualitative data collected allow us to synthesise the contribution of sport tourism and the SDGs in the islands, based on Fig. 28.3 below. The challenge here is to aim at preservation objectives through the exceptional natural heritage character of the islands while allowing its sustainable socio-economic development.

Fig. 28.3 Sport tourism and the SDGs in the islands. (Source: World Tourism Organization, 2019)

Conclusions

This chapter contributes to the existing literature on sport tourism through theoretical advances, novel methodologcial approaches and critical discussions, all through the lens of islands as eco-territories. Our results show that the model of sustainable island development represents new spaces and issues applied to sport tourism. There is no single model applicable due to the specificity of each island. Each territory is individualised according to its history, identity, constraints and resources. It can be stated with Weaver (2017), there is no island model for sustainable tourism, and as such, it is the case study that allows the best understanding of the relationship between sport tourism and the sustainable development in islands.

The SDGs applied to sport tourism in the islands also allow us to envisage a model of development compatible with the existing resources and environmental constraints. Island territories can be conducive to study sustainable development, especially with both the synergies and antagonism effects that sustainable development generates through sport tourism events and practices. At the same time, small-scale sport tourism activities are seen as development opportunities for island communities; these activities are part of sport tourism products and services. Ultimately, approaches to sustainable sport tourism depend on the socio-cultural context of the territory and are linked to the overall development context of the island. The results highlighted best practices and implications that can help other similar destinations to develop more sustainable activities in line with the SDGs. Faced with their vulnerabilities, islands have no choice but to embrace the path towards sustainable development. The promotion of sport tourism within these eco-territories must demonstrate a holistic commitment to genuine sustainable development, as well as a strategic plan to achieve the sustainable development goals articulated as part of the United Nations' 2030 Agenda.

References

Agathe, H. (2016). *Economic history and development prospects in Rodrigues* (p. 19). Unpublished paper.

Baldachino (2014), Godfrey Baldacchino, "Small Island States: Vulnerable, Resilient, Doggedly Perseverant or Cleverly Opportunistic?", Études caribéennes [En línea], 27–28 | Avril-Août 2014, Publicado el 14 agosto 2014, consultado el 19 mayo 2024. URL: http://journals.openedition.org/etudescaribeennes/6984; DOI: https://doi.org/10.4000/etudescaribeennes.6984

Bertile, W. (2013). *Mascareignes and Seychelles, Creole archipelagos of the Indian Ocean*, Indian Ocean Studies [Online], 49–50 accessed 19 April 2019. https://doi.org/10.4000/oceanindien.1811.

Bessy, O., & Naria, O. (2004). L'origine géographique des participants au Grand-Raid de La Réunion, M@ppemonde, n°74.

Bessy, O., & Naria, O. (2005). Les enjeux des loisirs et du tourisme sportif de nature dans le développement durable de l'île de la Réunion. In P. Bouchet, C. Sobry, & (dir) (Eds.), *Management et marketing du sport: du local au global* (pp. 307–339). Presses Universitaires du Septentrion, collection Sports et Sciences Sociales.

Bessy, O (dir). (2008). *Sport, loisir, tourisme et développement durable des territoires* (p. 175). PUS.

Beeharry, Y., Makoondlall-Chadee, T., & Bokhoree, C. (2014). Policy analysis for performance assessment of integrated coastal zone management initiatives for coastal sustainability. *APCBEE procedia, 9*, 30–35.

Bouchet P., & Bouhaouala, M. (2009). Tourisme sportif, *Téoros* [Online], 28-2 | 2009, Online since 28 April 2016, connection on 16 October 2022. http://journals.openedition.org/teoros/438

Bouchet, P., & Gay, J.-C. (1998). Les Hauts de La Réunion conquis par les loisirs. *Mappemonde, 58*, 31–37.

Bouchet, P., & Sobry, C. (2019). Sports tourism: Contemporary issues and new trends on a global market. In M. Desbordes, P. Aymar, & C. Hautbois (Eds.), *The global sport economy* (pp. 295–314, 330p). Routledge..

Butler, R. (Ed.). (2006). *The tourism area life cycle: Conceptual and theoretical issues* (Vol. 2). Channel view publications.

Cazes-Duvat V. (2001), «Évaluation de la vulnérabilité des plages à l'érosion: application à l'archipel des Seychelles», Géomorphologie, 1, p. 31–40.

Christie, T., & Crompton, D. (2003). *Etude du Secteur Tourisme*, Africa Region Working Paper Series No. 63, Republic of Madagascar, 155 p.

Dehoorne and Saffache (2008), Olivier Dehoorne y Pascal Saffache, "Le tourisme dans les îles et littoraux tropicaux : ressources et enjeux de développement", Études caribéennes [En línea], 9-10 | Avril-Août 2008, Publicado el 15 agosto 2008, consultado el 19 mayo 2024. URL: http://journals.openedition.org/etudescaribeennes/852; DOI: https://doi.org/10.4000/etudescaribeennes.852

Erik, S. (2004) Globalisation or 'glocalisation'? Networks, territories and rescaling, *Cambridge Review of International Affairs*, 17:1, 25–48, https://doi.org/10.1080/0955757042000203632

Gay, J.-C. (1999). Les balbutiements du tourisme mahorais. *Travaux & documents.*, Université de La Réunion, *11*, 137–152.

Gay (2001) L'Île-hôtel, symbole du tourisme maldivien. Les Cahiers d'Outre-Mer. Revue de géographie de Bordeaux, n° 213, p. 26–52.

Gay, J. C. (2017). Tourism in the Tropical Islands. Fagnoni E. *Les espaces du tourisme et des loisirs*, Armand Colin, 2017, 978-2-200-62008-0. ffhalshs-03583673e.

Giffinger, R. (2015). Smart City Concepts: Chances and Risks of Energy Efficient Urban Development. In M. Helfert, K.-H. Krempels, C. Klein, B. Donellan, & O. Guiskhin (Éd.), *Smart Cities, Green Technologies, and Intelligent Transport Systems* (p. 3–16). Cham, Springer International Publishing.

Government of Seychelles (2018), "Tourism Master Plan (2020–2023), 198 p.

Gowreesunkar, Vanessa & Soteriades, Marios. (2015). Entertainment of leisure tourists in island destinations: evidence from the island of Mauritius. *African Journal of Hospitality Tourism and Leisure*. 4.

Higham, J., & Hinch, T. (2006). Sport and tourism research: A geographic approach. *Journal of Sport & Tourism, 11*(1), 31–49. https://doi.org/10.1080/14775080600985267

HTL. (2019). *Schéma Régional de Développement du Tourisme et des Loisirs de Mayotte*. Département de Mayotte.

INSEE Analyses (2019), Les inégalités de niveau de vie se sont creusées, Juillet 2020, N°25, INSEE, https://www.insee.fr/fr/statistiques/4622454-

Jauze, J.-M. (1998). *Rodrigues la troisième ile de Mascareignes* (p. 272). L'Harmattan.

Jauze, J.-M., & Guebourg, J.-L. (2005). *Inégalités et spatialité dans l'Océan Indien* (p. 420). L'Harmattan.

Jauze, J. -M. (dir). (2019). *Define Indianoceania for me* (p. 303). O.I.E.S, Faculté des Lettres et des Sciences Humaines, Université de La Réunion, Epica Editions.

Klöck, C., & Fink, M. (Eds.). (2019). *Dealing with climate change on small islands: Towards effective and sustainable adaptation*. Göttingen University Press. https://doi.org/10.17875/gup2019-1208

Kundur, S. K., & Murthy, K. (2013). Environmental impacts of tourism and management in Maldives. *International Journal of Environmental Sciences, 2*(1), 44–50.

Magnan, A. (2005). *Tourisme, développement et dynamique territoriale dans l'archipel des Maldives et à l'île Maurice (océan Indien)*. Thèse de doctorat de géographie, Université Paul Valéry, Montpellier III: Montpellier, 462 p.

Melo, R., Sobry, C., & Rheenen, V. (2021). *Small scale sport tourism events and local sustainable development; a cross-national comparative perspective*. Springer.

Ministry of Tourism Maldivian. (2019). *Tourism yearbook 2019* (p. 52).

Morgan, H., Bush, A., & McGee, D. (2021). The contribution of sport to the sustainable development goals: Insights from commonwealth games associations. *Journal of Sport for Development*. Retrieved from https://jsfd.org/2021/09/01/the-contribution-of-sport-to-the-sustainable-development-goals-insights-from-commonwealth-games-associations/

Naria, O. (2011). Eléments de réflexion sur les formes de développement des territoires touristiques et sportifs de nature dans les Hauts de La Réunion: entre marquage culturel et enjeux de développement durable. In J.-M. Jauze & (dir) (Eds.), *Les Hauts: Terres de tradition et d'avenir* (pp. 243–270). Université de La Réunion, Océan Edition, Saint-André.

Naria, O., & Sherwin, K. (2011). Tourism, nature sports and sustainable development in the Seychelles. In O. Dehoorne (Ed.), *Etudes Caribéennes.*, The changing world of coastal, island and tropical tourism (Vol. 19, pp. 46–67).

Naria, O., Van Rheenen, D., Melo, R., & Sobry, C. (2022). Sport tourism, Island territories and sustainable development: Approach to a methodological framework for research based on the insular context of the Indian Ocean. In B. Mazza & C. Sobry (Eds.), *Sport tourism and its territorial development and opportunities* (pp. 19–40). Cambridge Scholars Publishing.

ORSTOM. (1986). *Les sols de l'île Rodrigues* (89 p). MSIRI Mauritius and ORSTOM.

Pébarthe, H. (2001). *Littoral mauricien et tourisme: quelles perspectives de développement et de gestion intégrées pour le sud-est de l'île ?* CNRS-UMR PRODIG, 95, 2 Grafigéo 2-901560-44-X.

Picard, D. (2015). Making ecotourism sustainable: Refocusing on economic viability. Lessons learned from the "Regional strategic action plan for coastal ecotourism development in the South Western Indian Ocean". *Journal of Sustainable Tourism*. https://doi.org/10.1080/09669582.2015.1019512

Pluss Ch. (1986). *Au Fil des îles. Propos sur les Maldives et le tourisme,* PhD thesis, Paris, Université Paris-VII, 456 p.

Rajeriarison, P. (2013). *Union des Comores: note de politique sur le tourisme* (p. 39). World Bank.

Région Réunion. (2018). *Actualisation du schéma de développement et d'aménagement touristique de la Réunion* (p. 537). SDATR.

Réunion National Park. (2013). *La Charte du parc national de La Réunion* (p. 202). Graphica Edition.

Robert, R. Caze, V. (2000). Etude quantitative de la fréquentation des espaces de tourisme et de loisirs à La Réunion, in Les Rencontres du Conseil Régional de La Réunion, 60 p.

Schéma d'Aménagement Régional (1995) SAR : Schéma d'Aménagement Régional de La Réunion; Approuvé par décret n°95-1169 du novembre, Rapport AGORAH, St-Denis.

Serviable, M. (1983). *Le tourisme aux Mascareignes* (182). Centre Universitaire de la Réunion.

Singaravelou, S. (1989). Le rôle du tourisme dans l'économie mauricienne. In Iles et tourisme en milieux tropical et subtropical. *Actes du XIe Colloque SEPANRIT*, Bordeaux, 9-10 avril 1987, 265–278.

Silva, C., Kastenholz, E., & Abrantes, J. L. (2013). Place-attachment, destination image and impacts of tourism in mountain destinations. *Anatolia, 24*(1), 17–29. https://doi.org/10.1080/13032917.2012.762312

Small Island Developing States. (2013). *World Heritage*, No. 66, UNESCO.

Taglioni, F. (2009). Le tourisme dans le sud-ouest de l'océan Indien: une durabilité à géométrie variable. In X. L. Vo Sang (Ed.), *Le tourisme durable, un instrument d'aide à la lutte contre la pauvreté?* (pp. 154–169). Agence universitaire de la francophonie and Van Lang University.

Teelucksingh, S., & Watson, P. (2013). Linking tourism flows and biological biodiversity in Small Island Developing States (SIDS): Evidence from panel data. *Environment and Development Economics, 18*(4), 392–404. https://doi.org/10.1017/S1355770X13000120

UN. (2019). *The least developed countries report 2018* (p. 190).

UN (2015), the Sustainable Development Goals were taken from the United Nations website www.un.org/sustainabledevelopment/ and the targets for each Goal are those included in General Assembly resolution 70/1 entitled "Transforming our world: the 2030 Agenda for Sustainable Development"..

UNDP. (2019). *Human development report* (p. 351).

UNESCO (2013), Tracking Key Trends in Biodiversity Science and Policy. Based on the proceedings of a UNESCO International Conference on Biodiversity Science and Policy, edited the United Nations Educational, Scientific and Cultural Organisation.

Weaver, D. B. (2017). Core-periphery relationships and the sustainability paradox of small Island tourism. *Tourism Recreation Research, 42*(1), 11–21.

Wise, N., & Harris, J. (2017). *Sport, events, tourism and regeneration*. Routledge.

World Bank. (2013). *Comoros tourism sector review: Discovering the tourism potential of natural wonders*. Washington, DC https://openknowledge.worldbank.org/handle/10986/16707

World Bank. (2019). *World development report 2019: The changing nature of work*. Washington, DC: World Bank. https://doi.org/10.1596/978-1-4648-1328-3

World Tourism Organization. (2019). *Sport tourism and the sustainable development goals (SDGs)*. UNWTO. https://doi.org/10.18111/9789284419661

Chapter 29
A Comparative Study of Small Islands, Sport Tourism, and Sustainable Development: A Local to Global Perspective

Derek Van Rheenen, Olivier Naria, Ricardo Melo, and Claude Sobry

Introduction

In order to propose a multi-dimensional analysis of the 21 cases studies featured in this comparative research project, we have condensed the 17 United Nations Sustainable Development Goals (SDGs) under four main themes: the social, economic, and environmental dimensions of sustainability, which make up the three articulated pillars of the Brundtland Report (e.g. the triple bottom line), to which we add governance or the political dimension of sustainability. This dimension, absent from past analyses, has become pivotal at the crossroads of multiple global issues (Meadowcroft, 2007) and of crucial importance in the study of small island states or territories. This international research project on sport tourism and sustainable development has shown that islands are at the heart of tomorrow's social and environmental challenges and may therefore be seen as a bellwether for global sustainability.

D. Van Rheenen (✉)
Cultural Studies of Sport in Education, University of California Berkeley, Berkeley, CA, USA
e-mail: dvr@berkeley.edu

O. Naria
Department STAPS, Université de La Réunion, Le Tampon, île de la Réunion, France
e-mail: olivier.naria@univ-reunion.fr

R. Melo
Coimbra Education School, Polytechnic Institute of Coimbra, Coimbra, Portugal

C. Sobry
Faculty of Sport and Physical Education, Université of Lille, Lille, Nord, France
e-mail: Claude.sobry@univ-lille.frxt

D. Van Rheenen et al. (eds.), *Sport Tourism, Island Territories and Sustainable Development*, Sports Economics, Management and Policy 24,
https://doi.org/10.1007/978-3-031-51705-1_29

In this chapter, we will present the contribution of sport tourism to the achievement of the SDGs, based on a comparative analysis from the diversity of chapters in this book. Based on this synthesis, we formulate an analytical rubric outlining the role of sport tourism and the 17 articulated SDGs. We envision this rubric serving as a useful tool to help key stakeholders better understand sport tourism as a vehicle for sustainable development in island territories. We address each of the four dimensions of sustainability—environmental, economic, social and political—as a basis of comparison across the 21 island case studies. After highlighting the limitations of our research design, we recommend a series of practical proposals at this dynamic intersection.

Sport Tourism and Sustainable Development in Small Island Ecoterritories

It is important initially to acknowledge that the contribution of sport tourism to the achievement of the global SDGs can only be achieved by the presence and planning of the sport tourism production or supply on the island territory. However, intentional, successful implementation requires an analysis of the sport tourism ecosystem and the unique ecoterritories comprising these island states.

As we have discussed throughout this book, we use the concept of ecoterritories as unique geographic ecosystems with diverse political economies and sociocultural histories. The concept of ecoterritories provides a more flexible tool for analysis at the micro- or meso-level, between islands and islets, bi-island, multi-island or archipelagos, of very diverse topographies, with the common characteristic of being surrounded by water on all sides.

Island ecoterritories likewise rely on a moral mandate for change amidst a changing climate and global context, a focus on a critical consciousness or ecopedagogy. A moral mandate requires political will and a local to global commitment to social and environmental justice. As such, this critical framework draws attention to not only the economic life of a community but also its decision-making processes, a focus on people in relationship to one another and in relation to all of the natural world.

The particular configurations of islands make them privileged laboratories for analysing the limits and potential scalability of sport tourism with respect to the SDGs. If the economic effects are prioritized at the expense of the social and environmental issues of these ecoterritotries, then sustainable development will not be realized. Similarly, the relationship between stakeholders and local governance will guide the future potentialities of island spaces. In this analytical effort, we argue that the environmental issues inherent to sport tourism is essential to preserve the unique biodiversity, promote resource management, and strengthen resilience to climate change, while simultaneously promoting economic growth and socio-cultural exchanges.

Environmental Perspectives in Comparison

For the purpose of this book, it is impossible to discuss the intersection of sport tourism and sustainable development in small island states or territories without addressing the environmental context of this dynamic intersection. The very existence of particular forms of sport tourism, such as nature sports, depends on the unique characteristics of the island context, such as its geography, geomorphology, and biodiversity. As contributors throughout this book have emphasized, there is a potentially powerful relationship between sport tourism and the environment. Sport tourism relies on the natural landscape of the islands while these activities potentially engender an appreciation of these resources. An appreciation and awareness of these unique ecosystems leads in turn to the promotion of environmental preservation and conservation initiatives. A commitment to ecotourism is a decision made by individual consumers but also by commercial entities, government stakeholders (from local to global agencies), international bodies, and non-governmental organisations (NGOs). Thus, before addressing the economic potentialities of sport tourism at this unique site of analysis, we must first acknowledge the physical (e.g. natural) characteristics of small island states.

Islands, while a diverse collection of physical and socio-economic contexts, share important environmental commonalities, such as being surrounded by water in the form of seas, oceans and other marine waterways. This unique characteristic creates both opportunities and vulnerabilities within these ecoterritories. The relative isolation of island states, for example, can make travel to and from these places an obstacle to global (even regional) visibility and development. Regarding tourism, particularly for those islands thousands of kilometres from a mainland, transportation has served as a significant barrier to growth. It was not until the mid- to-late twentieth century when affordable long-range commercial air travel and the construction of island infrastructure, such as airports, provided a viable venue for market growth within these remote locations. Many islands in the Atlantic, Indian, and Pacific oceans exemplify this historical development. Still today, many islands are only reachable by boat, particularly those in island archipelagos comprising hundreds of islands. In these instances, there is often one or more island hubs where international travel provides access to these territories. It should be noted, however, that even in highly populated and developed economies, such as Germany, this is the case. For example, the island of Helgoland can only be accessed via sea. These limitations impact the economic potential of sport tourism within these sites.

Of the 17 sustainable development goals (SDGs) articulated in the United Nations' 2030 Agenda for Sustainable Development, four are commonly categorized as primarily environmental goals, while the others map onto the economic and social pillars of sustainability (Mangikia & Sklarew, 2023; Sachs et al., 2022). These goals, SDG #12, Responsible Consumption and Production, SDG # 12, Climate Action, SDG# 14, Life Below Water, and SDG #15, Life Above Land, are critical to any discussions of sustainable development in small island states or territories.[1] Given the environmental vulnerability of small islands, particularly low-lying, less developed islands, climate action (SDGs# 13) must be a priority for the future. This goal, in particular, is interconnected with rising sea levels, placing many of these territories at existential risk. Climate action, in addition to the health and well-being of the island inhabitants (e.g. a reduction in poverty and the basic needs of food, water, and health care, [SDGs #1, #2, #3 and #4, respectively]), must take precedence over any discussion regarding the development and diversification of a particular industry, such as sport tourism. And yet, such mature and emerging markets may also help to provide opportunities to address some of these basic needs, while promoting economic, social, and environmental sustainability, mechanisms or drivers for responsible growth and progress. The political role of small island developing states (SIDS) has likewise grown more important in global efforts to confront planetary crises such a climate change. In this sense, small island states or territories represent a potential laboratory for innovation.

This relative geographic isolation and the necessary mode of transportation to and from islands run counter to sustainable development goal (SDGs) # 13, action to curb climate change. The CO_2 emissions associated with travel to these places pose a dilemma for stakeholders in search of a multi-dimensional approach towards sustainable development. The promotion of ecotourism to these locations sends an odd, perhaps hypocritical, message to potential travellers: fly halfway across the world to our pristine island where we offer a "responsible" sustainable destination, despite your decision to travel here produces significant carbon emissions and runs counter to the sustainable development goals we espouse. Such examples highlight the complexity of promoting sustainable development both locally and globally, especially when seeking to incorporate the economic, social, and environmental dimensions of sustainability. This particular belief may be exaggerated, however. The emergence of "flygskam" or flight shame as a recent phenomenon refers to an increasing number of citizens concerned with their individual impact of global warming as a result of flying commercially (Chiambaretto et al., 2021; Gössling et al., 2020; Mkono, 2020). Scholars have found, however, that flight shame can be partially explained by a distorted public perception of the environmental impact of air transport, finding that this form of transportation accounts for 2–3% of global carbon emissions (Baledon & Kosby, 2018; Larsson et al., 2019; Staples et al., 2018).[2]

---

[1] In truth, these pillars of sustainable development goals are porous as heuristic categories and should be understood as such. For example, within the cake model of SDGs proposed by Rockström and Sukhdev (2016), SDG# 6, *Clean Water and Sanitation* was categorized as an environmental goal but has been reframed as a social sustainable goal based on its objective to provide safe drinking and sanitation to all, just as SDG# 2, *Zero Hunger*, is seen as primarily a social goal within this global initiative.

[2] This discussion demonstrates that air travel is a controversial topic in climate debates. While there is disagreement about the extent of a frequent flyer's individual contribution to climate change, this calculation is more difficult for countries who are only expected to count domestic aviation emissions; international flights fall into a separate category, known as "bunker fuels." These flights are not included in any country's emission data, which limits the incentive for these countries to reduce them (Ritchie et al., 2020).

Climate stories tend to blame "all of us" and imply that there is no hope for real climate change, as we have not done enough soon enough. It draws attention away from the real culprits of resource extraction, deforestation, carbon emissions, and global warming, such as agribusiness, livestock farming, and fossil fuel companies. As climate scholar and author Genevieve Guenther (2018) argues, "The *we* responsible for climate change is a fictional construct, one that's distorting and dangerous. By hiding who's really responsible for our current, terrifying predicament, [the pronoun] *we* provides political cover for the people who are happy to let hundreds of millions of other people die for their own profit and pleasure."

And it the wealthiest people and the wealthiest nations, the haves rather than the have nots, who bear the greatest responsibility for these emissions and disastrous results. For example, the United States has emitted far more CO_2 than any other country in the world, accounting for a quarter of all global emissions since the middle of the eighteenth century (Ritchie et al., 2020). China is now the biggest producer of greenhouse gases and the second largest economy in the world (Plumer et al., 2022). And sadly, it is often the poorest countries that bear the greatest relative burden of climate change, based on their socio-political and historical geography and their environmental vulnerability.

The complexity of these real and perceived trade-offs may resonate in particular with low-lying islands, whose very existence is in jeopardy as a result of climate change and the corresponding sea-level rise. For example, Kiribati, one of the world's Least Developed Countries (LDC) and a low-lying small island developing state (SIDS) in the Pacific, is particularly vulnerable to environmental disasters and shocks. As a result, this island archipelago remains in the ideation stage related to sport tourism development, as other government priorities take precedence, such as combatting food insecurity (SDGs # 2), improving the health of its peoples (SDGs # 3), and climate change mitigation and adaptation (SDGs # 13). Similarly, the Maldives archipelago, one of the lowest lying and vulnerable island states in the world have projected that 200 of its nearly 1200 islands will be submerged by 2100. From an environmental perspective, the Maldives have utilized land reclamation processes to develop an adaptive response to climate change. Land reclamation is a common practice in many archipelagic regions (Wittesaele, 2021; Zhu et al., 2022). This operation consists of dredging sand from shallow lagoons to add usable land to these low-lying islands. While this engineering process facilitates new development and infrastructural projects, the process has negative environmental impacts, such as damaging coral reefs and negatively impacting plant and animal ecosystems.

Island isolation also serves as a geographical mechanism for environmental preservation. On the one hand, limited access from the outside has helped to keep these natural spaces unspoiled or at least less spoiled by the carbon and literal footprints of human beings. But the physical separation of islands to other land masses (e.g. continents) has created unique and diverse ecosystems of flora, fauna, and zoological species. This biodiversity can be found within the marine and coastal environment, where coral reefs, marine fauna, coastal forests, and mangrove swamps proliferate. This biodiversity extends on land as well. On the Union of Comoros Islands, for example, there have been more than 2000 terrestrial plants recorded. Madagascar, the planet's oldest island when it broke away from the Indian continental plate 88 million years ago, boasts one of the most biodiverse hotspots in the world. More than 80% of the island nation's plants and animals, including more than 100 different varieties of lemur, exist nowhere else in the world. Referred to as the "eighth continent," 90% of all plants on Madagascar and 85% of all animals are endemic to the island (Black et al., 2000; Tyson, 2013).

Needless to say, this impressive biodiversity is at risk due to the anthropogenic pressures of population growth, overfishing, over-tourism, and climate change. In some instances, these pressures on the environment have led to the depletion of these natural resources, thus endangering many of the species within these diverse ecosystems. Given these troubling trends, the idea of further developing sport tourism on these islands may seem counterintuitive. As such, a critical theme emerged within nearly all island case studies featured in this collection. Despite the economic promise associated with sport tourism as a niche and emerging market, the development of these particular activities and events in sensitive island states or territories highlights the dilemma faced by territorial stakeholders, particularly those in charge of governance and the Destination Management Organization (DMO), such as Destination Sustainability Management Groups (see Azores, for example). The question arises: should we promote sport tourism for its economic benefits (and corresponding SDGs) or should we prioritize the preservation of these vulnerable territories? This dilemma may lead to contradictions in actions related to such development, as those with the greatest power often enact policies and guidelines for both economic development and environmental sustainability (not to mention the preservation of social-cultural heritage and traditions inherent to these indigenous contexts).

And yet, as Aussant reminds us in his chapter on Tahiti, we must move beyond the simplistic dichotomy in which sport tourism development is viewed as *either* an assumed exploitation of precious natural resources and a corresponding degradation of the environment *or* as an assumed means of promoting an environmental awareness and consciousness that a priori leads to environmental preservation and conservation. An additional problematic assumption is that economic development must negatively impact the social-cultural and environment wealth within these island territories. Conversely, there is an inverse assumption that environmental protection regulations will automatically stymie economic growth.

This process of demystifying simple either-or propositions underlies the contribution of this book. Fortunately, we have the potential as human beings to learn from our mistakes and (hopefully) not make them again. We can also learn from others who have charted a similar course, but with a different approach and perhaps more favourable outcomes. The value of such comparative analyses is that we can articulate models or blueprints for future success. These blueprints include identifying and involving key stakeholders, while also ensuring governmental support to safeguard legislative frameworks. This requires an effective enforcement mechanism to protect the land and preserve public spaces, thereby improving the sustainability of the industry (Baldacchino & Caruana, this volume; Caruana & Sultana,

2021). For example, as Fadda notes relative to the niche boardsports market in Sardinia, "as the coastal areas in which boardsports are practiced are public property, they require public intervention. In this regard, the intervention of local government authorities is fundamental in planning such activity in the territory" (p. 136). Governments must also enhance and enforce environmental code regulations, require permits and hold accountable those in violation of these regulations. For example, French Polynesia tightened its environmental regulations in 2018, forbidding whale watching in lagoons, bays, and passes, while requiring a distance outside of these areas of 100 m and 30 m for whales and dolphins, respectively. The relative level and enforcement of such regulations varies considerably from one island to the next.

While there are examples in which economic growth is in fact pitted against environmental preservation as a zero-sum proposition (see, for example, Baldacchino & Caruana's discussion of Malta and the establishment of national nature parks to preserve public space in response to the proposed development of a golf course and university campus), there are likewise examples of environmental preservation enhancing an island's brand and sustainable economic growth.

As the cases of Singapore and the Azores attest, the promotion of environmental sustainability goals (SDGs 12, 13, 14, 15) need not compromise economic development and growth. In order to achieve this critical balance, however, adept international planning and management is required, in addition to an effective marketing and/or branding strategy.

Sustainable Development Policies: Government Plans, Charters and Initiatives

The United Nations' attention to small island developing states over the past 30 years and the international organization's promotion of the 2030 Agenda for Sustainable Development, with its articulated sustainable development goals (SDGs), has made a substantial impact on these territories and their national strategic plans. There appears to be a continuum of success among small island states when it comes to their development and implementation of sustainable development policies and plans. This has much to do with the physical characteristics of these islands (e.g. low-lying, relative distance from the mainland, etc), their geopolitical status, and current state of economic development. As such, the wealth of a nation will inform its possibilities and priorities. This is a critical factor in the relative investment and promotion of sustainable development goals and strategies. However, whether more or less developed, all island states or territories require a plan of action towards these ends. Without a plan or roadmap, it will be impossible to set a course towards genuine sustainability. Many states prioritize certain SDGs and leave out others in their national strategies. As Forestier and Kim (2020) have argued, "such cherry-picking defies the 'integrated and indivisible' nature of SDGs and could negatively impact overall progress on sustainable development globally" (p. 126).

And yet, many of these islands have specific plans relative to sustainable development, such as the Environmental Charter and subsequent 15-year National Environmental Action Plan (NEAP) in Madagascar, the Sustainable Development Plan for Rodrigues (SIDPR), Mauritius's 2017 National Biodiversity Strategy, and the Sustainable Development Plan (PADD) of the General Development Scheme (SAGE) in French Polynesia. Another example, the Emerging Comoros Plan, has helped to build a broad national consensus around the 2030 Agenda for Sustainable Development, articulating 68 distinct goals tied to the multidimensional understanding of sustainability. Thus, some of these territorial plans set optimistic but tangible key performance indicators (KPI) relative to sustainable development. As a part of the Azores Sustainability Charter, many of these environmental KPIs have been reached, including a decrease in per person greenhouse emissions from 2018 to 2021 and an increase to 35% of electricity produced from renewable sources.

The success of these indicators has led to international acclaim, recognizing nations, cities, and territories with a growing list of sustainability awards. In 2018, the Azores won first place as "Best of the Atlantic" category by Green Destinations.[3] The following year, the Azores was named the first archipelago in the world to be designated as a sustainable tourist destination by EarthCheck Sustainable Development Programme. In the Pacific, the "Green City" of Singapore was ranked 4th worldwide by the Sustainable Cities Index in 2018 for its impressive efforts of promoting sustainability as a driving force behind its national brand.

The establishment of nature reserves, parks, and sanctuaries are a major step in the direction of promoting environmental sustainability. In some instances, the entire island is considered a nature reserve or sanctuary, such as Helgoland (Germany) and Pulau Ubin (Singapore). These designations set in motion a set of policies and protocols, often developed and enforced by a strong government agency. For example, the Natural Park in Mayotte was created in 2010 under the direct authority of the Prefect and the Directorate of Food, Agriculture and Forestry. In 2021, a National Nature Reserve of the Forests of Mayotte (2800 ha) was established, in part to respond to the growing popularity of nature sports and sport tourism on the island. Similarly, the development of Marine Conservation Areas or Marine Protected Areas (MPAs) in Zanzibar seeks to combine economic opportunities while working towards biodiversity conservation and species protection. The establishment of these national parks are often heralded as an important step towards the prioritization of environmental sustainability, while remaining an attractive space to participate in an expanding market. These parks and reserves also create the spaces often needed to develop certain sport tourism activities, such as nature sports.

[3] Based in Leiden, the Netherlands, Green Destinations is a global organization that supports and measures sustainable destinations, businesses, and their communities.

Some islands see themselves as models for sustainable development while others acknowledge that sport tourism is in its infancy and look to other islands as providing a positive blueprint for their own development. For example, Madagascar looks to its neighbour in the Indian Ocean, La Réunion Island, as an inspiration for sustainable development. In the Mediterranean Sea, Malta's Marine Natural 2000 encompasses 18 sites and covers 4100 km^2 as part of its Special Areas of Conservation (SACs) and Special Protected Areas (SPAs), designated by EU directives for the protection of marine habitats and species, such as sea birds.

In the Pacific Ocean, Singapore may be an ambitious model for other islands to emulate, although the city state has been able to pursue this sustainable path as a result of its wealth and geopolitical position. Nonetheless, the plan Singapore has pursued has paid off in economic dividends and global recognition, branding itself "the ultimate sustainable destination." The island in the Pacific, 379 kilometres from Malaysia and 3742 km from mainland China, boasts 400 parks, four nature reserves, and smaller islands designed as nature sanctuaries and leisure parks. The city state's land management plan likewise includes a Park Connector Network (PCN), a 300 km island-wide network of green corridors linking the main parks and reserves, as well as a 36 km Coast-to-Coast Trail, which winds through a variety of parks and nature reserves. These integrated sustainability efforts are developed by Singapore's National Parks Board (NPB), a statutory agency under the Ministry of National Development to manage all facets of this strategic development initiative. The Board holds significant power within the national governance structure.

A Consortium of Stakeholders in the Promotion of Environmental Sustainability

The role of governmental agencies in implementing strategic plans and enforcing legislative regulations are indeed critical, but other stakeholders are important as well in the successful implementation and enforcement of environmental sustainability within these island territories. Many of the contributing authors highlighted this diverse group of stakeholders working at this nexus, including local communities, tourists, tourism operators, politicians, environmental associations, and non-governmental organizations (NGOs). These stakeholders have the potential to create public-private-community partnerships in support of local sustainable development. In many locations, governments seek to incentivize private entrepreneurship as a means to promote economic sustainability and the sustainable development goals (SDGs) articulated by the United Nations.

In juxtaposition to financial incentives to encourage sustainable businesses to invest in these islands, efforts towards environmental sustainability often come at a cost to both tourists and local residents. Numerous contributors from islands across the globe described visitor or tourist fees imposed for use of the natural spaces enjoyed as a context for sport tourism activities. The costs of these environmental taxes often varies by season, charging more during the peak season, often connected to the carrying capacity of certain locations. These luxury fees or taxes help to preserve these spaces for future use, a foundational tenet of sustainable development. This form of revenue has become commonplace, with minimal resistance from these nature sport enthusiasts. Many of these tourists value environmental protection and preservation and are likewise willing to pay to participate in activities enjoyed within these protected and well-maintained areas. Additionally, some of these fees are voluntary, asking the consumer to pay an add-on fee, appealing to travellers' growing environmental consciousness and even climate shame.[4]

In some instances, local residents are asked to support environmental protection through taxes, citing, in part, the benefits of hosting sport tourism activities or events within these areas. In the Irago Peninsula in Japan, for example, the area has hosted an annual triathlon for many years, which has been identified as an opportunity to attract a large crowd over a short period of time, thereby enhancing the local economy. When beaches along the coast were lost year after year due to erosion from the waves, the local Aichi Prefecture constructed jetties and breakwaters to mitigate against coastal erosion. In explaining the necessity for an environmental improvement project, the local government sought to convince residents to approve taxpayer funds in support of the project. The government used the tradition of this sporting event and the potential for future sport tourism activities as a means to elicit local support, despite relatively limited participation in the local race over time.

Though both locals and tourists are asked to invest in environmental sustainability connected to sport tourism, there exist tensions between these stakeholders in many of the islands highlighted in this book. These tensions often pit locals against tourists and tourist providers. In Zanzibar, for example, local seaweed farmers, mostly women, compete for shoreline areas also used for swimming, kitesurfing, and kayaking. In this same location, it has been reported that kayak providers have cut down mangrove forests to provide a more efficient outlet to the sea, negatively impacting the local environment. Jet skis may pose the greatest threat to the coastal environment, including noise pollution both above and below the sea, beach erosion, degradation of shallow zone marine life habitats, and negative impacts on marine biodiversity, such as dolphins, whales, sea turtles, and sea birds. These vessels threaten locals and tourists alike, such as farmers, fishermen, swimmers and

[4]The commercial transportation industry has begun to offer these consumer "opportunities" to support specific carriers or national rail services in an effort to contribute to reducing CO2 emissions. As one international airline carrier notes to passengers checking in online, "Flying is not sustainable. At KLM, we invest in a series of initiatives to help reduce our carbon footprint. And you can also contribute by joining our CO2 Impact Programme." These initiatives include a reforestation programme, introducing operational improvements, utilizing more fuel-efficient aircraft, and enhanced air-rail connections. Perhaps the most impactful initiative in the airline industry today is replacing fossil-based fuel with Sustainable Aviation Fuel (SAF), currently much more expensive than regular fossil fuel. Airlines ask their passengers to help foot the bill to increase the purchase of SAF.

snorkellers who share the waters not far from the shoreline. To make matters worse, there appear to be no regulatory policy frameworks in Zanzibar to monitor some of these emerging sport tourism activities.

In Tahiti, these tensions go beyond local surfers and tourists jockeying for the next set of waves. Residents assert that tourist providers are willing to damage the local environment in order to turn a profit. They cite examples of diving providers who drop anchor and destroy coral beds while whale watching boats disrupt these majestic creatures' habitat and push them away from the location. These feelings extend beyond environmental concerns, stirring up anger at these tourists' perceived disdain and contempt for the local, indigenous traditions and their land. These social-cultural feelings, prompted by what locals see as environmental damage caused by sport tourism, likewise stir up feelings of territorial dispossession and the vestiges of colonialism. One remedy for combatting these tensions is to intentionally develop sport tourism opportunities to ensure that residents and visitors alike can participate in harmony rather than in competition of space and island prioritization. Within this framework of inclusion, some island rules have sought to offer locals the first opportunity to enjoy the natural resources, such as the best breaks in the waves (Sotomayor & Barbieri, 2016).

As island territories develop in their relationship to sport tourism, many of these mature destinations, as well as emerging markets, seek to diversify their range of these activities. In some instances, this has meant a shift from a focus on a niche sport tourism market to one that offers opportunities across all parts of the island territory and at varying times of the year. This strategy has been cited as a means to curb the challenges with seasonality in these island territories. The diversification of sport tourism supply has led to the promotion of both "green" and "blue" sports, with a corresponding focus on activities on the land and along the coastline, respectively. Green sports, tied to the concept of a green economy (Barbier, 2011; Kasztelan, 2017; Mikhno et al., 2021), include hiking, biking, and climbing, while blue sports, connected to the concept of the blue economy (Lee et al., 2020; Smith-Godfrey, 2016; Voyer et al., 2018), include surfing, swimming, snorkelling, diving, and kayaking, among others.

This has meant that islands offer more than simply "sun, sea and sand" tourism, a lesson learned long ago in Malta. This small archipelago in the Mediterranean has very few sandy beaches, forcing it to develop activities off the rocky sections of coastline, where scuba diving and snorkelling are popular among both locals and tourists alike. But so too have non-beach-based activities emerged in more rural areas of the island, ranging from hiking, running, and cycling to climbing, abseiling, and geocaching. This supports the assertion made by Marrero Rodrigues and Santana Talavera (2018), who highlight the historical process of "complexing the sun and beach through the incorporation of a diverse set of natural and cultural activities and products" (p. 291).

Many islands reported strong connections between local stakeholders and non-governmental organizations. For example, several islands reported the presence of well trained and effective local NGOs focused on both marine and terrestrial conservation. In Rodrigues and Mauritius, for example, these NGOs have made a significant contribution to natural reserves management and conservation. Well-established collaborations between local authorities, tourism operators, and these NGOs help to monitor and remedy the degradation of marine biodiversity and fragile ecosystems as a result of diving and whale watching. These environmental groups include the Marine Megafauna Conservation Organisation (MMCO), Reef Conservation in Mauritius and Shoals Rodrigues, to name but a few. The diversification of sport tourism activities has also highlighted the unique and specific conservation efforts based on the natural context within which these activities occur. The number of these environmental NGOs may likewise grow as the islands pursue a strategy of sport tourism diversification. Their existence and proliferation should not allow local and national governmental agencies to abdicate their responsibilities towards environmental regulations and enforcement within these island ecoterritories. Additionally, these government agencies must balance economic growth with both social and environmental sustainability efforts, seeking methods to integrate the multiple dimensions of sustainability.

Economic Perspectives in Island Comparison

From an economic point of view, the case studies in this book clearly demonstrate that sport tourism can help to respond to SDGs 1, eradicating poverty and creating wealth in a given territory; SDGs 2, creating jobs that allow for remuneration and thus obtaining purchasing power; SDGs 3, enabling all active participants, local or otherwise, to improve their health through physical activity; SDGs 4, developing the necessary training that must be put in place in order to manage this new form of tourism, sport tourism, and SDGs 9, creating the corresponding island infrastructure for the promotion of all types and dimensions of sport tourism. While noted in the previous section, SDGs 14 and 15, protection of aquatic and terrestrial flora and fauna, must be included, as it is clear that the development of sport tourism is conceivable only by pursuing a policy of environmental preservation in parallel with economic development. One must therefore adopt a socio-economic approach, underscoring a commitment to a critical ecopedagogy. This approach allows for a more harmonious process of sustainable development rather than pursuing economic growth at the detriment of these other dimensions.

Sport Tourism as an Economic Driver for Island Development

The islands studied within this collection can be divided into three broad groups: (1) those which have integrated sport tourism as a factor of development for a substantial period of time, sometimes several decades (Malta, Ruegen, Barbados, Kalymnos, etc.); (2) those which perceive the interest of developing this niche tourism segment and have begun the process (Azores, Reunion, Sardinia, Singapore, etc.); and (3) those, few in number, for which sport tourism is not yet included in their development projects and strategies (Madagascar, Kiribati).

The difficulty encountered by most authors in finding reliable statistical or/and local data should be stressed from the outset. Because of a lack of institutional recognition, sport tourism is not one of the documented types of tourism listed by the World Tourism Organization. Additionally, because of the multiple and diverse array of definitions describing these activities, sport tourism data are drowned out among other sports or tourism data. However, some small islands have little data altogether, especially regarding sport. The data is often produced at the international level by large organizations, which offer only approximate numbers since they do not have a reliable or uniform mechanism for collecting these data at the local level.

It appears from the case studies carried out in this book that, as with sport (cf. Sobry, 2003), sport tourism develops only under certain conditions, the first being that the basic needs of the local population are met. In an area where hunger and clean drinking water are recurring problems, local authorities cannot afford to invest in recreational facilities. They devote their limited resources to meeting local demand for sport, which is generally modest.

The use of sport tourism as a means of development is essentially based on international tourism and it can be analysed as part of a large value chain, one factor inducing another in a spiral which, under ideal conditions, proves to be virtuous. A positive spiral phenomenon develops, combining the Keynesian multiplier and the accelerator effects, respectively.

While in most cases it is essential to seek to satisfy the basic needs of local people before turning to the international market, in the case of sport tourism, we have a paradox: sport tourism is initially focused towards an international market. In terms of tourism in general or sport tourism specifically, the needs of the locals are most often different from the expectations of international visitors. Local spending on sporting activities geared to an outside market is non-existent or very limited. In order to attract international tourists, local decision-makers, private and public, value the local natural advantages, the sea and the sand in the first place, but also the topography of the island. Local reception areas, homestay rooms, modest hotels are sufficient for a first generation of "explorer" tourists, who are looking for "undiscovered" sites or "hotspots," generally unknown to the tourism industry. The development of sport tourism as a sustainable economic activity comes only in the second phase of island development.

In most cases, the exploitation of one of the island's natural resources is initiated by a private, often foreign, actor, usually copied by others who in turn try to take advantage of the niche. The virtuous circle takes place when the local authorities take over. A double collaboration often develops thereafter between private actors (coopetition, a conceptual ideal existing between cooperation and competition (Pickel-Chevalier, 2015) and between private and public actors. On several occasions, this double synergy is either stressed or regretted because it does not exist or is insufficient. In Malta, "Government support, stakeholder consultation, site and activity management via 'public–private' partnerships, and stringent accreditation and licensing, will help lift this niche industry to the respect and significance that it deserves." (p. 135). Conversely, in Sardinia, it is deplored that "in terms of networking with other actors, frequent collaborations were observed with other tourism businesses, but fewer with public institutions."

The role of the public sector is varied and essential. It must facilitate, while supervising, private investment, equipment and equipment rental companies, guides, hosting companies, etc. The purpose of this framework is not to allow structures or activities to destroy or damage the environment that will attract tourists. It is observed that sport tourists are sensitive to the environment, probably more than traditional tourists. The framework concerning the environment, often fragile in and around the islands, can go further by allowing public decision-makers to set up and develop protected areas, national parks, marine wildlife protection areas, protection of endangered species, etc. Sport tourism can make it possible to go even further by promoting the rehabilitation of certain places such as, for example, the resettlement of corals in areas that have suffered. In most cases, the development of sport tourism is an opportunity for local authorities to implement environmental protection policies but also to preserve certain social and cultural elements. It is about preserving global authenticity in a fragile environment. It should be noted that, in rare cases, this opportunity is not seized and that, while everything seems in place for tourists to be satisfied, some elements are missing, such as correct topographical maps or the lack of maintenance of hiking trails, an essential component of tourism promotion. This aspect is to be considered from a socio-economic point of view since investments, mostly public, must be secured. Political decisions must be made in order to preserve and organize the context in which sport tourism as a source of income can be sustainably developed.

The public sector plays a critical role in developing structures to meet the needs of tourists. These structures can be quite general, such as the creation of airports and the development of modernized transportation infrastructure, including marinas, roads, public transport, etc. The creation and development of these infrastructures must be linked to the regulation of the number of tourists who wish to visit these territories. There are certain infrastructures specific to sport tourism that are not supported by the private sector, such as the maintenance of hiking trails, certain heavy structures such as sailing bases, etc. Heavy equipment is rarely geared towards competition, at least high-level competition. Indeed, few cases are studied that try to develop sport tourism by relying solely on international competitions (Sardinia, Barbados, Sri Lanka, Irago, Singapore).

Sport tourists have several peculiarities. According to the WTO, these tourists spend more than the majority of tourists. This induces important revenues for local actors, including the public sector through taxes. They need equipment that they rent or buy on site and utilize instructors or guides. In addition, sport tourists often travel with friends or family, who may seek out other activities. These secondary practices, sports or not, also require structures and staff capable of accommodating this demand. We must remember that tourism is like a chain; its quality depends on its weakest link. The development of sport tourism therefore induces that of other sectors of activity. This larger enterprise requires training individuals and businesses to meet the multiple new demands, such as instructors, security and rescue personnel, travel agencies, specialized stores in sports equipment, hotels and catering services, etc. The establishment of training centres is still primarily a public sector role.

Some islands have chosen to organize sporting events in order to attract tourists. The analysis made on Barbados in the Caribbean Sea is particularly interesting regarding the expected and assumed effects and the observable outcomes. The authors point out that "hosting non-national sporting events could be used to boost economic growth in small developing countries like Barbados." Such a choice "requires effective planning and management of the niche" and "It may be prudent to target key sporting events that can attract tourists, where the burden of infrastructure spending is minimal and significant private investment." They argue that these events "can increase visitor spending, public spending and employment and generate revenue." However, they conclude that "our analysis found no measurable evidence that sports tourism and hosting non-domestic sporting events have enhanced economic growth in small developing nations like Barbados." This conclusion could be applied to all island territories organizing sporting events, mostly non-domestic sporting events, with the aim of attracting tourists. There are few studies analysing the impact of a non-domestic sporting event on the economic activity of a small island. We can assume a positive direct effect, if some conditions are met, as the cited authors point out. We can also see the search for indirect and induced effects by using the event organized as a communication vector to attract other tourists to this particular site. This would allow many tourist resorts in early summer to launch their season and thus draw attention to their businesses connected to an upcoming sport tourism event.

On several occasions, coordination between the public and private sectors was stressed by the authors, as was the need for long-term plans or strategies for successful implementation. On the contrary, many authors also deplored their lack of strategic planning within these island contexts. These multi-year plans, put in place by governments, are designed to set targets, processes, and regulations. Sometimes it is the private actors who self-organize. If this coordination and planning is not organized, however, conflicts arise between key stakeholders, negatively impacting the potential socio-economic development of these emerging markets.

Necessary Conditions for the Development of a Positive Socio-Economic Spiral or Multiplier Effect

In order for a positive spiral to take place, it appears that the origin of capital and skilled personnel are essential. Although this point is only rarely raised, it seems that the financing of investments must originate locally or through the aid granted by international organizations for a real socio-economic development to appear. If the investors are essentially private and foreign, only offering low-skilled and low paying positions to the local population, much of the profits will leave the territory. If, on the contrary, investments are made by indigenous people, whether or not using the aid of international structures, the development process may be slower, but the local authorities retain control of these domestic investments with all of the corresponding induced effects.

As the authors of the study on Mauritius note, "Sport tourism is a major contributor to economic growth and economic development as it allows the emergence of new entrepreneurs in various fields and the distribution of income across the country and the population." To support this assertion, it is necessary to analyse the various elements involved in this development.

We distinguish between the direct effects of sport tourism on the development of the small island territories and the indirect and induced effects of the jobs created, as well as new income distributed through direct and indirect effects.

A descriptive approach to the Keynesian multiplier and accelerator helps to understand the importance of combining the different factors discussed. The multiplier is a relationship between a change in investment in a territory and the change in national income. Both have an impact on economic activity and therefore on employment. The volume of initial investment may increase either because entrepreneurs' expectations are optimistic or because the State has increased its expenditure. While expectations are optimistic, entrepreneurs are inclined to increase their investments, which translates into orders for capital goods from other companies specializing in these types of products. To meet the demand, the latter are led to hire additional staff, hence distributing new revenues. Some of that income is spent on consumption, thereby increasing the demand for consumer goods, and so on. Demand is a determining factor in the investment, which triggers the accelerator effect. In this case, a change in the capital stock appears to be the result of a change in demand, correlated with the change in national income. A change in national income on investment is therefore affected through consumption.

To illustrate our point, and starting from the assumption that investments have been made in a territory to welcome and attract tourists, we can observe the following mechanism (Table 29.1).

The money spent by tourists comes mainly from outside the observed territory. This injection of money into the economy will circulate from one sector to another and will thereby generate additional activity and additional revenues and taxes, a cycle favourable to new investments. At each iteration, the amount re-injected into the system decreases (taxes, imports, savings), acknowledging the "leaks" of Kahn

Table 29.1 How tourist spending flows through the economy

Tourists spend for:	Second round of expenditures:	Ultimate beneficiaries (a partial list):
Lodging Food Beverages Entertainment Clothing Gifts and souvenirs Personal care, medicines, cosmetics Photography Recreation Tours, sightseeing, guides and local transportation Miscellaneous	Wages and salaries Tips and gratuities Payroll taxes Commissions Music and entertainment Administrative and general expenses Professional services Purchase of food and beverage supplies Purchase of goods for resale Purchase of materials and supplies Repairs and maintenance Advertising, promotion and publicity Utilities Transportation Licenses Insurance premiums Rental of facilities and equipment Interest and principal payments of borrowed funds Income and other taxes Replacement of capital assets Return to government	Accountants Appliance repairpersons Architects Artisans and craftspeople Arts and crafts suppliers Athletes Attorneys Auto servicepersons Bakers Bank workers Butchers Carpenters Cashiers Charities Cinema and video makers/distributors Clerks Clothing manufacturers Cooks Cultural organizations Dairies Dentists Department store owners/workers Doctors Education providers Electricians Engineers Farmers Fisherpersons Freight forwarders Furniture makers Gardeners Gift shop operators Government workers Grocers Health care providers Housekeeping staff Insurance workers Laundry service providers Manufacturing workers Office equipment suppliers Painters Petrol stations Plumbers Porters Printers and publishers Recreation equipment, sales/rental Resort owners, operators and workers Restaurant owners, operators Road maintenance workers Sign makers Transportation workers Utilities, providers of and repairpersons Waiters and waitresses Wholesale suppliers

Leakage: When the private or public sector purchases goods or services from sources outside the community, that money is no longer subject to the multiplier effect and the economic benefits leak out of the community.

Source: Adapted from World Tourism Organisation

(1931); but when the effects of the initial injection are completed, the impact on the observed economy is several times greater than the initial injection, hence the concept of multiplier. We thus have a virtuous circle and the phenomenon of economic growth is initiated.

It is important to emphasize how this mechanism impacts employment. On the one hand, due to the increase in public and private spending, whether it is investments or spending by tourists, the number of jobs increases, fuelling the need for skills development among workers. In the case of territories with endemic underemployment, in which a new niche of activity is established, it can only be net job creation.

On the other hand, in the field of tourism the factors required to welcome tourists under the best possible conditions are significant and multiple. To welcome tourists with unqualified staff quickly disqualifies the destination concerned. To achieve a satisfactory level of qualification for staff from local populations, it is essential to set up training protocols. We thus have an effect on both quantitative and qualitative local employment. Such an approach is conceivable only in the long term, with a planning of the multiple steps to be taken by setting clear objectives with adequate means, supported and led by a stable government.

It is clear from these few elements how public and private expenditure must be articulated to create direct, indirect, and induced effects. We also better understand how the development of sport tourism can respond to SDGs 1, 2, 3, and 4, as well as the implementation of SDGs 9, 14, and 15, depending on the political will implemented through long-term development plans.

Since data on the impact of sport tourism on small islands are currently limited or non-existent, the question of territorial analysis must be addressed in an effort to assess this impact from a socio-economic point of view. Adopting a tool proposed by C. Sobry et al. (2016), the aim is to apply to sport tourism the same approach as that used for the definition of the so-called Vilnius system of categorization. This basic tool helps to promote sport tourism research and to improve sport tourism development. It is therefore necessary to create a series of indicators based on the definition of sport tourism.

This definition offers a key element in the organization of these indicators and thus of the results obtained. The definition adopted by IRNIST (Sobry et al., 2016) was used to develop this assessment tool to assess the socio-economic characteristics of sport tourism activities. According to the direct and indirect socio-economic activities induced by sport tourism, the socio-economic indicators of sport are divided into central and extensive parts.

The central or core represents the direct role of sport tourism from a socio-economic perspective, while the extension identifies its indirect role (Fig. 29.1). As long as sport tourism activities exist, they must develop alongside corresponding social and economic activities. These activities may either be sport tourism activities themselves, such as ticket sales to attend a sporting event, or they may be part of other economic activities driven by sport tourism, for example the hosting needs generated by Olympic tourism.

If it took many years for sport to be perceived not simply as a consumer of wealth but as an economic activity in its own right, creating jobs and wealth, we can hope that this awareness will be more readily understood for sport tourism. As the case studies provided in this book clearly demonstrate, sustainable economic development must coincide with a sensitivity to the local population and a commitment to the preservation and rehabilitation of the environment.

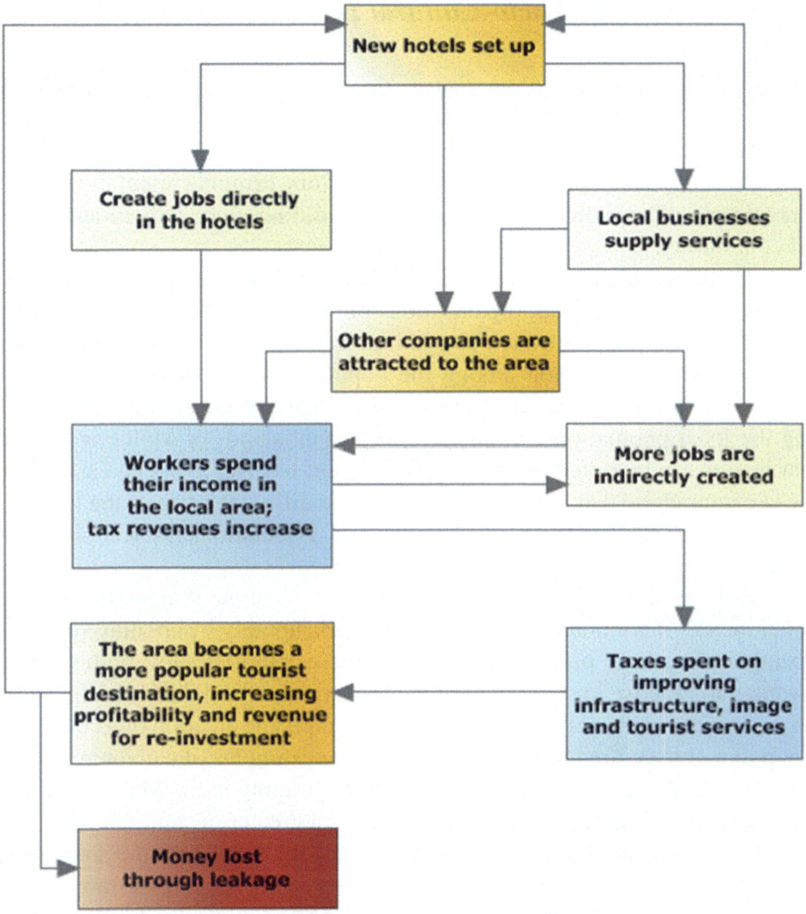

Fig. 29.1 The sport tourism multiplier effect. (Source: Barcelona Field Studies Center. Geographyfieldwork.com)

Socio-Cultural Perspectives in Comparison

Chapters in the book have addressed several contributions of sport tourism to socio-cultural perspectives, focusing in particular on SDGs 3—Ensure healthy lives and promote well-being for all at all ages; SDGs 5—Achieve gender equality and empower all women and girls to access employment and income; SDGs 11—Make cities and human settlements inclusive, safe, resilient, and sustainable; SDGs 16—Promote peaceful and inclusive societies, ensure access to justice for all, and build inclusive institutions; and SDGs 17—Strengthen the means of implementation and reinvigorate the global partnership for sustainable development. Despite the contribution of sport tourism to these particular SDGs, there remain significant differences in the degree to which the SDGs are being achieved in these various island territories.

Multiple Roles in the Socio-Cultural Balance of Populations

A first observation relevant to the contribution of sport tourism to the SDGs is the extent to which it contributes to the socio-cultural balance of island territories. The practice of sport tourism plays a role in the socio-cultural balance of the islands because it is at the same time unifying for local actors, bringing populations together and integrating marginalized groups, but also building a collective identity of the island community.

In the Caribbean island of Barbados, government ministries, hotel providers, and private and non-governmental organizations on the islands develop national water- and land-based tournaments such as cycling, trail running, sailing, and fishing to foster socio-cultural cohesion among the populations. The results show that the organization of tourist and sporting events encourages local fraternity and solidarity among the local populations. In the Mediterranean islands of Malta and Sardinia, similar results can be stated due to the role of sport tourism as social catalyst in the bonds of camaraderie that are formed between the different strata of the local population through contact with the stakeholders of associations, NGOs, and service companies.

The results of the analysis of the chapters also illustrate that sport tourism is a potential means of communication for local actors, notably by creating social links. For example, for a large proportion of islands, these activities allow both young men and women to practise in synergy in order to overcome the difficulties linked to the rough or extreme course of these sports. This prompts the creation of a collaborative or group dynamic centred around an activity. The examples of canyoning in Cilaos in Réunion, climbing in Kalymnos, or parachute jumping in the Maldives allow for group cohesion and motivate individuals seeking a greater meaning in life.

As can be observed in Rodrigues, such relationships between the socio-cultural balance and the integrated role of local actors, show that private initiatives were at the origin of the development of diving and sailing activities, due to the weak presence of public authorities in the tourism sector at that time. The results show that joint action between private partners of different activities has allowed the structuring of these activities in Mourouk Ebony. Mr. Percy Gontran, CEO of Mourouk Ebony, and Sydney Caré, CEO of Cotton Bay, were decisive in managing the link with the external diving instructors from France and South Africa. This shared promotional work helped attract international clients. This dynamic has been progressively appropriated by the public authorities by forcing the sharing of benefits for the locals, by creating a real partnership and social dynamic around these activities. This structuring of sport tourism products has helped to reduce inequalities and the inclusion of local populations. Roads and accommodation structures, small collectives, gites and hotels, have improved the general accessibility of the island to sport tourists.

Sport tourism also brings people together, showing the importance of their contribution to the achievement of relevant SDGs targets. Results are visible with the gathering today of young people from various neighbourhoods. The activities allow young people who are not used to meeting each other to build relationships. They encourage exchanges between young and old, between men and women, indigenous populations and tourists, and even between young people from different islands and archipelagos. As a place of cultural integration and social cohesion, these practices play a role in connecting the lives of these young people. Similar sentiments were expressed by the role that sport tourism plays in bringing people together in the Canary Islands. Indeed, foreign practitioners in maritime activities have become sport tourism entrepreneurs between tour operators and their networks in the countries of origin. In Mauritius, sport tourism activities, often run by local councils, often provide opportunities for dialogue with people from different cultures. These encounters are a source of education, social exchange, and mutual understanding between locals and tourism stakeholders.

Other results demonstrate that these activities may potentially integrate specific groups. In Réunion Island, sport tourism activities contribute to the social enhancement of young and marginalized groups. In this context, these activities play an important cultural role by being considered as an emblematic activity of community identification. These young people, most of whom come from modest backgrounds, come to share the values of effort by symbolically appropriating the Réunionese heritage in the national park. The immersion in nature and the search for harmony with the elements that these sports require, make them aware of the beauty and richness of the places in their native National Park.

Additional results have shown that sport tourism can contribute to the construction of a collective identity for indigenous populations. Faced with mass unemployment, island populations are confronted in their quest for identity with the problem of the dual belonging between the norms of Creole tradition (endogenous) and those of Western modernity (exogenous). In this context, the practice of nature sports plays a significant cultural role. These activities can participate in the process of identity building, by conveying, in their diversity, plural values in which young islanders find themselves. Because they can generate original relationships with oneself, others, but also with nature across time and space, sport tourism can be viewed as a positive vector at the socio-cultural level.

Contributions to Health Education, Healthy Living and Life Values

At the intersection of sport tourism and the SDGs, it has been shown that the practice of sport tourism can be beneficial to health education, healthy living, as well as socializing positive life values (SDGs 3). Indeed, sport tourism has a deterrent effect on young people; through their interest in sport, they can avoid falling prey to social ills such as alcoholism, gambling, and drugs. Land, water, and air activities organized in the islands, such as the Dodo Ferney Trail in Mauritius, the Grand-Raid in Réunion, the Singapore Ultra Trail, the Lanzarote International Regatta, the Madeira

Island Ultra Trail, the O'Neil Deep Blue Contest in the Maldives, attract young people to these activities. Furthermore, the results from Madagascar reveal that the extent of sport tourism is part of the employment generated from the often informal sector, which allows the vast majority of Malagasy people, without education and training, to develop new skills and obtain a subsistence income.

In some islands, sport tourism opportunities have targeted the integration of young people and island women. In Kalymnos, for example, local people can work as instructors and guides in the climbing and diving sectors. The income generated by these activities allows them to be paid more than the average salary in the country. Professional integrity and autonomy is a pillar of this SDGs. In La Palma (in the Canary Islands), public actors have supported social inclusion through economic employment activity as a priority by targeting activities such as maintenance, cleaning of the pathway network, packaging, repair, and fire watch for young islanders. In Rodrigues, young people and women are involved in sport tourism activities. They are a means of providing supplementary income activities. There are several trail and regatta events which are very popular among the population. The success of the autonomy of young people and women in Rodrigues can be explained mainly by the role of the operators of sport tourism activities, who mainly work in rural guesthouses and bed-and-breakfasts to welcome tourists. This situation is accentuated by the fact that the majority of operators in the sector are local. The findings of the Zanzibar analysis show that these activities create jobs for young people living in the marine conservation areas (MCAs), where most of these tourism and sports activities take place. The organized events of Jambiani Beach soccer, the Pemba Tourist Bonanza, or the Ngalawa race ensure the empowerment of communities along the islands' coastline.

Thus, the creation of local island enterprises has proven successful in achieving youth and women's empowerment. In Mauritius, contributions to the SDGs are reinforced by the commitment of the local Ministry of Youth Empowerment, Sports and Recreation to promote the destination "Mauritius" as an "active sports island." In Réunion, the results show that sport tourism activities are a vector for environmental education, social cohesion, and health, both in terms of individual and collective responsibility. The immersion in nature and the search for harmony with the elements that these sports require make people aware of the beauty and richness of these native places. The practice of sport tourism encourages, firstly, an active and sensitive discovery of the heritage, and secondly, an appreciation and defence of this island resource. It thus appears to be a stimulant to the construction of an eco-citizenship as part of an expanding global ecopedagogy. The role played by local institutions and associations, such as the Office National des Forêts, the Commissariat à l'Aménagement des Hauts, the Maison de la Montagne et de la Mer and the Direction Régionale de l'Environnement, in raising awareness and educating people about the environment is exemplary in this respect. Sport tourism can also help achieve educational and personal development objectives (camaraderie, tolerance, respect...) for young people. The results in Fuerteventura (Canary Islands) show that water sports have allowed the development of small local businesses, which have created jobs in management and services, thus allowing the development of tolerance and respect between locals and foreign visitors. The activities have promoted health and well-being among island communities while strengthening social cohesion.

These significant results reflect a comprehensive consideration of several SDGs, but there remain limitations to the integration of women in sport tourism activities (SDGs 5). In particular, there is no explicit reference made to training management and microfinancing opportunities for women's service delivery activities. The results also show that the wealth created for the local population is sometimes limited, with problems of technical expertise, leading to mostly precarious and seasonal employment. These aspects favour certain forms of social exclusion, sometimes petty crime, especially for young people who have remained on the fringe of these opportunities.

The role of sport tourism in achieving the varied SDGs and the governance of the actors are intimately linked, as their achievement requires coordination, collaboration, and shared responsibility between all of the stakeholders in these sport tourism activities.

The Challenges of Local Governance in Comparison

The results of this comparative study demonstrate the essential role of stakeholders in setting island territories in motion to achieve the SDGs. Decision-makers play a vital role in the territorial animation of the ecosystem of sport tourism actors. These actors stimulate initiatives and encourage the construction of fruitful partnerships. However, most contributors to this book emphasized the importance of improving the role of stakeholder governance as a key element in achieving the SDGs.

In general, the conclusions of this study illustrate a real general awareness of the roles and responsibilities of key stakeholders in the coordination and planning of the strategy for the promotion of the territories. Several results underscore the need for enhancing the tourism image of destinations in order to achieve the SDGs, and to define a robust feature of future programming and strategic intent for sport tourism. They also contribute to the improvement of the living environment and favour the creation of a sustainable supply for the practice of local populations. The structure of this supply is very uneven, depending on the island context. Much remains to be done to improve the reception of "sport" tourists and their companions through quality facilities and services. The poor coordination between state services, local authorities, private groups, etc. is a real obstacle to the operationality of projects. For example, the duration of requests for work to improve the accessibility of paths serving the sites and practice areas within the protected areas of the national park in Réunion, in the reserves of Mayotte or French Polynesia have taken several years to achieve due to regulations, decrees, and laws.

In concert, local actors develop the attractiveness of their island territories for sport tourists on different scales (e.g. coast, mountain, island, archipelago). They support investment policies by mobilizing stakeholders for the sustainable sport tourism project, depending on the uniqueness of the specific location. Faced with competition from other island destinations at the end of the 1990s, the conclusions of the case study of the Canary Islands show that a primary focus on golf and sailing must be repositioned around a more diverse array of sport tourism activities, such as trail riding, mountain biking, surfing, rallies, etc.

Findings from Madagascar show that hiking has become a priority for initiatives to structure the island's highlands, such as the Ultra-Trail des Ô Plateaux (UTOP), which symbolizes the efforts made by private and public actors. Moreover, hiking represents an accessible and therefore inexpensive sport, more easily accessible to the population. This activity has created a collective dynamic among decision-makers and has provided opportunities for encounters and collaborations between isolated villagers and sport tourists from very different cultural, economic and social backgrounds.

Moreover, it is now common for local elected representatives to participate in the construction of the sport and tourist opportunities established on the islands. In Tahiti, the government of French Polynesia has supported sport tourism organizations by opening training courses to professionalize the sector, as shown by the creation of a Polynesian diving certificate or the communication campaigns of the Tahiti Tourism Economic Interest Group. Other examples include the growing market of canyoning on the islands of the Azores. Partnerships have enabled the territory to be positioned as a sport tourism destination for discovery or challenge. The sustainable development policies implemented in the territory has been a collective effort between numerous decision-makers, including the local government, environmentalists (structure for the protection of the marine environment of the North-East Atlantic, Natura 2000 network), tourism organizers (world council for sustainable tourism, structure for the management of the sustainability of tourist destinations), NGOs (Azores Trail Run), and the Desnível adventure sports association. These multiple stakeholders have collaborated to develop sites, train local technicians, and communicate sport tourism opportunities broadly.

Data indicates that the majority of the case studies involved multiple stakeholders in the territory, although there were significant regional island variations. In Barbados, for example, the authors acknowledge the inadequacy of the coordination of actors in the planning and efficient management of the sport tourism sector so as to achieve a return on investment. These organizational difficulties include a lack of institutional development, an inefficient communication network, as well as the ambiguous positioning of the island in terms of its regional or global brand as a potential tourist destination. Conversely, in Kalymnos, the municipal council has brought together associations and volunteers to secure and organize sporting events with strong tourist interest in both climbing and diving, two well-developed niche markets.

Initiatives to Organize Stakeholder Cooperation to Be Improved Over the Long Term

Although there are still marked island differences, the case studies from around the world indicate that initiatives have improved cooperation on sport tourism between the islands. In Malta, successful results have been developed by actors in the diving sector. Indeed, the structuring of a niche market has been relatively successful through the support of PADI clubs and associations, in conjunction with the Malta Tourism Authority and the NGO Nature Trust. Public–private partnerships have enabled this niche industry to contribute to the sustainable development dynamic.

The qualitative data obtained from the analysis of governance partnerships shows that they could help to improve the development of sport tourism in several areas. For example, in Sardinia, the ambivalence of the regional government to promote sport tourism events has led to it not initiating a strategic planning approach. Private stakeholders would like to see greater ambition in promoting Sardinia as a winter sport tourism destination, with the implementation of strategies for deseasonality, economic growth and respect for the environment. These stakeholders' demands seek to increase the services to board sports enthusiasts in order to strengthen an emerging tourism segment in an environmentally sustainable way.

Findings from the analysis of the chapters also indicate that strategic proposals and action plans for sustainable development can be formulated for the structuring of sport tourism production and consumption. The results of the case studies highlight and reinforce the potential contribution of sport tourism to the SDGs. At the level of the International Labour Organization (ILO), the results show that in the post-Covid management of SIDS, sport tourism presents a sector with new opportunities to approach sustainable development. The capacity of stakeholders to innovate represents a dynamic cross-fertilization around SIDS with the organization of an annual forum to generate synergies for the future of their economies and social policies.

In the Azores, despite the momentum, several avenues are being considered to continue the growth of the sector. Commitment to training, awareness-raising in school sports, better knowledge of demand, and the establishment of carrying capacities for the most sensitive areas have been agreed upon. In the Canaries, actors have clearly positioned themselves for a governance as an adaptation to international competition marked by the diversification of the market, whereby sport tourism provides a niche segment in the same way as gastronomy, heritage culture, etc.

These results mask some island variations. For example, in Comoros, findings demonstrate that the Comoros 2030 Emerging Plan is sustainable. This plan details the key ideas and large-scale projects that will stimulate the structural transformation of the country's economy in favour of the development of tourism and sport tourism. It is a pragmatic tool for mobilizing the archipelago to achieve its development while ensuring its regional integration and contributing to the well-being of its population. It integrates the achievement of the SDGs by 2030 and the African Union's Agenda 2063 (African Union, 2015).

In Zanzibar, the results show the shortcomings of stakeholder governance that can have a negative impact on the attractiveness of the islands. These activities are not well planned and coordinated to maximize profits, user experience, and sustainability issues. Problems of cohabitation of the space with different users cause conflicts between jet ski pilots, swimmers, divers, and fishermen. The study found that there was no policy framework regulating the use of jet skis, nor was there effective institutional coordination between the main stakeholders. In Kiribati, findings identified that the governance of sport tourism reflects the level of development of the island. The condition of structuring this sector will depend on the capacity of the actors to provide the islands of the archipelago with the primary needs in terms of functional facilities such as roads, airports, public facilities (e.g. island infrastructure).

In French Polynesia, the added value of sport in the tourism supply is currently low and the expected profit margins are almost non-existent. Achieving the objectives of sustainable tourism remains to be seen. It requires relevant choices and a set of voluntary and targeted measures on the part of stakeholders. These measures must include voluntary actions in the framework of public and private policies that support the SDGs to be achieved.

The results from this comparative study highlight several promising practices or implications on the possibilities and capacities of local and global actors in implementing sport tourism as a vehicle to achieve the SDGs. These results may help similar destinations to develop more sustainable activities in line with the SDGs. Sport tourism contributes to influencing the attractiveness of island destinations with variable impact gradients due to the level of touristicity of the islands or the maturity of the sustainable tourism policies of the territories. However, these results must be qualified with regard to the transversal limits to the global tourism development of the islands. These include, for example, the cost of air transport and accommodation, strong tourism competition in the island regions, absence of a real tourism policy for the islands, lack of tourism information and promotion, heterogeneity of the island territories, etc.

Faced with their natural and anthropogenic vulnerabilities, small island states serve as laboratories to experiment with innovative solutions to achieve the stated SDGs. These ecoterritories must develop sport tourism practices utilizing an ecopedagogical approach, committed to a broad-based global sustainability platform.

An Assessment Tool for the Contribution of Sport Tourism to the Global SDGs

This book concludes by proposing a series of reflections and critical analyses on the contribution of sport tourism to potential models of sustainable development applied to the world's island territories. These models must acknowledge the diversity of geopolitical, socio-economic, and environmental contexts of the islands where tourism takes shape.

Based on the comparative results of the case studies included in this collection, an analytical and assessment tool (see Fig. 29.2 below) was developed in order to propose a novel methodology for studying the positive and negative impacts of sport

Fig. 29.2 An analytical and assessment tool to understand the contribution of sport tourism as a means of sustainable development in the islands

tourism on sustainable development. The model defines the multiple and integrated dimensions of sustainability, inclusive of the environmental, economic, social, and political (governance structures) inherent to such analyses. The reading of this analytical heuristic must also be based on the presentation of the insular contexts, specific to each island or archipelago, while also being sensitive to the regional and international pressures placed on the national and local contexts of these island ecoterritories.

This analytical model responds to the need to develop key performance indicators for measuring sport policies mentioned in several international framework documents for sustainable development, notably the Kazan Action Plan (UNESCO, 2017), the United Nations Action Plan on Sport for Development and Peace (UN, 2016), and in regionalized studies (Lindsey & Chapman, 2017; University of the South Pacific, 2018; The Commonwealth, 2019). Thus, this tool has several purposes.

First, this analytical heuristic allows for an ecosystem-based assessment of the effects of sport tourism on islands and for appropriate measures to be taken to maximize benefits and minimize negative impacts. Secondly, islands, because of their environmental fragility and their often high dependence on tourism, are particularly concerned by the implementation of the SDGs. This framework ensures that island sport tourism is aligned with a long-term sustainable development vision. Thirdly, this tool provides a structured framework for assessing sport tourism in islands. It can be used by policymakers, destination managers, sport organizations, and local stakeholders to help identify potential areas for improvement, assess risks and opportunities, monitor performance, and make informed decisions in real time. Ultimately, this framework promotes sustainability by assessing existing practices, identifying best practices, and encouraging the adoption of intentional ecoterritorial measures.

In order to propose the most operational tool possible, the analytical rubric suggests the salient criteria to be filled in at the scale of the island territories. This tool was built around the axes of sustainable development by positing four key questions (see Table 29.2 below):

Table 29.2 Island or archipelago under analysis (available data from 2019)

Island or archipelago under analysis (name)	
Capital (city name)	
Geographical size of the island [if an island archipelago indicates the size of the island where the capital is situated/if about an island indicate the size of the island] (number in km^2)	
Population (number in millions)	
Share of population under 20 years old (%)	
Share of population above 65 years old (%)	
Unemployment rate (%)	
Unemployment rate of population under 20 years old (%)	
Average annual income (US Dollar, per capita)	
Sport participation index (%)	

As depicted in the upper left quadrant regarding the political dimension, how does the structure of governance include, rather than exclude, local stakeholder perspectives and priorities in the promotion of sport tourism as a vehicle for sustainable development? (e.g., public-private-community partnerships).

As illustrated in the upper right quadrant regarding the environmental dimension, does sport tourism foster a critical ecopedagogy and contribute to environmental preservation, conservation, and rehabilitation within island ecoterritories?

In the lower left quadrant of the model, regarding the economic dimension, to what extent does sport tourism promote sustainable economic growth and address the socio-economic disparities of island territories?

Finally, in the lower right quadrant regarding the social dimension, does sport tourism contribute to the socio-cultural health, well-being, and preservation of indigenous island territories?

In order to most effectively answer these key questions, the analysis of sport tourism in the islands through the prism of the SDGs can be facilitated by the use of relevant indicators. Thus, each of the sustainable development dimensions proposes a series of indicators with a double objective, the first being to highlight the island potential of sport tourism development to achieve the SDGs, the second being to monitor the actual evolution of these changes as well as to measure the commitment of stakeholders to these articulated goals.

In conclusion, these indicators need to be adapted to the specificities of the islands and their own sustainable development objectives (Sachs et al., 2022). It is also important to set up a timely and consistent monitoring and evaluation system to measure progress. Adjustments, if necessary, should be made promptly to achieve these SDGs. The ultimate goal of this pragmatic tool is to promote sport tourism production and consumption that contributes to local development, while preserving the environment and respecting the social and cultural values of indigenous island communities (Fig. 29.2).

As discussed in the introduction, the purpose of this book is to compare experiences of using sport tourism as a means of developing small island states or territories. The reader has observed that the editors have taken every precaution to make the various chapters as coherent as possible. A precise methodology has been provided to the authors so that they may follow a common framework. This methodology asked project contributors to provide island data, including demographic, geographic, and topographic statistics and descriptions, as well as social-historical factors (see tables below). We have sought to compare and analyse these data in order to obtain conclusions and recommendations with empirical rigour (Tables 29.2, 29.3, 29.4, 29.5, 29.6 and 29.7).

Table 29.3 Characteristics of the archipelago (if more than one island) or island (available data from 2019)

Island or archipelago typology (mono-island, multi-islands, archipelago)	
Number of islands [if an archipelago] (number)	
Geographical size of the archipelago [if an archipelago] (number in km^2)	
Political status of the island or archipelago (independent, autonomous region, islands under the authority of a metropolis...)	
Main colonizing country (name)	

Table 29.4 Indicators of island or archipelago type (available data from 2019)

Location [ocean or sea] (name)	
Type of geomorphology of islands (continental island, oceanic island, reef island, etc.)	
Inhabited islands (number)	
Main means of transportation to and from the island or archipelago (number and type)	
Seasonality (number of seasons)	
Conservation areas [national parks, natural parks, regional parks, protected areas, etc.] (name, number, and type)	
Marine conservation area(s) (number and names)	
World heritage site(s) (number and names)	

Table 29.5 Global indicators (available data from 2019)

Gross domestic product (GDP) per capita (US Dollars, in millions)	
Gini index (0.0–1.0)	
Principal vulnerabilities (environmental threats, natural hazards, health challenges, security risks, etc.)	

Table 29.6 Island tourism statistics (available data from 2019)

Tourists (number)	
Number of cruise passengers (number)	
Average length of stay (number of days)	
Total revenues (US Dollars, in millions)	
Tourism revenue as direct contribution (% of total GDP)	
Domestic travel agencies (number)	
Tourist beds (number)	

Study Limitations

In order to obtain the data sought, a global call for contributors invited academics of all disciplines, specifically those familiar with the local context under study. These two elements play a role in the limitations encountered by the comparative work presented in this international research project. Multi-disciplinarity has many advantages: among these advantages are an openness to other research methods and

Table 29.7 Sport tourism on islands (available data from 2019)

Local sport tourism companies (number)	
Jobs in the sport tourism sector (number)	
Total sector revenue (US Dollars, in millions)	
Sites and spaces of sport tourism activities and events (number)	
Planning documents of sport tourism sites and locations [2010–2020] (number, name, and year)	
Participation rates of tourists who engage in sport tourism activities [in national or local tourism survey] (%)	
Main sport tourism activity # 1 (name)	
Sites and spaces of sport tourism activity # 1 (number)	
Sport tourism companies of activity # 1 (number)	
Revenue in sport tourism activity # 1 (US Dollar, in millions)	
Sport tourists in activity # 1 (number)	
Sport tourism activity # 1 project (name, location, and cost)	
Events organized in activity # 1 (number)	
Main sport tourism activity # 2 (name)	
Sites and spaces of sport tourism activity # 2 (number)	
Sport tourism companies of activity # 2 (number)	
Revenue in sport tourism activity # 2 (US Dollar, in millions)	
Sport tourists in activity # 2 (number)	
Sport tourism activity # 2 project (name, location, and cost)	
Events organized in activity # 2 (number)	

techniques, other epistemologies or ways of knowing, and a celebration of diverse research paradigms. There would not be so many maps and diagrams in this book if geographers had not contributed their expertise to sociologists, ethnologists, and economists. By the same token, however, each chapter has been written by one or several authors with a shared disciplinary training. The focus and findings are often reflective of a particular specialization and rarely interpreted through a multidisciplinary lens and consortium of scholars. This suggests that even with the most precise instructions, the case study results may be somewhat biased, interpreted from a particular theoretical and methodological perspective.

Though this international, comparative approach is undeniably enriching and provides a contribution to the extant literature, it is still necessary to take into account the varying island cultures which largely influence the authors' emic or insiders' points of view and analyses. The same figures, which may be scientifically indisputable, may also be the subject of somewhat different analyses depending on the region of the world from which they come. And yet, this variation likewise exists within these same regions, such as oceans, seas, and island archipelagos. For international scholars embarking on the study of these island contexts, a similar caution must be considered, as an etic or outsider perspective may not capture the sociocultural nuances and understandings of these island territories. Fortunately, this book offers an integrated analysis, inclusive of both an emic and etic perspective.

This same international framework leads to working with authors who do not have access to the same quality of data sources, such as common statistics utilized in some but not all parts of the world. Local culture and local needs may explain these differences. If some countries have very few statistics on tourism and none on sport tourism, it may be for economic reasons, given the lack of financial and human resources to develop a service or department dedicated to such data collection. This lack of reliable and easily comparable data may also be because local interests are varied based on the historical context and more pressing issues of the day. This may mean that this sector of activity, while important to us as researchers, is not considered a priority from the perspective of decision-makers and/or government bodies.

In any case, regardless of the island under study, specific data on sport tourism are still relatively rare, most often included in tourism statistics in general, if they exist at all. Researchers in these territories are required to do substantial investigative work to obtain data that is often only approximate. Sport tourism, a recent field of scientific study, is at a similar point as the sports economy was in the 1970s and 1980s in most Western countries, before the concerted interest in sport from an economic perspective. One working definition of sport, known as "Vilnius definition" is a tool for assessing the economic impact of sport in the various countries of the European Union. Several titles concern the transport (land, water, air) of athletes to competitions and sport tourists to sporting events and to a sports holiday destination. Under the heading restaurant/hotels are listed the expenses of sportspeople during their competitions and training as well as those of tourists attending a sporting event or during their sports holidays. There is a direct linkage between the economy of sport and that of sport tourism, providing an existing mechanism to measure the impact of sport tourism within a given economy.

Nothing as structured as the Vilnius system exists at the level of international bodies concerning sport tourism. Only a first approach was carried out at the academic level (Sobry et al., 2016), inspired by this European definition of sport. However, in the United States, The National Association of Sports Commissions (NASC), the governing body of the $11.4 billion sport tourism industry, has been transformed since January 2020 into the Sports Events and Tourism Association (Sports ETA). As such, the United States has a tool for knowledge collection and analyses of both the sports market and that of the sport tourism industry. The connections between this organization and certain universities make it possible to develop tools for evaluating the economic effects of events and sports equipment considered as elements for the development of sport tourism. No other country has such a structure at this time.

The summaries and recommendations proposed here are therefore dependent on the multiple variables noted above. Part III of the book has endeavoured to make the best use of the data at our disposal, acknowledging that some errors of omission and biases may be included in the analyses. IRNIST will attempt in future research to contribute to this field of investigation with humility; we look forward to the time when there exists a common and therefore comparable set of sport tourism indicators to better inform the work of academics and governmental and non-governmental organizations around the world.

The final section of this culminating chapter offers several recommendations. We hope that these proposals will be useful to decision-makers, a diverse group of stakeholders who play an essential role in the implementation of policies and measures favourable to sustainable sport tourism within island territories worldwide.

Recommendations and Futures Implications

Sport tourism activities in small island states or territories must confront the multiple dimensions of sustainability recognized within this book. As one author noted in their chapter in this collection, "SIDS are eager to engage innovative thinking and planning. They often organize in collaboration with their university leaders to imagine the future of their economies and their social policies, ensuring that the two are aligned. Such an approach towards resource management and preservation is unique in the world today and should be put to concrete use beyond its academic merit (p. 560)."

Based on our work, this field of study holds promise and the possibility for future action. Naturally, any prospective approach must take into account the specificity and complexity of the ecosystem in which it is embedded. That is, each island is characterized by its own unique characteristics, seeking to propose solutions to the diversity of existing challenges and situations. These recommendations are provided at a time when many islands in the world understand the global impact of climate change, income disparity and social inequality. They therefore seek novel models of sustainable development that encompass the framework of the 2030 agenda for the sustainable development of these unique ecoterritories.

1. *Expand action research on sport and island tourism in partnership with global organizations and alliances*

There are numerus efforts to better facilitate the development of policies and decision-making by actors in this field of study at several scales:

At the level of the Small Island Developing States (SIDS), organization and the Alliance of Small Island States (AOSIS), Institute for Small Islands, Islands and Small States Institute, International Small Islands Studies Association, the Island Innovation Agency, etc.

At the Indian Ocean level: the Indian Ocean Commission (IOC), Indian Ocean Rim Association, the Vanilla Islands cluster, Université Rurale de l'Océan Indien (UROI), Agence Française de Développement/Océan Indien, universities, etc.

At the Pacific Ocean level: the International Pacific Forum (PIF), Pacific Community (SPC), Pacific Islands Development Forum (PIDF), universities, etc.

At the Caribbean Sea level: the Caribbean Community and Common Market (CARICOM), Caribbean Development Bank (CDB), Association of Caribbean States (ACS), Organization of Eastern Caribbean States (OECS), AFD/Atlantic, universities, etc.

These different actors involved at the global, regional, national, and local levels are working towards the implementation of an ethic of sustainability for economic development, the protection of indigenous populations and the preservation of the environment. It will therefore be necessary to extend this research agenda to deepen the knowledge and achieve the multiple objectives of sustainable development in the islands.

2. *Support an ecoterritorial approach to sport tourism to inspire innovative solutions for the development of these territories*

   Within this action research, we recommend utilizing an ecoterritorial approach to (re)imagine solutions to respond to the problems of a territory. This approach must operate at a coherent and functional scale, seeking coordination and partnership at the various levels of the decision-makers to achieve common objectives. These objectives will align with the sustainable development goals across multiple dimensions, notably in terms of environmental, economic, and social impacts. The development of sport tourism practices must therefore be enacted in a reasoned and respectful way, in accordance with the resources and constraints of the local island context. This approach likewise adopts a critical ecopedagogy towards social and environmental justice and relational equity. The relationship between sport tourism and sustainable development is not a simple proposition; it depends on the goodwill of multiple stakeholders whose work in synergy is more necessary now than ever to move towards a more promising and viable future.

3. *Create a Global Sport Tourism Observatory (and islands) with international institutions*

   This proposal will serve as a tool for the establishment of a Global Sport Tourism Observatory. This observatory aims to collect, store, and present relevant data on destinations and sports events in order to facilitate analysis. This approach of creating a tool to evaluate the socio-economic impact of sport tourism can be one of the factors behind the creation of a sport tourism observatory. The geographical observatory of sport tourism has several interests.

   This proposed global observatory will allow for trend analyses through the study of sport tourism data such as the most popular destinations, major sports events, emerging sports activities, etc. It will provide tourism professionals, sport event organizers and policymakers with valuable information to make informed decisions regarding the development of sport tourism. It will help promote destinations by highlighting their assets and showcasing available sports activities, which can help attract more sport tourists.

   A Geographic Information System (GIS) of sport tourism could be realized to better understand and manage the functioning of a system marked by diversity, complexity, and variability over time. The principle of an interactive tool with an open data platform would give access to a regularly updated database as well as documenting the evolution of the supply and demand of sport tourism in these territories. This would allow scholars, practitioners, and policymakers to respond to the transformations that are taking place in the short and long term in the sport tourism market. This observatory would also provide elements of comparison with other sources that are essential for a more sustainable approach to territorial policy.

The observatory could build on the findings of Sobry et al. (2016) who proposed a system of socio-economic indices for sport tourism. With this tool, it would be possible to calculate these indicators from the database of the sport and tourism administration by analysing and cross-referencing them with each of the 17 SDGs.

This observatory will also improve the communication of sport tourism activities from the global to the local level. It is necessary to improve the communication of sport tourism at the territorial level by grouping more information points, publishing adapted guides and signposting places. The aim is to promote island destinations through sport tourism products, using digitalization, applications, or social networks. For example, an initiative with bloggers might be an interesting next step. Each blogger can discover a different sport tourism area of focus and then become an ambassador through the experiential value spread to the larger global community.

4. *Establish a global development charter and an implementation plan for the practice of sustainable sport tourism for the islands with key performance indicators (KPI)*

The aim of this recommendation is to propose a global development charter for sport tourism in the islands, giving priority to small, light structures for the reception of sportsmen and women and for the organization of services, using local materials, local wood, solar panels, management towards zero waste, or water recycling. The general idea is to include concrete strategies to minimize environmental impacts, while promoting the conservation of natural resources, foster the participation of local communities and encourage sustainable economic models. It will also involve the thematic edition of documents for sustainable sport tourism with regard to each activity concerned in the territories.

This global charter will also propose an implementation plan for the promoted practice of sustainable sport tourism with key performance indicators (KPI) around:

(a) Supporting sport tourism for education and health

This effort will promote the development of the practice of activities in schools as well as the support of educational projects enhancing sport tourism activities through the protection of natural resources, waste reduction, and the preservation of local fauna and flora. It will also integrate sport tourism activities into the intersection of health and wellness issues by offering activities adapted to different audiences. It is also about fighting against diabetes and obesity through the practice of these activities. Finally, it is to contribute to the well-being of the population, in particular young people, by reconnecting them to nature through these natural activities.

(b) Promoting sport tourism as a creator of local jobs and a vector of social and solidarity

We propose initiating insertion projects in the sector because of the endemic unemployment in the islands, thereby deploying green jobs on spaces, sites, and practice courses in order to favour the maintenance of supporting service providers who invest in the reception of people with disabilities in order to favour the inclusion of all peoples.

(c) Establishing a closer relationship with cultural heritage and the environment

The aim here is to create combined products at the interface of sport, culture, and the environment. The idea is to offer products that combine the heritage skills of local guides and the technical skills of professionals, in order to make these activities powerful tools for raising tourists' awareness of environmental and social issues.

(d) Creating multi-sport and multi-site products of varying difficulty and duration to stimulate local job creation

We recommend adapting to the different levels and time availability of sport tourists. The products created must allow all participant to discover the multiple facets of the islands by practising the activities most adapted to the specificity of the territories in question. One-day or multi-day tours can be set up. In the same way, mono-activity products designed on the basis of a daily change of site on the islands, by traveling in the atolls or archipelagos (Mauritius, Rodrigues, Mayotte, Seychelles...) are to be imagined.

(e) Evaluating sport tourism development policies on the basis of indicators used to achieve the SDGs.

This proposal focuses on the evaluation of sport tourism policies in order to observe and compare the evolution of actions undertaken over several years. The evaluation will be a comparison between territories having undertaken a similar approach and/or a temporal comparison for the same territory. It will be a matter of defining indicators that will make it possible to assess progress in achieving the SDGs and to orient actions according to the principles of sustainable development.

5. *Improve the governance of decision-makers and increase close collaboration between stakeholders*

This recommendation aims to better coordinate the actors or stakeholders (institutional, political, associative, economic...) in charge of sport tourism and sustainable development. Each institution often works in isolation and singularly within its sector, whereby sport actors do not equally prioritize tourism and sustainable development. Conversely, tourism and environmental actors often do not sufficiently integrate sport into their issues. It will be necessary to integrate representatives of the different institutions concerned in the representative councils of each of them by creating the conditions for a common work on this theme.

It is also necessary to propose participatory approaches involving local populations and tourism industry stakeholders to find solutions collaboratively. The tourism industry, including policymakers, local businesses, tour guides, equipment suppliers, local accommodation providers, etc., need to consult more with indigenous communities, especially women, youth, and people with disabilities, to involve them more in the decision-making process and formalization of projects, so that the socio-economic benefits create more jobs and improve their living conditions.

6. *Participate in joint projects related to development in different environments and in obtaining funding for sport tourism projects*

This proposal promotes the coherence of sport tourism policies and legal and institutional frameworks to achieve environmental objectives within the framework of sustainable development. It will solicit international donors to support the financing of sport tourism projects at all levels. Many governments offer grants to support research projects related to sustainable development and tourism. We will need to identify the appropriate agencies, organizations, foundations, institutions or programs to apply for funding, clearly explaining our proposed study and its potential contribution to sustainable development goals at all governmental levels.

7. *Integrate sport tourism into the smart destination concept to enhance the experiential value of participants and the adoption of sustainable practices*

   This recommendation aims to integrate sport tourism into the concept of intelligent destinations in order to improve the experiential value of the participants. It will concern the incorporation of new information technologies at the service of the sport tourism customers in the islands, characterized by the ability to use updated and effective data in the creation of an island-smart-destination application. This can include intelligent transportation systems, mobile applications for booking sport tourism activities, sensors to monitor weather conditions and pollution levels, free Wi-Fi access points, charging stations for electric vehicles, etc. The goal is to provide travellers with personalized services and information, as well as increased connectivity and sustainability.

   It will also involve the publication of a guide to best practices in sport tourism in island territories. The aim is to place the environment at the heart of any development project, considering the natural environment not as an unlimited resource or place of consumption but truly as a partner to be respected[5] . It is about using sport tourism activities as a relevant lever for environmental education, a guide of best practices in promoting sustainable development. Such educational efforts seek to promote responsible behaviors, such as reducing one's impact on biodiversity, preserving water quality, reducing pollution and thereby minimizing the impacts on marine ecosystems. It could be drafted by public authorities and distributed to practitioners by the various actors.

8. *Create an ecolabel for sport tourism and sustainable development in the territories*

   This ecolabel will distinguish sustainable sport tourism professionals (service providers, associations, companies, etc.) and will guarantee that visitors will find eco-responsible service providers. It will also concern the creation of a label for sport tourism, demonstrating the commitment of a tourist destination or a sports event to sustainable practices. These certifications offer external recognition and reassure travellers of a genuine commitment to sustainability.

[5] Scholars have conceptualized the environment as a stakeholder (Nunkoo et al., 2010; Nunkoo & Ramkissoon, 2010; Song & Kang, 2023 cite several) to be included in all planning and implementation of sustainable development strategies. The question has been, and continues to be, who advocates for nature and away from an anthropocentrist preoccupation with economic growth and profit/income generation? We advocate for a shared ecopedagogy, moral, and ethical commitment to social and environmental justice globally.

All of the recommendations presented here must be integrated into the framework of a global tourism development policy that takes into account the economic, social, and environmental dimensions of the island territories. This development plan must be designed in close collaboration with the multiple stakeholders in order to guarantee the success of the approach. These recommendations must be integrated into master plans at the level of each island and/or regional island organizations that will set the strategic orientations for the next decade, with clear, precise, quantifiable objectives and the practical means to achieve them.

For more than a decade now, IRNIST has provided international expertise in sport tourism and sustainable development through numerous publications, symposiums, and conferences. IRNIST offers to intensify its role as an expert in sport tourism and sustainable development in collaboration with several networks of actors. These networks include international actors who contribute to the promotion of sustainable development, such as the World Tourism Organization (WTO), the International Labour Organization (ILO), the World Bank, the United Nations Development Programme, the United Nations International Children's Fund (UNICEF), the United Nations Educational, Scientific and Cultural Organization (UNESCO), the European Union, the United Nations Organization (UNO), etc.

Secondly, these actors include global sport networks to demonstrate that sport tourism participates in sustainable development and to strengthen living communities at the level of UN Sport, Sport and Development, Peace and Sport Organization, Play International, Impact 2024, the International Olympic Committee (IOC), Sport4impact, Commonwealth Sport, Social Role of Sport, international federations, leagues, and regional sports committees.

Integration into these international networks would allow IRNIST to conduct research activities in the field of strategic observation of sport tourism within the framework of an interdisciplinary approach to sustainable development in connection with policymakers, NGOs, civil society, and associations in order to promote and strengthen field interventions. The research territories would be island, mountain, coastal, and continental territories. The aim will be to improve scientific knowledge, to structure and capitalize on the knowledge of the sector on their dynamics and interactions, as well as to facilitate shared access to data and to develop exchanges with researchers/decision-makers in the territories. It will also be necessary to participate in or organize various consultation meetings to this effect in order to present the scientific approach of the results obtained in this work, as well as to publicize the results during events such as the Fourth U.N. SIDS Conference held in Antigua and Barbuda in May 2024. It will also co-construct a research partnership with stakeholders for seminars and symposia, scholarships or PhDs, produce periodic reports on sport tourism trends and statistics from the data collected, or conduct in-depth analyses.

Finally, IRNIST wishes to provide training and curricular materials to support the creation of jobs in this sector and to further professionalize the field. Training is one of the essential means to ensure the success of the sustainable planning of sport tourism in the islands. The aim is to consolidate the sector's professions by improving skills at each level of intervention, from animation to project management, from strategic and operational marketing to human resources management, from grant applications to the search for funding partners. It is also a question of developing specialized training to better professionalize the actors who design and implement policies and products related to sport tourism. Continuous training sessions for actors would be set up by IRNIST in this spirit. These could be MOOCs (Massive Online Open Courses), webinars, certification, master classes, etc. This training would aim at acquiring a double competence, both technical and managerial, all in an effort to realize the 17 articulated sustainable development goals.

These recommendations seek to complement the comparative analyses of the 21 island case studies included in this international research project. As a group of international scholars, the majority of whom identify as indigenous island peoples, our aim is to contribute to the solutions rather than simply highlighting the challenges, indeed crises, facing small island territories today. In order to meet the future regarding the sustainable development of the planet, we offer a specific lens through which to promote positive change. This lens of sport tourism promises the conception and implementation of developing relational equity among peoples and in harmony with nature. The case studies in this volume, analysed from an ecoterritorial perspective and utilizing a comparative methodology, have allowed us to delve deeply into the practical while retaining an unwavering commitment to social and environmental justice.

References

African Union. (2015). *Agenda 2063: The Africa we want*. African Union Commission, 24 p. https://au.int/en/agenda2063/overview

Baledón, M. S., & Kosoy, N. (2018). "Problematizing" carbon emissions from international aviation and the role of alternative jet fuels in meeting ICAO's mid-century aspirational goals. *Journal of Air Transport Management, 71*, 130–137.

Barbier, E. (2011). The policy challenges for green economy and sustainable economic development. In *Natural resources forum* (Vol. 35, No. 3, pp. 233–245). Blackwell Publishing Ltd.

Black, R., Petersen, A., Pinter, G. G., & Reid, R. (2000). The eighth continent: Life, death, and discovery in the lost world of Madagascar. *American Scientist, 88*(5), 452.

Caruana, S., & Sultana, T. (2021). Exploring scuba diving tourism sector in Malta and its sustainable impact on the Island. In V. Katsoni & C. Van Zyl (Eds.), *Culture and tourism in a smart, globalized and sustainable world* (pp. 85–103). Springer.

Chiambaretto, P., Mayenc, E., Chappert, H., Engsig, J., Fernandez, A. S., & Le Roy, F. (2021). Where does Flygskam come from ? The role of citizens' lack of knowledge of the environmental impact of air transport in explaining the development of flight shame. *Journal of Air Transport Management, 93*, 102049.

Forestier, O., & Kim, R. E. (2020). Cherry-picking the sustainable development goals: Goal prioritization by national governments and implications for global governance. *Sustainable Development, 28*(5), 1269–1278.

Gössling, S., Humpe, A., & Bausch, T. (2020). Does 'flight shame' affect social norms? Changing perspectives on the desirability of air travel in Germany. *Journal of Cleaner Production, 266*, 122015.

Guenther, G. (2018). Who is the *We* in 'We are causing climate change?' *Slate*. https://slate.com/technology/2018/10/who-is-we-causing-climate-change.html. Accessed 22 Dec 2022.

Kahn, R. F. (1931). The relation of home investment to unemployment. *The Economic Journal, 41*(162), 173.

Kasztelan, A. (2017). Green growth, green economy and sustainable development: Terminological and relational discourse. *Prague Economic Papers, 26*(4), 487–499.

Larsson, J., Elofsson, A., Sterner, T., & Åkerman, J. (2019). International and national climate policies for aviation: A review. *Climate Policy, 19*(6), 787–799.

Lee, K. H., Noh, J., & Khim, J. S. (2020). The Blue Economy and the United Nations' sustainable development goals: Challenges and opportunities. *Environment International, 137*, 105528.

Lindsey, I., & Chapman, T. (2017). *Enhancing the contribution of sport to the sustainable development goals*. The Commonwealth.

Mangukiya, R. D., & Sklarew, D. M. (2023). Analyzing three pillars of sustainable development goals at sub-national scales within the USA. *World Development Sustainability, 2*, 100058.

Marrero Rodriguez, J., & Santana Talavera, A. (2018). La estructura de oportunidades empresariales en los destinos turísticos de sol y playa. El caso de Fuerteventura (Canarias). *Cuadernos de Turismo, 42*, 285–304. https://doi.org/10.6018/turismo.42.12

Meadowcroft, J. (2007). Who is in charge here? Governance for sustainable development in a complex world. *Journal of Environmental Policy & Planning, 9*(3–4), 299–314.

Mikhno, I., Koval, V., Shvets, G., Garmatiuk, O., & Tamošiūnienė, R. (2021). Green economy in sustainable development and improvement of resource efficiency. *Central European Business Review*. https://doi.org/10.18267/j.cebr.252

Mkono, M. (2020). Eco-anxiety and the flight shaming movement: Implications for tourism. *Journal of Tourism Futures*. https://doi.org/10.1108/JTF-10-2019-0093

Nunkoo, R., & Ramkissoon, H. (2010). Small Island urban tourism: A residents' perspective. *Current Issues in Tourism, 13*(1), 37–60.

Nunkoo, R., Gursoy, D., & Juwaheer, T. D. (2010). Island residents' identities and their support for tourism: An integration of two theories. *Journal of Sustainable Tourism, 18*(5), 675–693.

Organisation des Nations unies pour l'éducation, la science et la culture [UNESCO]. (2017). Plan d'action de Kazan, document final de la sixième Conférence internationale des ministres et hauts fonctionnaires responsables de l'éducation physique et du sport (MINEPS VI), Kazan, Russie. Récupéré sur: https://unesdoc.unesco.org/ark:/48223/pf0000252725

Pickel-Chevalier, S. (2015). Can equestrian tourism be a solution for sustainable tourism development in France? *Loisir et Société/Society and Leisure, 38*(1), 110–134. https://doi.org/10.1080/07053436.2015.1007580

Plumer, B., Friedman, L., Bearak, M., & Gross, J. (2022, November 19). In a first, rich countries agree to pay for climate damages in poor nations. *New York Times*. https://www.nytimes.com/2022/11/19/climate/un-climate-damage-cop27.html#:~:text=the%20main%20story-,In%20a%20First%2C%20Rich%20Countries%20Agree%20to%20Pay%20for%20Climate,about%20how%20it%20would%20work. Accessed 22 Dec 2022.

Ritchie, H., Roser, M., & Rosado, P. (2020). *CO_2 and greenhouse gas emissions*. Our World in Data.

Rockström, J., & Sukhdev, P. (2016). *How food connects all the SDGs*. Stockholm Resilience Centre, 14.

Sachs, J., Lafortune, G., Kroll, C., Fuller, G., & Woelm, F. (2022). *Sustainable development report 2022: From crisis to sustainable development: The SDGs as roadmap to 2030 and beyond*. Cambridge University Press.

Smith-Godfrey, S. (2016). Defining the blue economy. *Maritime Affairs: Journal of the National Maritime Foundation of India, 12*(1), 58–64.

Sobry, C. (2003). *Socioéconomie du sport; les structures sportives ébranlées par le libéralisme économique*. De Boeck.

Sobry, C., Liu, X., & Li, J. (2016). Sport tourism: Contribution to a definition and categorization. *Acta Touristica, 28*(1), 7–26.

Song, J., & Kang, J. (2023). The relationship between marine sports tourist destinations, social responsibility, and environmentally responsible behavior. *Sustainability, 15*(10), 7739.

Sotomayor, S., & Barbieri, C. (2016). An exploratory examination of serious surfers: Implications for the surf tourism industry. *International Journal of Tourism Research, 18*(1), 62–73.

Staples, M. D., Malina, R., Suresh, P., Hileman, J. I., & Barrett, S. R. (2018). Aviation CO2 emissions reductions from the use of alternative jet fuels. *Energy Policy, 114*, 342–354.

The Commonwealth. (2019). *Measure the contribution of sport, physical education and physical activity to the goals of sustainable development* (103 p). The Commonwealth.

Tyson, P. (2013). *Madagascar: The eighth continent: Life, death and discovery in a lost world.* Bradt Travel Guides.

U.N. (21 July 2016). United Nations Action Plan on Sports for Development and Peace. Retreived at chrome-extension://efaidnbmnnnibpcajpcglclefindmkaj/https://www.un.org/development/desa/dspd/wp-content/uploads/sites/22/2018/06/14.pdf

University of the South Pacific, Oceania National Olympic Committees (ONOC) Government of Fiji & Government of Samoa. (2018). *Maximising the contribution of sport to economic and social development of sport in Pacific Island countries: The case of Fiji and Samoa preliminary report*. University of the South Pacific.

Voyer, M., Quirk, G., McIlgorm, A., & Azmi, K. (2018). Shades of blue: What do competing interpretations of the blue economy mean for oceans governance? *Journal of Environmental Policy & Planning, 20*(5), 595–616.

Wittesaele, C. (2021). Art, land reclamation and green governmentality in Indonesia: Teja Astawa's Dewa Murka and Tita Salina's 1001st Island – The Most Sustainable Island in archipelago. *Journal of Southeast Asian Studies, 52*(2), 309–335.

Zhu, J., Yao, Y., Li, L., & Guan, W. (2022). Impacts of land reclamation on hydrodynamics in Zhoushan archipelago. In *Third international conference on geology, mapping and remote sensing (ICGMRS), Zhoushan, China, 2022* (pp. 74–77).

Chapter 30
The United Nations (UN), the International Labor Organization (ILO) and the Small Islands Developing States (SIDS)

Giovanni di Cola

Introduction

Small Island Developing States (SIDS) are a group of 38 United Nations (UN) member states and 20 non-member states (seven of them are Least Developed Countries LDCs 1)—grouped in three regions: the Caribbean, the Pacific, as well as the Africa, Indian Ocean, and South China Sea (AIS) region. At the regional level, SIDS are supported by inter-governmental organizations, primarily the Caribbean Community (CARICOM), the Pacific Islands Forum (PIF), and the Indian Ocean Commission (IOC).

SIDS are formed by a diverse group of countries and territories, including large geographical spaces, cultural diversity and socio-economic differences between and within the Pacific, Caribbean, and AIS regions. The aggregate population of all the SIDS is 65 million, slightly less than 1% of the world's population with a total GDP of USD 733 billion, of which Singapore accounts for half. GDP per capita in SIDS—on average weighted by population—is USD 11,200, ranging widely from about USD 64,600 in Singapore to USD 800 in Guinea Bissau.

SIDS share similar challenges to sustainable development, exacerbated by the COVID-19 pandemic. Such challenges include geographic remoteness, small economies, limited domestic resources, low economic diversification, high marginal costs of essential services and products, cost of transport, heavy dependence on imports, limited external financial investments, vulnerability to environmental, economic, and other external shocks and crises (i.e. natural disasters, pandemics, and climate change). All the above may positively or negatively impact the SIDS populations.

G. di Cola (✉)
International Labour Organisation (ILO) Special Adviser, Genève, Switzerland
e-mail: dicola@ilo.org

D. Van Rheenen et al. (eds.), *Sport Tourism, Island Territories and Sustainable Development*, Sports Economics, Management and Policy 24,
https://doi.org/10.1007/978-3-031-51705-1_30

However, the SIDS are the quintessence of resilience and anti-fragility. They have a very rich cultural heritage. Their music (Calypso, Reggae, Soca, etc.), for example, is known all over the world. The Nobel Prizes per capita is staggering: St Lucia received two Nobel Prizes in Economy and Literature and Trinidad and Tobago received one in Literature for a total population of 1.6 million peoples.

The level of research in science and social sciences and university education is widely appreciated and recognized in both the Pacific and West Indies universities. The quality of health service education is such that most of the nurses certified are immediately hired in the USA, Canada, the UK and in other Commonwealth countries (creating a deficit estimated at 40% of the demand of nurse services in the Caribbean labour market). The success in Olympic sports, particularly in athletics, is outstanding. In cricket, the famous West Indies Cricket Team, which did not lose a match between 1980 and 1995, is still admired for their fast bowling technique as documented in Fire in Babylon (Cricket's Greatest team in history: Fire in Babylon 1.—video Dailymotion). The victories of the team contributed to create a sense of civic achievement and pride in fighting for equality across the Caribbean following their independence.

The diplomacy of the SIDS and the capacity to represent their population at the multilateral level is one of the finest despite the limited number of embassies and representations around the world. The reserves of gas and oil in the Caribbean Sea, mostly untapped, are finally being explored, providing good prospects offshore from Venezuela and Trinidad and Tobago (350 million cubic feet of gas, January 25, 2023, Reuters), and Guyana and Suriname (S&P Global Commodity Insights). Important hydroelectric infrastructures projects are being carried out in Guyana (Amaila Plant, Kumu, and Moco Moco hydro projects), successfully providing opportunities to exploit green energies.

Caribbean economies were among the worst hit by the COVID-19 pandemic, especially due to their high dependence on tourism and limited fiscal flexibility. And yet, they became the centre of all attention in leading discussions over climate change resilience in Sharm El Sheikh (UNCCC COP 27) and by formulating an important package of projects for a total of USD 32 million through the UN SDGs Fund for SIDS, being implemented by the UN country teams in all SIDS.

Background

The majority of SIDS are Middle-Income Countries (MICs) with very specific needs and capacities. However, the impact of COVID-19 has reduced the capacity of the local actors to engage effectively in existing frameworks, such as advances in the digital economy, skills anticipation, and climate change response as part of the SIDS agenda. Despite these challenges, the resilience of the SIDS is part of their DNA and of their culture. It is an asset for all climate change-related policies and programmes and this includes the slow but affirmative response to the pandemic.

Politically, the SIDS are a vocal group, particularly at the level of the UN in New York and through the various multi-bilateral fora to which they contribute positively. In the UN setting, with the exception of the New York-based UN General Assembly and the UN HQs Secretariat, the SIDS find themselves working between and with distinct regional groups. As the SIDS grouping is interregional and covers three regions, many diplomatic efforts are displayed to ensure the queries and needs of the SIDS are heard by the international community. Since Samoa 2014, the capacity to build consensus around the SIDS situation and demand has grown exponentially.

Against this background, describing pathways towards sustainability, such as those articulated in the SAMOA Plan, is of critical importance to understand how SIDS are forging blue and green economies, rescheduling debt for climate change and preserving nature and to unlock sustainable recovery while achieving SDGs towards resilient, inclusive societies and more diversified economies.

As an example, the ECOSOC (Social and Economic Council of the UN) Segment of 23 April 2021, devoted to SIDS, highlighted the following issues:

- The impact of the pandemic hit the SIDS economies hard with a minus 16–18% of the GDP in 2020 for the Caribbean countries. Additionally, the eruption of the Volcano La Sufrière in St Vincent and the Grenadines also placed the country in a vulnerable situation.
- The Multidimensional Vulnerability Index (MVI) is a useful tool to access concessional financing, since the Gross National Income (GNI) of middle-income countries, although highly indebted, does not provide for such access.
- Equitable financing, debt relief, vulnerability assessment, climate action, and COVAX must be prioritized under such challenging conditions.

Due to their relatively small jurisdictions, SIDS are well suited for testing and operationalizing new and innovative policies with very specific advisory services. SIDS are often eager to serve as "incubators" for developing actionable and scalable tools aimed at strengthening institutions, skills, and leadership capabilities for coherent SAMOA Pathway and SDG implementation.

International Frameworks Supporting SIDS

The SAMOA Pathway was adopted in 2014 in Apia, Samoa. It was the third international UN conference on SIDS after the:

- Barbados Programme of Action (1994)—enabling actions for SIDS to achieve sustainable development.
- Mauritius Strategy (2005)—engaging further implementation of the Barbados Plan of Action.
- The SAMOA (SIDS Accelerated Modalities of Action) (2014) Pathway for the sustainable development of SIDS is built around five priority areas: (a) sustainable,

inclusive and equitable economic growth with decent work for all, (b) climate change, (c) bio-diversity, (d) human health and social development, and (e) partnership development.

The SAMOA Pathway is a dedicated, internationally agreed-upon programme of action for small island developing States (SIDS) for the decade 2014–2024. The SAMOA Pathway was the outcome of the Third International Conference on Small Island Developing States held in Samoa in 2014. The theme of the Conference was "The Sustainable Development of Small Island Developing States Through Genuine and Durable Partnerships."

The above framework included responses to concrete questions, such as (1) What is the SAMOA Pathway? (2) What are the Key Priority Areas in the SAMOA Pathway? Sustainable, Inclusive, and Equitable Economic Growth with Decent Work for All; Sustainable Energy, Oceans and Seas; Climate Change; Food Security and Nutrition; Water and Sanitation; Sustainable Transportation; Sustainable Consumption and Production; Management of Chemicals and Waste; Including Hazardous Waste; Health and Non-Communicable Diseases; Gender Equality and Women's Empowerment; Biodiversity; Social Development; Invasive Alien Species; Means of Implementation and Partnerships.

In 2019, five years after the adoption of the SAMOA Pathway, the international community reviewed progress towards the implementation of the SAMOA Pathway goals. SIDS, development partners, the UN system, civil society, the private sector, and other relevant actors had the opportunity to take stock of progress being made to achieve the priority areas of the SAMOA Pathway in support of the sustainable development of SIDS. In that regard, a series of regional meetings took place for SIDS in three island regions, identifying gaps in the implementation as well as exchanging best practices and lessons learned.

It should be noted that the SAMOA Pathway offers a more open framework than the previous plans or strategies (Barbados and Mauritius) to articulate regional and national programmes with international agenda items.

Perhaps because of this flexibility, the SIDS have actively contributed to the:

- Sendai Framework for Disaster Risk Reduction (March 2015)
- Addis Ababa Action Agenda on Financing for Development (July 2015)
- UN 2030 Agenda for Sustainable Development, (SDGs) (September 2015)
- Paris Agreement COP 21 (December 2015)

The UN Quadrennial Comprehensive Policy Review (QCPR) in 2020 emphasized the need to improve the UN system's support to Least Developed Countries (LDCs), Landlocked Developing Countries (LLDCs), and Small Island Development States (SIDS).

The UN Office of the High Representative for Least developed, Landlocked and Small islands developing countries (UN-OHRLLS) since 2001 mobilized and coordinated the United Nations system and other stakeholders towards supporting implementation, follow-up, and monitoring of the Barbados Programme of Action (BPoA), the Mauritius Strategy, and the SAMOA Pathway at the country, regional, and global levels.

As part of the UN Reform, the Regional and Multi-Country Offices Reviews have been put in place at the level of SIDS (in the Pacific and in the Caribbean) to better respond to long-standing concerns of SIDS and LDCs and UN presence in the 41 countries covered by multi-country resident coordinator offices.

In March 2021, as part of the initiative Financing for Development in the Era of COVID-19 and Beyond, the UN Secretary General urged governments to (a) provide concessional financing, especially for SIDS and LDCs; (b) allocate drawing Rights; and (c) extend the World Bank's Debt Suspension Initiative until June 2022, especially for SIDS that have been gravely affected by crisis. (Meeting of Heads of State and Government on the International Debt Architecture and Liquidity [29 March 2021] and WB–IMF Spring meetings [9 April 2021]).

The above chronology demonstrates the SIDS' success in further enhancing their agenda to accelerate their strategies coherently.

In the Aftermath of the SAMOA Pathway and Future Directions

In the framework of the UN Reform, with regard to Multi-Country Offices (MCOs), all UN agencies are committed through an agreed-upon framework called Commitment Tracker to contribute to the following Action Areas:

More Tailored Support to SIDS

> Support RCs, RCOs, and UNCTs in MCOs in developing system-wide Country Implementation Plans the UN supports and establishing Priority Theme Groups to pull expertise and resources around priority challenges affecting SIDS.

Enhanced and Aligned Physical Presence in SIDS

> Increasing or strengthening on-the-ground capacities in MCO-serviced countries and territories.
> Take concrete steps to establish capacities in the new North Pacific MCO.
> Take concrete steps to review and delegate additional authorities to staff at country level.

Strengthened and More Visible Policy and Technical Resources

> Take concrete steps to strengthen policy and technical capacities available to countries and territories serviced by MCOs, including at the regional and global levels.
> Establish SIDS focal points, as far as possible, at sub-regional, regional, and/or HQ levels.

Integration and Participation in Intergovernmental Fora and Processes

Enhance support to countries and territories serviced by MCOs in their preparation for and participation in intergovernmental forums and UN-led processes.

Strengthening National Data Systems, Data Collection and Usage

Increase entity support to the collection of data and its use for evidence-based policymaking in issue areas prioritized by countries and territories serviced by MCOs.

Improved Access to Finance

Strengthen policy support for financing for development and provide instruments and tools addressing the specific financing needs of countries and territories serviced by MCOs, in particular SIDS, including country case studies and sub-regional SDG financing conferences.

Fostering South-South Cooperation and Cross-Regional Learning

Enhance the promotion of south-south and triangular cooperation modalities to allow countries and territories serviced by MCOs to benefit from the experiences of countries with similar development settings, as well as to share their own good practices.

The Multidimensional Vulnerability Index (MVI)

The COVID-19 pandemic exposed the vulnerabilities of SIDS, prompting their long-standing call to facilitate access to concessional resources. The economic fallout suffered by SIDS showed the urgent need for measures that better reflect the challenges these countries face. In fact, the Gross National Income (GNI) per capita excludes the majority of SIDS from concessional development financing. It is still considered by the majority of donors and financial institutions as the most effective criterion for eligibility and financing allocations. In reality, such an approach does not respond timely and effectively to exogenous shocks.

"In response to paragraph 8 (a) of General Assembly Resolution 75/215, which called upon the Secretary-General to inter alia provide recommendations on the potential development, finalization and use of a Multidimensional Vulnerability Index (MVI) for SIDS, the Secretariat conducted a series of consultations during 2021, which examined the work of UN and Non UN System entities that were developing, had worked on or had developed a multidimensional vulnerability index or vulnerability measures for SIDS. Based on these consultations, the Secretary General presented a set of recommendations in his report A/76/211 confirming inter alia that:

(i) It is possible to develop an MVI, within a short time span, based on a set of key guiding principles;
(ii) That an MVI has several possible uses including for access to concessional resources; and
(iii) In order to build consensus, the development of the MVI should be a member State driven process, which should be finalized by a high-level expert panel, headed by two eminent persons, one of whom should be from a SIDS, both appointed by the President of the General Assembly (PGA), tasked with finalizing the index."

"The Panel's work stems largely from more than three decades of advocacy from small island developing States (SIDS), on the need for the international community, including IFIs and MDBs, to better understand their particular vulnerabilities and its resulting needs. In accordance with the Secretary-General's guidance, the Panel is however cognizant that the vulnerabilities of all developing countries must be captured in the MVI to ensure credibility and comparability. Vulnerability is increasingly being recognized as a hindrance to sustainable development. Developing countries, including countries in special situations such as Least Developed Countries (LDCs), Landlocked Developing Countries (LLDCs) and Small Island Developing States (SIDS), have unique inherent and vulnerabilities that hinder their ability to overcome economic, environmental, and social shocks and stressors."

In its interim report of August 2022, the Panel considered that "the MVI must be designed as an index of both structural vulnerability and resilience."

The Principle of Universality should be considered as well as the three dimensions of vulnerability (economic, environmental, and social) and included in the index to assess proper risks to sustainable development.

The central node is the quality and availability of data which are critically lacking in SIDS countries. They should not be immediately correlated with GNI per capita and should focus on quality indicators.

The human-centred approach is paramount in selecting indicators based on accessible data. The approach is innovative since the GNI dogma is shaken by the importance of looking at needs and possible responses and then linking them with resources.

Such resources should be planned ahead of external shocks, which are looming at SIDS at all times.

The exposure of SIDS to external shock is centre stage in the conceptual framework proposed by the Panel and is underpinned by vulnerability and resilience as responses to environmental, social, and economic shocks.

The panel indicated that social vulnerability and structural resiliencies are important to eliminate redundancy towards the identification of appropriate quality indicators.

ILO and the SIDS

The overall mandate of the ILO is to promote decent work for all and advance social justice—central features of sustainable development. Several of these issues were highlighted as priorities by the SIDS in the Barbados Outcome Document of August 2013 and SAMOA Pathway of September 2014.

They fall directly within the mandate of the ILO and its priorities for a human-centred approach in accelerating the Decade of Action for the UN 2030 Agenda, achieving the SDGs while recovering from COVID-19.

These include:

- Employment Promotion: Assistance in formulating plans, programmes, and policies for the promotion of sustainable micro, small, and medium enterprises; competitiveness, productivity, and jobs; youth employment and green jobs; women empowerment; skills development and labour market information systems; effective labour migration management.
- Social Protection: Moving towards broader and deeper coverage of social security schemes and improve working conditions and safety and health at work.
- International Labour Standards: Promotion and realization of fundamental principles and rights at work (child and forced labour, non-discrimination, freedom of association and collective bargaining).
- Social Dialogue: Strengthening the representative organizations of employers and workers, tripartite labour councils, labour ministries and administrations, to enhance their influence on the range of policies and laws that affect the workplace.

ILO has established strategic alliances with SIDS regional organizations, such as: the Caribbean Community (CARICOM and CARIFORUM and the Association of the Caribbean States (ACS), the Organization of Eastern Caribbean States (OECS), the Pacific Islands Forum (PIF), and the Indian Ocean Commission (IOC); Alliance of Small Island States (AOSIS)—an intergovernmental which is carrying out advocacy for SIDS and influencing international environmental policy; UN Inter-Agency Consultative Group for SIDS to ensure coherent and coordinated support to SIDS; SIDS National Focal Points Network (2018)—that strengthens coherence and coordination of SIDS-related sustainable development issues at the national, regional, and global levels.

ILO south-south and Triangular Cooperation (SSTC projects between SIDS-to-SIDS, country-to-country, regional and inter-regional projects include employment services, women economic empowerment, employment policies, skills, informality, Occupational Safety and Health (OSH), labour migration, Technical and Vocational Education and Training (TVET), sustainable tourism, and Disaster Risk Reduction, among others. They are adapted to address the impacts of COVID-19 as well as post-COVID-19 recovery strategies towards a better normal through decent work.

The ILO has a permanent presence in the SIDS with sub-regional offices located in Fiji in the Pacific and Trinidad and Tobago in the Caribbean. In Africa, the ILO has DWT/CO offices in Dakar, Antananarivo, and Yaoundé that provide support to

SIDS in the region. In addition, the ILO has project staff working in several of the SIDS.

Through the ILO Decent Work Team and Office for the Caribbean, the ILO participates in the UN Country Teams of Belize, Jamaica, Barbados, Trinidad and Tobago, Guyana and Suriname. It has contributed to the development and implementation of two with a third one under preparation. DWCPs aligned with UN Multi-Country Sustainable Development Framework (MSDF) for the Caribbean which defines how the UN will collectively provide support to the 18 English and Dutch speaking countries and Overseas Territories in the Caribbean for: an Inclusive, Equitable, and sustainable economic growth; Prosperous Caribbean; A Healthy Caribbean; A Safe, Cohesive, and Just Caribbean; A Sustainable and Resilient Caribbean.

The ILO Decent Work Team for East and South-East Asia and Pacific Office has contributed to the development and implementation of a number of Decent Work Country Programmes (DWCPs) aligned with UN Pacific Strategy covering 14 countries of Cook Islands, Fiji, Federated States of Micronesia, Kiribati, Nauru, Niue, Palau, Republic of Marshall Islands, Samoa, Solomon Islands, Tokelau, Tonga, Tuvalu, and Vanuatu for addressing: climate change, disaster, resilience and environmental protection; gender equality; sustainable and inclusive economic empowerment; equitable basic services; governance and community engagement; human rights. The ILO also supports technically the Multi-country Pacific Strategy 2018–2022.

Moreover, the ILO co-chairs several regional issue-based coalitions, such as: coalitions on equitable growth in Latin America with the Economic Commission for Latin America and the Caribbean and the UNDP; on harnessing demographic dividends, gender, and youth for development in Africa (for LDCs) with the UNFPA and the WHO; and on inclusion and empowerment in Asia and the Pacific with the United Nations Children's Fund.

Conclusions

The UN commitment tracker, programmes, and mandates are consistent with the overall SIDS frameworks and the SAMOA Pathway. However, the integration of SIDS issues into strategies, narratives, and institutional mechanisms may continue to be reviewed and further improved. There is a need to increase capacity to formulate plans and deliver services and projects in a timely manner, while attracting funding on a regular basis.

In SIDS countries, prospects for employment remain bleak as the economies continue to recover slowly from the effects of the pandemic. Given various development challenges compounded by COVID-19, delivering policies and programmes through very strong development lenses is important for the SIDS countries. The UN needs a SIDS Development Agenda and sub-regional Action Plans (Pacific, Caribbean specific) with predictable funding sources.

Further alignment between UN agency programmes may improve delivery and increase efficiency in attracting foreign investment into development plans through partnerships also led by ACS, Pacific, AIMS, CARICOM, and other regional frameworks.

The ILO has supported technically sub-regional policy frameworks based on social tripartite dialogue. There are outstanding examples of tripartite social dialogue in the SIDS also at the macro-economic levels, with involvement in the implementation of IMF structural adjustment plans. In Samoa, the ILO supported tripartite panels organized by sub-regional social partners and Member States. This is quite an exception for UN Summits which have their own formal framework, with delegations from Ministries of Foreign Affairs rather than technical departments like the Ministry of Labour and social partners (employers and workers' organizations). It shows the innovative approach to which SIDS are, and continue to be, receptive.

Fostering policy coherence with Bretton Woods Institutions with regard to financing for development and innovative sources of funding and mechanisms on fresh concessional financing should go hand-in-hand with the formal adoption of the MVI. A new general allocation of IMF Special Drawing Rights should continue into 2023, especially for SIDS that have been gravely affected by crisis. The Panel on MVI indicated that some action in that regard should be taken promptly by Member States and relevant partners of SIDS.

SIDS UN Multi-Country Office(s) (MCOs) will provide a hub of information on a particular region's value proposition, expert personnel, capacities, skills, and knowledge that could help to accelerate identification and provision of services in support of SIDS. MCOs have the potential of becoming strategic in some of the regional initiatives in particular energy and blue economy related to bring global expertise for contextualization of SIDS regional/national frameworks.

Supporting the SIDS on policy coherence for sustainable development to achieve the goals of the SAMOA Pathway, including the implementation of the UNCCC recommendations from the COP 27 in Sharm el-Sheikh, is an opportunity to maintain focus on SIDS with possible implementation of new planning tools

The SIDS are eager to engage innovative thinking and planning. They often organize in collaboration with their university leaders to imagine the future of their economies and their social policies, ensuring that the two are aligned. Such an approach towards resource management and preservation is unique in the world today and should be put to concrete use beyond its academic merit.

References

Caribbean future forum Archives – CARICOM.

CARIBBEAN FUTURES FORUM | The Futures Forum.

Energy Diplomacy in Guyana and CARICOM, Oil Now, February 2, 2023.

Good-practices-in-sstc-for-sustainable-development-in-sids%20(5).pdf

SAMOA Pathway (3).pdf (SIDS ACCELERATED MODALITIES OF ACTION [S.A.M.O.A.] Pathway DOCUMENT • A/RES/69/15 – SIDS Accelerated Modalities of Action (SAMOA) Pathway [Arabic] [Chinese] [English] [French] [Russian] [Spanish).

Sands, S. R., Ingraham, K., & Salami, B. O. (2020). SIDS Multi Country Operations (MCO) Commitment Tracker 20 January 2021 (Internal UN document) Caribbean nurse migration—A scoping review. *Human Resources for Health, 18*, Article number: 19.

The Challenge of Small Island Developing States: Barbados, Mauritius, Samoa and beyond | ACS-AEC.

United Nations. (2022, August). MVI Interim Report | Department of Economic and Social Affairs (un.org).

United Nations Conference on Environment & Development (Barbados Plan of Action) 1994.

Chapter 31
Epilogue—Reflections at the End of the Journey

Olivier Naria

Planet Earth is our shared island, let's join forces to protect it.

Ban Ki-Moon, Secretary-General of the United Nations

The world has become complex, the problems so immense as we have embarked on a global exploration and expansion with a corresponding ethical responsibility to preserve the planet. A multitude of stakeholders and decisionmakers confront an evolving context marked by uncertainty and unpredictability.

The challenge for today's governments is not only to respond to everyday emergencies but also to lay the foundation for defining a new paradigm of sustainable development. This paradigm offers a new model of society that is more resilient and sustainable, based on experimentation and innovation. Its success can only be achieved over the long term, an aspirational future and path forward. Confetti strewn about the seas and oceans, islands and archipelagos are reflections of this complex world. They are directly exposed to numerous vulnerabilities, the most visible effects of which are global warming, but above all, they have to deal with the legacy of complex economic and social contexts. After decolonization and assimilation to the metropolises, the islands are now subject to the dictatorship of urgency, as they are far from decision-making centers and lag far behind structurally, accentuated by demographic pressure. While meeting the challenges of housing, employment, education, health, mobility, modernization of production facilities, islands and archipelagos have been forced to embark on a process of accelerated social, economic, and environmental change, which has taken mainland territories several decades to achieve. As a result, they are in the vanguard of international environmental, social, and economic policies.

O. Naria (✉)
Department STAPS, Université de La Réunion, Le Tampon, ïle de la Réunion, France
e-mail: olivier.naria@univ-reunion.fr

D. Van Rheenen et al. (eds.), *Sport Tourism, Island Territories and Sustainable Development*, Sports Economics, Management and Policy 24,
https://doi.org/10.1007/978-3-031-51705-1_31

Against this backdrop, the Université of Réunion has positioned itself as a catalyst for social and environmental transformation, in line with its societal responsibilities for the training of its students. The President has appointed a group of teacher-researchers and staff from the Université de la Réunion as Sustainable Development and Social Responsibility (SDRS) referents, to take up the challenge of creating an Eco-campus in the only French and European university in Indianoceania, and to respond to the challenges of adapting our island campus to tomorrow's world.

It was at this time that my IRNIST colleagues and I began an international research project, which culminated in this book. Thanks in part to the already-established links of the research network, manuscripts poured in from all over the world, materializing the interest in opening up to the world for our Indian Ocean university.

Since the end of 2021, I have been responsible for the pillar of regional, national, and international cooperation around Sustainable Development and have thus established exchanges and partnerships on cross-cutting issues linked to the development of our Eco-campus. The various meetings of this group show just how complex it is to develop benchmark campuses in tropical ecologies, as they reflect the image of island evolution, a complex process involving many interdependent factors. Setting the eco-campus agenda is in line with the sustainability approach to island societies, since the assessment of resources and needs must be based on community involvement, so that everyone takes ownership of the ambition from the earliest stages of the process. Designing the eco-campus infrastructure according to the principles of environmental sustainability and energy efficiency, integrating elements such as low-energy buildings, remain goals that can only be achieved in a global and local context, focused on environmental issues and sustainable practices.

In addition to the contacts made with the Vice-Chancellor of the University of Seychelles, the Dean of the Faculty of Letters at the University of Comoros, and the first Eco-campus agreement between the Mauritius University of Technology and our university, I have also maintained relations, thanks to Professor Van Rheenen, between our school and the University of California at Berkeley, which was named the most sustainable university among 700 higher education institutions worldwide, due to its focus on environmental sustainability as well as social and governance objectives (Wu, 2022).

In April 2023, as part of an Erasmus+ program, I was able to learn more about UC Berkeley's efforts to make the campus as sustainable as possible in the short and long terms. In addition, the campus, and multiple members comprising its diverse community, developed a plan to achieve specific sustainable goals, such as zero waste, carbon neutrality, and other tangible institutional efforts. These sustainability goals included engagement in social and environmental impact research, teaching and training, such as those conducted at the CoolClimate Network, the Institute for Parks, People and Biodiversity and the Student Environmental Resource Center (SERC). During my time at Berkeley, I was able to highlight the leadership of the US institution to Marc Fisher, the university's Vice President for Sustainability and Board of Trustees. I also presented to Kira Stoll, Chief Sustainability & Carbon

Solutions Officer of the Sustainable Development Program, the challenges of the Université de La Réunion's ecological and environmental transition program (UR, 2022) with a view to international cooperation between our institutions.

Whether we are on an island in the Indian Ocean, the Caribbean, or an archipelago in the Pacific, in a region of California in the USA or in a rural province in Asia, whatever part of the world we are in, today we have a burning obligation to provide answers to the challenges of sustainable development for our populations. Despite the implementation of several decades of programs and action plans by international and local organizations, why are there still so many inequalities in these island and archipelagic territories?

Inequalities can be deeply rooted and complex, requiring holistic approaches adapted to each context. There is a need to address these issues together with local populations marked by their mixed origins, multi-belonging, and cultural complexity. This painstaking work needs to be carried out with a sustainable pedagogy, so that the local population can change their mentality, so that they are at the center of these concerns and true agents of change. The "smart governance" of sustainable public policies on islands needs to be catalyzed on the ground by citizens, certainly on a more human scale. It is also a question of spreading the seeds of training among a new generation of young people in the fertile soil of the islands, so that they can take on more responsibility for the committed transition of island societies, conveying inclusive and sustainable approaches to ensure more equitable and balanced development.

It therefore becomes crucial to implement effective policies and measures to meet these challenges, paying particular attention to the resilience of islands and archipelagos in the face of climate change, the preservation of biodiversity, economic sustainability, food security, and the protection of the most vulnerable populations. It will also regularly assess progress in terms of environmental, social, and economic sustainability, in order to identify areas for improvement and adjust action steps accordingly.

Faced with these complex interactions, islands and archipelagos have demonstrated their agility, adaptability, and resilience (Pagney Bénito-Espinal, 2022) by promoting innovative, nature-based solutions encompassing integrated approaches that take into account environmental, geomorphological, climatic, socio-economic, and political dimensions (Baldacchino, 2023). Examples include, but are not limited to, eco-neighborhoods on the islands of Samso in Denmark, Ravine-Blanche on Réunion Island, the village of Graciosa on Lanzarote in the Canary Islands, the island of Bali, Bon Air caribéen in Martinique and many others, which demonstrate successful transformations in urban recycling, climate resilience, and ecological and demographic transition, to test and support innovative solutions and processes. In the same vein, some islands have developed electrically-assisted bicycles or electric car rentals for island populations (Saint-Pierre de la Réunion); others have linked their territory with an electric tramway (Mauritius). Solar and photovoltaic villages are springing up all over the islands, in El Hierro on the Canaries, Ta'u in the Solomons, Koh Rong in Cambodia, and Manapany les Hauts on Réunion Island. The introduction of participatory budgets for citizen-run projects is further evidence of the societal transformations taking place. Recourse to territorial innovation and experimentation thus becomes a key factor if islands are to become territories of innovation, bringing together the challenges of sustainability, the acceptability of local populations toward these achieving sustainable development objectives, combining the capacity for innovation to deploy solutions to meet the challenges of adapting to climate change (Carayannis et al., 2012).

Meeting formats in the form of conferences and forums through international cooperation with the global initiatives of intergovernmental and governmental agencies, civil society, non-governmental organizations (NGOs), economic players, and island stakeholders represent real opportunities to advance new ideas and stimulate the achievement of sustainable development goals. The adaptation of applied research to major island issues (Mcmahon & Baldacchino, 2023) represents an important lever in strategies for living with the rapid changes in small island ecosystems. A prime example is the work of the University of Malta's Institute of Islands and Small States (ISSI), who have introduced innovative concepts and methodologies (Moncada et al., 2021).

After this period of incubation, during which small areas of many islands served as laboratories for testing the integration of sustainable development at neighborhood or village level, it is now time to advance scale and deploy these innovations in order to consolidate the trajectories of growth and regional development throughout the island and archipelago territories of a new world. This is undoubtedly one of the best ways to encourage the development of coherent, relevant projects, in line with local island realities, so as to address the needs of local populations and the changes taking place in our society. A land of contrasts and cross-fertilization, the islands have not yet managed to resolve all of the inequalities produced by development logics that have left part of the population on the margins. We believe that the future of island territories will depend on the ability of local decisionmakers to create meaning and coherence, and on the ability of local populations to come up with original solutions rather than seeking them elsewhere.

This suggests that change is inevitable for many island and archipelagic territories, but that it is also possible to make a real difference in our collective efforts. As Peter Drucker, Professor Emeritus of Management at New York University (NYU), puts it, "the best way to prevent the future is to create it." To focus these aspirational efforts, vectors of involvement and transformation, we must foster innovation by integrating multiple epistemologies and methodologies, facilitating the exchange of ideas and enabling the implementation of broader, global solutions.

These efforts can be focused on the growing sport tourism market and the pursuit of sustainable development objectives in these territories, as small-scale initiatives present significant challenges in terms of local impact, demonstration effect, learning, replicability, and commitment to indigenous populations. This is certainly one of the best ways of transforming atypical spaces into dynamic eco-territories. This book has sought to provide a conceptual framework and a methodological tool for advancing these sustainable development goals. Time is a luxury that islands and archipelagos can no longer afford. Instead, it is time to act.

References

Baldacchino, G. (2023). *The success of small states in international relations: Mice that roar?* (280 pp). Routledge.

Carayannis, E. G., Thorsten, D., Barth, D., & Campbell, F. J. (2012). The quintuple innovation helix model, global warming as a challenge and driver for innovation. *Journal Innovation and Entrepreneurship, 1–2*, 1–12.

McMahon, E., & Baldacchino, G. (2023). Special section: Considering suitable research methods for islands. *Geographical Research, 61*, 93–95.

Moncada, S., Baldacchino, G., & Briguglio, L. (2021). The importance of academic research in studying islands and small states. *The Parliamentarian, 102*(1), 48–51.

Pagney Bénito-Espinal, F. (2022). *Îles en devenir des Antilles et d'ailleurs, AIHP-GEODE* (356 p). Schoelcher, Martinique Éditions Karthala.

Université de La Réunion. (2022). *Programme de Transition Écologique et Environnementale de l'Université de La Réunion*. Vers La première Université ultramarine engagée dans la transition énergétique et environnementale, Saint-Denis, 57 p.

Wu, B. (2022). University of California, Berkeley ranked first in first-ever QS Sustainable World University Rankings. *U.C. Berkeley News*. https://sustainability.berkeley.edu/news/uc/Berkeley/wins/1/first-ever/QS/Sustainability/World/University/Rankings. Accessed 10 Nov 2022.

SPRINGER NATURE

GPSR Compliance

The European Union's (EU) General Product Safety Regulation (GPSR) is a set of rules that requires consumer products to be safe and our obligations to ensure this.

If you have any concerns about our products, you can contact us on ProductSafety@springernature.com

In case Publisher is established outside the EU, the EU authorized representative is:

Springer Nature Customer Service Center GmbH
Europaplatz 3
69115 Heidelberg, Germany

The manufacturer's authorised representative in the EU is Springer Nature Customer Service Centre GmbH, Europaplatz 3, 69115 Heidelberg, Germany. If you have any concerns regarding our products, please contact ProductSafety@springernature.com

Printed and bound by CPI Group (UK) Ltd, Croydon, CR0 4YY
01/10/2026
02225904-0019